T0388552

Microbial Enzymes in Production of Functional Foods and Nutraceuticals

This book is a valuable reference that discusses green technologies, like enzyme technologies, to meet the ever-growing demand of nutraceuticals and functional foods. Microorganisms like bacteria (lactic acid bacteria, Bacillus species), yeasts, and filamentous fungi have been exploited for food preparations globally. *Microbial Enzymes in Production of Functional Foods and Nutraceuticals* discusses how to use them commercially. Chapters include enzyme sources, processing, and the health benefits of microbial enzymes. Other interesting Chapters include the application of metagenomics and the molecular engineering of enzymes. This book is useful for students, academicians, and industry experts in food science and applied microbiology.

Microbial Enzymes in Production of Functional Foods and Nutraceuticals

Edited by
Amit Kumar Rai
Ranjna Sirohi
Luciana Porto de Souza Vandenberghe
Parameswaran Binod

CRC Press is an imprint of the
Taylor & Francis Group, an informa business

First Edition published 2023
by CRC Press
6000 Broken Sound Parkway NW, Suite 300, Boca Raton, FL 33487-2742

and by CRC Press
4 Park Square, Milton Park, Abingdon, Oxon, OX14 4RN

CRC Press is an imprint of Taylor & Francis Group, LLC

ISBN: 978-1-032-31756-4 (hbk)
ISBN: 978-1-032-31757-1 (pbk)
ISBN: 978-1-003-31116-4 (ebk)

DOI: 10.1201/9781003311164

Typeset in Times
by KnowledgeWorks Global Ltd.

Contents

Preface

The book entitled *Microbial Enzymes in Production of Functional Foods and Nutraceuticals* is part of a book series that is published with the partnership of Taylor & Francis Group and the Biotech Research Society, India (BRSI). Application of microbial enzymes in the functional food industry is gaining popularity due to their advancement in production-specific nutraceuticals using novel microbial enzymes with unique properties. Microbial enzymes have been applied on different food substrates to produce nutraceuticals with a wide range of health benefits as well as reduction of antinutritional factors. These enzymes are also applied for recovery of nutraceuticals from agri-food processing by-products. Due to the increasing demand for enzymes in the functional food industry, they are produced using native microorganisms as well as genetically modified candidates with improved properties. Enzymes produced using *Bacillus* species, lactic acid bacteria, and fungi have been characterized and applied for production of a wide range of nutraceuticals.

This book covers a broad range of topics related to microbial enzymes, their sources, and application in the production of functional foods. The contribution of experts from the scientific community in specific areas of microbial enzymes from different parts of the globe is compiled in this book. The book has four sections including (1) introductory and perspective chapter, (2) sources of microbial enzymes for nutraceutical production, (3) specific microbial enzymes and their application in functional foods production, and (4) technologies for improved enzyme systems for nutraceutical production. The biocatalysts that have been applied for the production of value-added nutraceuticals mainly include proteases, and specific carbohydrate active enzymes (pullulanases, β-galactosidases, β-glucosidase, xylanase, etc.) are discussed. There are specific chapters on different microbial enzymes for the production of value-added molecules. The value-added molecules produced and modified using microbial enzymes include bioactive peptides, galactooligosaccharides, resistant starch, fructooligosaccharides, and free polyphenols. There are separate chapters on the application of microbial enzymes on the reduction/transformation of antinutritional factors and lactose. The bioactive compounds produced on enzymatic treatment are responsible for specific health benefits including antioxidant, anticancer, antihypertensive, cardioprotective, and prebiotic activities that are discussed in different chapters. The strategic approach of applying enzyme technologies for the production of high-value nutraceuticals from agri-food processing waste is also described. The scientific information on mining genes from metagenomes and genetic modification of microbial enzymes for improving their catalytic capacity providing economical alternatives to chemical industrial processes for functional food production is provided in this book. Due to increasing global demand for functional foods with specific health benefits, microbial enzymes are gaining popularity. This book on microbial enzymes provides comprehensive information on current enzyme technologies related to the production of functional foods.

We sincerely appreciate the contribution of scientific experts from different countries for sharing their knowledge on microbial enzyme technologies for the functional food industry and health sector during the pandemic period. We strongly believe that this book provides enriched scientific information that will be useful for researchers, students, academicians, and industry experts in food biotechnology and applied microbiology. We are grateful to reviewers for their sincere efforts and acknowledge them for their contribution in the critical review of the chapters. We sincerely acknowledge Prof. Ashok Pandey, Chief Mentor, BRSI, and Managing Editor of the series and the BRSI for providing us the opportunity to prepare this book and guiding us during the editing and publication process for bringing this book in the final form. We thank the team of Taylor & Francis Group including Dr. Gagandeep Singh, Senior Publisher, Ms. Madhurima Kahali, Editor II (Life Sciences),

Ms. Neha Bhatt, Editorial Assistant, CRS Press, and the entire team at CRC Press/Taylor & Francis Group for their consistent support during the publication process.

Editors

Amit Kumar Rai

Ranjna Sirohi

Luciana Porto de Souza Vandenberghe

Parameswaran Binod

About the Authors

Dr Amit Kumar Rai is Scientist (C) at the Institute of Bioresources and Sustainable Development, Mizoram Node, Aizawl, India. He completed his doctorate from CSIR – Central Food Technological Research Institute, Mysore. His major area of interest is food biotechnology for the production of nutraceuticals and functional foods rich in bioactive peptides and isoflavones using microorganisms. He has developed bioprocess for production of milk and legume-based bioactive peptides enriched fermented products that can be beneficial during oxidative stress and hypertension. He has filed 4 patents, 102 publications, including 73 research and review papers, 3 books, and 26 book chapters to his credit.

Dr Ranjna Sirohi is an Assistant Professor with the University of Petroleum and Energy Studies, Uttarakhand, India. She has served as a post doctorate researcher at Korea University, Seoul, South Korea. She completed her doctorate in Process and Food Engineering from G.B. Pant University of Agriculture and Technology, Pantnagar, India. She has worked as researcher at École Polytechnique Fédérale de Lausanne, Switzerland. Her major research interests are bioprocess technology, food and food waste valorization, waste to wealth, and biofuels. She has 94 publications including 60 research and review papers, 13 book chapters, 7 popular articles, and 14 conference proceedings to her credit.

Dr Luciana Porto de Souza Vandenberghe is a Full Professor with the Department of Bioprocess Engineering and Biotechnology, Federal University of Paraná, Curitiba, Brazil. She is a member of the Bioprocess Engineering and Biotechnology Graduation Program at UFPR. Dr. Vandenberghe obtained her PhD in Génie de Procédés Industriels – Biotechnologie from Université de Technologie de Compiègne (2000), France. Her areas of interest include bioprocess engineering and biotechnology and industrial microbiology, with focus on valorization of solid and liquid agro-industrial sub-products through submerged and solid-state fermentation for biomolecules production including industrial enzymes, organic acids, biopolymers, and plant growth hormones. She has published 136 papers, 60 book chapters, and 25 deposed patents.

Dr Parameswaran Binod is a Principal Scientist with the Microbial Processes and Technology Division of CSIR – National Institute for Interdisciplinary Science and Technology, Trivandrum, India. He obtained his PhD in Biotechnology from the University of Kerala, Thiruvananthapuram, India. He worked as a post-doctoral Fellow at the Korea Institute of Energy Research, Daejeon, South Korea, and later joined as scientist at CSIR – National Institute for Interdisciplinary Science and Technology, Thiruvananthapuram, India. His research interests include biomass to fuels and chemicals, biopolymers, and enzyme technology. He has more than 200 publications with h-index of 48. His name is listed in the world's top 2% of scientists for the whole career as per the study by Stanford University and Elsevier in the year 2020.

About the Authors

Contributors

Ali Muhammed Moula Ali
Department of Food Science and Technology
School of Food-Industry
King Mongkut's Institute of Technology Ladkrabang
Bangkok, Thailand

Cristina Álvarez
Advanced Biofuels and Bioproducts Unit
Department of Energy
CIEMAT
Madrid, Spain

Embalil Mathachan Aneesh
Centre for Research in Emerging Tropical Diseases (CRET-D)
Department of Zoology
University of Calicut
Malappuram, Kerala, India

Ammini Naduvanthar Anoopkumar
Centre for Research in Emerging Tropical Diseases (CRET-D)
Department of Zoology
University of Calicut
Malappuram, Kerala, India

Aline B. Argenta
Department of Chemical Engineering
Federal University of Paraná
Curitiba, Paraná, Centro Politécnico, Curitiba, Brazil

Mukesh Kumar Awasthi
College of Natural Resources and Environment
Northwest A & F University
Yangling, Shaanxi, China

Palanisamy Athiyaman Balakumaran
Microbial Processes and Technology Division
CSIR – National Institute for Interdisciplinary Science and Technology, (CSIR-NIIST)
Thiruvananthapuram, Kerela, India
and
Academy of Scientific and Innovative Research, (AcSIR)
Ghaziabad, Uttar Pradesh, India

Rwivoo Baruah
Department of Microbiology and Fermentation Technology
CSIR – Central Food Technological Research Institute
Mysuru, Karnataka, India

Sri Charan Bindu Bavisetty
Department of Fermentation Technology
School of Food-Industry
King Mongkut's Institute of Technology Ladkrabang
Bangkok, Thailand

Parameswaran Binod
Microbial Processes and Technology Division
CSIR – National Institute for Interdisciplinary Science and Technology (CSIR-NIIST)
Thiruvananthapuram, Kerela, India
and
Academy of Scientific and Innovative Research
Ghaziabad, Uttar Pradesh, India

Dão Pedro de Carvalho Neto
Federal Institute of Education
Science and Technology of Paraná
(IFPR), Londrina, Paraná, Brazil

Kalivarathan Divakar
Department of Biotechnology
Sri Venkateswara College of Engineering
Sriperumbudur, Tamil Nadu, India

Leticia Schneider Fanka
Department do Bioprocess Engineering and Biotechnology
Federal University of Paraná
Curitiba, Paraná, Brazil

Vivek Kumar Gaur
Centre for Energy and Environmental Sustainability
Lucknow, Uttar Pradesh India
and
School of Energy and Chemical Engineering
UNIST. Ulsan, Republic of Korea

Krishna Gautam
Centre for Energy and Environmental Sustainability
Lucknow, Uttar Pradesh, India

Cristina González-Fernández
Biotechnological Processes Unit
IMDEA Energy
Móstoles, Spain
and
Department of Chemical Engineering and Environmental Technology
School of Industrial Engineering
Valladolid University, Valladolid, Spain
and
Institute of Sustainable Processes
Valladolid, Spain

Luis Daniel Goyzueta-Mamani
Vicerrectorado de Investigación
Universidad Católica de Santa María
Urb. San José s/n—Umacollo, Arequipa, Peru

Pallavi Gupta
Bioscience and Biotechnology Department
Banasthali University
Jaipur, Rajasthan, India

Prakash M. Halami
Department of Microbiology and Fermentation Technology
CSIR – Central Food Technological Research Institute
Mysuru, Karnataka, India

Subrota Hati
Dairy Microbiology Department
SMC College of Dairy Science
Kamdhenu University
Anand, Gujarat, India

Leonardo Wedderhoff Herrmann
Department do Bioprocess Engineering and Biotechnology
Federal University of Paraná
Curitiba, Paraná, Brazil

Shahrim Ab Karim
Department of Foodservice Management
Faculty of Food Science and Technology
University Putra Malaysia, Serdang
Selangor Darul Ehsan, Malaysia

Susan Grace Karp
Department do Bioprocess Engineering and Biotechnology
Federal University of Paraná
Curitiba, Paraná, Brazil

Anand Krishnan
Cardiology Department
Sree Chitra Tirunal Institute for Medical Sciences and Technology
Thiruvananthapuram, Kerela, India

Mohammed Kuddus
Department of Biochemistry
University of Hail
Hail, Kingdom of Saudi Arabia

Reena Kumari
Institute of Bioresources and Sustainable Development
Regional Centre
Gangtok, Sikkim, India

Laya Liz Kuriakose
Department of Food Technology
TKM Institute of Technology
Kollam, Kerala, India

Luiz Alberto Junior Letti
Department do Bioprocess Engineering and Biotechnology
Federal University of Paraná
Curitiba, Paraná, Brazil

Ashwini Manjunath
Department of Microbiology
Faculty of Life Sciences
JSS Academy of Higher Education and Research
Mysuru, Karnataka, India

Clara Matte Borges Machado
Department do Bioprocess Engineering and Biotechnology
Federal University of Paraná
Curitiba, Paraná, Brazil

Aravind Madhavan
Mycobacterium Research Group
Rajiv Gandhi Center for Biotechnology
Jagathy, Thiruvananthapuram, Kerala, India

Nur Maiyah
Department of Fermentation Technology
School of Food-Industry
King Mongkut's Institute of Technology Ladkrabang
Bangkok, Thailand

Walter José Martínez-Burgos
Department do Bioprocess Engineering and Biotechnology
Federal University of Paraná
Curitiba, Paraná, Brazil

Wahyu Haryati Maser
Department of Food Science and Technology
School of Food-Industry
King Mongkut's Institute of Technology Ladkrabang
Bangkok, Thailand

Fernanda Kelly Mezzalira
Department do Bioprocess Engineering and Biotechnology
Federal University of Paraná
Curitiba, Paraná, Brazil
and
Department of Chemistry and Biology
Universidade Tecnológica Federal do Paraná
Curitiba, Paraná, Brazil

Gilberto Vinicius de Melo Pereira
Department do Bioprocess Engineering and Biotechnology
Federal University of Paraná
Curitiba, Paraná, Brazil

Ariane Fátima Murawski de Mello
Department do Bioprocess Engineering and Biotechnology
Federal University of Paraná
Curitiba, Paraná, Brazil

María José Negro
Advanced Biofuels and Bioproducts Unit
Department of Energy
CIEMAT
Madrid, Spain

Maria Giovana Binder Pagnoncelli
Department of Chemistry and Biology
Universidade Tecnológica Federal do Paraná
and
Department do Bioprocess Engineering and Biotechnology
Federal University of Paraná
Curitiba, Paraná, Brazil

Ashok Pandey
Centre for Innovation and Translational Research
CSIR – Indian Institute for Toxicology Research (CSIR-IITR), Lucknow, India
and
Centre for Energy and Environmental Sustainability
Lucknow, Uttar Pradesh, India
and
Sustainability Cluster
School of Engineering
University of Petroleum and Energy Studies
Dehradun, Uttarakhand, India

Pratyusha Patidar
School of Biosciences and Bioengineering,
Indian Institute of Technology Mandi,
Kamand, Mandi, Himachal Pradesh, India

Rafaela de Oliveira Penha
Department do Bioprocess Engineering and Biotechnology
Federal University of Paraná
Curitiba, Paraná, Brazil

Kantiya Petsong
Department of Food Technology
Faculty of Technology
Khon Kaen University
Khon Kaen, Thailand

Dave Jaydeep Pinakin
Department of Food Science and Technology
School of Food-Industry
King Mongkut's Institute of Technology
Ladkrabang
Bangkok, Thailand

Muthu Suryia Prabha
Multi Disciplinary Research Unit (MDRU), Tirunelveli
Medical College, Tirunelveli, Tamil Nadu, India

Fernanda Guilherme Prado
Department do Bioprocess Engineering and Biotechnology
Federal University of Paraná
Curitiba, Paraná, Brazil

Sangeeta Prakash
School of Agriculture and Food Sciences
University of Queensland
Queensland, Australia

Amit Kumar Rai
Institute of Bioresources and Sustainable Development
Regional Centre
Sikkim, India

Mousumi Ray
Department of Microbiology and Fermentation Technology
CSIR – Central Food Technological Research Institute
Mysuru, Karnataka, India

Carolina Mene Savian
Department do Bioprocess Engineering and Biotechnology
Federal University of Paraná
Curitiba, Paraná, Brazil

Thamarys Scapini
Department do Bioprocess Engineering and Biotechnology
Federal University of Paraná
Curitiba, Paraná, Brazil

Agnes de Paula Scheer
Department of Chemical Engineering
Federal University of Paraná
Centro Politécnico, Curitiba, Paraná, Brazil

Theeraphol Senphan
Program in Food Science and Technology
Faculty of Engineering and Agro-Industry
Maejo University
Chiang Mai, Thailand

Poonam Sharma
Department of Bioengineering
Integral University
Lucknow, Uttar Pradesh, India

Alexander da Silva Vale
Department do Bioprocess Engineering and Biotechnology
Federal University of Paraná
Curitiba, Paraná, Brazil

Raveendran Sindhu
Department of Food Technology
TKM Institute of Technology
Kollam, Kerela, India

Carlos Ricardo Soccol
Department do Bioprocess Engineering and Biotechnology
Federal University of Paraná
Curitiba, Paraná, Brazil

Divyang Solanki
School of Agriculture and Food Sciences
University of Queensland
Brisbane, Australia

Tulika P. Srivastava
School of Biosciences and Bioengineering,
Indian Institute of Technology Mandi,
Kamand, Mandi, Himachal Pradesh, India

Elia Tomás-Pejó
Biotechnological Processes Unit
IMDEA Energy
Móstoles, Spain

Kim Kley Valladares-Diestra
Department do Bioprocess Engineering and Biotechnology
Federal University of Paraná
Curitiba, Paraná, Brazil

Jalaja Vidya
Department of Botany and Biotechnology
Milad E-Sherif Memorial College (MSM)
Kayamkulam, Alappuzha, Kerala, India

Jéssica Aparecida Viesser
Department do Bioprocess Engineering and Biotechnology
Federal University of Paraná
Curitiba, Paraná, Brazil

Yesodharan Vysakh
Cardiology Department
Sree Chitra Tirunal Institute for Medical Sciences and Technology
Thiruvananthapuram, Kerela, India

Adenise Lorenci Woiciechowski
Department do Bioprocess Engineering and Biotechnology
Federal University of Paraná
Curitiba, Paraná, Brazil

Section I

Introduction and Perspective

1 Microbial Enzymes for Production of Functional Foods and Nutraceuticals

Kalivarathan Divakar[1], Muthu Suryia Prabha[2], Parameswaran Binod[3,4], and Palanisamy Athiyaman Balakumaran[3,4]

[1] Department of Biotechnology, Sri Venkateswara College of Engineering, Sriperumbudur, Tamil Nadu, India

[2] Multi Disciplinary Research Unit (MDRU), Tirunelveli Medical College, Tirunelveli, Tamil Nadu, India

[3] Microbial Processes and Technology Division, CSIR – National Institute for Interdisciplinary Science and Technology (CSIR-NIIST), Thiruvananthapuram, Kerala, India

[4] Academy of Scientific and Innovative Research (AcSIR), Ghaziabad, Uttar Pradesh, India

CONTENTS

1.1 INTRODUCTION

The beneficial effect imparted by the dietary components on the health and its awareness among consumers increased the market requirements and supply of functional foods and nutraceuticals (Rai et al., 2019). Regular foods, apart from their normal nutritional characteristics, deliver several benefits to the health like the deterrence of specific diseases (Nesse et al., 2011; Rai et al., 2017a; Sanjukta et al., 2021). Enzymes from different origins (Pokora et al., 2013; Ruthu et al., 2014; Shu et al., 2018) and microbial fermentation augment the bioactive component in foods (Sanjukta et al., 2021).

DOI: 10.1201/9781003311164-2

Bioprocessing in food industries through hydrolysis, mediated by enzymes and fermentation, has propelled the growth of functional foods against specific diseases (Fernandez-Lucas et al., 2017).

Catalytically superior enzymes find immense application in the food industry and appeal to researchers worldwide for bioprocess development based on enzymatic process (Speranza et al., 2018). Enzymes on one side deactivate the antinutritional property of the food, while on the other side promote stabilization, solubilization, and bioavailability of bioactive components (Draelos, 2008). Enzymes from different origins have been successfully deployed for (1) reducing the antinutritional components (Bhagat et al., 2016; Porres et al., 2006), and (2) producing bioactive molecules through hydrolytic or transformative strategies (Kuo et al., 2018). Besides the augmentation of bioactive molecules in functional foods, the enzymatic process has also been successfully deployed in the mining of nutraceuticals with a high value from the by-products of food processing industries (Hathwar et al., 2012; Rai et al., 2013). Bioactive molecules released in functional foods as a result of enzymatic bioprocessing possess unique properties like anticancer potential, immunomodulation, and scavenging free radicals (Agyei et al., 2018; da Silva et al., 2019; Delgado-Garcia et al., 2019; Fujimoto et al., 2009; Morais et al., 2013; Park and Oh, 2010).

Asparaginase, lipase, phytase, protease, and carbohydrate-active enzymes are the most popular enzymes known to improve the bioactivity of functional foods. Most of the enzymes used in the food industry are derived from microbes due to several advantages like ease of optimization, enzyme engineering, and better control of the microbial process. Enzymes from animal and plant origin have also been applied in the food processing industries (Aburai et al., 2019; Afsharnezhad et al., 2019; Lafarga et al., 2016; Sripokar et al., 2019; Timon et al., 2019; Zhang et al., 2017).

1.2 ROLE OF ENZYMES IN THE PRODUCTION OF FUNCTIONAL FOODS AND NUTRACEUTICALS

1.2.1 Role of Protease

The release of bioactive peptides to enhance the functional properties of foods is primarily mediated by protease enzymes through the hydrolysis of peptide bonds (Agyei et al., 2018; Chourasia et al., 2020a; Speranza et al., 2018). Different varieties like aspartic protease, glutamic protease, serine protease, and metalloprotease have immense applications in food industries. Properties like anticancer potential, immunomodulation, and scavenging of free radicals are imparted by bioactive peptides, produced as a result of protease activity (Rai and Jeyaram, 2015; Sanjukta et al. 2021). The size of the peptides and sequence of the amino acids generated as a result of protease-mediated hydrolysis results in different activities on health. Proteases isolated from microbes thriving in extreme conditions have been known to possess better catalytic properties and stability than the enzymes extracted from animal/plant sources (Chourasia et al., 2020b). Hypertension and other related disorders have been treated using the inhibitory peptides of Angiotensin-I-converting enzyme (ACE) (Abedin et al., 2022; Chourasia et al., 2022; Rai et al., 2017b). The use of rennet from microbes for the production of cheese from cow milk with enriched antioxidant potential and goat milk hydrolysis using ACE have shown the potential role of protease for functional food products (Chourasia et al., 2020c; Timon et al., 2019). Food industries have started applying the fish protein hydrolysate using papain as a functional feed with high antioxidant potential (Fernandez-Lucas et al., 2017). Allergenicity in milk-derived products induced by β-lactoglobulin has been significantly lowered through the serine protease trypsin, derived from animals (Mao et al., 2019). Rennet derived from microbes like *Aspergillus* sp. and *Rhizomucor* sp. are now used as an alternative to animal-derived rennet by dairy industries worldwide to meet 33% of cheese production (Afroz et al., 2015).

1.2.2 Oligosaccharides Production through Enzymes

Oligosaccharides and polysaccharides play a vital part in maintaining human physiology and also play a major role in disease prevention and reducing the pathogenic effects. Carbohydrates like glycoproteins and gangliosides are well known for their health benefits. Bioactive carbohydrates with a low glycaemic index were used for making fortified foods and pharmaceutical applications. Fructooligosaccharides (FOS), Xylooligosaccharides (XOS), Galactooligosaccharides (GOS), Transgalactooligosaccharides (TOS), Chitin oligosaccharides (Na-COS), and Mannan-oligosaccharides (MOS) are prebiotics and dietary fibres with health benefits (Pereira-Rodríguez et al., 2012; Picazo et al., 2018; Martins et al., 2019). Conversion of the substrate inulin and sucrose to FOS is primarily mediated by inulinases and fructosyl transferases (Vega and Zuniga-Hansen., 2014). Na-COS are reported for their inhibitory effect of oxidative stress in live cells (Ngo et al., 2008). Mannan oligosaccharides (MOS) produced from pretreated coconut meal substrate using a recombinant mannanase was reported to have epithelial tight junction enhancing activity.

Enzymes from fungi like *Aspergillus* sp. and *Penicillium* sp. are commonly known to be used for the production of FOS (Dominguez et al., 2014; Muniz-Marquez et al., 2016; Sangeetha et al., 2004). *Aureobasidium pullulans* β-fructofuranosidase mediated transfructosylation is a widely used protocol for the production of FOS (Jung et al., 1987). Endo-inulinase is a significant enzyme used for FOS production next to fructofuranosidase. Many microbes including *Aspergillus* sp., *Xanthomonas* sp., and *Pseudomonas* sp. are known to produce endo-inulinase (Picazo et al., 2018). FOS production through endo-inulinase isolated from *Pseudomonas* sp. reached a higher yield of 80% compared to the yield of 65% with commercially available endo-inulinase (Hyun Kim et al., 1997; Singh et al. 2016). Immobilized enzymes have also been routinely applied for the production of FOS by utilizing molasses as the substrate and the reaction through a membrane reactor (Ghazi et al., 2006).

1.2.3 Role of Asparaginase

Plant-derived vegetables, such as potatoes, and grains serve as an important raw material for the production of fried and baked foods. The Maillard reaction is a type of chemical reaction that happens between sugars and asparagine and produces acrylamide in fried foods (Xu et al., 2016). A high concentration of acrylamide is toxic to health and hence strategies like reducing the temperature, pH level, and use of additives are applied to the process (Palazoglu and Gokmen, 2008; Zyzak et al., 2003). Asparaginase acts as an alternative to physical methods by converting asparagine into products like aspartic acid and ammonia. It helps to reduce the acrylamide precursor burden in baked and fried foods (Hendriksen et al., 2009). Research studies show that *A. oryzae*-derived asparaginase (Dias et al., 2017) and *Aquabacter*-derived asparaginase reduced the acrylamide concentration by 72% and 88%, respectively (Sun et al., 2015). Acrylaway, the commercial name of the asparaginase enzyme produced by Novozymes from *A. oryzae* and preventase from *Aspergillus niger* are the commercial enzymes used for the preparation of acrylamide-free products (Speranza et al., 2018).

1.2.4 Role of Phytase

Phytases are known to promote phytic acid dephosphorylation and generate products that possess better solubility and reduced chelating range, and have a lower impact on the absorption of minerals (Greiner and Konietzny, 2006). The salt form of phytic acid is known to impart properties like complex formation with minerals such as potassium, calcium, manganese, zinc, iron, and magnesium that occur in foods. Phytase plays a key role in the development of functional foods, especially for people susceptible to mineral deficiency (Jain and Singh, 2016). Research reports provide evidence for the development of foods with low phytate content through the process of phytate dephosphorylation mediated by phytase (Gupta et al., 2015). Phytase treatment reduced oxalate and phytate content in foods and suppressed the development of kidney stones (Israr et al., 2017). Likewise, treatment of wholegrain bread with phytase augmented the mineral and fibre bioavailability and reduced the phytate levels without compromising the nutritional characteristics of the bread (Bokhari et al., 2012; Dulinski et al., 2016; Garcia-Mantrana et al., 2014; Iglesias-Puig et al., 2015). Consumption of rice, wheat, and sorghum treated with phytase was reported to enhance the bioavailability of zinc and iron (Abid et al., 2017; Nielsen and Meyer, 2016). In another study, the consumption of phytase-treated sorghum by rats improved the digestibility of its phosphate content and decelerated phosphate excretion (Schons et al., 2011).

1.3 ENZYMES IN THE FOOD INDUSTRY

Enzymes such as lipase, amylase, phytase, invertase, and protease find immense uses in the food industry (Kumari et al., 2021). Chemical or enzyme-mediated modifications alter the physical properties of food products. Chemical-mediated modifications have several disadvantages like unfavourable reactions, trouble in downstream processing, and application of nonspecific reagents. Enzymes, on the other hand, possess unique properties like specific catalysis and rapid reactions to improve the functional properties of the food products.

Size of proteins and their conformation are closely associated with the functional properties of proteins like solubility, gelation, and emulsification. Hydrolysed protein-derived peptides are classified based on their molecular weight. Smaller peptides are used in pharmaceutical industries, whereas medium and large peptides are used in clinical and functional care products (Chourasia et al., 2020a,c). Fruit juices, infant food products, beverages, and therapeutic products are fortified with protein-derived hydrolysates and impart several health benefits to the body. *Bacillus*- and *Aspergillus*-derived protease have been used in the preparation of ACE inhibitors for the treatment of hypertension (Erdmann et al., 2008). Fishery resources derived from protein hydrolysates have been used as ingredients and supplements in functional foods and beverages. The backbones of fish hydrolysed by protease enzymes exhibited excellent antioxidative functions (Sheriff et al., 2014).

Food additives have been prepared by utilizing the biocatalytic properties of lipase. Heterologous expression of *Candida antarctica* lipase in methylotrophic yeast *Pichia pastoris* produced vitamin A esters with a yield of 80%. Many esters of ascorbic acid and several antioxidants were synthesized using lipase. Cheese ripening and augmentation of cheese flavour is enhanced through the fine-tuning of fatty acid chain length mediated by lipases (Ghosh et al., 1996). Like lipase, arabinofuranosidases are also known to induce hydrolysis and the subsequent release of arabinose. L-arabinose is used in food industries as a food additive due to its unique properties like low metabolic absorption rate, antiglycaemic nature, and sweeter taste (Raweesri et al., 2008). Arabinofuranosidases also improve the hydrolysis of pentosans present in the dough, promoting the formation of free pentoses with improved solubility and enhanced acidification mediated by lactic acid bacteria (Gobbetti et al., 2000).

Agro-residues consist of three main components namely cellulose, hemicellulose, and lignin. These components are predominantly comprised of phenolic compounds and polysaccharides which can be hydrolysed by lignocellulosic enzymes. Lignocellulosic enzymes comprise nearly 25% of the world's commercial enzyme market used in the juice industries. A blend of different enzymes at predefined ratios is used for augmenting the extraction process, the processing output, and preparation of solids-free fruit juices (Toushik et al., 2017). Endo-β-D glucanase, exo- β-D-glucanase, and β-glucosidase constitute the three enzymes of the cellulase complex. Endo- and exo-glucanase convert cellulose to oligosaccharides followed by the hydrolysis of oligosaccharide molecules to glucose by β-glucosidase. Alteration of the glycosylated compounds is known to augment the aroma of wine through the action of β-glucosidase. Bitterness in citrus fruit juices tends to lower the product quality. Therefore, a mixture of β-glucosidase and pectinases enzymes is used in beverage industries to improve the taste, aroma, and properties of juices (Baker and Wicker, 1996). An enzymatic cocktail containing enzymes like hemicellulase, cellulase, and β-glucosidase is known to improve the extraction of oil from olives, decreasing the viscosity and accelerating the cloud stability.

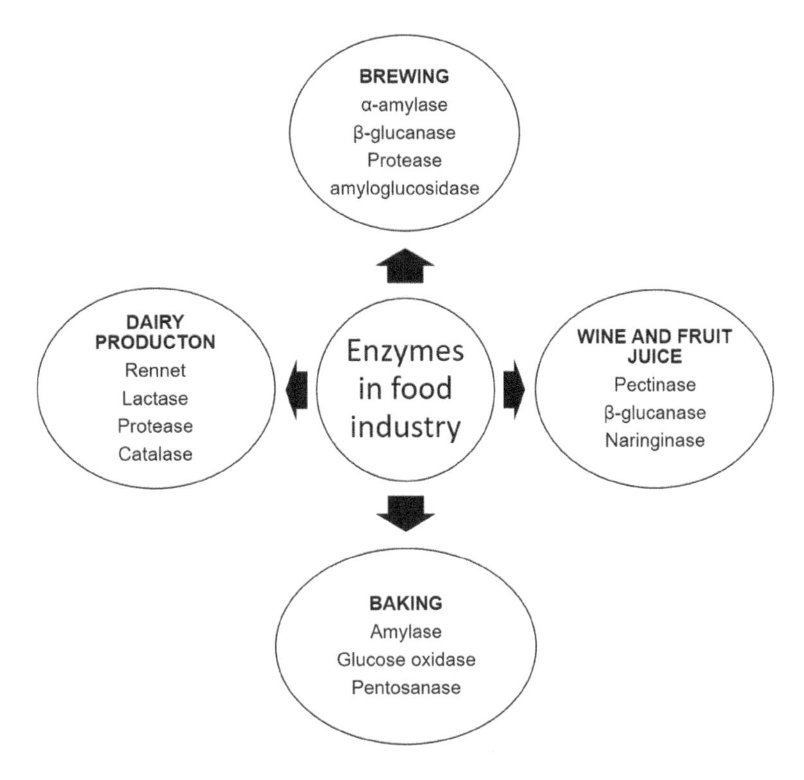

FIGURE 1.1 Schematic representation of enzymes used in different food industries.

Likewise, the nutrition composition of the wheat-based diet is improved through endo-1,4-β-xylanase isolated from *Thermomyces* sp. (Francesch et al., 2012). Similarly, enzymes that degrade the cell wall components of plants comprise mainly lignin peroxidase, laccase, and manganese peroxidases. Laccases are copper-containing oxidoreductases capable of both lignin degradation and clarification of fruit juice and wine through the removal of unwanted phenolic compounds that contribute to turbidity and haziness in juices (Rodríguez and Toca, 2006). The release of aromatic compounds is favoured through the action of lignin peroxidase and manganese peroxidase (Lesage-Meessen et al., 1996). Enzymes used in different food industries are mentioned in Figure 1.1.

1.4 MICROBIAL SOURCE OF ENZYMES USED IN FUNCTIONAL FOODS

Functional food using microbes may be prebiotics, probiotics, or a combination of both; or the natural enzymes extracted from microbes or enzymes, or secondary metabolites which are produced as a result of genetic engineering. Enzymes are used as part of the biochemical operations such as the biotransformation and stabilization of foods like baked goods, fruit juices, dairy, wine, and brewed beverages. Traditionally, several enzymes have been used for the production of compounds such as xylose oligosaccharides and bioactive peptides (Ottens and Chilamkurthi, 2013; Tahergorabi et al., 2012). Microbes produce enzymes naturally for their metabolism, for their survival, or as secondary metabolites for their survival in competition from antagonism or symbiotic relationships. These microbial enzyme usages are mostly in functional foods used for improving a person's natural digestive tract microflora or in food for people intolerant to certain kinds of food products. Using microorganisms for the fermentation of food is an age-old process employed since ancient times. Fermentation methods are still used to produce a variety of foods that provide additional health benefits. Microbial enzymes are used a lot in the food industries as they are more stable than plant and animal enzymes. These enzymes of microbial origin can be produced cost-effectively with less time and space. Microbes can be used to produce a large number of enzymes than animals and plants which are not recovered economically from plants and animals, easy, cost-effective and consistent in production. Microorganisms such as bacteria, yeast, fungi, and their enzymes are widely used in several food preparations for improving taste and texture, and they offer huge economic benefits to industries (Divakar et al., 2010; Priya et al., 2014; Sindhu et al., 2022). The application of microbial enzymes and their sources are shown in Table 1.1.

TABLE 1.1
Microbial Enzymes Used for Functional Food Processing and Their Sources

Enzyme	Microbial Source	Applications	Reference
α-Amylases (EC 3.2.1.1)	*A. oryzae, Bacillus* sp., *Bacillus licheniformis*	Bakery products, brewing industry Clarification of fruit juices, starch liquefaction	(Sindhu et al., 2017) (Zhang et al. 2019).
Xylanase	*Streptomyces* sp., *Bacillus* sp. and *Pseudomonas* sp.	Bakery products, beer-making industry	(Sanghi et al., 2010) (Dervilly, 2002)
Glucose and hexose oxidase (EC1.1.3.4 and 1.1.3.5)	*Aspergillus* sp., *Hansenula polymorpha*	Bakery products for better dough quality	(Hellmuth and van den Brink, 2013)
Glucoamylase (EC3.2.1.3)	*Aspergillus niger* and *Aspergillus awamori, Rhizopus oryzae*	Bakery products, production of high-glucose syrups and high-fructose syrups	(James et al., 1996)
Protease	*Aspergillus usami Bacillus* sp.	Brewing, meat tenderization, coagulation of milk, bread quality improvement	(Razzaq et al., 2019)
Lactase β-galactosidase	*Kluyveromyces lactis Kluyveromyces fragilis Lactobacillus* sp.,	Prebiotic food ingredients, lactose intolerance	(Boon et al., 2000) (Jurado et al., 2002)
Lipase	*Penicillium camemberti, Aspergillus niger* or *A. oryzae, Candida* sp.,	Polyunsaturated fatty acids (PUFA) production, short chain fatty acid esters, flavouring agents	(Gerits et al., 2014) (Castejón and Señoráns, 2020)
Phospholipases	*Fusarium oxysporum,*	Production of oils, dairy industry and in the manufacture of several bakery items, enhancing the cheese flavour, lipolyzed milk fat production for use in butter as flavour enhancer, etc.	(Law, 2009) (De Maria et al., 2007) (Zhao et al., 2010)
Pectinase	*Bacillus, Erwinia, Kluyveromyces, Aspergillus, Rhizopus, Trichoderma, Pseudomonas, Penicillium, and Fusarium*	Paper bleaching, food industry, remediation, juice industry	(Saadoun et al., 2013) (Pedrolli et al., 2009)
Laccases (EC 1.10.3.2)	*Bacillus licheniformis Bacillus vallismortis*	Polyphenol removal from wine, baking	(Giardina et al., 2010) (Morozova et al., 2007)
Inulinases	*Pseudomonas sp., Aspergillus sp., Bacillus sp., Yarrowia sp., Penicillium sp.,* and *Xanthomonas sp.,*	FOS production	(Chourasia et al., 2020a)

Following are the characteristics that any microorganism should possess so that they can be used as a source of enzymes in food processing:

- Higher growth rate under inexpensive medium
- High enzyme yield
- Minimal production of undesirable contaminant enzymes
- Minimal fermentation/production cost and time
- Easy purification and recovery process

1.5 ENGINEERING ENZYMES FOR FUNCTIONAL FOODS

Enzymes used in food industries do not always meet the demands of harsh processing conditions like temperature, pH, and salinity and provide the expected results. Engineered enzymes, on the other hand, impart special properties like improved stability, substrate specificity, catalytic performance, and multifunctional properties (Chourasia et al., 2020b). Several hurdles and safety issues circumvent the use of engineered enzymes for functional foods, yet its fruitful outcomes in relevant industries and consumers have prompted several research studies on recombinant enzymes. Methylotrophic yeast *Pichia pastoris* and *E.coli* are known to be the common hosts used for the production of engineered enzymes (Zhang et al., 2019). α-amylases isolated from *Bacillus licheniformis* was engineered by replacing its His residue with Asp residue. This novel amylase exhibited higher saccharification and activity at pH 4.5 (Zhang et al., 2019). Aglycone production was enhanced through the site-directed mutagenesis of β-glucosidase from *Thermotoga maritima* (Sun et al., 2014). Cleavage of β-1,4 glycosidic linkages in xylan was enhanced by endo-β-1,4-xylanase through specific mutation. This resulted in higher production of XOS from wheat straw (Faryar et al., 2015). Mutagenesis of *Rhizopus oryzae*-derived α-amylase at histidine present at 286th position changed the characteristics of the enzyme to withstand a lower pH and higher temperature of around 60°C compared to the wild-type enzyme (Li et al., 2018). Replacement of lysine with alanine or leucine in β-1,4 glucanase drastically augmented its ability to withstand high temperatures, halotolerance, and effectively hydrolysed β-1,4 glucans in barley and rice (Lee et al., 2017). The catalytic efficiency of invertase from *Saccharomyces cerevisiae* to convert sucrose to glucose and fructose was enhanced through the substitution of hydrophilic residues with hydrophobic residues (Mohandesi et al., 2017). Likewise, similar modifications enhanced the transfructosylation ability of invertase for efficient production of FOS (Marin-Navarro et al., 2015). Rational designing of β-galactosidase from *Thermotoga maritima* enhanced the production of GOS through its improved transglycosylating activity (Talens-Perales et al., 2016). The ability of β-galactosidase to process milk at a temperature as low as 8°C was accelerated through directed evolution (Rentschler et al., 2016). Thermal tolerance of lipase from *Malassezia globosa* was enhanced to promote diacylglycerol synthesis (Gao et al., 2014). Similarly, T1 lipases were genetically modified to improve the production short-chain fatty acids in milk to enhance the flavour of milk products (Tang et al., 2017). Protease from *Aspergillus* sp. was truncated to stimulate its activity for enhanced production of beneficial peptides, oil processing at optimum conditions (Ke et al., 2012). Higher catalytic activity and fibrinolytic performance were observed in nattokinase subjected to site-directed mutagenesis (Weng et al., 2015).

Caldicellulosiruptor sp.-derived cellobiose epimerase effectively produced lactulose from lactose without the by-product epilactose (Shen et al., 2016). Lactulose is known to confer health benefits and is also consumed as a prebiotic product. Low-calorie sweetener namely D-psicose was produced from D-fructose through D-psicose 3-epimerase-mediated isomerization (Zhang et al., 2016). *Yersinia*-derived phytase was engineered by site-specific mutagenesis to improve its thermotolerance, resistance to trypsin, and enhanced hydrolysis of phytic acid (Niu et al., 2017). Production of the prebiotic fructose dianhydride from inulin was enhanced by engineered inulin fructotransferase derived from *Streptomyces* sp. (Yu et al., 2017).

1.6 PRODUCTION AND RECOVERY OF ENZYMES

Enzymes have numerous applications in the food industry; various enzymes are produced on different scales depending on their utility rate. Speciality enzymes are produced on a small scale, whereas general-use food-grade enzymes are produced on a large scale, in tons of kilograms as per industry requirements. Other than a few special enzymes which can be produced only through plant and animal cells, most of the industrial enzymes are produced by microorganisms. About 90% of the enzymes available in the market are of microbial origin owing to their ease of maintenance and production and purification processes (Rebello et al., 2019; Sharma and Upadhyay, 2020). Microbial enzymes possessing the required functionality can be produced on a large scale through fermentation. The production of enzymes by microbes is very dynamic; it requires the optimization of fermentation conditions to produce the enzymes with the desired property at the desired quantity. The cost of enzyme-deriving processes basically depends on the cost of the enzyme. The cost of the enzyme depends on the raw material used for production, the fermentation conditions, and recovery processes. For example, in the production of the lignocellulosic enzyme, the raw material contributes to 28% of its total production cost (Ferreira et al., 2018). To reduce the cost of production of food-grade enzymes, low-cost substrates such as agricultural waste materials are used. In addition to this two or more enzymes are co-produced using single microbial fermentation, for which a single microbial strain able to produce multiple enzymes is preferred. This drastically reduces the time and cost of fermentation (Divakar et al., 2017; 2018; Suryia Prabha et al., 2015). Recombinant DNA techniques are applied for engineering the strains to increase the expression of enzymes thus improving the yield. Engineering the microbial strains and producing enzymes using microorganism is the easiest way when compared to other sources. GRAS-certified ("generally recognized as safe" certification by USFDA) microbes are preferred for the production of food-grade enzymes to reduce the risk of pathogenic strains or toxins, or undesirable toxic metabolites being present in the products (Binod et al., 2019). Screening of enzymes from GRAS organisms and optimizing the production process at a larger scale is generally considered a tedious task. To replace this, the genes responsible for enzymes produced by other microbes are cloned and expressed in GRAS microbes used for the production of food-grade enzymes/products. *Bacillus* strains are generally used for the production of enzymes because of ease of manipulation and production, and high product yield. Recombinant *Bacillus* strains are reported for production of several food grade enzymes including lipases, proteases, amylases, pullulanase, and cellulases (Cai et al., 2019; Kumari et al., 2021; Phukon et al., 2022). The purification of an enzyme depends on the type of microbe and the secretion processes adopted by the microbes (Ferreira et al., 2018). The enzyme production process is broadly classified as upstream bioprocessing and downstream bioprocessing. The upstream processing involves strain selection, strain improvement, selection of suitable raw materials, and media optimization. Batch or fed-batch fermentation conditions are used for the production of enzymes. Solid-state fermentation is also applied in enzyme production, especially for enzymes produced by mycelia using low-cost agricultural waste materials as substrates (Xu et al., 2021). After the upstream and fermentation processing, the enzymes are recovered through a series of downstream processing steps. The number and nature of the recovery process depends on the purity of the enzyme required at end. The downstream processing involves cell separation through centrifugation or membrane separation, purification of enzymes using simple precipitation, membrane separation, and chromatographic techniques, with the final finishing steps including freeze drying and storage (Binod et al., 2019; Divakar et al., 2017; Sharma et al., 2019). Generally continuous centrifuges are used in the separation of bacterial cells, but, for actinomycetes and mould filtration methods, plate and frame filters and rotary filters are preferred (Illanes, 2008). For purification of intracellular enzymes, cells are separated and lysed using mechanical methods like homogenization and ultrasonic cell lysis. Chemical and enzymatic methods are usually not preferred for cell lysis in enzyme purification, because chemical treatment may denature the target enzymes and the presence of trace chemicals in food products may lead to toxicity that may complicate the approval processes. The enzymatic method of cell lysis is not preferred due to the cost of the process involved and it cannot be applied for low-cost enzyme production on a large scale (Illanes, 2008).

1.7 ENZYMATIC BIOTRANSFORMATION FOR THE PRODUCTION OF NUTRACEUTICALS

Biotransformations using enzymes have been adopted for centuries to make foods and food products. Morden food technological industries use food-grade enzymes for the production and modification of food products and beverages. In addition to this, enzymatic biotransformation methods are used for the pre-treatment process in food product manufacturing units, in the selective modification of metabolites to improve the quality and nutritional value of food products. Several bioactive compounds are produced from fruits using enzymatic biotransformation. In addition to the improvement of bioactive molecules/metabolites in foods, the enzymatic bioprocess has been applied to improve the recovery and yield of high-value nutraceuticals in the food processing industries. Extraction of bioactive compounds from the fruits and vegetables are achieved through enzymatic methods. Extraction and biotransformation of phenolic compounds from citrus fruits were employed. Pectinase, cellulase, naringinase, lipases, esterases, tannase, and β-glucosidase are used independently or in combination to perform several biotransformation reactions. Citrus fruit juice extracts were treated with enzymes to change the phenolic profile, produce flavanones, and improve antioxidant activity (Ruviaro et al., 2019).

Food-derived biologically active proteins and peptide molecules are produced by the enzymatic action of several proteases. Peptides produced by enzymatic hydrolysis of casein and whey protein using microbial proteases have been reported as inhibitors of ACE. The protease enzyme papain, which degrades the Gln-Pro-rich allergenic wheat proteins, has been used for processing hypoallergenic flour. β-Galactosidase was reported to convert lactose to allolactose, which is used in the production of lactose-free milk products. Inulin is used for FOS biotransformation and production. Endo-inulinases enzymes were reported from several bacterial and fungal strains *Pseudomonas* sp., *Aspergillus* sp., *Bacillus* sp., *Yarrowia* sp., *Penicillium* sp., and *Xanthomonas* sp (Chourasia et al., 2020b). During the backing process at high temperatures, the amino acid asparagine reacts with the reducing sugars via the Maillard reaction that produces acrylamide. Acrylamide is a carcinogen; presence of this carcinogen in food is lethal to cells. Since asparagine is the main precursor of acrylamide, reducing the concentration of asparagine present in the food before the baking process could reduce the accumulation of toxic product. This can be achieved by the enzymatic hydrolysis of asparagine into aspartic acid and ammonia y enzyme L-asparginase (Meghavarnam and Janakiraman, 2018).

In many citrus plants, biotransformation of the flavonoids happens naturally and, through this, glycosidic conjugates such as O-rutinosides or O-neohesperidosides are formed. These glycosylation reactions are catalysed by glycosyltransferases. Several glycosyltransferases are reported from various plant species. Apart from natural processes, the biotransformation of plant metabolites under in-vitro conditions using enzymes is attracting more attention. Enzymes are reported to convert glycosylated platycosides in *Platycodi radix* into deglycosylated platycosides. Biotransformation of Platycodin, platycoside E by cellulose, laminarinase, β-Glucosidase has been reported (Kil et al., 2019). The bioavailability of molecules present in the food products can be enhanced by enzymatic pre-treatment processes; for example, the bioavailability of flavonoids hesperidin and hesperetin can be enhanced by enzymatic biotransformation of Hesperetin to Hesperetin-7-O-glucoside by naringinase (Lee et al., 2012). Eicosapentaenoic acid (EPA), and docosahexaenoic acid (DHA) rich omega-3 fatty acids are produced by a biotransformation reaction catalysed by several lipase enzymes. DHA/EPA groups are present in one of the positions in the triglycerides, the other positions are filled with other fatty acids. The specific enhancement of DHA/EPA in the glycerol moiety can be achieved by selective lipolysis of fatty acid chains present in the triglycerides. For example, fish oils contain DHA in the sn2 position of the glycerol backbone; positions sn1 and sn3 occupied by other fatty acids can be selectively hydrolysed by lipases to produce polyunsaturated fatty acids (PUFA) (Castejón and Señoráns, 2020).

1.8 APPLICATIONS OF ENZYMES IN FUNCTIONAL FOODS AND NUTRACEUTICALS

The functional modification of foods using chemical and mechanical methods has its disadvantages ranging from the high cost of processing to the side effects of using chemicals. The mechanical methods such as ultrasonication, pulse-field electric, and hydrostatic pressure require high energy, are difficult to operate, and increase the processing cost. Extraction methods involving active enzyme ingredients are considered an environmentally friendly approach for the processing of food products and their raw materials. Enzymes play a key role in the extraction of pigments, flavours, and metabolites from agricultural raw materials. Food industries including agriculture processing industries or meat processing industries use enzymes for improving the functional and nutritional value of food products. Using enzyme for the preparation of functional foods does not require harsh industrial processing. Lipolytic enzymes are used for recovering phytosterols having high lipid content. Low-molecular-weight carbohydrates are used as functional constituents in food products due to their physiochemical properties but they are non-digestible. Enzymes such as phospolipases (A and C) are used in the extraction of lecithin and other active metabolites, without using organic solvents. In a few cases, the presence of wild-type enzymes reduce the usability of the food grains or its products. Modification of these sources could improve their availability for the production of processed food. For example, rice bran (a byproduct obtained from the rice milling industry) is known to be rich in nutraceuticals, but is not utilized because of rancidity caused by lipases and lipooxigenases present in the rice bran. Naringinases are used for treating citrus pulp to remove bitterness in processed fruit juices and pectinases are used as a stabilizing agent in clarifying fruit juice. The antioxidant potential in the wheat bran was increased by enzymatic treatment of wheat bran with α-amylase. Also, α-amylase is used for milk coagulation. Bacterial amylases are widely used in bread making, which improves yield and shelf life, and decreases the crumb firmness. The viscosity of the dough is reduced by hydrolyzing the starch present in wheat flour which increases the bread yield. In addition to this, amylases are used for the production of sugar syrups. Bromelain present in pineapple is known to act against cancer cells by improving the cytotoxicity of monocytes and also it acts as an immunomodulating agent. These non-digestible oligosaccharides need enzymatic pre-processing to make them more usable. The chemical inhibition of the above-mentioned enzymes will lead to side effects if a trace amount of chemical molecules is present in it after processing. Alcalase enzyme treatment was reported to inhibit the lipase, lipooxigenase, and endogluconase activity. This reduced the oxidation of free fatty acids and improved the availability of soluble fibre content (Vallabha et al., 2015). Glucose oxidase acts as an oxygen scavenger and improves the shelf life of beverages; it also reduces the loss of food colour which is mainly caused by oxidation/interaction with oxygen. Microbial transglutaminase was used for the restructuring of meat. The restructuring of meat was catalysed by the formation of high molecular weight polymers by the transglutaminase enzyme through covalent cross-linking of amino acids glutamine and lysine. In addition to this transglutaminase can modify the textural property of meat. The exogenous lysine is used for producing low-salt meat products (Sorapukdee and Tangwatcharin, 2018). Apart from this the enzyme-modified products are used as cryopreservants and improve the shelf life of the products. The peptides formed through hydrolysis of gelatin by proteases were reported to act as cryo-preservatives for marine-based products (Karnjanapratum and Benjakul, 2015). A list of several enzymes, their function in the formation of functional foods/ingredients, and their pharmaceutical application is given in Table 1.2. Enzymes used for food processing and their benefits are shown in Figure 1.2.

TABLE 1.2
Effect of Enzymes on Production of Functional Products and Their Biological Activity

Enzyme	Enzymatic Product / Functional Food	Biological Activity	Reference
Candida antarctica lipase A	Highly pure (~90%) DHA concentrate from alga oil	Reduce the prevalence of cardiovascular diseases	(Akanbi et al., 2012)
Lipase	Omega 3 concentrates useful for food fortification	Prevented non-melanoma skin cancer	(Gunathilake et al., 2021) (Castejón and Señoráns, 2020)
β-Galactosidases	Production of oligosaccharide prebiotics (GOS, TOS)	Stimulating the growth and activities of probiotic bacteria in the gut	(Pereira-Rodríguez et al., 2012)
Protease	Production of functional peptides	ACE inhibition properties Immunomodulatory properties	(Lafarga et al., 2016)
Carbohydrase / Cellulases	Enzyme-assisted carbohydrate and oligosaccharide extraction	Protective effects against high glucose-induced oxidative stress	(Lee et al., 2010)
κ-carrageenase	κ-carrageenan oligosaccharides	Anticoagulant, antioxidant, anti-inflammatory activities	(Zhu et al., 2018)
Alcalase	Seed protein hydrolysates	Antibacterial activity	(Osman et al., 2021)

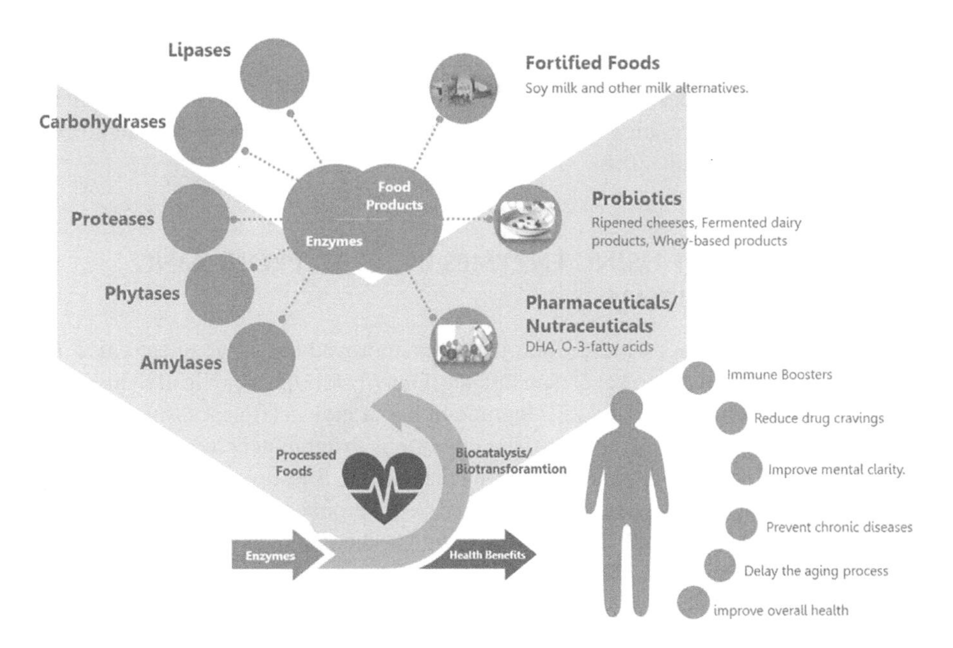

FIGURE 1.2 Enzymes in processing food products for pharmaceuticals/nutraceuticals and their health benefits.

1.9 CHARACTERISTICS OF ENZYMES TO BE USED FOR FOOD PROCESSING AND NUTRACEUTICALS

Enzymes used for food processing are selected from various sources, such as plants, animals, and microbes. The sources should support a high yield of enzymes and have very few side effects. This can be achieved by a series of purification and recovery processes. However, the source material should possess certain characteristics such as non-pathogenicity, being free of toxic substances/metabolites, or free of carcinogenic metabolites, and should not be modified to toxic form under various process conditions. For the enzymes derived from plant material the source should not produce any other toxic metabolite even in trace concentration. For microbial sources it should be a GRAS organism, it should produce enzymes at a high rate, and be easy to modify and scale up. Beside the source material, the enzyme itself should possess certain characteristics which suit the process conditions required for optimal processing of food products. The enzyme should possess operational stability and should be stable under minor operational fluctuations. Enzymes which are stable under wide pH and temperature ranges are most preferred. The characteristics of stable enzyme include the following:

- Low production cost
- No side effects from the formation of by-products
- A good shelf life at ambient conditions
- Stability under a wide range of ambient conditions
- Resistance against alterations by mild chemical or biological reactions

1.10 CURRENT LIMITATIONS IN USING ENZYMES

Currently available enzymes are limited due to specific requirements of enzymes of each individual food industry. Various operating conditions are followed in the food industries for the processing of various food products. The industrial needs are currently met using the available enzymes or industrial processes are modified to suit the existing commercial enzymes. There is always, therefore, an inherent lag in meeting industrial requirements. An approach to finding an enzyme that meets industrial requirements needs to be adopted. In the production of pineapple-based jelly, fresh pineapple extracts cannot be used because the bromelain (a proteolytic enzyme present in the fresh pineapple) inhibits the gel formation by hydrolyzing the gelatin, a key ingredient required for the formation of jelly (Ahmad et al., 2018).

1.11 FUTURE SCOPE OF USING ENZYMES IN FOOD PROCESSING AND NUTRACETUCIALS

The recent increase in the usage of molecular tools has improved the usage of enzymes in several food industries. Enzymes tuned to act on specific nutrients/metabolites can lead to meeting industrial needs. The native enzymes often fail/degrade in harsh industrial processing. Further modification of enzymes to meet the harsh industrial conditions through genetic engineering methods could improve enzyme usage in food processing industries. Site-directed mutagenesis (SDM) was used for improving the specificity of an enzyme to produce a high yield of product and also to improve the pH or thermal stability of enzymes. The enzymes are tuned to improve the catalytic rate through SDM approach. Gene sequence optimization was also adopted to improve the expression, production and activity of the enzyme several times higher than the wide type strains. In this codon bias of the host cell is optimized to improve the heterologous production of active enzymes. Based on knowledge of currently available well-studied enzymes, enzymes are designed through a directed evolution approach. Through which several sets of genes coding the enzymes are generated, the one with the highest activity and specificity or selectivity were chosen (Zhang et al., 2019). Currently, the enzymes used for food industries are derived from the known microbial strain cultivated under laboratory conditions. The genetically modified enzymes are again derived for the cultivable microbiome. However, screening of enzymes from uncultivable microbiome could give the space for exploring more enzymes with unusual activity and stability. Metagenomics is a tool used for screening enzymes that can be stable and active under extreme environmental conditions. Metagenomic approaches could advance the possibility of screening the enzyme with diverse characteristics from uncultivable microbiomes (Boddu et al., 2022; Boddu and Divakar, 2018; Sindhu et al., 2022).

1.12 CONCLUSION

Currently, enzymes are considered as an inherent part of any bioprocessing industry. However, the production, purification and usage of enzymes under industrial process conditions is always considered a challenging task. Among the industrial enzymes, the enzymes used in the food industry share the major market. Production, storage of enzymes and improving the shelf life is further considered as challenging. The feasibility of enzymes to be used for biotransformation of food-grade products and key metabolites needs to be analysed. Advancements in genome engineering tools can add up the opportunity for using enzymes in food processing industries and the production of functional foods and nutraceuticals.

REFERENCES

Abedin, M.M., Chourasia, R., Phukon, L.C., Singh, S.P., & Rai, A.K. 2022. Characterization of ACE inhibitory and antioxidant peptides in yak and cow milk hard chhurpi cheese of the Sikkim Himalayan region. *Food Chemistry: X* 13: 100231.

Abid, N., Khatoon, A., Maqbool, A., Irfan, M., Bashir, A., Asif, I., Shahid, M., Saeed, A., Brinch-Pedersen, H., & Malik, K.A. 2017. Transgenic expression of phytase in wheat endosperm increases bioavailability of iron and zinc in grains. *Transgenic Research* 26: 109–122.

Aburai, N., Maruyama, S., Shimizu, K., & Abe, K. 2019. Production of bioactive oligopeptide hydrolyzed by protease derived from aerial microalga *Vischeria Helvetica. Journal of Biotechnology* 294: 67–72.

Afroz, Q.M., Khan, K.A., Ahmed, P., & Uprit, S. 2015. Enzymes used in dairy industries. *International Journal of Applied Research* 1: 523–527.

Afsharnezhad, M., Shahangian, S.S., & Sariri, R. 2019. A novel milk-clotting cysteine protease from Ficus johannis: Purification and characterization. *International Journal of Biological Macromolecules* 121: 173–182.

Agyei, D., Akanbi, T.O., & Oey, I. 2018. Enzymes for use in functional foods. *Enzymes in Food Biotechnology*, Elsevier Inc. Available from: http//doi.org/10.1016/b978-0-12-813280-7.00009-8.

Ahmad, I.Z., Tabassum, H., Ahmad, A., Kuddus, M. 2018. Food Enzymes in Pharmaceutical Industry: Perspectives and Limitations. In: Kuddus, M. (eds) *Enzymes in Food Technology*. Springer, Singapore.

Akanbi, T.O., Barrow, C.J., & Byrne, N. 2012. Increased hydrolysis by Thermomyces lanuginosus lipase for omega-3 fatty acids in the presence of a protic ionic liquid. *Catalysis Science and Technology* 2: 1839.

Baker, R.A. & Wicker, L. 1996. Current and potential applications of enzyme infusion in the food industry. *Trends in Food Science and Technology* 7: 279–284.

Bhagat, J., Kaur, A., & Chadha, B.S. 2016. Single step purification of asparaginase from endophytic bacteria *Pseudomonas oryzihabitans* exhibiting high potential to reduce acrylamide in processed potato chips. *Food and Bioproducts Processing* 99: 222–230.

Binod, P., Papamichael, E., Varjani, S., & Sindhu, R. 2019. Introduction to green bioprocesses: Industrial enzymes for food applications. *Energy, Environment, and Sustainability* 1: 1–8.

Boddu, R.S., Ajay Prabhakar, K., & Divakar, K. 2022. Metagenomic bioprospecting of uncultivable microbial flora in soil microbiome for novel enzymes. *Geomicrobiology Journal* 39: 97–106.

Boddu, R.S. & Divakar, K. 2018. Metagenomic insights into environmental microbiome and their application in food pharmaceutical industry. *Microbial Biotechnology* 1: 23–38.

Bokhari, F., Derbyshire, E., Li, W., Brennan, C.S., & Stojceska, V. 2012. A study to establish whether food-based approaches can improve serum iron levels in child-bearing aged women. *Journal of Human Nutrition and Dietetics* 25: 95–100.

Boon, M.A., Janssen, A.E.M., & van 't Riet, K. 2000. Effect of temperature and enzyme origin on the enzymatic synthesis of oligosaccharides. *Enzyme and Microbial Technology* 26: 271–281.

Cai, D., Rao, Y., Zhan, Y., Wang, Q., & Chen, S. 2019. Engineering *Bacillus* for efficient production of heterologous protein: Current progress, challenge and prospect. *Journal of Applied Microbiology* 126: 1632–1642.

Castejón, N. & Señoráns, F.J. 2020. Enzymatic modification to produce health-promoting lipids from fish oil, algae and other new omega-3 sources: A review. *New Biotechnology* 57: 45–54.

Chourasia, R., Abedin, M.M., Phukon, L.C., Sahoo, D., Singh, S.P., & Rai, A.K. 2020c. Biotechnological approaches for the production of designer cheese with improved functionality. *Comprehensive Reviews in Food Science and Food Safety* 20: 960–979.

Chourasia, R., Padhi, S., Phukon, L.C., Abedin, M.M., Singh, S.P., & Rai, A.K. 2020a. A potential peptide from soy cheese produced using lactobacillus delbrueckii WS4 for effective inhibition of SARS-CoV-2 main protease and S1 glycoprotein. *Frontiers in Molecular Biosciences* 7: 413.

Chourasia, R., Phukon, L.C., Abedin, M.M., Sahoo, D., & Rai, A.K. 2022. Production and characterization of bioactive peptides in novel functional soybean chhurpi produced using Lactobacillus delbrueckii WS4. *Food Chemistry* 387: 132889.

Chourasia, R., Phukon, L.C., Singh, S.P., Rai, A.K., & Sahoo, D. 2020b. Role of enzymatic bioprocesses for the production of functional food and nutraceuticals. *Biomass, Biofuels, Biochemicals* 1: 309–334.

da Silva, R.R., Conceicao P.J.P., Menezes, C.L.A., *et al.* 2019. Biochemical characteristics and potential application of a novel ethanol and glucose-tolerant β-glucosidase secreted by *Pichia guilliermondii* G1.2. *Journal of Biotechnology* 294: 73–80.

De Maria, L., Vind, J., Oxenbøll, K.M., Svendsen, A., & Patkar, S. 2007. Phospholipases and their industrial applications. *Applied Microbiology and Biotechnology* 74: 290–300.

Delgado-Garcia, M, Flores-Gallegos, A.C., Kirchmayr, M., *et al.* 2019. Bioprospection of proteases from *Halobacillus andaensis* for bioactive peptide production from fish muscle protein. *Electronic Journal Biotechnology* 39: 52–60.

Dervilly, G. 2002. Isolation and characterization of high molar mass water-soluble arabinoxylans from barley and barley malt. *Carbohydrate Polymers* 47: 143–149.

Dias, F.F.G., Bogusz Junior, S., Hantao, L.W., Augusto, F., & Sato, H.H. 2017. Acrylamide mitigation in French fries using native L-asparaginase from *Aspergillus oryzae* CCT 3940. *LWT Food Science and Technology* 76: 222–229.

Divakar, K., Priya, J.D.A., & Gautam, P. 2010. Purification and characterization of thermostable organic solvent-stable protease from *Aeromonas veronii* PG01. *Journal of Molecular Catalysis B: Enzymatic* 66: 311–318.

Divakar, K., Priya, J.D.A., Selvam, G.P., *et al.* 2018. Detection of Multiple Enzymes in Fermentation Broth Using Single PAGE Analysis. *Methods in Molecular Biology* 133–138.

Divakar, K., Suryia Prabha, M., Devi, G.N., Gautam, P., Nandhinidevi, G., & Gautam, P. 2017. Kinetic characterization and fed-batch fermentation for maximal simultaneous production of esterase and protease from *Lysinibacillus fusiformis* AU01. *Preparative Biochemistry & Biotechnology* 47: 323–332.

Divakar, K, Suryia Prabha, M., & Gautam, P. 2017. Purification, immobilization and kinetic characterization of G-x-S-x-G esterase with short chain fatty acid specificity from *Lysinibacillus fusiformis* AU01. *Biocatalysis and Agricultural Biotechnology* 12: 131–141.

Dominguez, A.L., Rodrigues, L.R., Lima, N.M., & Teixeira, J.A. 2014. An overview of the recent developments on fructooligosaccharide production and applications. *Food and Bioprocess Technology* 7 (2): 324337.

Draelos, Z.D. 2008. The cosmeceutical realm. *Clinics in Dermatology* 26: 627–632.

Dulinski, R., Cielecka, E.K., Pierzchalska, M., Byczynski, Ł., & Zyła, K. 2016. Profile and bioavailability analysis of myo-inositol phosphates in rye bread supplemented with phytases: A study using an in vitro method and Caco-2 monolayers. *International Journal of Food Sciences and Nutrition* 67: 454–460.

Erdmann, K., Cheung, B.W., & Schröder, H. 2008. The possible roles of food-derived bioactive peptides in reducing the risk of cardiovascular disease. *The Journal of Nutritional Biochemistry* 19: 643–654.

Faryar, R., Linares-Pasten, J.A., Immerzeel, P., *et al.* 2015. Production of prebiotic xylooligosaccharides from alkaline extracted wheat straw using the K80R-variant of a thermostable alkali-tolerant xylanase. *Food and Bioproducts Processing* 93: 1–10.

Fernandez-Lucas, D., Castaneda, D., & Hormigo, D. 2017. New trends for a classical enzyme: Papain, a biotechnological success story in the food industry. *Trends in Food Science and Technology* 68: 91–101.

Ferreira, R.D.G., Azzoni, A.R., & Freitas, S. 2018. Techno-economic analysis of the industrial production of a low-cost enzyme using E. coli: The case of recombinant β-glucosidase. *Biotechnology for Biofuels* 11: 1–13.

Francesch, M., Perez-Vendrell, A.M., & Broz, J. 2012. Effects of a mono-component endo-xylanase supplementation on the nutritive value of wheat-based broiler diets. *British Poultry Science* 53: 809–816.

Fujimoto, Y., Hattori, T., Uno, S., Murata, T., & Usui, T. 2009. Enzymatic synthesis of gentiooligosaccharides by transglycosylation with β-glycosidases from *Penicillium multicolor. Carbohydrates Research* 344: 972–978.

Gao, C., Lan, D., Liu, L., Zhang, H., Yang, B., & Wang, Y. 2014. Site-directed mutagenesis studies of the aromatic residues at the active site of a lipase from *Malassezia globose. Biochimie* 102: 29–36.

Garcia-Mantrana, I., Monedero, V., & Haros, M. 2014. Application of phytases from bifidobacteria in the development of cereal-based products with amaranth. *European Food Research and Technology* 238: 853–862.

Gerits, L.R., Pareyt, B., Decamps, K., & Delcour, J.A. 2014. Lipases and their functionality in the production of wheat-based food systems. *Comprehensive Reviews in Food Science and Food Safety* 13: 978–989.

Ghazi, I., Fernandez-Arrojo, L., De Segura, A.G., Alcalde, M., Plou, F.J., & Ballesteros, A. 2006. Beet sugar syrup and molasses as low-cost feedstock for the enzymatic production of fructo-oligosaccharides. *Journal of Agricultural and Food Chemistry* 54: 2964–2968.

Ghosh, B.C. & Singh, S. 1996. Effect of different levels of titrable acidity in curd from buffalo milk at stretching on Mozzarella cheese. *Journal of Food Science and Technology* 33: 70–72.

Giardina, P., Faraco, V., Pezzella, C., Piscitelli, A., Vanhulle, S., & Sannia, G. 2010. Laccases: A never-ending story. *Cellular and Molecular Life Sciences* 67: 369–385.

Gobbetti, M., Ferranti, P., Smacchi, E., Goffredi, F., & Addeo, F. 2000. Production of Angiotensin-Iconverting-enzyme-inhibitory peptides in fermented milks started by *Lactobacillus delbrueckii subsp. bulgaricus* SS1 and *Lactococcus lactis* subsp. *cremoris* FT4. *Applied and Environmental Microbiology* 66 (9): 3898–3904.

Greiner, R., & Konietzny, U. 2006. Phytase for food application. *Food Technology and Biotechnology* 44: 125–140.

Gunathilake, T., Akanbi, T.O., & Barrow, C.J. 2021. Lipase-produced omega-3 acylglycerols for the fortification and stabilization of extra virgin olive oil using hydroxytyrosyl palmitate. *Future Foods* 4: 100045.

Gupta, R.K., Gangoliya, S.S., & Singh, N.K. 2015. Reduction of phytic acid and enhancement of bioavailable micronutrients in food grains. *Journal of Food Science and Technology* 52: 676–684.

Hathwar, S.C., Rai, A.K., Modi, V.K., & Narayan, B. 2012. Characteristics and consumer acceptance of healthier meat and meat product formulations—a review. *Journal of Food Science and Technology* 49: 653–664.

Hellmuth, K. & van den Brink, J.M. 2013. Microbial production of enzymes used in food applications. In: *Microbial Production of Food Ingredients, Enzymes and Nutraceuticals*. Elsevier, pp. 262–287.

Hendriksen, H.V., Kornbrust, B.A., Østergaard, P.R., & Stringer, M.A. 2009. Evaluating the potential for enzymatic acrylamide mitigation in a range of food products using an asparaginase from *Aspergillus oryzae*. *Journal of Agricultural and Food Chemistry* 57: 4168–4176.

Hyun Kim, D., Jin Choi, Y., Koo Song, S., & Won Yun, J. 1997. Production of inulo-oligosaccharides using endoinulinase from a *pseudomonas sp*. *Biotechnology Letters* 19: 369372.

Iglesias-Puig, E., Monedero, V., & Haros, M. 2015. Bread with whole quinoa flour and bifidobacterial phytases increases dietary mineral intake and bioavailability. *LWT Food Science and Technology* 60: 71–77.

Illanes, A. 2008. Enzyme biocatalysis: Principles and applications. *Enzyme Biocatalysis: Principles and Applications*: 1–391.

Israr, B., Frazier, R.A., & Gordon, M.H. 2017. Enzymatic hydrolysis of phytate and effects on soluble oxalate concentration in foods. *Food Chemistry* 214: 208–212.

Jain, J. & Singh, S.B. 2016. Characteristics and biotechnological applications of bacterial phytases. *Process Biochemistry* 51: 159–169.

James, J., Simpson, B.K., & Marshall, M.R. 1996. Application of enzymes in food processing. *Critical Reviews in Food Science and Nutrition* 36: 437–463.

Jung, K.H., Lim, J.Y., Yoo, S.J., Lee, J.H., & Yoo, M.Y. 1987. Production of fructosyl transferase from *Aureobasidium pullulans*. *Biotechnology Letters* 9: 703–708.

Jurado, E., Camacho, F., Luz6n, G., & Vicaria, J. 2002. A new kinetic model proposed for enzymatic hydrolysis of lactose by a β-galactosidase from *Kluyveromyces fragilis*. *Enzyme and Microbial Technology* 31: 300–309.

Karnjanapratum, S., & Benjakul, S. 2015. Cryoprotective and antioxidative effects of gelatin hydrolysate from unicorn leatherjacket skin. *International Journal of Refrigeration* 49: 69–78.

Ke, Y., Huang, W.-Q., Li, J., Xie, M., & Luo, X. 2012. Enzymatic characteristics of a recombinant neutral protease I (rNpI) from *Aspergillus oryzae* expressed in *Pichia pastoris*. *Journal of Agricultural and Food Chemistry* 60: 12164–12169.

Kil, T.G., Kang, S.H., Kim, T.H., Shin, K.C., & Oh, D.K. 2019. Enzymatic biotransformation of balloon flower root saponins into bioactive platycodin d by deglucosylation with *Caldicellulosiruptor bescii* β-glucosidase. *International Journal of Molecular Sciences* 20.

Kumari, M., Padhi, S., Sharma, S., Phukon, L.C., Singh, S.P. & Rai, A.K. 2021. Biotechnological potential of psychrophilic microorganisms as the source of cold-active enzymes in food processing applications. *3 Biotech* 11: 1–18.

Kuo, H.P., Wang, R., Huang, C.Y., Lai, J-T., Lo, Y-C., & Huang, S-T. 2018. Characterization of an extracellular β-glucosidase from *Dekkera bruxellensis* for resveratrol production. *Journal of Food and Drug Analysis* 26: 163–171.

Lafarga, T., Aluko, R.E., Rai, D.K., O'Connor, P., & Hayes, M. 2016. Identification of bioactive peptides from a papain hydrolysate of bovine serum albumin and assessment of an antihypertensive effect in spontaneously hypertensive rats. *Food Research International* 81: 91–99.

Law, B.A. 2009. Enzymes in Dairy Product Manufacture. In: *Enzymes in Food Technology*. Wiley-Blackwell, pp. 88–102.

Lee, J.M., Moon, S.Y., Kim, Y-R., Kim, K.W., Lee, B-J., & Kong, I-S. 2017. Improvement of thermostability and halostability of β-1,3-1,4-glucanase by substituting hydrophobic residue for Lys 48. *International Journal of Biological Macromolecules* 94: 594–602.

Lee, S.-H., Heo, S.-J., Hwang, J.-Y., Han, J.-S., & Jeon, Y.-J. 2010. Protective effects of enzymatic digest from Ecklonia cava against high glucose-induced oxidative stress in human umbilical vein endothelial cells. *Journal of the Science of Food and Agriculture* 90: 349–356.

Lee, Y.S., Huh, J.Y., Nam, S.H., Moon, S.K., & Lee, S.B. 2012. Enzymatic bioconversion of citrus hesperidin by Aspergillus sojae naringinase: Enhanced solubility of hesperetin-7-O-glucoside with in vitro inhibition of human intestinal maltase, HMG-CoA reductase, and growth of *Helicobacter pylori. Food Chemistry* 135: 2253–2259.

Lesage-Meessen, L., Delattre, M., Haon, M., *et al.* 1996. A two-step bioconversion process for vanillin production from ferulic acid combining *Aspergillus niger* and *Pycnoporus cinnabarinus. Journal of Biotechnology* 50: 107–113.

Li, S., Yang, Q., Tang, B., & Chen, A. 2018. Improvement of enzymatic properties of Rhizopus oryzae α-amylase by site-saturation mutagenesis of histidine 286. *Enzyme and Microbial Technology* 117: 96–102.

Mao, Y., Krischke, M., Hengst, C., & Kulozik, U. 2019. Influence of salts on hydrolysis of β-lactoglobulin by free and immobilised trypsin. *International Dairy Journal* 93: 106–115.

Marin-Navarro, J., Talens-Perales, D., & Polaina, J. 2015. One-pot production of fructooligosaccharides by a *Saccharomyces cerevisiae* strain expressing an engineered invertase. *Applied Microbiology and Biotechnology* 99: 2549–2555.

Martins, G.N., Ureta, M.M., Tymczyszyn, E.E., Castilho, P.C., & Gomez-Zavaglia, A. 2019. Technological Aspects of the Production of Fructo and Galacto-Oligosaccharides. Enzymatic Synthesis and Hydrolysis. *Frontiers in Nutrition* 6: 1–24.

Meghavarnam, A.K. & Janakiraman, S. 2018. Evaluation of acrylamide reduction potential of L-asparaginase from Fusarium culmorum (ASP-87) in starchy products. *LWT* 89: 32–37.

Mohandesi, N., Haghbeen, K., Ranaei, O., Arab, S.S., & Hassani, S. 2017. Catalytic efficiency and thermostability improvement of Suc2 invertase through rational site-directed mutagenesis. *Enzyme and Microbial Technology* 96: 14–22.

Morais, H.A., Silvestre, M.P.C., Silveira, J.N., Silva, A.C.S., Silva, V.D.M., & Silva, M.R. 2013. Action of a pancreatin and an *Aspergillus oryzae* protease on whey proteins: Correlation among the methods of analysis of the enzymatic hydrolysates. *Brazilian Archives of Biology and Technology* 56: 985–995.

Morozova, O. V., Shumakovich, G.P., Shleev, S. V., & Yaropolov, Y.I. 2007. Laccase-mediator systems and their applications: A review. *Applied Biochemistry and Microbiology* 43: 523–535.

Muniz-Marquez, D.B., Contreas, J.C., Rodriguez, R., Mussatto, S.I., Teixeria, J.A., & Aguilar, C.N. 2016. Enhancement of fructosyltransferase and fructooligosaccharides production by *A. oryzae* DIA-MF in Solid-State Fermentation using aguamiel as culture medium. *Bioresouce Technology* 213: 276–282.

Nesse, K.O., Nagalakshmi, A.P., Marimuthu, P., & Singh, M. 2011. Efficacy of a fish protein hydrolysate in malnourished children. *Indian Journal of Clinical Biochemistry* 26: 360–365.

Ngo, D.N., Kim, M.M., & Kim, S.K., 2008. Chitin oligosaccharides inhibit oxidative stress in live cells. *Carbohydrate Polymers* 74: 228–234.

Nielsen, A.V., & Meyer, A.S. 2016. Phytase-mediated mineral solubilization from cereals under in vitro gastric conditions. *Journal of the Science of Food and Agriculture* 96: 3755–3761.

Niu, C., Yang, P., Luo, H., Huang, H., Wang, Y., & Yao, B. 2017. Engineering of *Yersinia* phytases to improve pepsin and trypsin resistance and thermostability and application potential in the food and feed industry. *Journal of Agricultural and Food Chemistry* 65: 7337–7344.

Osman, A., Enan, G., Al-Mohammadi, A.R., *et al.* 2021. Antibacterial Peptides Produced by Alcalase from Cowpea Seed Proteins. *Antibiotics* 10: 870.

Ottens, M. & Chilamkurthi, S. 2013. Advances in process chromatography and applications in the food, beverage and nutraceutical industries. In: *Separation, Extraction and Concentration Processes in the Food, Beverage and Nutraceutical Industries*. Elsevier, pp. 109–147.

Palazoglu, T.K., & Gokmen, V. 2008. Reduction of acrylamide level in French fries by employing a temperature program during frying. *Journal of Agricultural and Food Chemistry* 56: 6162–6166.

Park, A.R., & Oh, D.K. 2010. Galacto-oligosaccharide production using microbial β-galactosidase: Current state and perspectives. *Applied Microbiology and Biotechnology* 85: 1279–1286.

Pedrolli, D.B., Monteiro, A.C., Gomes, E., & Carmona, E.C. 2009. Pectin and pectinases: Production, characterization and industrial application of microbial pectinolytic enzymes. *The Open Biotechnology Journal* 3: 9–18.

Pereira-Rodríguez, Á., Fernández-Leiro, R., González-Siso, M.I., Cerdán, M.E., Becerra, M., & Sanz-Aparicio, J. 2012. Structural basis of specificity in tetrameric Kluyveromyces lactis β-galactosidase. *Journal of Structural Biology* 177: 392–401.

Phukon, L.C., Chourasia, R., Padhi, S. *et al.* 2022. Cold-adaptive traits identified by comparative genomic analysis of a lipase-producing Pseudomonas sp. HS6 isolated from snow-covered soil of Sikkim Himalaya and molecular simulation of lipase for wide substrate specificity. *Current Genetics* 68: 375–391.

Picazo, B., Flores-Gallegos, A.C., Muñiz-Márquez, D.B., *et al.* 2018. Enzymes for fructooligosaccharides production: Achievements and opportunities. In: *Enzymes in Food Biotechnology*. Elsevier Inc. Available from: https://doi.org/10.1016/b978-0-12-813280-7.00018-9.

Pokora, M., Eckert, E., Zambrowicz, A., *et al.* 2013. Biological and functional properties of proteolytic enzyme-modified egg protein byproducts. *Food Science & Nutrition* 1: 184–195.

Porres, J.M., Aranda, P., Lo´pez-Jurado, M., & Urbano, G. 2006. Nutritional evaluation of protein, phosphorus, calcium and magnesium bioavailability from lupin (*Lupinus albus var. multolupa*)-based diets in growing rats: Effect of α-galactoside oligosaccharide extraction and phytase supplementation. *British Journal of Nutrition* 95: 1102–1111.

Priya, J.D.A., Divakar, K.K., Prabha, M.S., Selvam, G.P., & Gautam, P., 2014. Isolation, purification and characterisation of an organic solvent-tolerant Ca2+-dependent protease from Bacillus megaterium AU02. *Applied Biochemistry and Biotechnology* 172: 910–932.

Rai, A.K., Bhaskar, N., & Baskaran, V. 2013. Bioefficacy of EPA DHA from lipids recovered from fish processing wastes through biotechnological approaches. *Food Chemistry* 136: 80–86.

Rai, A.K., & Jeyaram, K. 2015. Health benefits of functional proteins in fermented foods. In: Tamang JP (ed) *Health benefits of fermented foods and beverages*. CRC Press. Pp. 455–474.

Rai, A.K., Pandey, A., & Sahoo, D. 2019. Biotechnological potential of yeasts in functional food industry. *Trends in Food Science & Technology* 83: 129–137.

Rai, A.K., Sanjukta, S., Chourasia, R., Bhat, I., Bharadwaj, P.K., & Sahoo, D. 2017a. Production of bioactive hydrolysate using protease, β-glucosidase and α-amylase of *Bacillus spp.* isolated from kinema. *Bioresource Technology* 235: 358–365.

Rai, A.K., Sanjukta, S., & Jeyaram, K. 2017b. Production of angiotensin I converting enzyme inhibitory (ACE-I) peptides during milk fermentation and their role in reducing hypertension. *Critical Reviews in Food Science and Nutrition* 57: 2789–2800.

Raweesri, P., Riangrungrojana, P., & Pinphanichakarn, P. 2008. α-L-Arabinofuranosidase from Streptomyces sp. PC22: Purification, characterization and its synergistic action with xylanolytic enzymes in the degradation of xylan and agricultural residues. *Bioresource Technology* 99: 8981–8986.

Razzaq, A., Shamsi, S., Ali, A., *et al.* 2019. Microbial proteases applications. *Frontiers in Bioengineering and Biotechnology* 7.

Rebello, S., Balakrishnan, D., Anoopkumar, A.N., *et al.* 2019. Industrial enzymes as feed supplements—Advantages to nutrition and global environment. *Energy, Environment, and Sustainability*: 293–304.

Rentschler, E., Schwarz, T., Stressler, T., & Fischer, L. 2016. Development and validation of a screening system for a β-galactosidase with increased specific activity produced by directed evolution. *European Food Research and Technology* 242: 2129–2138.

Rodríguez, S. & Toca, J.L. 2006. Industrial and biotechnological applications of laccases: A review. *Biotechnology Advances* 24: 500–513.

Ruthu, Murthy, P.S., Rai, A.K., & Bhaskar, N. 2014. Fermentative recovery of lipids and proteins from freshwater fish head waste with reference to antimicrobial and antioxidant properties of protein hydrolysate. *Journal of Food Science and Technology* 51: 1884–1892.

Ruviaro, A.R., Barbosa, P. de P.M., & Macedo, G.A. 2019. Enzyme-assisted biotransformation increases hesperetin content in citrus juice by-products. *Food Research International* 124: 213–221.

Saadoun, I., Dawagreh, A., Jaradat, Z., & Ababneh, Q. 2013. Influence of culture conditions on pectinase production by streptomyces sp. (Strain J9). *International Journal of Life Science and Medical Research* 3: 148–154.

Sangeetha, P.T., Ramesh, M.N., & Prapulla, S.G. 2004. Production of fructosyl transferase by *Aspergillus oryzae* CFR 202 in solid-state fermentation using agricultural by-products. *Applied Microbiology and Biotechnology* 65: 530–537.

Sanghi, A., Garg, N., Gupta, V.K., Mittal, A., & Kuhad, R.C. 2010. One-step purification and characterization of cellulase-free xylanase produced by alkalophilic *Bacillus subtilis* ash. *Brazilian Journal of Microbiology* 41: 467–476.

Sanjukta, S., Padhi, S., Sarkar, P., Singh, S.P., Sahoo, D. and Rai, A.K. 2021 Production, characterization and molecular docking of antioxidant peptides from peptidome of kinema fermented with proteolytic Bacillus spp. *Food Research International* 141: 110161.

Schons, P.F., Ries, E.F., Battestin, V., & Macedo, G.A. 2011. Effect of enzymatic treatment on tannins and phytate in sorghum (Sorghum bicolor) and its nutritional study in rats. *International Journal of Food Science and Technology* 46: 1253–1258.

Sharma, H. & Upadhyay, S.K. 2020. Enzymes and their production strategies. *Biomass, Biofuels, Biochemicals*, Advances in Enzyme Catalysis and Technologies: 31–48.

Sharma, V., Ayothiraman, S., & Dhakshinamoorthy, V. 2019. Production of highly thermo-tolerant laccase from novel thermophilic bacterium Bacillus sp. PC-3 and its application in functionalization of chitosan film. *Journal of Bioscience and Bioengineering* 127: 672–678.

Shen, Q., Zhang, Y., Yang, R., *et al.* 2016. Enhancement of isomerization activity and lactulose production of cellobiose 2-epimerase from *Caldicellulosiruptor saccharolyticus. Food Chemistry.* 207: 60–67.

Sheriff, S.A., Sundaram, B., Ramamoorthy, B., & Ponnusamy, P. 2014. Synthesis and in vitro antioxidant functions of protein hydrolysate from backbones of *Rastrelligerkanagurta* by proteolytic enzymes. *Saudi Journal of Biological Sciences* 21: 19–26.

Shu, G., Huang, J., Bao, C., Meng, J., Chen, H., & Cao, J. 2018. Effect of different proteases on the degree of hydrolysis and angiotensin I-Converting enzyme-inhibitory activity in goat and cow milk. *Biomolecules* 8: 101.

Sindhu, R., Binod, P., Madhavan, A., *et al.* 2017. Molecular improvements in microbial α-amylases for enhanced stability and catalytic efficiency. *Bioresource Technology* 245: 1740–1748.

Sindhu, R., Pandey, A., & Binod, P. 2022. *Gut microbes Role in production of nutraceuticals. Current Developments in Biotechnology and Bioengineering*: 273–299.

Singh, R.S., Singh, R.P., & Kennedy, J.F. 2016. Recent insights in enzymatic synthesis of fructooligosaccharides from inulin. *International Journal of Biological Macromolecules* 85: 565–572.

Sorapukdee, S. & Tangwatcharin, P. 2018. Quality of steak restructured from beef trimmings containing microbial transglutaminase and impacted by freezing and grading by fat level. *Asian-Australasian Journal of Animal Sciences* 31: 129–137.

Speranza, P., Lopes, D.B., & Martins, I.M. 2018. Development of functional food from enzyme technology: A review. *Enzymes in Food Biotechnology.* Elsevier Inc. Available from: https://doi.org/10.1016/b978-0-12-813280-7.00016-5.

Sripokar, P., Benjakul, S., & Klomklao, S. 2019. Antioxidant and functional properties of protein hydrolysates obtained from starry triggerfish muscle using trypsin from *albacore tuna* liver. *Biocatalysis and Agricultural Biotechnology* 17: 447–454.

Sun, A., Xu, X., Lin, J., Cui, X., & Xu, R. 2015. Neuroprotection by saponins. *Phytotherapy Research* 29: 187–200.

Sun, H., Xue, Y., & Lin, Y. 2014. Enhanced catalytic efficiency in quercetin-40-glucoside hydrolysis of *Thermotoga maritima* β-glucosidase a by site-directed mutagenesis. *Journal of Agricultural and Food Chemistry* 62: 6763–6770.

Suryia Prabha, M., Divakar, K., Deepa Arul Priya, J., Panneer Selvam, G., Balasubramanian, N., & Gautam, P. 2015. Statistical analysis of production of protease and esterase by a newly isolated Lysinibacillus fusiformis AU01: Purification and application of protease in sub-culturing cell lines. *Annals of Microbiology* 65: 33–46.

Tahergorabi, R., Beamer, S.K., Matak, K.E., & Jaczynski, J. 2012. Functional food products made from fish protein isolate recovered with isoelectric solubilization/precipitation. *LWT – Food Science and Technology* 48: 89–95.

Talens-Perales, D., Polaina. J., &Marín-Navarro, J., 2016. Structural dissection of the active site of thermotoga maritima β-Galactosidase identifies key residues for transglycosylating activity. *Journal of Agricultural and Food Chemistry* 64 (14):2917–2924.

Tang, Q., Lan, D., Yang, B., Khan, F.I., & Wang, Y. 2017. Site-directed mutagenesis studies of hydrophobic residues in the lid region of T1 lipase. *European Journal of Lipid Science and Technology* 119: 1600107.

Timon, M.L., Andres, A.I., Otte, J., & Petron, M.J. 2019. Antioxidant peptides (< 3 kDa) identified on hard cow milk cheese with rennet from different origin. *Food Research International* 120: 643–649.

Toushik, S.H., Lee, K.T., Lee, J.S., & Kim, K.S. 2017. Functional applications of lignocellulolytic enzymes in the fruit and vegetable processing industries. *Journal of Food Science* 82: 585–593.

Vallabha, V.S., Indira, T.N., Jyothi Lakshmi, A., Radha, C., & Tiku, P.K. 2015. Enzymatic process of rice bran: A stabilized functional food with nutraceuticals and nutrients. *Journal of Food Science and Technology* 52: 8252–8259.

Vega, R., & Zuniga-Hansen, M.E. 2014. A new mechanism and kinetic model for the enzymatic synthesis of short-chain fructooligosaccharides from sucrose. *Biochemical Engineering Journal* 82: 158–165.

Weng, M., Deng, X., Bao, W., *et al.* 2015. Improving the activity of the subtilisin nattokinase by site-directed mutagenesis and molecular dynamics simulation. *Biochemical and Biophysical Research Communications* 465: 580–586.

Xu, F., Oruna-Concha, M.J., & Elmore, J.S. 2016. The use of asparaginase to reduce acrylamide levels in cooked food. *Food Chemistry* 210: 163–171.

Xu, X., Song, Z., Yin, Y., *et al.* 2021. Solid-State Fermentation Production of Chitosanase by Streptomyces with Waste Mycelia of *Aspergillus niger*. *Advances in Enzyme Research* 9: 10–18.

Yu, S., Zhang, Y., Zhu, Y., Zhanf, T., Jiang, B., & Mu, W. 2017. Improving the catalytic behavior of DFA I-forming inulin fructotransferase from *Streptomyces davawensis* with site-directed mutagenesis. *Journal of Agricultural and Food Chemistry* 65: 75797587.

Zhang, W., Jia, M., Yu, S., *et al.* 2016. Improving the thermostability and catalytic efficiency of the D-Psicose 3-Epimerase from *Clostridium bolteae* ATCC BAA-613 using site-directed mutagenesis. *Journal of Agricultural and Food Chemistry* 64: 3386–3393.

Zhang, Y., Geary, T., & Simpson, B.K. 2019. Genetically modified food enzymes: A review. *Current Opinion in Food Science* 25: 14–18.

Zhang, Y, Gu, H, Shi, H, Wang, F, & Li, X. 2017. Green synthesis of conjugated linoleic acids from plant oils using a novel synergistic catalytic system. *Journal of Agricultural and Food Chemistry* 65: 5322–5329.

Zhao, X., Shi-Jian, D., Tao, G., *et al.* 2010. Influence of phospholipase A2 (PLA2)-treated dried egg yolk on wheat dough rheological properties. *LWT – Food Science and Technology* 43: 45–51.

Zhu, B.W., Xiong, Q., Ni, F., Sun, Y., & Yao, Z. 2018. High-level expression and characterization of a new κ -carrageenase from marine bacterium *Pedobacter hainanensis* NJ-02. *Letters in Applied Microbiology* 66: 409–415.

Zyzak, D.V., Sanders, R.A., Stojanovic, M., *et al.* 2003. Acrylamide formation mechanism in heated foods. *Journal of Agricultural and Food Chemistry* 51: 4782–4787.

Section II

Sources of Microbial Enzymes for Nutraceutical Production

2 Enzymes from Lactic Acid Bacteria for Nutraceuticals Production

Mousumi Ray[1], *Ashwini Manjunath*[2], *Rwivoo Baruah*[1], *and Prakash M. Halami*[1]

[1] Department of Microbiology and Fermentation Technology, CSIR – Central Food Technological Research Institute, Mysuru, Karnataka, India

[2] Department of Microbiology, Faculty of Life Sciences, JSS Academy of Higher Education and Research, Mysuru, Karnataka, India

CONTENTS

2.1 INTRODUCTION

Nutraceutical is a modern and scientific term, a portmanteau of "nutrition" and "pharmaceutical". The term is basically used to specify some products that may be synthesized from herbal products, may be dietary supplements (nutrients), specific diets, and different processed foods like cereals, soups, and beverages. Nutraceutical products contain both medicinal properties and nutritional requirements (Kalra, 2003). Unlike pharmaceuticals, nutraceuticals are substances that generally

DOI: 10.1201/9781003311164-4

don't have patent protection. The use of both substances is almost same, in that both the pharmaceutical and nutraceutical products might be used to treat or inhibit various disorders, though only pharmaceutical compounds have governmental sanction (Chauhan et al., 2013).

Nutraceuticals are the foods or nutritional supplements used for health progression other than fulfilling the basic nutritional requirements (Kalra, 2003). When a dietary supplement contains one or more dietary components, it is considered a good food product. A bgood dietary supplement is considered a food product when it comprises one or more dietary components. These dietary constituents are vitamins, minerals, amino acids, medicinal herb or other botanical substances used as dietary supplements in order to increase the total intake of daily requirements (Nasri et al., 2014). Nowadays, people are more focused on healthy diet and there is a growing need for probiotics as nutraceutical foods via tablets, capsules, sachets, and other fermented probiotic food and beverage products. Therefore, they are used as dietary supplements (Bharti et al., 2020). Since long lactic acid bacteria (LAB) have been known for their extensive and significant advanced uses in the food, pharmaceutical and chemical industries (Das and Goyal, 2012). Most of the LAB or their metabolites that are safe to use, proven to cure various health problems, and used as nutraceutical products are available commercially (Radhika et al., 2011).

Nutraceuticals synthesized by LAB have several beneficial health effects and they are also used as low-calorie sweeteners and food additives. They are involved in nucleotide biosynthesis and act as antioxidant, antitoxic, antiobesity, anti-plaque agents and anticancer agents (Beena Divya et al., 2012). Hence, it confers various health benefits like preventing obesity, cardiovascular diseases, gastrointestinal disorder, diabetes, colon cancer, arthritis, and various other ailments. Additionally, LAB are also responsible for the synthesis of various metabolites like enzymes for nutraceuticals production that are essentially used in several food products and the pharmaceutical industry (Anandharaj et al., 2014; Bharti et al., 2020; Kumari et al. 2021; Chourasia et al. 2022a; Hugenholtz and Smid, 2002).Therefore, in this chapter, we will present LAB as a source for nutraceuticals and discuss how LAB are essential to produce nutraceuticals and nutraceutical foods that derive different health benefits. The chapter will also describe the metabolic engineering of LAB to produce several metabolites as nutraceuticals on an industrial scale. In this way, we can better understand the effective production of metabolites as nutraceuticals used as, or as part of, food products with such benefits to consumer health. In the rest of this chapter we will present several examples where metabolic engineering of homofermentative LAB have led or will lead to the efficient production of food ingredients (nutraceuticals) or food products with a beneficial effect for the consumer.

2.2 LACTIC ACID BACTERIA FOR NUTRACEUTICALS

LAB have widely been used as precise starter cultures in the fermented food production industry (Hati et al., 2013). Additionally, their part as biocontrol agents to counter the infections caused by spoilage and food-borne bacteria has also been projected in order to reduce the quantity of preservatives in food (Johnson-Green, 2018). Currently, the potential of LAB as probiotics and also as nutraceuticals has a greater impact on the food supplement and pharmaceutical industry (Das et al., 2012; Bharti et al., 2020). Even though it is promising to use LAB as a microbial cell factory in the production of functional food, their metabolic descriptions are often underexplored and most products underexploited. However, the naturally occurring biologically active compounds that promote medical benefits in humans are nutraceuticals (Das et al. 2012). Several nutraceuticals recognized as probiotics, prebiotics, and secondary metabolites that are produced by LAB have importance commercially (Bharti et al., 2020). Being the gut microbe, LAB is known to be responsible for the production of different types of nutraceuticals. Gut microbes are recognized to yield nutraceuticals, hence, there is a growing need to know the gut microbiome and the nutraceutical relative interface in the production of nutraceuticals by LAB (Balakumaran et al., 2022).

Nutraceuticals are biologically dynamic elements that it might be considered as food or a fragment of a food that offers health benefits in the treatment of disease (Sachdeva et al., 2020). Most

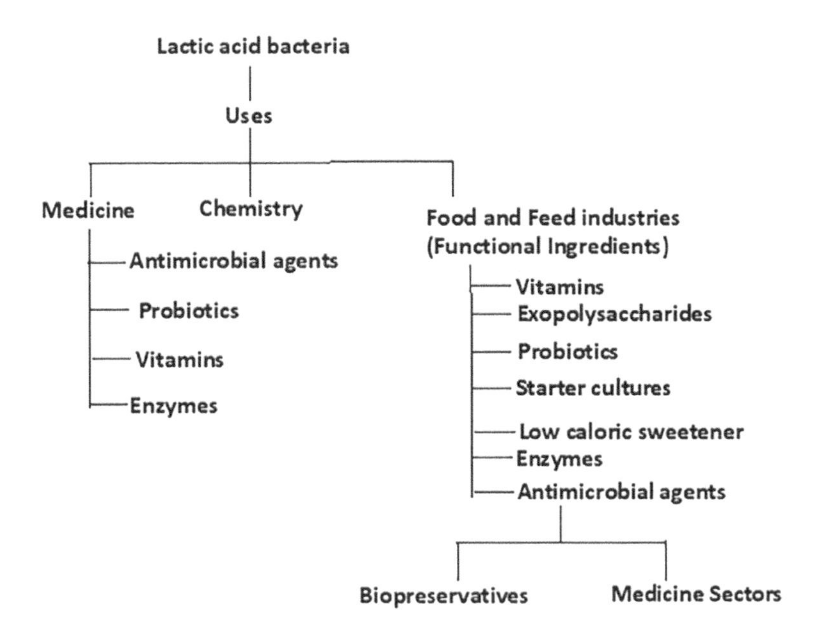

FIGURE 2.1 The uses of lactic acid bacteria and their functional ingredients as nutraceuticals.

diseases are the result of an unhealthy diet and lifestyle, for example, diabetes, obesity, and heart-related problems. But nutraceuticals can play an essential role in the prevention of disease that in turn promotes good health. Several ingredients of plants like polyphenols, lycopene, carotenoids, and catechins are also used as nutraceuticals. These nutraceuticals support the body's defence system and enhance immunity against diseases. They are effective in the prevention of arthritis and cancer, for instance (Johnson-Green et al., 2018). Additionally, LAB play a vital role in the production of pharmaceuticals, chemicals, nutraceuticals, and other useful products (Figure 2.1) (Florou-Paneri 2013).

2.3 LAB AS A SOURCE OF ENZYMES

Most of the LAB belonging to the enzymatic group are capable of influencing the quality and other organoleptic properties of feeds and foods. Several enzymes are released by the LAB into the gastrointestinal tract. These enzymes exert potential synergetic effects on digesting food and reduce the symptoms of imperfect absorption of food substances by the small intestine (Florou-Paneri et al., 2013). Additionally, these LAB act as a source for the enzyme extract preparations that are capable of functioning under various environmental conditions during fermentation (Tamang, 2011). The most studied LAB is the one isolated from a wine source or fermented food, such as yoghurt or cheese, and is used for its enzymatic action (Matthews et al., 2004; Mtshali, 2007).

Most of the commonly associated LAB species in fermented foods include *Pediococcus* sp. and *Lactococcus* sp. (Matthews et al., 2004). The enzymes, viz. amylases produced by the LAB, are used in sourdough technology to perfect the natural texture of the bread (Mtshali 2007). Also, LAB contributes to the flavour and aroma of the fermented food products. The cheese sensory quality has been improvised by the peptidases produced by the *cremoris* subsp. of *Lactococcus lactis* (Florou-Paneri et al., 2013). Likewise, lipolysis and proteolysis might increase the flavour of several varieties of cheese (Guldfeldt et al., 2001). The LAB strains of Spanish Genestoso cheese were examined for their enzymatic action. *Lactococcus* sp. was found with a high level of dipeptidase action along with *Enterococcus* sp. with enterolytic action (González et al., 2010). The enzymes produced by the LAB play a major role in the preparation of wine. The enzymatic activity of LAB adds to the aroma and flavour of the wine together with the flavours obtained from the alcoholic fermentation of grapes. The growth of LAB is observed in wine even during malolactic fermentation followed by the alcoholic fermentation. The broad range of secondary alterations made by the LAB enhances the flavour and taste of the wine (Matthews et al., 2004).

LAB causes acidification of the unprocessed materials by producing organic acids, essentially lactic acid. In addition, they also produce acetic acid, ethanol, aroma compounds, bacteriocins, exopolysaccharides (EPS), and numerous enzymes. Enzymes like proteases, peptidases, polysaccharide degrading enzymes, ureases, lipases, amylases, esterases, and phenoloxidases are synthesized by LAB (Padmavathi et al., 2018). These derivatives effectively improve product shelf life, microbial safety, and texture, and also contribute to the pleasing sensory profile of the end product. On the subject of dairy fermentations, starter cultures play a variety of roles. LAB, *Propionibacterium*, surface-ripening bacteria, yeasts, and moulds are some of the most common starter cultures used to make fermented milk products (Hati et al., 2013; Chourasia et al., 2022a).

2.4 PRODUCTION OF NUTRACEUTICALS USING LAB SPECIES

The majority of nutraceuticals produced through the fermentation process of LAB have shown to be beneficial to human health (Table 2.1). These enzymes such as pullulanases, proteinases and peptidases, amylases, phytases, lipases, proteases, lipases and esterases, produced by LAB will now be discussed.

TABLE 2.1
Microbial Enzymes Produced by Lactic Acid Bacteria

Enzymes	Lactic Acid Bacteria	Sources	References
Proteolytic enzymes	*Lactococcus lactis* ssp. cremoris	Cheese	Hati et al., 2013
Proteinases and peptidases	LAB	Fermented milk	Kumar et al., 2017
Amylase	*Lactobacillus fermentum and Lactobacillus sp G3_4_1TO2*	Milk, curd and bovine colostrum	Padmavathi et al., 2018
Lipases, proteases	*L. casei*, two strains of *L. paracasei*, two strains *of L. plantarum*, and two strains of *P. acidilactici*	Several dairy products	García-Cano et al., 2019
Lipolytic and esterase	*Enterococcus faecium* and *Enterococcus durans*	Biotopes of Algeria and Mauritania	Aravindan et al., 2007
Pullulanase	*Lactococcus lactis*	Fermented cereal beverages, sourdoughs, and cooked fish products	Petrova et al., 2013
Phytase	*L. salivarius, L. reuteri,*	Wholewheat bread	Sharma and Shukla, 2020
β-glucosidase	LAB	Intestinal and food fermenting LAB	Michlmayr and Kneifel, 2014
β-galactosidase	*Lactobacillus plantarum* Ln4 and G72, and *L. rhamnosus* KCTC12202BP	Kimchi	Son et al., 2017
Glutamate decarboxylase	*Lactobacillus* spp., *Lactococcus* spp. and *Streptococcus* spp.	Kimchi, Thai fermented sausage *nham*, fermented Asian fish products and cheese	Yogeswara et al., 2020

2.4.1 Proteolytic Enzymes of LAB and Production of Bioactive Peptides

The proteolytic system of LAB converts protein substrates into free small peptides and amino acids to carry out a variety of physiological processes such as controlling intracellular pH, generating metabolic energy, resisting stress, and biosynthesizing protein (Fernández and Zúñiga, 2006). This proteolytic system of LAB is a suitably designed mechanism that consists of endopeptidases, peptide transporters and cell-envelope proteinases (Savijoki et al., 2006). The genome of LAB encrypts this protein system into clusters of gene that display differences linked with their capacity for amino acid biosynthesis among LAB strains (Venegas-Ortega et al., 2019). Few proteins encrypt bioactive sequences that are released by the process of hydrolysis; these by-products obtained by the degradation process by either microbial or non-microbial enzymatic reactions are known as bioactive peptides (Meisel and FitzGerald, 2003). Due to the proteolytic system of LAB, they release these peptides during the process of fermentation (Liu et al., 2010). The LAB contains three chief components of proteolytic system: (1) cell wall-bounded proteinase that catalyses the extracellular milk protein (casein) degradation into oligopeptides, (2) transporters of peptides that take the peptides inside the cell, and (3) several intracellular peptidases that break down the peptides into smaller peptides and amino acids. Caseins in particular are rich in proline, so the LAB contains several proline peptidases to break down peptides rich in proline (Liu et al., 2010). Further conversion of amino acids into numerous flouring agents like alcohols, aldehyde and esters were carried out by several microbes (Liu et al., 2010).

The isolation of lactobacilli is considered on the basis of various physiological and genetic aspects from various ecological places. A previous study has found huge differences among the strains as well as species at the physiological and genetic levels (Sun et al., 2015). It is reported that LAB from various sources differ in their process of protein hydrolysis (Jensen et al., 2009). This supports the evidence that lactobacilli can yield a huge diversity of bioactive peptides. Also, this highlights the cautious selection of LAB strains used for the production of bioactive peptides. The bioactive peptides are engrossing from the perspective of nutritive and health care values. Each species displays various proteinases inhibitor subdue that lead to a variety of proteolytic action. This highlights the high capability of *Lactobacillus* strains to yield novel bioactive peptides that are of primary interest (Raveschot et al., 2018). There are fermented dairy products stating functional health features linked with bioactive peptides (Korhonen and Pihlanto, 2006; Hafeez et al., 2014). These proteolytic enzymes play a vital role in the production of fermented dairy foods like cheese and yoghurt (Kieliszek et al., 2021). There are several studies indicating the ability of LAB to produce bioactive peptides using its proteolytic system. *Lactobacillus helveticus* used as a starter culture for Emmental cheese production produces Angiotensin-I-converting enzyme (ACE)-inhibitory bioactive peptides. These peptides help regulate blood pressure in patients with hypertension. Probiotic bacteria *Lb. rhamnosus* GG was reported to produce immunosuppressing peptides from casein hydrolysis (El-Sayed and Awad, 2019).

The *Lactobacillus* genus is the chief diverse group of LAB. LAB are involved in the fermentation of foods such as vegetable, meat, and dairy products in various food industries. Also, these LAB are used as good bio-preservatives and as a probiotic application due to their therapeutic qualities (Stefanovic et al., 2017; Chourasia et al., 2022b). As these LAB require various amino acids, they are known to be fastidious microbes. Therefore, to utilize amino acids from their environment, LAB uses its proteolytic system for their growth and survival through the enzymatic action known as cell envelope proteinases. This action releases the free amino acids and peptides in the fermentation media (Savijoki et al., 2006). Besides contributing modifications in the organoleptic features of the fermented products, these released peptides can show several biological activities (Hafeez et al., 2014). Various studies have reported the existence of bioactive peptides in ripened cheeses such as Swiss cheese, manchego, cheddar, Gouda (Pihlanto, 2013) or in yoghurts (Jin et al., 2016; Rutella et al., 2016). The best example of bioactive peptide obtained by *Lactobacillus* species are the antihypertensive tripeptides, Val-Pro-Pro and Ile-Pro-Pro produced from hydrolysis of casein by diverse *Lactobacillus helveticus* strains (Pihlanto, 2013). Also, the fermented dairy products comprising antihypertensive bioactive peptides produced by *Lactobacillus* strains are marketed already (Hafeez et al., 2014).

2.4.2 Alanine Dehydrogenase and the Production of Alanine

The enzyme alanine dehydrogenase is responsible for the conversion of pyruvate into L-alanine. Alanine is a commercially relevant food and pharmaceutical component produced by *Lactiplantibacillus plantarum* and *Lactococcs lactis* (Hugenholtz and Smid, 2002). L-alanine is the form of alanine primarily found in amino acid-rich constituents of medical supplements. These are also the substrate for most of the synthesis of biochemical reactions in food and pharmaceutical industries. The higher level of alanine dehydrogenase expression has been engineered in the lactate dehydrogenase (LDH)-deficient lactococcal cells to completely convert the pyruvate pool into alanine. This conversion is obtained by growing the cells in a high ammonium environment and by the functional alteration of the *alr* gene. This can be considered as an efficient production of alanine by the homofermentative metabolism of *Lactococcus lactis* (Hugenholtz et al., 2000). Additionally, alanine is used as a sweetener in low calorie foods as it has a lower caloric value (Hugenholtz and Smid, 2002).

2.4.3 Lipases

Lipases (EC 3.1.1.3) are serine hydrolases that catalyse the hydrolysis of triglycerides to glycerol and free fatty acids at the oil–water interface (Liu et al., 2006). Because of their capacity to catalyse a wide range of processes in both aqueous and nonaqueous conditions, microbial lipases have risen to prominence among biocatalysts. In most commercial applications, microbial lipases are chosen over plant and animal lipases because their enzymatic characteristics and substrate specificity are more diverse (Gupta et al., 2004). Lipases are used commercially to degrade lipids in more recent applications because they are inexpensive and adaptable catalysts. Lipases from LAB are extremely important as industrial biocatalysts, medicinal agents, and in other applications. In the food sector, they are used as flavouring agents. Intracellular or extracellular lipases are seen in LAB strains. LAB's capacity to execute complex fatty acid transformation reactions such as isomerization, hydration, dehydration, and saturation has various industrial applications. As a result, there is a worldwide focus on LAB lipase research (Ramakrishnan et al., 2015, 2016). Lipase, like many other enzymes, is utilised in the food industry to modify and break down lipid-polymers. Commercial lipase is often used to generate flavour in dairy products, as well as meat, baking, and alcoholic fermentation (Matthews et al., 2004).

Some of the possible industries where the lipase enzyme or LAB-producing lipases have been employed are as follows: (1) the dairy industry for the hydrolysis of milk fat and development of flavour in cheese (Collins et al., 2003); (2) meat processing for fat removal using the bio-lipolysis procedure (Hasan et al., 2006); (3) baking and confectioneries to increase the flavour content and shelf life of bakery items through esterification of short-chain fatty acid, while lipase catalyzation increases texture and suppleness (Laboret and Perraud, 1999); (4) The food industries uses immobilized lipase in the determination of quantitative triglycerol. As sensors for the quantitative detection of triacylglycerol, these immobilized lipases are rapid, efficient, precise, and cost effective, so this use is critical in the food business, particularly in the areas of fats and oils, beverages, soft drinks, pharmaceuticals, and clinical diagnosis (Aravindan et al., 2007). The main idea behind utilizing lipases as biosensors is to produce glycerol from triacylglycerol in the analytical sample and then quantify the released glycerol using a chemical or enzymatic approach. Because of the wide variety of LAB lipases, they can produce a variety of compounds with varied specificities that can be used for sensing (Ramakrishnan et al., 2019).

2.4.3.1 Nutraceutical Applications of Lipases

Polyunsaturated fatty acids (PUFAs) are extracted from animal and plant lipids and enriched by lipases extracted from microbes. The mono- and diacylglycerides of animal and plant lipids are used to make a variety of medications. Because of the metabolic benefits they offer, such as their antihypertensive, anticancer, antioxidant, antidepression, antiaging, and antiarthritis properties, PUFAs are widely exploited as food additives, medicines, and nutraceuticals. Immobilized lipases, extracted from *Limosilactobacillus reuteri*, are reportedly effective in the synthesis of nutraceuticals (Aravindan et al., 2007).

2.4.4 β-Glucosidase and Aglycone Production

The enzyme β-Glucosidase that belongs to the foremost group of glucosyl hydrolase brings about the cleavage of glycosidic bonds (genistin, diadzin, and glycitin) by hydrolysing β-1-4 linked glucose residues into its biologically active aglycones (genistein, daidzein, and glycitein) (Zouhar et al., 2001). *Lactobacillus* spp. shows the highest specific activity of β-Glucosidase and several studies indicate this by decrease in concentration of β-Glucosides during fermentation of soya milk. In a study with a mixed culture of *Lactobacillus delbrueckii* subsp. *bulgaricus* and *Streptococcus thermophiles* during soya milk fermentation, high β-Glucosidase activity was observed. This resulted in the high amounts of isoflavone aglycones (Hati et al., 2020). The enzyme β-Glucosidase has therapeutic values to treat Gaucher's disease caused due to the inadequacy of this enzyme in various tissues and cells (Mahajan et al., 2012). β-lucosidase as an immobilized enzyme is also extremely beneficial in the beverage indsutry as it can be utilized as a flavour enhancer for fruit juice and wine (Mahajan et al., 2012).

2.4.5 Pullulanase

The glycoside hydrolase family 13 subfamily 14 (GH13 14) pullulanases are the most frequent extracellular glycan-active enzymes in human gut lactobacilli. These enzymes have a catalytic module, two starch binding modules, a domain of unknown function, and a C-terminal surface layer association protein (SLAP) domain that is not seen in other bacteria. In the well-characterized *Lactobacillus acidophilus* NCFM, a regularly used probiotic, the specificity of *Lactobacillus acidophilus* Pul13 14 (LaPul13 14), a member of this type of pullulanase, and its role in branched-glucan metabolism were reported (Møller et al., 2017). On starch-derived branching substrates, growth tests with *L. acidophilus* NCFM demonstrated a preference for -glucans with short branches of roughly two to three glucosyl moieties over amylopectin with longer branches.

In the presence of β-glucans, cell-attached debranching activity was detectable, but glucose inhibited it. LaPul13 14 is required for debranching activity, which is lost in a mutant strain missing a functional LaPul13 14 gene. The preference for short-branched -glucan oligomers was validated by the hydrolysis kinetics of recombinant LaPul13 14, which matched the growth results. Surprisingly, this enzyme has the best catalytic efficiency and the lowest Km of any pullulanase ever discovered (Møller et al., 2017). The presence of other glucan binding sites besides the active site of the enzyme, which may contribute to the exceptional substrate affinity, was disclosed by inhibition kinetics by cyclodextrin. The enzyme was also reported with high thermostability and activity in the acidic pH range, indicating that it has adapted to the physiologically demanding conditions of the human gut (Møller et al., 2017).

Lactobacillus acidophilus NCFM, a commercial probiotic strain, is one of the best studied of this taxonomic group (Bull et al., 2013). According to the CAZy database, the genome of *L. acidophilus* NCFM (Altermann et al., 2005) encodes nine enzymes of glycoside hydrolase family 13 (GH13), which contains -glucan-active enzymes (Lombard et al., 2014). The only extracellular -glucan-active enzyme anticipated is a pullulanase-type -glucan debranching enzyme. This enzyme is multimodular, with an N-terminal carbohydrate binding module family 41 (CBM41) starch binding domain, a domain of unknown function, a CBM48 domain, a GH13 subfamily 14 (GH13 14) catalytic module (Lombard et al., 2014), and a C-terminal SLAP (Lombard et al., 2014: Marchler-Bauer et al., 2015) catalytic module. Hydrolysis of α-1,6-linked branches in glycogen, amylopectin, and other starch-derived glucans, as well as pullulan, are catalysed by pullulanases.

Enzyme activity against α-1,4 and α-1,6 glycosidic linkages are required for complete starch hydrolysis. One enzyme (amylopullulanase) or two enzymes (a-amylase and pullulanase type I) can demonstrate these two sorts of activities (Hii et al., 2012). (Depending on their source, amylolytic enzymes can have a wide range of specificity and characteristics (Sonnenburg and Bäckhed, 2016). Pullulanase type I (pullulan 6-glucanohydrolase, EC 3.2.1.41), a debranching enzyme that preferentially cleaves certain α-1,6 glycosidic linkages in starch, amylopectin, pullulan, and related oligosaccharides, is abundant in thermophilic and mesophilic bacteria (Møller et al., 2014). It was calculated that pullulan digestion by these bacteria's pullulanase resulted in a 100% conversion to maltotriose. Vishnu et al. (2006) identified an amylolytic enzyme with both amylase and pullulanase activity in *L. amylophilus* GV6 as a 90-kD protein (Waśko et al., 2011). For this strain, the amylase and pullulanase activities were found to be 0.439 U=g=min and 0.18 U=g=min, respectively, in semi-solid-state fermentation with wheat bran (Waśko et al., 2011). This paper is the first study on pullulanase type I from *Lactococcus lactis* (Waśko et al., 2011).

2.4.6 Phytase and its Uses

Phytase-producing LAB in the gut may have medicinal and nutritional implications that degrades the phytate or phytic acid. Phytase in human food increases the availability of important divalent mineral ions and the release of myo-inositols, which may be used to treat disorders such as Crohn's disease, Alzheimer's disease, and irritable bowel syndrome. Phytase (myo-inositol hexakisphosphatephosphohydrolases) hydrolyses the phosphoric monoester linkages in phytic acid, a key source of phosphorus in plant diets. Phytic acid has a high chelation with divalent minerals like Ca2+, Mg2+, Zn2+, and Fe2+.Although there was no data on exopolysaccharide (EPS) synthesis from *Pediococcus pentosaceus* KTU05-8, KTU05-9 and *P. acidilactici* (KTU05-7), among these strains, KTU05-8 and KTU05-9 had the highest extracellular phytase activity (Cizeikiene et al., 2015). A scientific study has reported on the probiotic activities of a phytase-producing *Pediococcus acidilactici* strain SMVDUDB2 isolated from a traditional fermented cheese product, Kalarei (Bhagat et al., 2020). Also, *Pediococcus pentosaceus strains* (CFR R38 and CFR R35) isolated from chicken intestine were found to possess phytate degrading ability (Raghavendra and Halami, 2009).

While LAB fermentation improves the textural properties of bread, it also improves bread mineral bioavailability and reduces phytate concentration (Milanović et al., 2020). Phytic acid binds to basic amino acids in proteins and functions as a chelator, making cations like Ca2+, Mg2+, Fe2+, and Zn2+ inaccessible for absorption in the digestive tract (Kumar et al., 2010). Phytase (myo-inositol hexakisphosphate phosphohydrolase), an enzyme that catalyses the progressive elimination of orthophosphates from this molecule, can break down phytic acid. This enzyme is found in a wide range of species, including plants, animals, and microbes (Zeller et al., 2015).

Endogenous phytases are present in cereal grain flours, but their levels are typically too low to result in considerable phytic acid reductions (Milanović et al., 2020). It has been discovered that sourdough LAB acidification promotes endogenous flour phytases, which, when combined with phytase-producing LAB and yeasts, can dramatically lower the phytic acid concentration of bread (Nuobariene et al., 2015). As a result, using phytase-producing microbes in baking has a lot of promise for increasing mineral bioavailability in cereal-based baked goods. Lactic acid fermentation has been shown to lower phytate concentration in plant-based diets while simultaneously improving mineral solubility. Extracellular phytase activity was proposed to be responsible for the observed drop in phytate concentration during lactic acid fermentation (Damayanti et al., 2017). Phytate dephosphorylation is begun by a class of enzymes known as phytases. Research has shown that LAB, such as that used for phytate degradation during cereal dough fermentation – *Lacticaseibacillus casei* DSM 20011 and *Lactiplantibacillus plantarum* W42, *L. plantarum, L. plantarum, L. reuteri* – exhibit phytase degrading activity (Damayanti et al., 2017). Many traditional foods in Yogyakarta, Indonesia that employed grain and legume as basic materials used the fermentation process, such as gembus tempeh (Glycine max), soyabean tempeh (Glycine max), lamtoro tempeh (Leucaenaleucocephala), and karabenguk tempeh (Leucaenaleucocephala) (Mucunapruriens) (Sarwono, 2010). In traditional grain and legume-based meals, LAB are one of the most common microbial fermentation processes (Damayanti et al., 2017). Ray et al. (2017) described that the *Bifidobacterium* sp. MKK4-fermented rice gruel was enriched with essential multivalent minerals which were associated with dephytinization of phytic acid by the microbial enzyme phytase. Phytate content in soyabean have been greatly reduced through fermentation using phytase-producing *L. plantarum* and resulted in improved mineral solubility (Theodoropoulos et al., 2018).

Thirteen phytase-producing LAB were isolated from 28 newborn faeces samples and phenotypically and prebiotically characterized. *Pediococcus acidilactici* BNS5B was identified phylogenetically as *Pediococcus acidilactici* BNS5B using both 16S rRNA and MALDI-TOF MS analysis. Isolate 5b exhibited potent probiotic potential with significant (p.01) phytase activity (4.55 U/mL) and was identified phylogenetically as *Pediococcus acidilactici* BNS. After 15 minutes of phytase treatment at 37°C, phytic acid from a modified diet (96.59%) and brown bread (88.89%) was significantly dephytinized by phytase from *P. acidilactici* BNS5B (Sharma and Shukla, 2020).

2.4.7 β-Galactosidase

The enzyme β-Galactosidase (-D-galactosidegalactohydrolases, EC 3.2.1.23) catalyses the cleavage of terminal galactosyl groups from the non-reducing ends of various galactosides (Thakur et al. 2022). The enzyme can be found in plants, animal organs, and microorganisms, among other locations in nature. Microorganisms are the preferred source of highly active and stable enzymes due to their ease of synthesis. Because of their capacity to hydrolyse lactose, the most abundant sugar in milk and its by-products, β-Galactosidases are now widely used in the food and pharmaceutical industries. This helps to alleviate the lactose intolerance problem, which affects a large portion of the human population, and consequently expands the market for these goods (Panesar et al., 2010). It also helps to overcome some of the technological challenges that come with lactose use in the food sector, such as low solubility, easy crystallization, and lack of sweetness (Ghosh et al., 2013; Vieira et al., 2013). Furthermore, inhibiting lactose's great capacity to absorb aromas and scents improves the sensory qualities of dairy products (Husain, 2010). The treatment of wastewater from the dairy industry is another major topic of β-galactosidase application. Because of the limited biodegradability of lactose, the disposal of whey and whey permeates generates major environmental issues. As a result, hydrolysed whey lactose is a viable solution to the pollution problem (Sen et al., 2012). β-Galactosidases, on the other hand, can catalyse galactoside production reactions under particular conditions such as high lactose concentration, low water activity, and high temperatures (Jovanovic-Malinovska et al., 2012).

Due to their strong prebiotic activity and potential favourable effects on human health, interest in this aspect of β-Galactosidase use is fast developing, particularly in the field of galacto-oligosaccharides (GOS) manufacture (Carević et al., 2015). LAB strains are also popular among the many accessible microbiological sources (yeasts, moulds, and bacteria) (Schwab et al., 2011). Lactic acid bacterial enzymes can be the alternative due to their GRAS (generally recognized as safe) status, which allows for unrestricted use of their enzymes without the need for extensive and costly purification (Carević et al., 2015). β-Galactosidase from LAB has been discovered to be a neutral enzyme that can hydrolyse sweet whey and milk lactose as well as produce GOS. Because of their GRAS status, LAB represents a varied group of commonly used industrial microorganisms (Carević et al., 2015). Although they are already used in large-scale lactic acid synthesis, their full potential in enzyme commercial production, particularly β-Galactosidase, has yet to be realized. *Bifidobacterium* species and *Streptococcus thermophiles* have been the most extensively studied strains thus far, with only a few -galactosidases from *Lactobacilus* sp. having been identified (Carević et al., 2015). Also β-Galactosidase of *Enterococcus faecium* MTCC5153 can be utilized in dairy industry for lactose hydrolysis thereby improving the ability of digestion (Badarinath and Halami 2011).

2.5 MISCELLANEOUS ENZYMES

LAB generate enzymes such as proteases, peptidases, polysaccharide degrading enzymes, ureases, lipases, amylases, esterases, and phenoloxidases. Amylases (E.C.3.2.1.1) are enzymes that catalyse the first hydrolysis of starch into short oligosaccharides by cleaving a-D 1, 4 glycosidic linkages (Magalhães and Souza, 2010). The acidic character of starch hydrolysed with amylase is overcome, and the high temperature is maintained, resulting in fructose syrup production. Amylases are secreted by a variety of organisms, including plants, mammals, bacteria, fungus, and yeast. Commercially available amylases today are derived from microbial sources. The advantage of using microbial amylases in industrial applications is that they are more stable than any other source and are simple to alter in order to generate enzymes with the desired properties in bulk manufacturing at a low cost (Padmavathi et al., 2018). *Lactobacillus amylovorus*, *Lactiplantibacillus plantarum*, *Lacticaseibacillus manihotivorans*, and *Limosilactobacillus fermentum* are examples of amylolytic LAB (Padmavathi et al., 2018).

LAB that are amylolytic can not only convert simple sugars to lactic acid but they can also break down starch, according to research conducted over the last three decades. While LAB have long been known to be involved in fermented indigenous foods, Nakamura's isolation of *Lactobacillus amylophilus* and *Lactobacillus amylovorus* in 1979 and 1981 was the first to demonstrate their ability to hydrolyse starch. *Lactobacillus, Lactococcus,* and *Streptococcus* are the genera that can directly utilise starch to make lactic acid, with the lactobacilli group being the most efficient, according to DNA tests of other bacteria isolated since then (Petrova et al. 2013). Aside from lowering the stages and costs involved in commercial lactic acid production, amylolytic LAB must produce lactic acid yields that are comparable to or better than those produced by bacteria utilized in commercial operations. In these industrial methods, more than 90% of the fermentable carbohydrates employed as substrate are transformed into lactic acid (Akoetey, 2015). The ability of amylolytic LAB to hydrolyse starch was linked to the synthesis of extracellular amylase enzymes, which were evaluated for starch hydrolysis using cell culture supernatants. As a result of genome sequencing, amylolytic lactic LAB produce enzymes that are classified as α-amylases, maltogenic amylases, amylopullulanases, pullulanases, neopullulanases, glycogen phosphorylases, and oligo 1, 6-glucosidases (Petrova et al., 2013). Genes coding for both cytoplasmic and extracellular amylases were transcribed with growth in starch media in *Lactococcus lactis* sp. *lactis* B84, though other genes for glycogen phosphorylase and amylopullulanase were found in the genome (Petrov et al., 2008).

Lacticaseibacillus paracasei B41, a novel strain isolated from a fermented beverage, was also discovered to have amylolytic activity (Petrova and Petrov, 2012). The amylolytic capacity was discovered to be owing to the amyl gene, which was induced in the presence of starch, maltose, or pullulan whereas glucose suppressed the gene's transcription. Their research also found that controlling the pH at 5.0 rather than 6.0 increased starch hydrolysis and lactic acid generation. From 40g starch, 37.3 g/L L (+) lactic acid was obtained, indicating a 93% conversion rate of substrate to lactic acid. Starch hydrolysis appeared to be the process's limiting stage. Bacterial growth was preferred in glucose as compared to other substrates. After the depletion of glucose, maltose and amylase production by *L. manihotivorans* OND32T was seen in mixed substrate fermentation.

Amylase synthesis required nitrogen sources as well (Guyot and Morlon-Guyot, 2001). The discovery of the genes responsible for amylolytic activity has sparked research towards genetically modifying other lactic acid species in order to expand the range of bacteria that can be used. The α-amylase amyA gene was inserted into *L. plantarum* by Okano et al. (2009). The ldhL1 gene, which converts pyruvate to L (+)-lactic acid, has been disrupted and replaced by the -amylase gene. As a result, the transformed bacteria could now manufacture pure D (+)-lactic acid from raw corn starch. A yield of 0.86g/g (g LA/g consumed substrate) was obtained after a 48-hour fermentation. Amylase catalyses the hydrolysis of starch resulting in production of glucose, maltose, and maltotriose (Ray, 2016). Amylases and lactic acid synthesis from starch appear to be two possible industrial uses of LAB, given the importance and availability of starchy biomass around the world. Amylases play a crucial role in starch degradation and are produced in large quantities by microorganisms, accounting for 25–33% of the global enzyme market (Fossi and Tavea, 2013). Amylases are now used in a variety of sectors, including clinical, medicinal, and analytical chemistry, as well as the textile, food, fermentation, paper, distillery, and brewing industries (Fossi and Tavea, 2013). The use of thermostable amylases in industrial processes has several advantages, including reduced contamination risk, lower external cooling costs, and increased diffusion rate (Fossi and Tavea, 2013).

As a result, a thermostable amylase-producing lactic acid bacterium could be a good choice for the food industry, particularly for creating high-density gruel from starchy raw materials like corn or wheat (Fossi and Tavea, 2013). This would necessitate a thorough understanding of the circumstances required to manufacture high-quality amylase and lactic acid. It is economically undesirable since it is a two-step process combining sequential enzymatic hydrolysis and fermentation. By combining the enzymatic hydrolysis of carbohydrate substrates and microbial fermentation of the resultant glucose into a single phase, the bioconversion of carbohydrate materials to lactic acid can be made significantly more effective. Many representative bacteria, including *Lactobacillus* and *Lactococcus* species, have been successfully used to produce lactic acid from raw starch materials (Fossi and Tavea, 2013). This, in combination with amylolytic LAB, may help to reduce the overall cost of fermentation. To make the procedure more cost effective, production strains that ferment starch to lactic acid in a single step must be developed.

2.5.1 Thermostable Amylases

Amylase's range of applications has expanded to include clinical, medicinal, and analytical chemistry, as well as its broad use in starch saccharification and the textile, culinary, brewing, and distilling sectors. Many enzymes offered for bulk industrial use have a high degree of thermostability. Because of their potential industrial applications, thermostable -amylases are of interest. They're used in a variety of sectors, including starch liquefaction, brewing, textile sizing, and paper and detergent manufacture (Ghorbel et al., 2009). The benefits of utilizing thermostable amylases in industrial operations include a lower risk of contamination and the cost of external cooling, improved substrate solubility, and a lower viscosity that allows for faster mixing and pumping. Although several thermostable -amylases have been isolated from *Bacillus* sp. and the variables determining their thermostability have been studied, the thermostability of amylases from LAB has received little scholarly attention. *Lactobacillus amylovorus*, *Lactiplantibacillus plantarum*, *Lactobacillus manihotivorans*, and *Limosilactobacillus fermentum* are LAB that have been examined for amylolytic activity (Fossi and Tavea, 2013).

2.5.2 Glutamate Decarboxylase (GAD) and GABA Production

Glutamate decarboxylase catalyses the reaction for the conversion of glutamate to γ-aminobutyric acid (GABA). LAB belonging to *Lactobacillus* spp., *Lactococcus* spp. and *Streptococcus* spp. produces GAD and is responsible for acid resistance in these bacteria. GABA is the most abundant neurotransmitter in the human brain. GABA has several beneficial properties such antidepressant, antihypertensive and has cholesterol-lowering activity (Yogeswara et al., 2020). GABA enriched functional foods have been of great interest in recent times. GABA-rich fermented foods include kimchi, Thai fermented fish plaa-som, fermented shrimp paste, fermented cassava, mulberry beer, fermented durian, kimchi cheese, and zlatar cheese (Sahab et al., 2020).

2.5.3 Enzymes for the Production of Tagatose

The carbohydrate that can replace the sucrose is tagatose. It has a sweeter taste than sucrose and is sweeter than erythritol, sorbitol, and mannitol. However, they are lower in calories as their degradation is very low in human body. The tagatose has been marketed recently as a prebiotic, sugar with a low calorie count (Hugenholtz et al., 2000). The enzymatic isomerization through the substrate arabinose is the viable process for the production of tagatose (Kim et al. 2001). The biochemical pathway that produces tagatose is the lactose degradation by LAB. During the metabolism of lactose by *Lactococcus lactis,* galactose fractions are broken down by tagatose 6-phosphate pathway (Hugenholtz et al., 2000). In this pathway, phospho-β-galactosidase hydrolyses lactose-phosphate followed by the degradation of glucoses via glycolysis. Later, galactose-phosphate isomerase converts galactose-phosphate into tagatose 6-phosphate, which is later phosphorylated to form tagatose-diphosphate by tagatose-phosphate kinase. Finally, tagatose-phosphate aldolase gets hydrolysed to produce glycolytic intermediates viz., glyceraldehyde phosphate and dihydroxyacetone-phosphate. *Lactococcus lactic* is used as a cell factory for the production of tagatose due to its ease for genetic manipulations (Hugenholtz and Smid, 2002).

2.5.4 Enzymes for the Production of Trehalose

Most microbes synthesize non-reducing sugars, one of which is trehalose. This is known as nutritional sugar and it is not fully digested in humans. This trehalose is featured to protect proteins against denaturation caused by various stress conditions such as freezing, drying, and heating. Trehalose is also used as preservative with various enzymes, membranes, human cells, and tissues (Guo et al., 2000; Hugenholtz et al., 2000; Hugenholtz and Smid, 2002). Trehalose is used as a preservative in most of the food products that can retain taste and freshness of the product for longer durations (Hugenholtz et al., 2000). The enzymes involved in the synthesis of trehalose are trehalose 6-phosphate synthase, which converts glucose 6-phosphate to trehalose 6-phosphate, and trehalose 6-phosphate phosphatase converts trehalose 6-phosphate to trehalose (Hugenholtz and Smid, 2002).

2.5.5 Production of Exopolysaccharides and Oligosaccharides

EPS are an important metabolite from LAB; they can function as nutraceuticals, prebiotic additives, and food hydrocolloids. They are of two types: (1) homopolysaccharides and (2) heteropolysaccharides. Both types of EPS are synthesized by carbohydrate active enzymes (CAzymes) like glycosyl hydrolase (GH) or glycosyl transferase (GT) (Baruah et al., 2022). α-glucans, a type of homopolysaccharides produced by LAB, are exclusively produced by an extracellular enzyme, glucansucrase, that uses sucrose as a substrate. Glucansucrase is produced by several LAB genera such as *Streptococcus, Leuconostoc, Lactobacillus, Limosilactobacillus, Onenococcus,* and *Weissela* species (Gangoiti et al., 2018).

Besides the production of EPS, these glucansucrase enzymes can be used to synthesize prebiotic oligosaccharides by introducing an acceptor sugar during the synthesis. Prebiotic oligosaccharides have been produced in different fruit juices to formulate a prebiotic fruit beverage by adding glucansucrase enzymes in orange, mango, apple, and pineapple juices (Baruah et al., 2017).

2.6 NEW STRATEGY TO MAKE ENGINEERED LAB FOR FOOD INDUSTRY

The subsequent chief application of LAB is in the food industryhese LAB are utilized at an industrial scale for the fermentation of raw food constituents, viz. vegetables and meat. LAB, particularly *Lactococcus lactis*, are known to be the absolute cell factory that can produce numerous significant nutraceuticals. In spite of having the potential ability to produce fermented food products, these

TABLE 2.2
Metabolically Engineered Lactic Acid Bacteria for Enzymes Responsible for Nutraceutical Production

Lactic Acid Bacteria	Modified Genes	Nutraceuticals	Importance/Findings	References
L. plantarum	*lac*L and *lac*M gene	β-Galactosidases and lactases	β-Galactosidases and lactases are important in the production of lactose-free dairy products	Halbmayr et al., 2008
L. brevis	*gad*A and *gad*B	GABA	The need for nutraceutical foods containing GABA is rising and most accepted for their health beneficial attributes and prevention of several diseases	Yogeswara et al., 2020
L. lactis	*gene (ldh)*	Lactate dehydrogenase	To convert sugar to lactate in *L. lactis*, Lactate dehydrogenase (LDH) acts as the last enzyme in the pathway	Platteeuw et al., 1995
L. lactis	*ilvBN gene*	α-acetolactate synthase (ALS)	ALS is one of the two enzymes that catalyses the conversion of pyruvate to α-acetolactate, the other being the acetohydroxy acid synthase	Hugenholtz, 1993
L. lactis	*UDP-glucose phosphorylase (GalU)*	Glycosyltransferases	EPS produced by *Lactococcus* are widely used by the dairy industry	Ruffing and Chen, 2006

LAB also have enormous capability in metabolic engineering approaches (Table 2.2). For example, production of nutraceuticals through the induced overexpression or/and disruption of related metabolic genes (Hugenholtz and Smid, 2002).

The disruption of enzyme alanine dehydrogenase in both *Lactococcus lactis* and *Lactiplantibacillus plantarum* has led in the creation of completely new kinds of metabolite such as mannitol (Hugenholtz et al., 2000). The fructose-6-phosphate reduces to form mannitol-1-phosphate which then forms into mannitol by dephosphorylation. By using this idea, strategies in metabolic engineering have been designed. For example, production of mannitol is induced in LAB, viz. *Lactococcus lactis,* has deficient or reduced activity of phosphofructokinase, therefore this organism has induced with phosphofructokinase activity to overproduce mannitol phosphate dehydrogenase (Wisselink et al., 2002). If the elimination of pyrolis is simplified, for example by the introduction of mannitol transporter in *Leuconostoc mesenteroides*, higher mannitol production is expected. Similarly, sorbitol production can also be induced in LAB species. Significant sorbitol production is observed in *Lactiplantibacillus plantarum* by the induction of gene encoding sorbitol dehydrogenase together with malate dehydrogenase (MDH) and lactate dehydrogenase (LDH) disruption. Also, by enhancing transportation and dephosphorylation of sorbitol, increased higher production of sorbitol can be obtained.

Lactococcus lactis is known to have potent cell factory that produce tagatose during carbon metabolism. It has broad collection of genetic tools, competent gene integration system and efficient system for the enzymes overproduction. The genes that code for tagatose pathway are found in the lactose operon (Hugenholtz et al., 2000). The disruption of genes *lac*D and/or *lac*C to produce either tagatose 1, 6-diphosphate or tagatose 6-phosphate was considered to induce tagatose metabolites accumulation in lactococcal cells. The gene *lac*D is disrupted using a two-step recombination process that involves incorporation of erythromycin resistance plasmid comprising *lac*F, *lac*C genes through single edge event, following the removal of *lac*D in a second impulsive double crossover recombinant event. The expected mutant can be obtained easily as 50% of cases are shown to have double crossover recombinant event. Though the growing rate of *lac*D mutant on glucose was similar to wild type, its rate of growth on lactose was strictly cooperated in the metabolically engineered strain. The increased production of unique metabolite, that is tagatose 1,6-diphosphate, has been observed in the genetically engineered strains when compared with wild type, viz. fructose and glucose phosphates. Thus, *Lactococcus lactis* was effectively altered into a cell factory that produces tagatose. For usage in applied (food) circumstances, nevertheless, strategies will have to established for the dephosphorylation of tagatose diphosphate and evacuation of the end product, tagatose.

2.7 ENZYMES FROM LAB AND THEIR MARKET POTENTIAL

Most of the enzymes produced by LAB have uses in medicine as diverse as they are in industry, and they are continually expanding. Proteolytic enzymes, which remove dead skin and burns, and fibrinolytic enzymes, which break up clots, are currently the most common medicinal applications of microbial enzymes. Lipases (EC 3.1.1.3) are the most common enzymes employed in organic synthesis, and they are utilized to make optically active alcohols, acids, esters, and lactones (Cambou and Klibanov, 1984; Saxena, 1999). The synthesis of (2R,3S)-3-(4-methoxyphenyl) methyl glycidate (an intermediary for diltiazem) and 3, 4-dihydroxylphenyl alanine (DOPA, for Parkinson's disease treatment) is aided by microbial lipases and polyphenol oxidases (EC 1.10.3.2) (Faber, 2011). The enzyme Superoxide dismutase and serrapeptases produced by *Lactiplantibacillus plantarum* are used as an anti-inflammatory. Also, the enzymes glutathione peroxidases and catalases produced by *Lactiplantibacillus plantarum* are used as antioxidants.

In the projected period of 2022–2029, the industrial enzymes market is expected to grow. According to Data Bridge Market Research, the market would increase at a compound annual growth rate of 5.8% over the forecast period. Industrial enzymes are essentially biological catalysts that speed up chemical reactions. The pharmaceuticals, chemicals, biofuels, food, diagnostics, detergents, feed, textile, paper, and leather industries, among others, employ them extensively. As the demand for sustainable solutions grows, so does the demand for industrial enzymes. The greatest difficulty facing fast-growing nations like India is delivering food and healthcare to their rising populations. India, an agriculture-based economy, is expected to grow at an annual rate of 7.9%, making it an appealing market for the industrial enzyme-based manufacturing industry. Although the Indian biotech sector accounts for only 2% of the global biotech market, it is garnering international attention due to investment opportunities and research output (Binod et al., 2013). Pharmaceutical enzymes account for about half of India's overall enzyme requirement (Binod et al., 2013).

Globally, the industrial enzymes market has grown at a rate of over 10% per year to nearly $173 million. Multinational businesses accounted for more than 60% of the enzymes market in India, while domestic producers accounted for the balance. Domestic enzyme consumption was at $110 million, with exports bringing in $32 million in revenue (Singh et al., 2016). Advanced Enzymes

Technologies Ltd, India's largest maker and exporter of enzyme products, had a nearly 30% market share in 2012–2013, second only to Denmark's Novozymes, which had a 44% market share. Rossari Biotech, Maps Enzymes, Lumis Biotech, and Zytex (CRISIL Research) were among the other notable manufacturers.

2.8 FUTURE PERSPECTIVE

The potential for industrial uses of microbial enzymes has risen dramatically in the twenty-first century and is continuing to rise, as enzymes provide enormous promise for many industries to satisfy the demands of a fast-growing population while also coping with natural resource exhaustion. It was demonstrated that enzymes have significant potential in a variety of industries, including food and dairy, pharmaceuticals, as well as nutraceuticals. Alternatively, these biomolecules might be used on a regular basis to meet the ever-increasing demand for food including nutritive and medicinal effect. Microbial enzymes are widely used in a variety of sectors to produce higher quality products at a faster rate of reaction with minimal pollution and lower costs.

2.9 CONCLUSIONS

Nutraceuticals can offer considerable health benefits particularly in the inhibition and/or treatment of various acute and chronic human disorders. Several factors like quality, safety, long-term adverse effects, toxicity, supplementation studies, and clinical trials in humans are major scientific aspects for its improvement and acceptance. Challenges are made to evade genetic ailments by using nutraceuticals in various forms of enzymes, probiotics, prebiotics, and fortified food. So-called modern foods established as a nutraceutical commercially must positively cross through the rigorous regulatory steps to deliver a significant impact on human health. It has already been proved that consumption of nutraceuticals (within its acceptable suggested dietary intakes) not only exerts health benefits but also will keeps ailments at bay. Administration of nutraceutical also allows individuals to maintain an overall good health for longer periods. Significant development of these useful nutraceuticals can be increased in fermented foods in two ways, moreover, by selecting high-producing LAB or by stimulating high production through engineering of the accountable metabolic pathways. The research is still carried out to improve novel fermented nutrifoods with enhanced nutritional value and significant health impacts. In the future, some of these nominated and created strains will certainly have an impact in the food industry.

REFERENCES

Akoetey, W. 2015. Direct fermentation of sweet potato starch into lactic acid by *Lactobacillus amylovorus*: The prospect of an adaptation process. Thesis. University of Arkansas, Fayetteville Retrieved from https://scholarworks.uark.edu/etd/41.

Altermann, E., Russell, W.M., Azcarate-Peril, M.A., Barrangou, R., Buck, B.L., McAuliffe, O., Souther, N., Dobson, A., Duong, T., Callanan, M., and Lick, S. 2005. Complete genome sequence of the probiotic lactic acid bacterium *Lactobacillus acidophilus* NCFM. *Proceedings of the National Academy of Sciences* 102: 3906–3912. doi:10.1073/pnas.0409188102.

Anandharaj, M., Sivasankari, B., and Parveen Rani, R. 2014. Effects of probiotics, prebiotics, and synbiotics on hypercholesterolemia: A review. *Chinese Journal of Biology*2014: 7. Article ID 572754. https://doi.org/10.1155/2014/572754

Aravindan, R., Anbumathi, P., and Viruthagiri, T. 2007. Lipase applications in food industry. *Indian Journal of Biotechnology* 6: 141–158. http://hdl.handle.net/123456789/3016.

Badarinath, V., & Halami, P.M. 2011. Purification of new β-galactosidase from Enterococcus faecium MTCC 5153 with transgalactosylation activity. *Food Biotechnology* 25(3): 225–239.

Balakumaran, P.A., Divakar, K., Sindhu, R., Pandey, A., and Binod, P. 2022. Gut microbes: Role in production of nutraceuticals. In: Current Developments in Biotechnology and Bioengineering (pp. 273–299). Elsevier.

Baruah, R., Deka, B., and Goyal, A. 2017. Purification and characterization of dextransucrase from *Weissella cibaria* RBA12 and its application in *in vitro* synthesis of prebiotic oligosaccharides in mango and pineapple juices. *LWT Food Science and Technology* 84: 449–456.

Baruah, R., Rajshee, K., and Halami, P.M. 2022. Exopolysaccharide producing microorganisms for functional food industry. In: Current Developments in Biotechnology and Bioengineering (pp. 337–354). Elsevier.

Beena Divya, J., KulangaraVarsha, K., MadhavanNampoothiri, K., Ismail, B., and Pandey, A. 2012. Probiotic fermented foods for health benefits. *Engineering in Life Sciences* 12(4): 377–390.

Bhagat, D., Raina, N., Kumar, A., Katoch, M., Khajuria, Y., Slathia, P.S., and Sharma, P. 2020. Probiotic properties of a phytase producing *Pediococcusacidilactici* strain SMVDUDB2 isolated from traditional fermented cheese product, Kalarei. *Scientific Reports* 10: 1926. https://doi.org/10.1038/s41598-020-58676-2.

Bharti, N., Kaur, R., and Kaur, S. 2020. Health benefits of probiotic bacteria as nutraceuticals. *European Journal of Molecular & Clinical Medicine* 7(07): 4797–4807.

Binod, P., Palkhiwala, P., Gaikaiwari, R., Nampoothiri, K.M., Duggal, A., Dey, K., and Pandey, A. 2013. Industrial enzymes: Present status and future perspectives for India: Present scenario and perspectives. *Journal of Scientific and Industrial Research* 72: 271–286.

Bull, M., Plummer, S., Marchesi, J., and Mahenthiralingam, E. 2013. The life history of *Lactobacillus acidophilus* as a probiotic: A tale of revisionary taxonomy, misidentification and commercial success. *FEMS Microbiology Letters* 349: 77–87. doi: 10.1111/1574-6968.12293.

Cambou, B. and Klibanov, A.M. 1984. Preparative production of optically active esters and alcohols using esterase-catalyzed stereospecific transesterification in organic media. *Journal of American Chemical Society* 106: 2687–2692. doi: 10.1021/ja00321a033.

Carević, M., Vukašinović-Sekulić, M., Grbavčić, S., Stojanović, M., Mihailović, M., Dimitrijević, A., and Bezbradica, D. 2015. Optimization of β-galactosidase production from lactic acid bacteria. *Hemijskaindustrija* 69(3): 305–312.

Chauhan, B., Kumar, G., Kalam, N., and Ansari, S.H. 2013. Current concepts and prospects of herbal nutraceutical: A review. *Journal of Advanced Pharmaceutical Technology & Research* 4(1): 4.

Chourasia, R., Kumari, R., Singh S.P., Sahoo, D., and Rai A.K. 2022a. Characterization of native lactic acid bacteria from traditionally fermented chhurpi of Sikkim Himalayan region for the production of chhurpi cheese with enhanced antioxidant effect. *LWT – Food Science and Technology* 154: 112801.

Chourasia, R., Phukon, L.C., Minhajul Abedin, M., Sahoo, D., and Rai, A.K. 2022b. Production and characterization of bioactive peptides in novel functional soybean chhurpi produced using Lactobacillus delbrueckii WS4. *Food Chemistry* 387: 132889.

Cizeikiene, D., Juodeikiene, G., Bartkiene, E., Damasius, J., and Paskevicius, A. 2015. Phytase activity of lactic acid bacteria and their impact on the solubility of minerals from wholemeal wheatbread. *International Journal of Food Sciences and Nutrition* 66: 736–742. doi: 10.3109/09637486.2015.1088939.

Collins, Y.F., McSweeney, P.L., and Wilkinson, M.G. 2003. Lipolysis and free fatty acid catabolism in cheese: A review of current knowledge. *International Dairy Journal* 13(11): 841–866.

Damayanti, E., Indriati, R., Sembiring, L., Julendra, H., and Sakti, A.A. 2017. Antifungal activities of lactic acid bacteria against Aspergillus flavus, A. parasiticus and Penicillium citrinum as mycotoxin producing fungi. In *Proceedings of the 16th AAAP Animal Science Congress* 2: 1742–1745.

Das, L., Bhaumik, E., Raychaudhuri, U., and Chakraborty, R. 2012. Role of nutraceuticals in human health. *Journal of Food Science and Technology* 49(2): 173–183.

Das, D. and Goyal, A. 2012. Lactic acid bacteria in food industry. In: Microorganisms in Sustainable Agriculture and Biotechnology (pp. 757–772). Springer.

El-Sayed, M. and Awad, S. 2019 Milk bioactive peptides: Antioxidant, antimicrobial and anti-diabetic activities. *Advances in Biochemistry* 7(1): 22–33.

Faber, K. 2011. Biotransformations in organic chemistry: a textbook (No. 660.634 F334B.). Springer: Heidelberg.

Fernández, M. and Zúñiga, M. 2006. Amino acid catabolic pathways of lactic acid bacteria. *Critical Reviews in Microbiology* 32(3): 155–183. doi: 10.1080/10408410600880643. PMID: 16893752.

Florou-Paneri, P., Christaki, E., and Bonos, E. 2013. Lactic acid bacteria as source of functional ingredients. In: Lactic Acid Bacteria – R & D for Food, Health and Livestock Purposes. IntechOpen.

Fossi, B.T. and Tavea, F. 2013. Application of amylolytic Lactobacillus fermentum 04BBA19 in fermentation for simultaneous production of thermostable alpha-amylase and lactic acid. In: Lactic Acid Bacteria-R & D for Food, Health and Livestock Purposes. IntechOpen.

Gangoiti, J., Pijning, T., and Dijkhuizen, L. 2018. Biotechnological potential of novel glycoside hydrolase family 70 enzymes synthesizing α-glucans from starch and sucrose. *Biotechnology Advances* 36(1): 196–207.

García-Cano, I., Rocha-Mendoza, D., Ortega-Anaya, J., Wang, K., Kosmerl, E., and Jiménez-Flores, R. 2019. Lactic acid bacteria isolated from dairy products as potential producers of lipolytic, proteolytic and antibacterial proteins. *Applied Microbiology and Biotechnology* 103(13): 5243–5257.

Ghorbel, R.E., Maktouf, S., Massoud E.B., Bejar, S., and Chaabouni, S.E. 2009. New thermostable amylase from Bacillus cohnii US147 with a broad pH applicability. *Applied Biochemistry and Biotechnology* 157: 5060.

Ghosh, M., Pulicherla, K.K., Rekha, V.P., Vijayanand, A., and Sambasiva Rao, K.R. 2013. Optimisation of process conditions for lactose hydrolysis in paneer whey with cold-active β-galactosidase from psychrophilic Thalassospira frigidphilosprofundus. *International Journal of Dairy Technology.* 66(2): 256–263.

González, L., Sacristán, N., Arenas, R., Fresno, J.M., and Tornadijo, M.E. 2010. Enzymatic activity of lactic acid bacteria (with antimicrobial properties) isolated from a traditional Spanish cheese. *Food Microbiology* 27(5): 592–597.

Guldfeldt, L.U. Sorensen, K.B., Stroman P., Behrndt, H., Williams, D., and Johansen, E. 2001. Effect of starter cultures with a genetically modified peptidolytic or lytic system on Cheddar cheese ripening. *International Dairy Journal* 11: 373–382.

Guo, N., Puhlev, I., Brown, D.R., Mansbridge, J., and Levine, F. 2000. Trehalose expression confers desiccation tolerance on human cells. *Nature Biotechnology* 18(2): 168–171.

Gupta, R., Gupta, N., and Rathi, P. 2004. Bacterial lipases: An overview of production, purification and biochemical properties. *Applied Microbiology and Biotechnology.* 64(6): 763–781.

Guyot, J.P., & Morlon-Guyot, J. 2001. Effect of different cultivation conditions on Lactobacillus manihotivorans OND32T, an amylolytic lactobacillus isolated from sour starch cassava fermentation. *International Journal of Food Microbiology* 67(3): 217–225.

Hafeez, Z., Cakir-Kiefer, C., Roux, E., Perrin, C., Miclo, L., and Dary-Mourot, A. 2014. Strategies of producing bioactive peptides from milk proteins to functionalize fermented milk products. *Food Research International* 63: 71–80. doi: 10.1016/j.foodres.2014.06.002.

Halbmayr. E., Mathiesen, G., Nguyen, T.H., Maischberger, T., Peterbauer, C.K., Eijsink, V.G., and Haltrich, D. 2008. High-level expression of recombinant β-galactosidases in *Lactobacillus plantarum* and *Lactobacillus sakei* using a sakacin P-based expression system. *Journal of Agricultural and Food Chemistry* 56(12): 4710–4719.

Hasan, F., Shah, A.A., and Hameed, A. 2006. Industrial applications of microbial lipases. *Enzyme and Microbial Technology* 39(2): 235–251.

Hati, S., Mandal, S., and Prajapati, J.B. 2013. Novel starters for value added fermented dairy products. *Current Research in Nutrition and Food Science Journal* 1(1): 83–91.

Hati, S., Ningtyas, D.W., Khanuja, J.K., and Prakash, S. 2020. β-Glucosidase from almonds and yoghurt cultures in the biotransformation of isoflavones in soy milk. *Food Bioscience* 34: 100542.

Hii, S.L., Tan, J.S., Ling, T.C., & Ariff, A.B. 2012. Pullulanase: role in starch hydrolysis and potential industrial applications. *Enzyme Research* 2012:921362. https://doi.org/10.1155%2F2012%2F921362

Hugenholtz, J. 1993. Citrate metabolism in lactic acid bacteria. *FEMS Microbiology Reviews* 12(1–3): 165–178.

Hugenholtz, J., Hols, P., Starrenburg, M.J., de Vos, W.M., and Kleerebezem, M. 2000. Metabolic engineering of *Lactococcus lactis* leading to high diacetyl production. *Applied Environmental Microbiology* 66: 4112–4114.

Hugenholtz, J. and Smid, E.J. 2002. Nutraceutical production with food-grade microorganisms. *Current Opinion in Biotechnology* 13(5): 497–507.

Husain, Q. 2010. β Galactosidases and their potential applications: A review. *Critical Reviews in Biotechnology* 30(1): 41–62.

Jensen, M.P., Vogensen, F.K., and Ardö, Y. 2009. Variation in aminopeptidase and aminotransferase activities of six cheese related *Lactobacillus helveticus* strains. *International Dairy Journal* 19: 661–668. doi: 10.1016/j.idairyj.2009.09.007.

Jin, Y., Yu, Y., Qi, Y., Wang, F., Yan, J., and Zou, H. 2016. Peptide profiling and the bioactivity character of yogurt in the simulated gastrointestinal digestion. *Journal of Proteomics* 141: 24–46. doi: 10.1016/j.jprot.2016.04.010.

Johnson-Green, P. 2018. Introduction to Food Biotechnology. CRC Press. https://doi.org/10.1201/9781420058383

Jovanovic-Malinovska, R., Fernandes, P., Winkelhausen, E., and Fonseca, L. 2012. Galacto-oligosaccharides synthesis from lactose and whey by β-galactosidase immobilized in PVA. *Applied Biochemistry and Biotechnology* 168(5): 1197–1211.

Kalra, E.K. 2003. Nutraceutical-definition and introduction. *AapsPharmsci* 5(3): 27–28.

Kieliszek, M., Pobiega, K., Piwowarek, K., and Kot, A.M. 2021 Characteristics of the proteolytic enzymes produced by lactic acid bacteria. *Molecules* 26(7): 1858.

Kim, P., Yoon, S.H., Roh, H.J., and Choi, J.H. 2001. High production of d-tagatose, a potential sugar substitute, using immobilized l-arabinose isomerase. *Biotechnology Progress* 17(1): 208–210.

Korhonen, H., and Pihlanto, A. 2006. Bioactive peptides: Production and functionality. *International. Dairy Journal* 16: 945–960. doi: 10.1016/j.idairyj.2005.10.012.

Kumar, M., Kumar, A., Nagpal, R., Mohania, D., Behare, P., Verma, V., Kumar, P., Poddar, D., Aggarwal, P.K., Henry, C.J., and Jain, S. 2010.Cancer-preventing attributes of probiotics: An update. *International Journal of Food Sciences and Nutrition* 61(5): 473–496.

Kumar, N., Raghavendra, M., Tokas, J., and Singal, H.R. 2017. Flavor addition in dairy products: Health benefits and risks. In: Nutrients in Dairy and their Implications on Health and Disease (pp. 123–135). Academic Press.

Kumari, M., Padhi, S., Sharma, S., Phukon, L.C., Singh, S.P., and Rai, A.K. 2021. Biotechnological potential of psychrophilic microorganisms as the source of cold-active enzymes in food processing applications. *3 Biotech* 11: 479.

Laboret, F. and Perraud, R. 1999. Lipase-catalyzed production of short-chain acids terpenyl esters of interest to the food industry. *Applied Biochemistry and Biotechnology* 82(3): 185–198.

Liu, C.H., Lu, W.B., and Chang, J.S. 2006. Optimizing lipase production of Burkholderia sp. by response surface methodology. *Process Biochemistry* 41(9): 1940–1944.

Liu, J., Huang, S.S., and Zhao, Z. 2010. Research on antioxidative capacity of lactic acid bacteria. *China Dairy Industry* 38(5): 38–41.

Lombard, V., GolacondaRamulu, H., Drula, E., Coutinho, P.M., and Henrissat, B. 2014. The carbohydrate-active enzymes database (CAZy) in 2013. *Nucleic Acids Research* 42: 490–495. doi:10.1093/nar/gkt1178.

Magalhães, O.P., & Souza, M.P. 2010. Application of microbial amylase in industry–A review. *Brazilian Journal of Microbiology* 41, 850–861.

Mahajan, P.M., Desai, K.M., and Lele, S.S. 2012. Production of cell membrane-bound α- and β-Glucosidase by *Lactobacillus acidophilus. Food Bioprocess Technology* 5: 706–718.

Marchler-Bauer, A., Derbyshire, M.K., Gonzales, N.R., Lu, S., Chitsaz, F., Geer, L.Y., Geer, R.C., He, J., Gwadz, M., Hurwitz, D.I., Lanczycki, C.J., Lu, F., Marchler, G.H., Song, J.S., Thanki, N., Wang, Z., Yamashita, R.A., Zhang, D., Zheng, C., and Bryant, S.H. 2015. CDD: NCBI's conserved domain database. *Nucleic Acids Research* 43: D222–D226. doi:10.1093/nar/gku1221.

Matthews, A., Grimaldi, A., Walker, M., Bartowsky, E., Grbin, P., and Jiranek, V. 2004. Lactic acid bacteria as a potential source of enzymes for use in vinification. *Applied and Environmental Microbiology* 70(10): 5715–5731.

Meisel, H. and FitzGerald, R.J. 2003. Biofunctional peptides from milk proteins: Mineral binding and cytomodulatory effects. *Current Pharmaceutical Design* 9(16): 1289–1295. doi: 10.2174/1381612033454847.

Michlmayr, H. and Kneifel, W. 2014 β-Glucosidase activities of lactic acid bacteria: Mechanisms, impact on fermented food and human health. *FEMS Microbiology Letters* 352(1): 1–10.

Milanović, V., Osimani, A., Garofalo, C., Belleggia, L., Maoloni, A., Cardinali, F., Mozzon, M., Foligni, R., Aquilanti, L., and Clementi, F. 2020. Selection of cereal-sourced lactic acid bacteria as candidate starters for the baking industry. *PLOS ONE* 15(7): e0236190.

Møller, M.S., Goh, Y.J., Rasmussen, K.B., Cypryk, W., Celebioglu, H.U., Klaenhammer, T.R., Svensson, B., and AbouHachem, M. 2017. An extracellular cell-attached pullulanase confers branched α-glucan utilization in human gut *Lactobacillus acidophilus. Applied and Environmental Microbiology* 83(12): e00402–17. doi: https://doi.org/10.1128/AEM.00402-17

Møller, M.S., Goh, Y.J., Viborg, A.H., Andersen, J.M., Klaenhammer, T.R., Svensson, B., and AbouHachem, M. 2014. Recent insight in α-glucan metabolism in probiotic bacteria. *Biologia (Bratisl)* 69: 713–721.

Mtshali, P.S. 2007. Screening and characterisation of wine related enzymes produced by wine associated lactic acid Bacteria. Doctoral dissertation, University of Stellenbosch.

Nasri, H., Baradaran, A., Shirzad, H., and Rafieian-Kopaei, M. 2014. New concepts in nutraceuticals as alternative for pharmaceuticals. *International Journal of Preventive Medicine* 5(12): 1487.

Nuobariene, L., Cizeikiene, D., Gradzeviciute, E., Hansen, Å.S., Rasmussen, S.K., Juodeikiene, G., and Vogensen, F.K. 2015. Phytase-active lactic acid bacteria from sourdoughs: Isolation and identification. *LWT-Food Science and Technology* 63: 766–772.

Okano, K., Yoshida, S., Yamada, R., Tanaka, T., Ogino, C., Fukuda, H., and Kondo, A. 2009. Improved production of homo-D-lactic acid via xylose fermentation by introduction of xylose assimilation genes and redirection of the phosphoketolase pathway to the pentose phosphate pathway in L-lactate dehydrogenase gene-deficient *Lactobacillus plantarum. Applied and Environmental Microbiology* 75(24): 7858–7861.

Padmavathi, T., Bhargavi, R., Priyanka, P.R., Niranjan, N.R., and Pavitra, P.V. 2018. Screening of potential probiotic lactic acid bacteria and production of amylase and its partial purification. *Journal of Genetic Engineering and Biotechnology* 16(2): 357–362.

Panesar, P.S, Kumari, S., and Panesar, R. 2010. Potential applications of immobilized β-galactosidase in food processing industries. *Enzyme Research*. doi: 10.4061/2010/473137.

Petrov, K., Urshev, Z, and Petrova, P. 2008. L (+)-Lactic acid production from starch by a novel amylolytic *Lactococcus lactis* subsp. lactis B84. *Food Microbiology* 25(4): 550–557.

Petrova, P., and Petrov, K.D. 2012. Direct starch conversion into L-(+)-lactic acid by a novel amylolytic strain of *Lactobacillus paracasei* B41. *Starch-stärke* 64(1): 10–17.

Petrova, P., Petrov, K., and Stoyancheva, G. 2013. Starch-modifying enzymes of lactic acid bacteria–structures, properties, and applications. *Starch-stärk*. 65(1-2): 34–47.

Pihlanto, A. 2013. Lactic fermentation and bioactive peptides. In: Lactic Acid Bacteria – R & D for Food, Health and Livestock Purposes, ed. M. Kongo (pp. 310–331). InTech Prepress. doi: 10.5772/51692.

Platteeuw, C., Hugenholtz, J., Starrenburg, M., van Alen-Boerrigter, I., and De Vos, W.M. 1995. Metabolic engineering of Lactococcus lactis: Influence of the overproduction of alpha-acetolactate synthase in strains deficient in lactate dehydrogenase as a function of culture conditions. *Applied and Environmental Microbiology* 61(11): 3967–3971.

Radhika, P.R., Singh, R.B., and Sivakumar, T. 2011. Nutraceuticals: An area of tremendous scope. *International Journal Research in Ayurveda Pharmacy* 2: 410–415.

Raghavendra, P., & Halami, P.M. 2009. Screening, selection and characterization of phytic acid degrading lactic acid bacteria from chicken intestine. *International Journal of Food Microbiology* 133(1–2): 129–134.

Ramakrishnan, V., Goveas, L.C., Halami, P.M., and Narayan, B. 2015. Kinetic modeling, production and characterization of an acidic lipase produced by *Enterococcus durans* NCIM5427 from fish waste. *Journal of Food Science and Technology* 52(3): 1328–1338.

Ramakrishnan, V., Goveas, L.C., Suralikerimath, N., Jampani, C., Halami, P.M., and Narayan, B. 2016. Extraction and purification of lipase from *Enterococcus faecium* MTCC5695 by PEG/phosphate aqueous-two phase system (ATPS) and its biochemical characterization. *Biocatalysis and Agricultural Biotechnology* 6: 19–27.

Ramakrishnan, V., Narayan, B., and Halami, P.M. 2019. Lipase of lactic acid bacteria. *Microbes for Sustainable Development and Bioremediation* 13: 313.

Raveschot, C., Cudennec, B., Coutte, F., Flahaut, C., Fremont, M., Drider, D., and Dhulster, P. 2018. Production of bioactive peptides by Lactobacillus species: From gene to application. *Frontiers in Microbiology* 9: 2354. https://doi.org/10.3389%2Ffmicb.2018.02354

Ray, R.C. 2016. Amylolytic lactic acid Bacteria: Microbiology and technological interventions in food fermentations. In: Fermented Foods, Part I. (pp. 143–160). CRC Press.

Ray, M., Ghosh, K., Har, P.K., Singh, S.N., and Mondal, K.C. 2017. Fortification of rice gruel into functional beverage and establishment as a carrier of newly isolated *Bifidobacterium* sp. MKK4. *Research Journal of Microbiology* 12: 102–117.

Ruffing, A. and Chen, R.R. 2006. Metabolic engineering of microbes for oligosaccharide and polysaccharide synthesis. *Microbial Cell Factories* 5(1): 1–9.

Rutella, G.S., Solieri, L., Martini, S., and Tagliazucchi, D. 2016. Release of the antihypertensive tripeptides valine-proline-proline and isoleucine-proline-proline from bovine milk caseins during in vitro gastrointestinal digestion. *Journal of Agricultural and Food Chemistry* 64: 8509–8515. doi: 10.1021/acs.jafc.6b03271.

Sachdeva, V., Roy, A., and Bharadvaja, N. 2020. Current prospects of nutraceuticals: A review. *Current Pharmaceutical Biotechnology* 21(10): 884–896.

Sahab, N.R., Subroto, E., Balia, R.L., and Utama, G.L. 2020. γ-Aminobutyric acid found in fermented foods and beverages: Current trends. *Heliyon* 6(11): e05526.

Sarwono, B. 2010. Usaha membuat tempe dan oncom. PT Niaga Swadaya.

Savijoki, K., Ingmer, H., and Varmanen, P. 2006. Proteolytic systems of lactic acid bacteria. *Applied Microbiology and Biotechnology* 71(4): 394–406.

Saxena, R.K. 1999. Microbial lipases: Potential biocatalysts for the future industry. *Current Science* 77: 101–115.

Schwab, C., Lee, V., Sørensen, K.I., and Gänzle, M.G. 2011. Production of galactooligosaccharides and heterooligosaccharides with disrupted cell extracts and whole cells of lactic acid bacteria and bifidobacteria. *International Dairy Journal* 21(10): 748–754.

Sen, S., Ray, L., and Chattopadhyay, P. 2012. Production, purification, immobilization, and characterization of a thermostable β-galactosidase from *Aspergillus alliaceus*. *Applied Biochemistry and Biotechnology* 167(7): 1938–1953.

Sharma, B., and Shukla, G. 2020. Isolation, identification, and characterization of phytase producing probiotic lactic acid bacteria from neonatal fecal samples having dephytinization activity. *Food Biotechnology* 34(2): 151–171.

Singh, R., Kumar, M., Mittal, A., & Mehta, P.K. 2016. Microbial enzymes: industrial progress in 21st century. *3 Biotech* 6(2): 1–15. https://doi.org/10.1007%2Fs13205-016-0485-8

Son, S.H., Jeon, H.L., Jeon, E.B., Lee, N.K., Park, Y.S., Kang, D.K., and Paik, H.D. 2017. Potential probiotic *Lactobacillus plantarum* Ln4 from kimchi: Evaluation of β-galactosidase and antioxidant activities. *LWT-Food Science and Technology* 85: 181–186.

Sonnenburg, J.L. and Bäckhed, F. 2016. Diet-microbiota interactions as moderators of human metabolism. *Nature* 535: 56–64.

Stefanovic, E., Fitzgerald, G., and McAuliffe, O. 2017. Advances in the genomics and metabolomics of dairy lactobacilli: A review. *Food Microbiology* 61: 33–49. doi: 10.1016/j.fm.2016.08.009.

Sun, Z., Harris, H.M.B., McCann, A., Guo, C., Argimón, S., Zhang, W., et al. 2015. Expanding the biotechnology potential of lactobacilli through comparative genomics of 213 strains and associated genera. *Nature Communications* 6: 8322. doi: 10.1038/ncomms9322.

Tamang, J.P. 2011 Prospects of Asian fermented foods in global markets. In The 12th Asian Food Conference (pp. 16–18).

Thakur, M., Rai, A.K. and Singh, S.P. 2022. An acid-tolerant and cold-active β-galactosidase potentially suitable to process milk and whey samples. *Applied Microbiology and Biotechnology* 106: 3599–3610

Theodoropoulos, V.C.T., Turatti, M.A., Greiner, R., Macedo, G.A., and Pallone, J.A.L. 2018. Effect of enzymatic treatment on phytate content and mineral bioacessability in soy drink. *Food Research International* 108: 68–73.

Venegas-Ortega, M.G., Flores-Gallegos, A.C., Martínez-Hernández, J.L., Aguilar, C.N., and Nevárez-Moorillón, G.V. 2019. Production of bioactive peptides from lactic acid bacteria: A sustainable approach for healthier foods. *Comprehensive Reviews in Food Science and Food Safety* 18(4): 1039–1051

Vieira, D.C, Lima, L.N, Mendes, A.A., Adriano, W.S., Giordano, R.C., Giordano, R.L., and Tardioli, P.W. 2013. Hydrolysis of lactose in whole milk catalyzed by β-galactosidase from *Kluyveromyces fragilis* immobilized on chitosan-based matrix. *Biochemical Engineering Journal* 15(81): 54–64.

Vishnu, C., Naveena, B.J., Altaf, M.D., Venkateshwar, M., & Reddy, G. 2006. Amylopullulanase—a novel enzyme of L. amylophilus GV6 in direct fermentation of starch to L (+) lactic acid. *Enzyme and Microbial Technology* 38(3–4): 545–550.

Waśko, A., Polak-Berecka, M., and Targoński, Z. 2011. Purification and characterization of pullulanase from *Lactococcuslactis*. *Preparative Biochemistry & Biotechnology* 41(3): 252–261.

Wisselink, H.W., Weusthuis, R.A., Eggink, G., Hugenholtz, J., and Grobben, G.J. 2002. Mannitol production by lactic acid bacteria: A review. *International Dairy Journal* 12(2-3): 151–161.

Yogeswara, I.B.A., Maneerat, S., and Haltrich, D. 2020. Glutamate decarboxylase from lactic acid bacteria—A key enzyme in GABA synthesis. *Microorganisms* 8(12): 1923.

Zeller, E., Schollenberger, M., Kühn, I., and Rodehutscord, M. 2015. Hydrolysis of phytate and formation of inositol phosphate isomers without or with supplemented phytases in different segments of the digestive tract of broilers. *Journal of Nutritional Science* 4: e1. pmid:26090091.

Zouhar, J., Vevodova, J., Marek, J., Damborsky, J., Su, X.D., and Bnobohaty, B. 2001. Insights into the functional architecture of the catalytic center of a maize beta-glucosidase *Zm-p60.l*. *Plant Physiology* 127: 973–985.

3 Fungal Enzymes for Applications in Functional Food Industry

Susan Grace Karp[1], Jéssica Aparecida Viesser[1], Maria Giovana Binder Pagnoncelli[1,2], Fernanda Guilherme Prado[1], Leticia Schneider Fanka[1], Walter José Martínez-Burgos[1], Fernanda Kelly Mezzalira[1,2], and Carlos Ricardo Soccol[1]

[1] Department do Bioprocess Engineering and Biotechnology, Federal University of Paraná, Curitiba, Paraná, Brazil

[2] Department of Chemistry and Biology, Universidade Tecnológica Federal do Paraná, Curitiba, Paraná, Brazil

CONTENTS

3.1 INTRODUCTION

According to data released by the Food and Agriculture Organization (FAO), around a tenth of the world population (approximately 800 million people) does not have enough food to maintain a healthy and active life. The data tend to indicate a worsening trend after the pandemic caused by SarsCoV-2, with up to 30% of the population denied access to a healthy diet (FAO, 2021). In addition, with the gradual increase in world population, this demand for food tends to intensify over the years, as areas destined for agricultural cultivation and cattle raising should not undergo expansion, in order to avoid advancing into environmental conservation areas. Another important issue to be pointed out is the food losses along the production chain, which reach a level of 30% of total production. All this wasted organic matter has contributed to the increase in global warming, due to the release of gases during the natural process of decomposition (UNEP, 2021). It has been made evident that the biorefinery concept will become a trend, making industrial processes more efficient and sustainable.

DOI: 10.1201/9781003311164-5

Within this scenario of transformation, changes in food behaviour have been a necessary path to ensure that what is being produced is enough to properly nourish the world's population (Ahmed et al., 2022). In this context, using a phrase attributed to Hippocrates in circa 400 BC, "Let food be thy medicine" when even then there was the knowledge that a healthy life is dependent on good nutrition, without any doubt, the path to be followed by the food industries is the search for healthy food with less waste. Japan was the first country to be concerned about the increase in the number of deaths and diseases caused by unhealthy habits, including food, establishing in 1981 the regulation of Food for Specified Health Uses (FOSHU). Since then, many industries have adhered to this new model of food and the term "functional food" has gained prominence and regulation in several countries (Birch and Bonwick, 2019; Iwatani and Yamamoto, 2019; Abedin et al., 2022).

Foods considered as functional are those which offer physiological and metabolic benefits, being able to reduce the risk of certain chronic diseases, in addition to their basic nutritional functions (El Sohaimy, 2012). Functional foods are the ones derived from plants or animals and can be categorized into groups based on their nature. Sometimes a whole food contains several ingredients which prove its beneficial properties (e.g. soybeans), in other cases the food contains an additional functional component, as in enriched, fortified, or enhanced foods. Food or food products can also be considered functional after removing some components or modifying them, for example, lactose-free products and food improved with bioavailability, like fermented soybeans (Ravindran, Sharma, and Jaiswal, 2016). Fermented foods have their anti-nutritional factors degraded due to endogenous enzymes, and can be an important source of functional foods (Nkhata et al., 2018). Also, bioactive compounds can be extracted from food and food waste using different methodologies, such as enzyme technologies (Wang et al., 2017).

Enzymes are highly specific biological catalysts that work under mild reaction conditions, and such characteristics make them preferrable to chemical reagents. The use of enzymes in the industrial sector is a trend that has been growing over the years, and a wide variety of products and ingredients can be obtained from the use of enzymes (Chapman, Ismail, and Dinu et al., 2018). Traditionally, in the food industry, enzymes are used in the transformation and stabilization of some products, such as the manufacture of bread, cheese, juice, and wine. Some non-digestible, biologically active oligosaccharides (fruto- and galacto-oligosaccharides) obtained by hydrolysis or by enzymatic synthesis can be incorporated into many products to improve one's health by stimulating the growth of a beneficial number of bacteria in the gut (Martins et al., 2019). Conversely, some components of several plant-based foods and milk, including α-galacto-oligosaccharides and lactose, have been the main cause of gastrointestinal disorders. The human digestive system does not secrete α-galactosidase, and most adult humans are not able to fully digest lactose. For this reason, the microbial α-galactosidases and β-galactosidases have important applications in therapeutic use and in the food industry in the production of lactose-free products. Microbial α-galactosidases are used in the beetroot and soybean processing industries as well (Mattar, Mazo, and Carrilho, 2012; Vandenberghe et al., 2019; Nyyssölä et al., 2020). Proteases are responsible for the production of bioactive peptides and, when introduced into the body, they perform health functions beyond nutritional function. Lipases are used to concentrate omega-3 fatty acids from oily fish (Agyei, Akanbi, and Oey, 2018).

Fungal enzymes are produced from fungi and are widely applied in food industries, since fungi are excellent enzyme producers and their application in food industries has been growing over the years. For example, pectinase, proteases, and amylases are already obtained from fungi. Many recent works have cited the use of fungi in the production of enzymes owing to their ability to secrete extracellular enzymes in greater quantity and their fast growth under low-cost media (Kumar, 2020; Takahashi et al., 2020; Ansari et al., 2021; Yahya et al., 2021).

This chapter presents the scientific advances and commercial applications related to fungal enzymes used in the segment of functional foods, namely, carbohydrate hydrolases, proteases, transglutaminases, and lipases. To evaluate the technological advances and innovations, a patent search was carried out, surveying the processes, products, or utility models that are being protected in this area.

3.2 CARBOHYDRATE HYDROLASES

Hydrolases are the major group of commercial enzymes, with around 75% of the total enzyme market (Li et al., 2012). The most important carbohydrate hydrolases, namely, amylases, cellulases, hemicellulases, and pectinases will be reviewed in this section.

3.2.1 AMYLASES

Amylases catalyze the breakdown of complex carbohydrates, representing an important group of industrial enzymes (Rana, Walia, and Gaur, 2013). Starch is a polysaccharide that corresponds to the natural substrate for amylase activity and consists of 75–80% amylopectin and 20–25% amylose (Sivaramakrishnan et al., 2007; Suriya et al., 2016; Zhang, Han, and Xiao, 2017). Based on their three-dimensional structures, reaction mechanisms, and amino acid sequences, amylases are divided into three classes: α-amylase, β-amylase, and glucoamylase (Janecek, Svensson, and MacGregor, 2014). Alpha-amylase (EC 3.2.1.1) hydrolyses the internal α-1,4-glycosidic bonds in starch producing malto-oligosaccharides. The β-amylase (EC 3.2.1.2) has the function of producing β-maltose from starch by hydrolysing the penultimate α-1,4-glucan bonds from the non-reducing end of the polymeric chains. Glucoamylase (EC 3.2.1.3) is the enzyme that produces glucose and in this process the α(1-4) glycosidic bonds are broken, as well as the α(1-6) glycosidic bonds at the non-reducing end of amylose and amylopectin (Sivaramakrishnan et al., 2007; Suriya et al., 2016).

Alpha-amylases are of great interest because they have the ability to hydrolyse starch into functional malto-oligosaccharides (Pan et al., 2017). Malto-oligosaccharide-forming amylase (MFAse, EC 3.2.1.133), a maltogenic α-amylase, is one of the fundamental enzymes with a dual catalytic function used to produce oligosaccharides. Oligosaccharides correspond to low molecular weight carbohydrates and have several health benefits due to their remarkable biopreservative and prebiotic properties. The many health benefits of functional oligosaccharides reflect in their great market demand and various applications in the food industry (Shinde and Vamkudoth, 2021).

Amylases are widely distributed among plants, animals, and microorganisms (Gopinath et al., 2017). Some of the great advantages of using microorganisms for the production of amylases are their mass productivity associated to economic fermentation processes and their easy manipulation to obtain enzymes with desired characteristics, with enzymes from fungal and bacterial sources dominating the applications in industrial sectors (Tanyildizi, Ozer, and Elibol, 2005).

Submerged fermentation (SmF) and solid-state fermentation (SSF) processes are used to produce amylases. SmF has been used particularly in the production of important industrial enzymes due to the ease of controlling different parameters such as pH, temperature, aeration, and oxygen transfer. On the other hand, SSF has the potential to resemble the microorganism's natural habitat, especially when the microorganism is a fungus, and therefore becomes a preferred environment for microorganisms to grow and produce value-added products (Orlandelli et al., 2012; Bastos et al., 2015). The properties of amylases, such as thermostability, pH profile, and pH stability, must be compatible with their application. These properties are often related to the growth conditions of the microorganism, which usually differ between bacteria and fungi. For example, most fungi grow in mildly acidic conditions, while bacteria require a neutral or alkaline pH (Reddy et al., 2003; Sundarram, Pandurangappa, and Murthy, 2014).

Microbial amylases can be produced by different species, but the production of α-amylase from bacteria may be cheaper and faster, which is desired for commercial applications. Bacterial α-amylases are mainly derived from the genus *Bacillus*. Some species can be mentioned as *Bacillus subtilis, Bacillus stearothermophilus, Bacillus amyloliquefaciens, Bacillus licheniformis, Bacillus coagulans, Bacillus polymyxa, Bacillus mesentericus, Bacillus vulgaris, Bacillus megaterium, Bacillus cereus, Bacillus halodurans*, and *Bacillus* sp. Ferdowsicous. These are good producers of thermostable α-amylase and have been widely used for the commercial production of the enzyme for various applications (Konsoula and Liakopoulou-Kyriakides, 2007; Hmidet et al., 2010; Prakash and Jaiswal, 2010).

Reports on fungal sources of amylases are limited to certain species of mesophilic fungi. Fungal sources are confined to terrestrial isolates, mainly *Aspergillus* spp. and *Penicillium* spp. The thermophilic fungus *Thermomyces lanuginosus* is considered an excellent producer of amylase due to its thermostability. The production of fungal amylases, for example from *Aspergillus niger*, is sometimes preferred due to the more accepted "generally recognized as safe" (GRAS) status of fungi, although some bacterial species such as *B. subtilis* are also GRAS (Jensen et al., 2002; Kathiresan and Manivannan, 2006; Kammoun, Naili, and Bejar, 2008).

Fungal amylases can be advantageous over bacterial amylases for presenting peculiar physicochemical characteristics and excellent potential for industrial applications. Resistance to organic solvents, thermostability, and pH-stability are some of these features. Besides, fungi are efficient enzyme producers, may not require excess water, are easily cultivated in SSF, and readily secrete the enzymes to the extracellular environment. Moreover, *A. niger*, *Aspergillus japonicus*, *Aspergillus terreus*, *Rhizopus* spp., *Penicillium* spp. and even *Trichoderma* spp. can be applied in the efficient production of amylases (Pasin et al., 2017; Kalia et al., 2021; Ünal et al., 2022).

3.2.2 Cellulases

Cellulases are enzymes responsible for the degradation of cellulose through the hydrolysis of β-1,4-glycosidic bonds. As a result, cellulose is converted into the simple sugar, glucose (Ejaz, Sohail, and Ghanemi, 2021). The complete enzymatic hydrolysis of cellulosic materials requires a complex of three cellulase enzymes, being endoglucanase or endo-1,4-β-D-glucanase (EG; EC 3.2.1.4), exoglucanase or exo-1,4-β-cellobiohydrolase (CBH; EC .2.1.3.2.1.91), and β-glucosidase (BG; EC 3.2.1.21), with each enzyme acting on different parts of the cellulose (Ejaz, Sohail, and Ghanemi, 2021).

Such enzymes act synergistically in the hydrolysis of cellulose. Endoglucanases attack randomly the O-glycosidic bonds, resulting in glucan chains of different lengths; exoglucanases act at the ends of the cellulose chain having β-cellobiose as the final product; and β-glycosidases act specially on β-cellobiose disaccharides producing glucose (Bayer, Morag, and Lamed, 1994; Sindhu, Binod, and Pandey, 2016). Cellulase production can be carried out by a large number of organisms, including bacteria, actinobacteria, filamentous fungi, plants, and animals. Some of these microorganisms used to synthesize cellulase enzymes are shown in Table 3.1. Microorganisms such as bacteria and fungi are pointed out as potential producers of cellulase; however, fungi are preferable because of their ability to penetrate and use various substrates (Srivastava et al., 2015). The cellulase enzyme obtained from aerobic fungi currently plays an important role in several industrial applications (Singh et al., 2021). The genera *Penicillium*, *Trichoderma*, and *Aspergillus* stand out as models for the production of cellulase on an industrial scale (Sajith et al., 2016; Passos, Pereira, and Castro, 2018).

Cellulases can be obtained by SSF or SmF techniques. SSF uses a solid material as substrate, usually an agro-industrial substrate, being characterized as a process that occurs in the absence of "free water". This system makes it difficult to control mass and heat transfers, but provides a natural environment for the growth of filamentous fungi (Hansen et al., 2015). On the other hand, in the SmF system, water is a major component. In the case of aerobic fermentations, the reaction medium can be agitated and receive an external supply of oxygen. SmF has become the most common process for the production of enzymes, due to its higher productivity and better operating capacity (Vaidyanathan et al., 1999; Florencio, Colli, and Farinas, 2017).

There are several industrial applications of microbial cellulases. In the food industry, the cellulase enzyme, when combined with pectinase and xylanase, acts as a macerating enzyme to improve product property and process performance (Juturu and Chuan, 2014; Toushik et al., 2017). Still, the use of this enzyme allows the increase of the benefits of antioxidant molecules, allowing to obtain a good quality and healthier product (Sharma and Yazdani, 2016).

TABLE 3.1
Microorganisms Presenting Cellulolytic Abilities

Genus	Species	Reference
Fungi		
Aspergillus	*A. niger, A. oryzae, A. fumigatus, A. nidulans, A. heteromorphus, A. acculeatus, A. terreus, A. flavus*	(Kuhad et al., 2011; Sajith et al., 2016; Kumar et al., 2019)
Fusarium	*F. solani, F. oxysporum*	(Kuhad et al., 2011)
Penicillium	*P. brasilium, P. octanus, P. decumbens, P. fumiculosum, P. janchinellum, P. pinophilum, P. echinulatum*	(Kuhad et al., 2011; Schneider et al., 2014)
Trichoderma	*T. reesei, T. longibrachiatum, T. harzianum, T. koningii, T. branchiatum, T. atroviride*	(Kuhad et al., 2011; Singh et al., 2011)
Bacteria		
Geobacillus	*Geobacillus* sp.	(Potprommanee et al., 2017)
Clostridium	*C. thermocellum, C. cellulolyticum, C. acetobutylium, C. papyrosolvens*	(Lamed et al., 1987)
Acetivibrio	*A. cellulolyticus*	(Lamed et al., 1987; Bayer, Morag, and Lamed, 1994; Kuhad et al., 2011)
Bacillus	*B. subtilis, B. pumilus, B. amyloliquefaciens, B. licheniformis, B. circulans, B. flexus*	(Kim and Kim, 1993; Kuhad et al., 2011)
Brevibacillus	*B. borstelensis*	(Ejaz et al., 2019)
Actinomycetes		
Thermomonospora	*T. fusca, T. curvata*	(Bayer, Morag, and Lamed, 1994; Kuhad et al., 2011)
Streptomyces	*S. drozdowiczii, S. lividans*	(Kuhad et al., 2011; Fujii et al., 2016)
Cellulomonas	*C. fimi, C. bioazotea, C. uda*	(Kuhad et al., 2011; Sadhu and Maiti, 2013; Ejaz et al., 2021)

3.2.3 Hemicellulases

Hemicellulose is a group of different polysaccharides with heterogeneous structures, such as mannans, xyloglucans, mixed-linkage β-glucans, and xylans. To break the bonds of these complex polymers, the action of multiple hemicellulases and auxiliary enzymes is necessary (Mamo, 2020). This enzyme is one of the most important hydrolases in terms of market size (Sarethy, Saxena, and Kapoor, 2011).

Hemicellulases are used in food processing and have the ability to improve the nutritional quality. Furthermore, the bioactive compounds, such as xylooligosaccharides (XOS) and mannooligosaccharides (MOS), have prebiotic, antioxidant, and potential immune modulating properties, and have been used as dietary supplements (Nawaz et al., 2018; Singh, Ghosh, and Goyal, 2018). XOS are recognized as non-digestible foods, because they are not absorbed in the human gastrointestinal tract. Consequently, these bioactive compounds have been used as a functional food material, for example, in dietary sweeteners for low-calorie foods.

The XOS can be obtained directly by enzymatic treatment of food with xylanase or mannanase. The nutritional quality can be improved because the use of xylanases modifies the dietary fibre. Diets rich in XOS have beneficial health effects, for example, lowering plasma cholesterol and the chances of developing type 2 diabetes. Xylanases improve the water level in fibre-enriched bread and pasta, promoting health benefits (Mcrae, 2017).

Most commercial hydrolases are produced by fungi such as *Trichoderma* and *Aspergillus* species, and some recent studies reported new promising sources of hemicellulases, like *Penicillium* and *Talaromyces*. *Trichoderma reesei* is the most used microorganism in the industrial production of cellulases. But recent studies showed several reasons why this species is not the best choice for the degradation of biomass rich in cellulases. Some species of *Penicillium* and *Talaromyces* are more efficient than *T. reesei* and give better yields of monosaccharides from biomass because they slowly release a battery of hemicellulases during the process (Eugenio et al., 2017; Yang et al., 2018).

Another hemicellulase producer cited in the literature is *Talaromyces amestolkiae*, a holomorphic genus with a very close taxonomical relatedness with *Penicillium*. Studies on the cultivation of *T. amestolkiae* with different carbon sources demonstrated that hemicellulases are produced as a response to the inducers, which did not happen in the control medium with glucose (Eugenio et al., 2017) because glucose is a strong repressor of the hemicellulolytic metabolism (Chávez, Bull, and Eyzaguirre, 2006). Some commercial enzyme products produced by fungi include cellulases derived from *A. niger* (Novozymes 188 – Novozymes; Biocellulase A, Biocellulose TRI, and Biocellulose A – Quest Intl.), *T. reesei* (Multifect CL and Accelerase® 1500 – Genencor; Celluclast 1.5 LFG®, Celtec2, and Celtec3 – Novozymes; Cellulase TRL – Solvay Enzymes), *T. reesei/Trichoderma longibrachiatum* (GC 880 – Genencor; Ultra-low microbial (ULM) – Iogen; Bio-feed beta L – Novozymes), and *Penicillium funiculosum* (Rovabio™ Excel – Adisseo), as reported by Vaishnav et al. (2018).

3.2.4 Pectinases

Pectins are categorized as soluble or insoluble fibres, which cannot be absorbed by the human digestive system, and are necessary in the food industry because of their ability to form substances which alter the texture and firmness of food products (Daniell et al., 2019; Moslemi, 2021). Structurally, pectin is a complex hetero-polysaccharide comprised of various monosaccharides, such as glucose, mannose, galactose, and arabinose (Kameshwar and Qin, 2018). However, pectinases are an enzyme group that digest pectin by modifying it to short polysaccharide fragments, and in this form pectin can be absorbed in human nutrition. Pectin is also a source of oligomers with prebiotic potential, which is interesting for the food and pharmaceutical industries (Babbar et al., 2016).

Pectinases can be efficiently produced by fungi belonging to the genus *Aspergillus*, using SmF and SSF (Ruiz et al., 2012). SSF offers the possibility of the use of agro-industrial residues as substrates for enzyme production, which is interesting because this reduces the cost of enzyme production and the environmental impact caused by these residues. Fungi are the microorganisms most adapted to SSF (Sandri and Moura, 2018). Pectic-oligosaccharides (POS) are prebiotic molecules that have gained attention as functional food ingredients (Kwan et al., 2021). These compounds are generally obtained from the partial depolymerization of pectin, usually from agro-residues through enzymatic hydrolysis (Ho, Lin, and Wu, 2017). These enzymes digest pectin to monosaccharides or oligosaccharides, and the final composition depends on the type of raw materials used as the carbon source (Gullón et al., 2013).

Pectinases are widely used in beverage industries. The application of these enzymes in juices results in the increased concentration of polyphenolic compounds (Kim, Park, and Rhee, 1990). In wine production, pectinases find applicability in improving stability, color extraction and increasing the anthocyanin content in red wine (Monsan and Paul, 1995). The addition of these compounds in beverages brings significant health benefits.

Studies have investigated the recovery of POS from citrus peel, lemon peel, sugar beet pulp, and orange peel by enzymatic hydrolysis (Gómez et al., 2016). The POS recovery with commercial pectinase from hydrolysates yielded high contents of arabinose and galactose. POS of this type could complement prebiotic growths, becoming candidates of health-promoting effects (Wongkaew et al., 2021).

3.3 PROTEASES

The peptide bonds between amino acids and protein molecules are cleaved by proteases (EC 3.4). Proteases consist of a group of hydrolytic enzymes that includes proteinases and peptidases

(Gurumallesh et al., 2019; Naveed et al., 2021). The classification of proteases is mainly based on the catalysed reaction and their site of action (Vandenberghe et al., 2020). Exopeptidases (EC 3.4.11-19) execute their enzyme activity by the hydrolysis of the C- and N-terminal points of the polypeptide chain, while endopeptidases (EC 3.4.21-99) cleave the internal portions of the polypeptide chain (Gurumallesh et al., 2019; Vandenberghe et al., 2020). They can also be classified according to the pH range, whether they have a higher activity in acidic (pH 2.0 to 6.0), neutral (pH 6.0 to 8.0), or alkaline (pH 8.0 to 13.0) values of pH (Souza et al., 2015).

Proteases are present in all biological systems, including microorganisms, animals, and plants (Gurumallesh et al., 2019). They are extremely important for metabolic and regulatory functions like processes of cell differentiation, cell division and migration, food proteins digestion, blood clotting, apoptosis, signal transduction, release of hormones, and retroviruses replication (Souza et al., 2015; Naveed et al., 2021). In the global market, proteases are an important group of industrial enzymes that reportedly comprise approximately 60% of the whole enzyme marketplace (Kumari et al., 2015). Proteases have a large variety of applications in medical, pharmaceutical, chemical, agricultural, and food industries (Gurumallesh et al., 2019; Naveed et al., 2021).

Microbial proteases are produced on a large scale by fermentation processes and contribute to nearly two-thirds of industrially essential protease (Gurumallesh et al., 2019). The microbial production of proteases presents many advantages if compared with the extraction from plants and animals. Microbial proteases are extracellular, have lower production costs, and present good stability and specificity (Gurumallesh et al., 2019; Vandenberghe et al., 2020). Frequently, fungal proteases are used in the modification of food proteins. Fungal proteases have superior advantages over the proteases produced by bacteria, and fungal species are more frequently characterized as GRAS strains (Naveed et al., 2021). Furthermore, proteases that are produced from fungi present high diversity, broader substrate specificity, stability under unfavourable conditions, and mycelium separation can be accomplished by the process of simple filtration (Rani, Rana, and Datt, 2012; Jisha et al., 2013).

Proteases are commonly used in the food industry for production of functional foods. The hydrolysis of proteins by proteases releases protein hydrolysates, peptides, and amino acids, resulting in enhanced functional and nutritional properties of foods through bioactive peptides liberation and alleviation of protein allergy (Chourasia et al., 2021). Bioactive peptides are small protein molecules with usually fewer than 20 units of amino acid residues and, when ingested, can provide several physiological benefits such as reduction in the risk of cardiovascular diseases, negative modulation of tumour growth, and reduction in blood sugar level (Aluko, 2012; Chew, Toh, and Ismail, 2018).

During food fermentation, bioactive peptides are produced by the action of fungi. Several fungal species are involved in the production of proteases associated with the release of bioactive peptides (Table 3.2). Most studies have used filamentous fungi as starter cultures. Filamentous fungi have a high potential to secrete functionally active enzymes, improving the functional properties of fermented foods such as tempeh, miso, douchi, and tofuyo (Kuba et al., 2003; Gibbs et al., 2004; Wang et al., 2008; Inoue et al., 2009). In addition, one of the advantages of using filamentous fungi is their ability to produce proteases in solid-state fermentation due to the low energy consumption during the process (Jisha et al., 2013; Gurumallesh et al., 2019).

Yeasts have also been reported as producers of bioactive peptides. Yeast-derived peptides revealed antimicrobial, antioxidant, antihypertensive, and antidiabetic activities, being mainly derived from commercial strains such as *Saccharomyces cerevisiae*, *Pichia fermentans*, and *Candida kefir* (Martínez-Medina et al., 2018; Mirzaei et al., 2021). Food-derived peptides by fungal protease-mediated hydrolysis have been largely applied as inhibitors of angiotensin-converting enzyme (ACE), a principal causative agent of hypertension (Table 3.2). Moreover, food-derived bioactive peptides have demonstrated other functional properties, including immunomodulatory, anti-inflammatory, anticancer, antithrombotic, opioid, and antioxidant activities (Aluko, 2012; Chourasia et al., 2021; Abedin et al., 2022).

Another application of fungal proteases in the functional food industry is associated with the release and purification of β-glucans. These are polysaccharides composed of D-glucose units

TABLE 3.2
Food-Derived Peptides Obtained by Fungal Protease-Mediated Hydrolysis or Fermentation

Microorganism	Enzyme	Substrate (Food)	Peptide	Bioactivity	Reference
Aspergillus oryzae *Monascus* sp.	-	Tofuyo	Ile-Phe-Leu Trp-Leu	ACE inhibitory	Kuba et al. (2003)
Rhizopus oligosporus	-	Tempeh	-	Antioxidant	Gibbs et al. (2004)
Aspergillus egypticus	Trypsinase	Fermented soybean	Phe-Ile-Gly	ACE inhibitory	Zhang et al. (2006)
Monascus pilosus	-	Fermented soybean	-	Antioxidant ACE inhibitory	Pyo and Lee (2007)
A. oryzae	-	Miso paste	Val-Pro-Pro Ile-Pro-Pro	Antihypertensive	Inoue et al. (2009)
A. oryzae	-	Douchi	-	Antioxidant	Wang et al. (2008)
Mucor Micheli ex fries	-	Mao-tofu	-	ACE inhibitory	Hang and Zhao (2012)
Aspergillus sojae	Corolase	Whey protein	-	ACE inhibitory	Morais et al. (2013a)
A. oryzae	Flavourzyme	Whey protein	-	ACE inhibitory	Morais et al. (2013b)
Pichia kudriavzevii *Kluyveromyces marxianus*	-	Kumis (fermented milk)	-	ACE inhibitory	Chaves-López et al. (2014)
Mucor miehei	Rennet	Casein	Glu-Ile-Val-Pro-Asn Asp-Lys-Ile-His-Pro-Phe Val-Ala-Pro-Phe-Pro-Gln	ACE inhibitory	Timón et al. (2019)

ACE: Angiotensin-converting enzyme.

linked by β-glycosidic bonds that can be found in cereals (e.g. oats, wheat, and barley) and in the cell wall of other plants, fungi, and bacteria (Ravindran, Sharma, and Jaiswal, 2016; Colosimo et al., 2021). Beta-glucans exist in close association with starch granules and several proteins in cereals, therefore, they need to be purified (Yang and Huang, 2021). This protein content is reduced by adding proteases through an enzyme-alkaline extraction method (Ravindran, Sharma, and Jaiswal, 2016; Karimi, Azizi, and Xu, 2019; Goudar et al., 2020; Nguyen et al., 2022).

A great potential of β-glucans with many health-promoting and prebiotic properties has been reported, including anti-inflammatory, anti-aging, anticancer, and immunomodulatory properties (Zhu, Du, and Xu, 2016). A diet rich in β-glucans helps to increase the peristalsis of the intestines and promote growth of healthy microflora (Ravindran, Sharma, and Jaiswal, 2016). Commercially, several proteases obtained from fungi are currently available on the market (Table 3.3). These data were compiled from reports by Ward, Rao, and Kulkarni (2009), Martínez-Medina et al. (2018) and Mora, Aristoy, and Toldrá (2018). *Aspergillus* has been a dominant fungal genus used for industrial protease production. *Aspergillus* species are capable of producing between 25 and 30 g/L of commercial extracellular enzyme (Ward, Rao, and Kulkarni, 2009).

Enzyme preparations usually contain an active enzyme, but they may also contain a mixture of two or more active enzymes (Ward, Rao, and Kulkarni, 2009). For instance, in the Flavourzyme preparation are identified three endopeptidases, two aminopeptidases, two dipeptidyl peptidases, and one amylase (Chew, Toh, and Ismail, 2018). The use of enzyme preparations in food processing helps in maximizing the quality and value of processed foods while optimizing the production processes (Chew, Toh, and Ismail, 2018; Mora, Aristoy, and Toldrá, 2018; Gurumallesh et al., 2019).

TABLE 3.3
Examples of Commercial Fungal Proteases, Applications, and Their Industrial Suppliers

Product Name (Enzyme)	Origin	Manufacturer	Application
Acid Protease A	*Aspergillus niger*	Amano Pharmaceuticals (Japan)	Protein hydrolysis
Acid Protease A2	*Aspergillus oryzae*	Amano Pharmaceuticals (Japan)	Protein hydrolysis
Acid Protease A-DS	*A. oryzae*	Amano Pharmaceuticals (Japan)	Dietary supplement
Acid Protease DS	*A. niger*	Amano Pharmaceuticals (Japan)	Dietary supplement
BakeZyme	*Aspergillus* sp.	DSM (Netherlands)	Baking
Corolase 7092	*A. oryzae*	Novo Nordisk (Denmark)	Protein hydrolysis
Corolase 7093	*A. oryzae*	Novo Nordisk (Denmark)	Protein hydrolysis
Corolase LAP	*Aspergillus sojae*	AB Enzymes (Germany)	Protein hydrolysis
Corolase PS	*A. oryzae*	Röhm GmbH (Germany)	Protein hydrolysis
Flavourzyme 1000M	*A. oryzae*	Novozymes (Denmark)	Protein hydrolysis
Fromase	*Rhizomucor miehei*	DSM (Netherlands)	Cheesemaking
Fungal-Protease	*Aspergillus* sp.	Novo Nordisk (Denmark)	Protein hydrolysis
		Genencor International (USA)	Baking
Maxiren	*Kluyveromyces lactis*	DSM (Netherlands)	Cheesemaking
Newlase F	*Rhizopus niveus*	Amano Pharmaceuticals (Japan)	Protein hydrolysis
Protease M	*A. oryzae*	Amano Pharmaceuticals (Japan)	Protein hydrolysis
Prozyme	*Aspergillus mellius*	Amano Pharmaceuticals (Japan)	Digestive aid
Suparen/Surecurd	*Cryphonectria parasitica*	DSM (Netherlands)	Cheesemaking

3.4 TRANSGLUTAMINASES

Transglutaminases (EC 2.3.2.13), also known as TGase or protein-glutamine ɣ-glutamyl-transferase, are a family of enzymes responsible for adding free amines into proteins by joining the glutamine residue (Fatima and Khare, 2018). Furthermore, these enzymes also catalyse the acyl transfer reaction, cross-linking between glutamine and lysine via a lysyl residue (i.e. transamidation), and deamidation into proteins (Romeih and Walker, 2017).

Animals, plants, and microorganisms are sources of transglutaminases (Akbari, Razavi, and Kieliszek, 2021). They are involved in many physiological processes, such as in coagulation, in antibacterial immune reactions, and photosynthesis (Kieliszek and Misiewicz, 2014). However, microbial transglutaminases (MTGase) have demonstrated to be cost effective and eco-friendly due to their Ca^{2+}-independency which minimizes the formation of by-products (mainly calcium-protein complexes), and have greater thermal stability (Kieliszek and Misiewicz, 2014; Fatima and Khare, 2018).

Enzymatic modifications by MTGases have been used in the food industry as a tool for improving the functional properties of food proteins, such as solubility, viscosity, elasticity, water holding capacity (WHC), emulsifying capacity, foaming, and gelation (Wang et al., 2018; Akbari, Razavi, and Kieliszek, 2021). In this way, modifications on the molecular structure of proteins improve texture and stability, without impacting the pH, colour, or flavour of foods, and moreover, make them more nutritious on account of the incorporation of essential amino acids (Grossmann et al., 2017; Fatima and Khare, 2018).

The cross-links generated by MTGases can modify physico-chemical properties of many food proteins, including milk caseins, whey proteins, wheat gluten, soybean globulins, and meat myosins (Martins et al., 2014). Commercially, MTGases used in the food industry are derived from wild-type strains belonging to the genus *Streptomyces*, mainly *Streptomyces mobaraense*

(Fatima and Khare, 2018; Miwa, 2020). Other strains were proven to have the ability to produce MTGases, such as *Bacillus firmus*, *Bacillus nakamurai*, *B. subtilis*, and *Pseudomonas putida* (Fatima and Khare, 2018; Akbari, Razavi, and Kieliszek, 2021). Recently, a novel MTGase from antartic *Penicillium chrysogenum* was isolated by Glodowsky et al. (2020). The partially purified MTGase was used as an additive to modify the rheology of a cold-set gelatin gel, achieving an increase in the gel strength of 32.25%.

Transglutaminases can be found in the cell wall of some yeasts, as observed in *S. cerevisiae* and *Candida albicans*, forming cross-link connections between structural glycoproteins (Iranzo et al., 2002; Mazáň and Farkaš, 2007; Reyna-Beltrán et al., 2018; Wang et al., 2018). Genetically modified yeasts have been reported as a good alternative for large-scale production of MTGases when compared with the conventional fermentation with wild strains (Wang et al., 2018). This is because the culture medium for *Streptomyces* sp. requires a large amount of expensive nutrients, which are not suitable for large-scale production (Kieliszek and Misiewicz, 2014). Thus, MTGase production by genetically modified microorganisms is more efficient due to the simple nutrition required and high productivity (Yokoyama, Nio, and Kikuchi, 2004; Wang et al., 2018).

For instance, Li et al. (2014) characterized a recombinant TGase from *Zea mays* expressed in *Pichia pastoris* and evaluated its effect on properties of different kinds of yoghurt. The optimized recombinant TGase (TGZo) was successfully expressed in *P. pastoris* and its specific activity was 0.889 U/mg. Moreover, the enzymatic cross-linking of milk proteins by TGZo improved the functional properties like texture and apparent viscosity in full and non-fat yoghurts. Therefore, this study indicated that TGZo could be used as a substitute for wild-type MTGase in the production of yoghurt.

Similarly, MTGase from *Streptomyces fradiae* was expressed in *P. pastoris* by Yang and Zhang (2019). The recombinant enzyme activity was approximately 0.70 U/mL. When used in meat products, the recombinant MTGase significantly improved the texture, slice integrity, and chewiness of minced pork. Liu et al. (2015) constructed a recombinant *Yarrowia lipolytica* strain that could efficiently secrete TGase from *Streptomyces hygroscopicus* with a specific activity of 7.8 U/mL. The results obtained in this study provided new possibilities for the efficient production of MTGases used in food processing.

3.5 LIPASES

In recent years, the knowledge about the effects of specific fatty acids, especially polyunsaturated fatty acids (PUFAs), has generated greater interest in the use of oils and fats to prevent chronic diseases. These lipids are being considered as functional foods (Castejon and Senorans, 2020; Szymczak et al., 2021). Currently, oil and fat modifications correspond to one of the main sectors of food processing industries; for example, fish oil, which is composed of PUFA, is being indicated both for the consumption of the whole fish itself, and the consumption of foods which contain fatty acid or concentrated fatty acid in capsules as a food supplement (Dong et al., 2019; Swapnil and Arpana, 2019; Coelho and Orlandelli, 2021). The world market has a significant variety of dietary supplements of PUFAs, mainly from the omega-3 and 6 families. Omega-3 fatty acids are being incorporated in infant formulas, bread, cereal products, milk and others (Caballero et al., 2014; Castejon and Senorans, 2019, 2020).

Omega-3 fatty acids are PUFAs that usually present 18–22 carbon chains, mainly with two or more double bonds. The omega-3 fatty acids are α-linolenic acid (ALA, C18:3n-3), stearidonic acid (STA, C18:4n-3), eicosapentaenoic acid (EPA, C20:5n-3), docosapentaenoicacid (DPA, C22:5n-3), and docosahexaenoic acid (DHA, C22:6n-3). The most important omega-3 fatty acids are DHA and EPA due to their health benefits, but PUFAs are moderately absorbed by the intestine as triglycerides and most promptly absorbed in the free-fatty acids form. Industrial production of DHA and EPA concentrates, which involves lipases, is a preferred and efficient method (Gámez-Meza et al., 2003; Agyei, Akanbi, and Oey, 2018).

Lipases (EC 3.1.1.3) comprise a group of enzymes capable of hydrolysing ester oils (triacylglycerols) into free fatty acids and glycerol at the lipid water interface.They are known to be able to change the fatty acid composition of food products (Dong et al., 2019; Castejon and Senorans, 2020; Chandra et al., 2020). These enzymes hydrolyse specific fatty acids at specific positions on the glycerol backbone of oil, and lipases which hydrolyse preferentially saturated and monounsaturated fatty acids are preferred to concentrate omega-3 fatty acids. Aiming to obtain high yields in the omega-3 concentration process, the fatty acid composition of the raw material and the enzymes to be selected must be known in order to guarantee the hydrolysis selectivity (Agyei, Akanbi, and Oey, 2018).

The lipases occur in plants and animals; however, the use of microorganisms for the production of enzymes has been driven by the vast microbial diversity. Microbial lipases stand out for their variety of available catalytic activities, versatility, high yield production, and ease of mass production. The cost of producing microbial lipases varies according to the costs of their fermentation and downstream process, where productivity and process stability are essential (Kumari et al., 2019; Szymczak et al., 2021; Phukon et al., 2022). Several studies have reported that filamentous fungal lipases are predominantly extracellular and highly active. The main genera of filamentous fungi which produce lipase are *Aspergillus, Rhizopus, Penicillium, Mucor, Ashbya, Geotrichum, Beauveria, Humicola, Rhizomucor, Fusarium, Acremonium, Alternaria, Eurotrium,* and *Ophiostoma*. Lipases are also produced by the species *Candida rugosa, Candida antarctica, T. lanuginosus*, and *Rhizomucor miehei*. Commercial lipases produced by filamentous fungi are obtained from the following species: *A. niger, C. rugosa, Humicola lanuginosa, Mucor miehei, Rhizopus arrhizus, Rhizopus delemar, Rhizopus japonicus, Rhizopus niveus*, and *Rhizopus oryzae* (Gopinath et al., 2013; Chandra et al., 2020; Cesario et al., 2021).

Fungal lipases can be easily obtained through SSF using oily raw material as a substrate. Oily substrates can be obtained from agro-residues, such as rice bran and wheat bran, and from residues from vegetable oils extraction, such as sunflower oil cake, palm kernel cake, babassu oil cake, and coconut oil cake. The fermentation process has factors which interfere in the production of the enzyme (pH, temperature, medium composition, agitation, aeration, type of inducer, and concentration), varying according to the needs of each fungi. Studies using the species of *Aspergillus flavus, P. chrysogenum*, and *Trichoderma harzianum* were carried out through SSF using similar agro-industrial residues. *A. flavus* showed maximum lipase production of 121.35 U/g of dry solid substrate which is five and nine times, respectively, more than *P. chrysogenum* and *T. harzianum*, respectively, using the same conditions of operation. *A. niger* presented a lipase activity of 121.53 U/g of dry solid substrate using the SSF technique with wheat bran as substrate. Scale-up studies for the production of lipase using other strains of *A. niger* were reported with a maximum activity of 745.7 U/g of dry solid substrate using tri-fermented substrates (Santhosh Kumar, 2015).

Some commercial products available on the market containing fungal lipase are Novozyme 435 (Novozymes), Lipomod 34MDP (Biocatalysts), Lipozyme RM IM (Novozymes), and Lipozyme TL IM (Novozymes). Novozyme 435 is a lipase produced by *C. antarctica* and immobilized on macroporous ion exchange acrylic resin. Lipomod 34MDP is a lyophilized free lipase from *C. rugosa*. Lipozyme RM IM is produced by submerged fermentation of *R. miehei* and immobilized on ion exchange resin, and the Lipozyme TL IM, also immobilized, is produced by submerged fermentation of *T. lanuginosus*.

3.6 TECHNOLOGICAL DEVELOPMENT

To evaluate the technological, and scientific advances, as well as the innovations in the area of fungal enzymes applied for the production of functional foods, a patent search was carried out, surveying the processes or utility models that are being protected in this area. The patent search was performed using the Derwent Innovations Index database with the following keywords in the topic: ((IP=(C12P*)) AND TS=(enzym*)) AND TS=(fung* OR macromycet* OR mushroom* OR mold* OR yeast*) AND (IP=(A23L-005* OR A23L-029* OR A23L-033* OR A23L-035) OR TS=(nutraceutic* food* OR functional food*)). The search was carried out from 1 January 2011 to 21 December 2021, totalling 328 patent documents found. Within the documents, 44% of the patents analysed were from China, indicating that this country can be considered the main holder of this technology. In addition, it was observed that countries such as the United States, Japan, France, and South Korea also had significant amounts of patent filings (Figure 3.1a). This could be explained by the fact that enzyme production in China has increased exponentially in the last decade, while the United States presents the largest market for these catalysts, with approximately 40% of the share (Guerrand, 2018).

In addition, it was observed that the number of deposits per year had increased from 2015 to 2019 (Figure 3.1b), demonstrating a growing interest in these technologies. It is interesting to highlight that there was a significant decrease in the amount of patent deposits in the years 2020 and 2021, which, according to Vandenberghe et al. (2021), occurs due to the average latency period of 18 months between the date of filing the patent and its publication.

It was observed that patents were classified into three major groups with the main international patent classification (IPC) codes A23L, C12N, and C12P (Figure 3.1c) according to the World Intellectual Property Organization (WIPO) classification (WIPO, 2022). As expected, 82% of

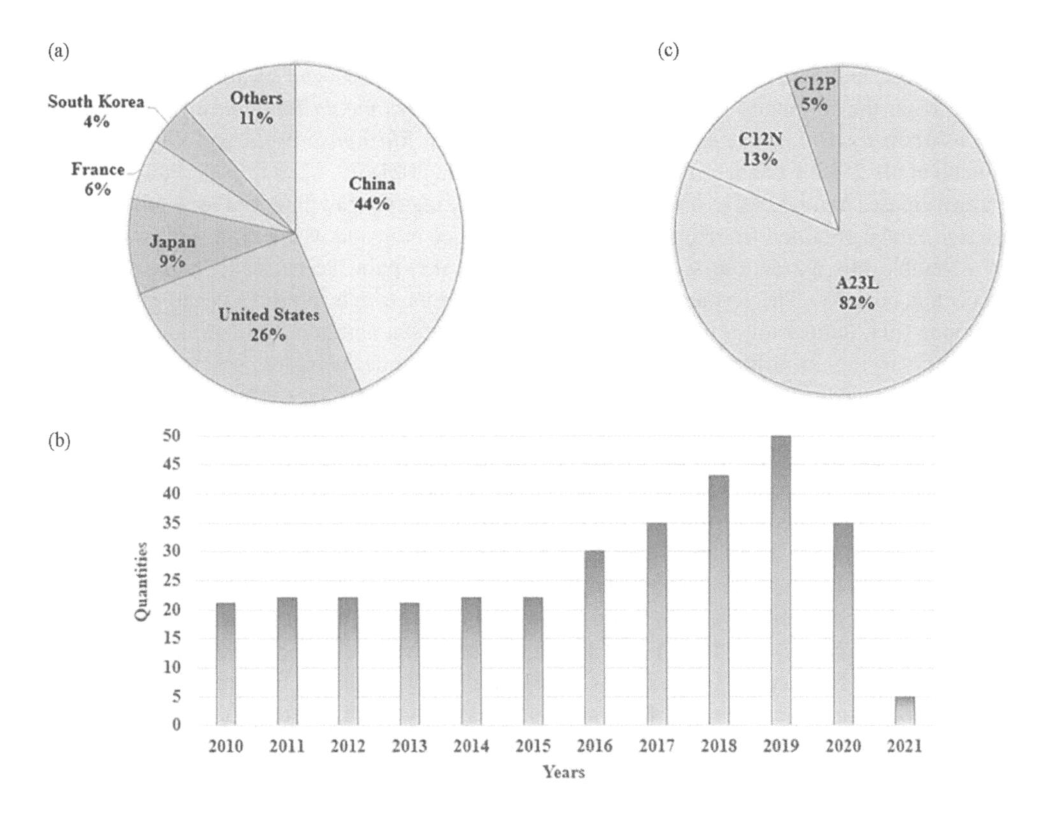

FIGURE 3.1 (a) Main countries that hold the technology of fungal enzymes applied in the production of functional foods; (b) Number of patents filed per year in the area of fungal enzymes applied in the production of functional foods; (c) Distribution of patent documents according to the main International Patent Classification code. **A23L:** Foods, foodstuffs, or non-alcoholic beverages, not covered by subclasses A21D or A23B-A23J; their preparation or treatment, e.g. cooking, modification of nutritive qualities, physical treatment (shaping or working, not fully covered by this subclass, A23P); preservation of foods or foodstuffs, in general. **C12N:** Microorganisms or enzymes; compositions thereof; propagating, preserving, or maintaining microorganisms; mutation or genetic engineering; culture media. **C12P:** Fermentation or enzyme-using processes to synthesize a desired chemical compound or composition or to separate optical isomers from a racemic mixture.

patent documents have A23L as the main IPC code associated with food. According to Guerrand (2018), 35% of enzyme production is destined for food, followed by cleaning agents with 25%. Some of these patent documents are described as follows.

Ogasawara et al. (2013), in the patent document represented by the application code WO2014142325A1, presented a new variant of the cellulase-producing fungus *T.reesei*, allowing the preferential production of cello-oligosaccharides. These compounds have physiologically active functions, as they are useful as raw materials in the formulation of functional foods (Karnaouri et al., 2019; Barbosa et al., 2020). According to the patent claims, the mutant strain also contains the egl1 gene, which is an endoglucanase gene. For the production of cello-oligosaccharides, the decomposition of the cellulosic substance starts from the use of cellulase generated by the mutant. Difeng et al. (2014), in patent CN104004813A, showed the process of preparing bioactive peptides from mushrooms (*Lentinula edodes*), which have antioxidant properties and can be used in the production of functional foods or products with high nutritional value for health. The process involves dissolving quantitative mushroom powder in distilled water with agitation, then ultrasonic disruption is performed. In addition, cellulase is used in a ratio of 100,000 to 200,000 international units per kilogram of fungal biomass powder. The protein is subsequently precipitated at a pH between 9.5 and 10.5 and a temperature of 5–15°C, and then separated by centrifugation and dried. Hao (2011), in patent CN102550802A, presented a process for extracting polypeptides and amino acids from craft beer residues, which can also be used as functional food additives. The process begins with the drying of the yeast, followed by the yeast biomass homogenization and treatment with high pressure. In addition, two extraction processes are carried out in parallel with enzymolysis processes to increase the permeability of the yeast cell walls, improving the extractive obtainment rate. Enzymolysis is performed with a mixture of proteases, namely, neutral protease, acid protease, trypsin, and papain of undefined origin and *B. subtilis* protease. Finally, the separation through filtration and drying of the amino acids and polypeptides is carried out. Kasahara et al. (2016), in the invention WO2017170919A1, presented a method for producing XOS. The process involves the hydrolysis of a biomass containing xylan and cellulose using cellulases from *Trichoderma* and *Aspergillus*. The enzyme used at the time of hydrolysis exhibits xylanase, cellobiohydrolase, and β-glucosidase activities, but does not substantially exhibit β-xylosidase activity. According to the inventors, the method produces XOS with high yield and purity in a simple, fast, economical way, and suppresses the breakdown of XOS into xylose. XOS are sugar oligomers of xylose units. They are recognized for their great prebiotic potential and nutritional benefits, promoting the growth of probiotic bacteria in the intestinal tract (De Freitas, Carmona, and Brienzo, 2019; Reque et al., 2019).

3.7 CONCLUSIONS

Enzymes are important tools in the obtaining of functional food products, owing to their ability to selectively remove anti-nutritional factors, increase the digestibility and bioavailability of nutrients, and extract bioactive compounds from several sources. Fungal enzymes are of particular importance because of the fungi's ability to secrete extracellular enzymes in greater quantity, their fast growth under low-cost media, and the GRAS status of several strains applied in the food industry. Several fungal enzyme preparations are commercially available for a great range of uses. The scientific literature search revealed recent advances in the production and application of fungal enzymes for food processing. Most examples are related to the use of carbohydrate hydrolases, proteases, transglutaminases, and lipases. Genetic tools are often applied to achieve high productivities and activities, and could be a valuable alternative in cases where there are no restrictions to their use in food products. The patent search revealed that the main technology holders in enzyme processes and products applied to the functional food industry are China, the United States, Japan, France, and South Korea. The number of new patent filings had a growing trend from 2015, showing that the maturity stage of the technology has not yet been reached and that the interest is increasing.

REFERENCES

Abedin, M. M., Chourasia, R., Phukon, L. C., Singh, S. P., & Rai, A. K. 2022. Characterization of ACE inhibitory and antioxidant peptides in yak and cow milk hard chhurpi cheese of the Sikkim Himalayan region. *Food Chemistry: X* 13: 100231.

Agyei, D., Akanbi, T. O., & Oey, I. 2018. Enzymes for use in functional foods. In M. Kuddus (Ed), *Enzymes in Food Biotechnology: Production, Applications, and Future Prospects.* Elsevier Inc. https://doi.org/10.1016/B978-0-12-813280-7.00009-8.

Ahmed, M. H., Vasas, D., Hassan, A., & Moltár, J. 2022. The impact of functional food in prevention of malnutrition. *PharmaNutrition* 19: 100288.

Akbari, M., Razavi, S. H., & Kieliszek, M. 2021. Recent advances in microbial transglutaminase biosynthesis and its application in the food industry. *Trends in Food Science and Technology* 110: 458–469.

Aluko, R. E. 2012. Functional foods and nutraceuticals. In R. H. Schmidt & G. E. Rodrick (Eds), *Food Science Text Series* (pp. 209–228). Wiley. https://doi.org/10.1002/047172159x.ch33.

Ansari, F., Samakkhah, S. A., Bahadori, A., *et al.* 2021. Health-promoting properties of *Saccharomyces cerevisiae var. boulardii* as a probiotic; characteristics, isolation, and applications in dairy products. *Critical Reviews in Food Science and Nutrition.* doi: 10.1080/10408398.2021.1949577. Online ahead of print.

Babbar, N., Dejonghe, W., Gatti, M., Sforza, S., & Kathy, E. 2016. Pectic oligosaccharides from agricultural by-products: production, characterization and health benefits. *Critical Reviews in Biotechnology* 36(4): 1–13.

Barbosa, F. C., Martins, M., Brenelli, L. B., *et al.* 2020. Screening of potential endoglucanases, hydrolysis conditions and different sugarcane straws pretreatments for cello-oligosaccharides production. *Bioresource Technology* 316: 123918.

Bastos, C. M. da S., Costa, S. T. C., Abreu-Lima, T. L. de, Zuniga, A. D. G., & Carreiro, S. C. 2015. Efeito das condições de cultivo na produção de amilase por duas linhagens de leveduras. *Revista Brasileira de Biociências* 13(3): 123–129.

Bayer, E. A., Morag, E., & Lamed, R. 1994. The cellulosome – a treasure-trove for biotechnology. *Trends in Biotechnology* 12: 259–265.

Birch, C. S. & Bonwick, G. A. 2019. Ensuring the future of functional foods. *International Journal of Food Science and Technology* 54(5): 1467–1485.

Caballero, E., Soto, C., Olivares, A., & Altamirano, C. 2014. Potential use of avocado oil on structured lipids MLM-type production catalysed by commercial immobilised lipases. *PLOS ONE* 9(9): e107749.

Castejon, N. & Senorans, F. J. 2019. Strategies for enzymatic synthesis of omega-3 structured triacylglycerols from *Camelina sativa* oil enriched in EPA and DHA. *European Journal of Lipid Science and Technology* 121(5): 1800412.

Castejon, N. & Senorans, F. J. 2020. Enzymatic modification to produce health-promoting lipids from fish oil, algae and other new omega-3 sources: A review. *New Biotechnology* 57: 45–54.

Cesario, L. M., Pires, G. P., Santos Pereira, R. F., *et al.* 2021. Optimization of lipase production using fungal isolates from oily residues. *BMC Biotechnology* 21(1): 65.

Chandra, P., Enespa, Singh, R., & Arora, P. K. 2020. Microbial lipases and their industrial applications: A comprehensive review. *Microbial Cell Factories* 19(1): 1–42.

Chapman, J., Ismail, A. E., & Dinu, C. Z. 2018. Industrial applications of enzymes: Recent advances, techniques, and outlooks. *Catalysts* 8(6): 238. MDPI AG. https://doi.org/10.3390/catal8060238.

Chaves-López, C., Serio, A., Paparella, A., *et al.* 2014. Impact of microbial cultures on proteolysis and release of bioactive peptides in fermented milk. *Food Microbiology* 42: 117–121. https://doi.org/10.1016/j.fm.2014.03.005

Chávez, R., Bull, P., & Eyzaguirre, J. 2006. The xylanolytic enzyme system from the genus *Penicillium. Journal of Biotechnology* 123: 413–433.

Chew, L. Y., Toh, G. T., & Ismail, A. 2018. Application of proteases for the production of bioactive peptides. In M. Kuddus (Ed), *Enzymes in Food Biotechnology: Production, Applications, and Future Prospects* (pp. 247–261). Elsevier Inc. https://doi.org/10.1016/B978-0-12-813280-7.00015-3.

Chourasia, R., Abedin, M. M., Phukon, L. C., Sahoo, D., Singh, S. P., & Rai, A.K. 2021. Biotechnological approaches for the production of designer cheese with improved functionality. *Comprehensive Reviews in Food Science and Food Safety* 20: 960–979.

Coelho, A. L. S. & Orlandelli, R. C. 2021. Immobilized microbial lipases in the food industry: A systematic literature review. *Critical Reviews in Food Science and Nutrition* 61(10): 1689–1703.

Colosimo, R., Mulet-Cabero, A. I., Cross, K. L., *et al.* 2021. β-Glucan release from fungal and plant cell walls after simulated gastrointestinal digestion. *Journal of Functional Foods* 83: 104543.

Daniell, H., Ribeiro, T., Lin, S., Saha, P., Mcmichael, C., Chowdhary, R., & Agarwal, A. 2019. Validation of leaf and microbial pectinases: Commercial launching of a new platform technology. *Plant Biotechnology Journal* 17(6): 1154–1166.

De Freitas, C., Carmona, E., & Brienzo, M. 2019. Xylooligosaccharides production process from lignocellulosic biomass and bioactive effects. *Bioactive Carbohydrates and Dietary Fibre* 18: 100184.

Difeng, R., Yudi, Y., Zhan, C., Qirong, K., Jun, L., & Jian, R. 2014. *Method for preparing shiitake bioactive peptide*. CN patent 104004813A.

Dong, Z., Liu, Z., Shi, J., *et al.* 2019. Carbon nanoparticle-stabilized pickering emulsion as a sustainable and high-performance interfacial catalysis platform for enzymatic esterification/transesterification. *ACS Sustainable Chemistry and Engineering* 7(8): 7619–7629.

Ejaz, U. *et al.* 2019. Methyltrioctylammonium chloride mediated removal of lignin from sugarcane bagasse for themostable cellulase production. *International Journal of Biological Macromolecules* 140: 1064–1072. doi: 10.1016/j.ijbiomac.2019.08.206

Ejaz, U., Sohail, M., & Ghanemi, A. 2021. Cellulases from bioactivity to a variety of industrial applications. *Biomimetics* 6(44): 2–11.

El Sohaimy, S. 2012. Functional foods and nutraceuticals-modern approach to food science. *World Applied Sciences Journal* 20(5): 691–708.

Eugenio, L. I. De, Líter, J. A. M., Domínguez, M. N., *et al.* 2017. Differential β-glucosidase expression as a function of carbon source availability in *Talaromyces amestolkiae*: A genomic and proteomic approach. *Biotechnology for Biofuels* 10(161): 1–14.

FAO. 2021. *The State of Food Security and Nutrition in the World 2021.* Food and Agriculture Organization of the United Nations. https://doi.org/10.4060/cb4474en.

Fatima, S. W. & Khare, S. K. 2018. Current insight and futuristic vistas of microbial transglutaminase in nutraceutical industry. *Microbiological Research* 215: 7–14.

Florencio, C., Colli, A., & Farinas, S. 2017. Current challenges on the production and use of cellulolytic enzymes in the hydrolyses of lignocellulosic biomass. *Química Nova* 40(9): 1082–1093.

Fujii, K. *et al.* 2016. Streptomyces abietis sp. nov., a cellulolytic bacterium isolated from soil of a pine forest. *International Journal of Systematic and Evolutionary Microbiology* 2013: 4754–4759. doi: 10.1099/ijs.0.053025-0

Gámez-Meza, N., Noriega-Rodríguez, J. A., Medina-Juárez, L. A., *et al.* 2003. Concentration of eicosapentaenoic acid and docosahexaenoic acid from fish oil by hydrolysis and urea complexation. *Food Research International* 36(7): 721–727.

Gibbs, B. F., Zougman, A., Masse, R., & Mulligan, C. 2004. Production and characterization of bioactive peptides from soy hydrolysate and soy-fermented food. *Food Research International* 37(2): 123–131.

Glodowsky, A. P., Ruberto, L. A., Martorell, M. M., Mac Cormack, W. P., & Levin, G. J. 2020. Cold active transglutaminase from antarctic *Penicillium chrysogenum*: Partial purification, characterization and potential application in food technology. *Biocatalysis and Agricultural Biotechnology* 29: 101807.

Gómez, B., Gullón, B., Yáñez, R., & Schols, H. 2016. Prebiotic potential of pectins and pectic oligosaccharides derived from lemon peel wastes and sugar beet pulp: A comparative. *Journal of Functional Foods* 20: 108–121.

Gopinath, S. C. B., Anbu, P., Arshad, M. K. M., *et al.* 2017. Biotechnological processes in microbial amylase production. *BioMed Research International* 2017: 1–9.

Gopinath, S. C. B., Anbu, P., Lakshmipriya, T., & Hilda, A. 2013. Strategies to characterize fungal lipases for applications in medicine and dairy industry. *BioMed Research International* 2013: 154549.

Goudar, G., Sharma, P., Janghu, S., & Longvah, T. 2020. Effect of processing on barley β-glucan content, its molecular weight and extractability. *International Journal of Biological Macromolecules* 162: 1204–1216.

Grossmann, L., Wefers, D., Bunzel, M., Weiss, J., & Zeeb, B. 2017. Accessibility of transglutaminase to induce protein crosslinking in gelled food matrices – Influence of network structure. *LWT – Food Science and Technology* 75: 271–278.

Guerrand, D. 2018. Economics of food and feed enzymes: Status and prospectives. Status and prospectives. In C. S. Nunes & V. Kumar (Eds), *Enzymes in Human and Animal Nutrition: Principles and Perspectives* (pp. 487–514). Elsevier Inc. https://doi.org/10.1016/B978-0-12-805419-2.00026-5.

Gullón, B., Gómez, B., Martínez-Sabajanes, M., Yánez, R., Parajó, J. C., & Alonso, J. L. 2013. Pectic oligosaccharides: Manufacture and functional properties. *Trends in Food Science & Technology* 30(2): 153–161.

Gurumallesh, P., Alagu, K., Ramakrishnan, B., & Muthusamy, S. 2019. A systematic reconsideration on proteases. *International Journal of Biological Macromolecules* 128: 254–267.

Hang, M. & Zhao, X. H. 2012. Fermentation time and ethanol/water-based solvent system impacted in vitro ACE-inhibitory activity of the extract of Mao-tofu fermented by Mucor spp. CYTA. *Journal of Food* 10:137–143. https://doi.org/10.1080/19476337.2011.601428

Hansen, G. H., Lübeck, M., Frisvad, J. C., & Lübeck, P. S. 2015. Production of cellulolytic enzymes from ascomycetes: Comparison of solid state and submerged fermentation. *Process Biochemistry* 50(9): 1327–1341.

Hao, G. 2011. *Method for extracting polypeptide and amino acid from waste beer yeast*. CN patent 102550802A.

Hmidet, N., Ali, N. E., Haddar, A., Kanoun, S., Alya, S., & Nasri, M. 2010. Alkaline proteases and thermostable a-amylase co-produced by *Bacillus licheniformis* NH1: Characterization and potential application as detergent additive. *Biochemical Engineering Journal* 47: 71–79.

Ho, Y., Lin, C., & Wu, M. 2017. Evaluation of the prebiotic effects of citrus pectin hydrolysate. *Journal of Food and Drug Analysis* 25(3): 550–558.

Inoue, K., Gotou, T., Kitajima, H., Mizuno, S., Nakazawa, T., & Yamamoto, N. 2009. Release of antihypertensive peptides in miso paste during its fermentation, by the addition of casein. *Journal of Bioscience and Bioengineering* 108(2): 111–115.

Iranzo, M., Aguado, C., Pallotti, C., Cañizares, J. V., & Mormeneo, S. 2002. Transglutaminase activity is involved in *Saccharomyces cerevisiae* wall construction. *Microbiology* 148(5): 1329–1334.

Iwatani, S. & Yamamoto, N. 2019. Functional food products in Japan: A review. *Food Science and Human Wellness* 8(2): 96–101.

Janecek, Š., Svensson, B., & MacGregor, E. A. 2014. α-Amylase: An enzyme specificity found in various families of glycoside hydrolases. *Cellular and Molecular Life Sciences* 71: 1149–1170.

Jensen, B., Nebelong, P., Olsen, J., & Reeslev, M. 2002. Enzyme production in continuous cultivation by the thermophilic fungus, *Thermomyces lanuginosus*. *Biotechnology Letters* 61: 41–45.

Jisha, V., Smitha, R. B., Pradeep, S., *et al.* 2013. Versatility of microbial proteases. *Advances in Enzyme Research* 01(03): 39–51.

Juturu, V. & Chuan, J. 2014. Microbial cellulases: Engineering, production and applications. *Renewable and Sustainable Energy Reviews* 33: 188–203.

Kalia, S., Bhattacharya, A., Prajapati, S. K. & Malik, A. 2021. Utilization of starch effluent from a textile industry as a fungal growth supplement for enhanced α-amylase production for industrial application. *Chemosphere* 279: 130554.

Kameshwar, A. K. S. & Qin, W. 2018. Structural and functional properties of pectin and lignin–carbohydrate complexes de-esterases: A review. *Bioresources and Bioprocessing* 5(1): 1–16.

Kammoun, R., Naili, B., & Bejar, S. 2008. Application of a statistical design to the optimization of parameters and culture medium for a-amylase production by *Aspergillus oryzae* CBS 819.72 grown on gruel (wheat grinding by-product). *Bioresource Technology* 99: 5602–5609.

Karimi, R., Azizi, M. H., & Xu, Q. 2019. Effect of different enzymatic extractions on molecular weight distribution, rheological and microstructural properties of barley bran β-glucan. *International Journal of Biological Macromolecules* 126: 298–309.

Karnaouri, A., Matsakas, L., Krikigianni, E., Rova, U., & Christakopoulos, P. 2019. Valorization of waste forest biomass toward the production of cello-oligosaccharides with potential prebiotic activity by utilizing customized enzyme cocktails. *Biotechnology for Biofuels* 12: 1–19.

Kasahara, T., Yamada, C., Hiroshi, K., & Katsunari, Y. 2016. *Method for producing xylo-oligosaccharide*. International Application Published under the Patent WO2017170919A1 Cooperation Treaty (PCT).

Kathiresan, K. & Manivannan, S. 2006. α-Amylase production by *Penicillium fellutanum* isolated from mangrove rhizosphere soil. *African Journal of Biotechnology* 5: 829–832.

Kieliszek, M. & Misiewicz, A. 2014. Microbial transglutaminase and its application in the food industry. A review. *Folia Microbiologica* 59(3): 241–250.

Kim, C. and Kim, D. 1993. Extracellular cellulolytic enzymes of bacillus circulans are present as two multiple-protein complexes. *Applied Biochemistry and Biotechnology* 42: 83–94.

Kim, S. Y., Park, P. S., & Rhee, K. C. 1990. Functional properties of proteolytic enzyme modified soy protein isolate. *Journal of Agricultural and Food Chemistry* 38(3): 651–656.

Konsoula, Z. & Liakopoulou-Kyriakides, M. 2007. Co-production of α-amylase and β-galactosidase by *Bacillus subtilis* in complex organic substrates. *Bioresource Technology* 98(1): 150–157.

Kuba, M., Tanaka, K., Tawata, S., Takeda, Y., & Yasuda, M. 2003. Angiotensin I-converting enzyme inhibitory peptides isolated from tofuyo fermented soybean food. *Bioscience, Biotechnology and Biochemistry* 67(6): 1278–1283.

Kuhad, R. C., Gupta, R. and Singh, A. 2011. Microbial cellulases and their industrial applications. *Enzyme Research* 2011. doi: 10.4061/2011/280696

Kumar, A. 2020. *Aspergillus nidulans*: A potential resource of the production of the native and heterologous enzymes for industrial applications. *International Journal of Microbiology* 2020: 894215.
Kumari, A., Kaur, B., Srivastava, R., & Sangwan, R. S. 2015. Isolation and immobilization of alkaline protease on mesoporous silica and mesoporous ZSM-5 zeolite materials for improved catalytic properties. *Biochemistry and Biophysics Reports* 2: 108–114.
Kumari, U., Singh, R., Ray, T., *et al.* 2019. Validation of leaf enzymes in the detergent and textile industries: launching of a new platform technology. *Plant Biotechnology Journal* 17(6): 1167–1182.
Kwan, Y., Kang, Y., Ram, B., Keun, S., & Hyuk, Y. 2021. Food Hydrocolloids Structural, antioxidant, prebiotic and anti-inflammatory properties of pectic oligosaccharides hydrolyzed from okra pectin by Fenton reaction. *Food Hydrocolloids* 118: 106779.
Li, S., Yang, X., Yang, S., Zhu, M., & Wang, X. 2012. Technology prospecting on enzymes: Application, marketing and engineering. *Computational and Structural Biotechnology Journal* 2(3): e201209017.
Li, H., Zhang, L., Cui, Y., *et al.* 2014. Characterization of recombinant *Zea mays* transglutaminase expressed in *Pichia pastoris* and its impact on full and non-fat yoghurts. *Journal of the Science of Food and Agriculture* 94(6): 1225–1230.
Liu, S., Wan, D., Wang, M., Madzak, C., Du, G., & Chen, J. 2015. Overproduction of pro-transglutaminase from *Streptomyces hygroscopicus* in *Yarrowia lipolytica* and its biochemical characterization. *BMC Biotechnology* 15(1): 1–9.
Mamo, G. 2020. Alkaline active hemicellulases. *Advances in Biochemical Engineering/Biotechnology* 172: 245–291.
Martínez-Medina, G. A., Barragán, A. P., Ruiz, H. A., *et al.* 2018. Fungal proteases and production of bioactive peptides for the food industry. In M. Kuddus (Ed), *Enzymes in Food Biotechnology: Production, Applications, and Future Prospects* (pp. 221–246). Elsevier Inc. https://doi.org/10.1016/B978-0-12-813280-7.00014-1.
Martins, I. M., Matos, M., Costa, R., *et al.* 2014. Transglutaminases: Recent achievements and new sources. *Applied Microbiology and Biotechnology* 98(16): 6957–6964.
Martins, G. N., Ureta, M. M., Tymczyszyn, E. E., Castilho, P. C., & Gomez-Zavaglia, A. 2019. Technological aspects of the production of fructo and galacto-oligosaccharides. Enzymatic synthesis and hydrolysis. *Frontiers in Nutrition* 6: 78.
Mattar, R., Mazo, D. F. de C., & Carrilho, F. J. 2012. Lactose intolerance: Diagnosis, genetic, and clinical factors. *Clinical and Experimental Gastroenterology* 5(1): 113–121.
Mazáň, M. & Farkaš, V. 2007. Transglutaminase-like activity participates in cell wall biogenesis in *Saccharomyces cerevisiae. Biologia* 62(2): 128–131.
Mcrae, M. P. 2017. Dietary fiber is beneficial for the prevention of cardiovascular disease: An umbrella review of meta-analyses. *Journal of Chiropractic Medicine* 16(4): 289–299.
Mirzaei, M., Shavandi, A., Mirdamadi, S., *et al.* 2021. Bioactive peptides from yeast: A comparative review on production methods, bioactivity, structure-function relationship, and stability. *Trends in Food Science and Technology* 118: 297–315.
Miwa, N. 2020. Innovation in the food industry using microbial transglutaminase: Keys to success and future prospects. *Analytical Biochemistry* 597: 113638.
Monsan, P. & Paul, F. 1995. Enzymatic synthesis of oligosaccharides. *FEMS Microbiology Reviews* 16: 187–192.
Mora, L., Aristoy, M. C., & Toldrá, F. 2018. Bioactive peptides. *Encyclopedia of Food Chemistry*: 381–389. https://doi.org/10.1016/B978-0-08-100596-5.22397-4
Morais, H. A., Silvestre, M. P. C., Amorin, L. L., *et al.* 2013a. Use of different proteases to obtain whey protein concentrate hydrolysates with inhibitory activity toward angiotensin-converting enzyme. *Journal of Food Biochemistry* 38: 102–109. https://doi.org/10.1111/jfbc.12032
Morais, H. A., Silvestre, M. P. C., Silveira, J. N., *et al.* 2013b. Action of a pancreatin and an Aspergillus oryzae protease on whey proteins: Correlation among the methods of analysis of the enzymatic hydrolysates. *Brazilian Archives of Biology and Technology* 56: 985–995. https://doi.org/10.1590/S1516-89132013000600014
Moslemi, M. 2021. Reviewing the recent advances in application of pectin for technical and health promotion purposes: From laboratory to market. *Carbohydrate Polymers* 254(24): 117324.
Naveed, M., Nadeem, F., Mehmood, T., Bilal, M., Anwar, Z., & Amjad, F. 2021. Protease—A versatile and ecofriendly biocatalyst with multi-industrial applications: An updated review. *Catalysis Letters* 151(2): 307–323.
Nawaz, A., Bakhsh, A., Irshad, S., Hossein, S., & Xiong, H. 2018. Fish and shell fish immunology the functionality of prebiotics as immunostimulant: Evidences from trials on terrestrial and aquatic animals. *Fish and Shellfish Immunology* 76: 272–278.

Nguyen, T. T. L., Flanagan, B. M., Tao, K., *et al.* 2022. Effect of processing on the solubility and molecular size of oat β-glucan and consequences for starch digestibility of oat-fortified noodles. *Food Chemistry* 372: 131291.

Nkhata, S. G., Ayua, A., Kamau, A. H., & Shingiro, J.-B. 2018. Fermentation and germination improve nutritional value of cereals and legumes through activation of endogenous enzymes. *Food Science & Nutrition* 6(8): 2446–2458.

Nyyssölä, A., Ellilä, S., Nordlund, E., & Poutanen, K. 2020. Reduction of FODMAP content by bioprocessing. *Trends in Food Science and Technology* 99: 257–272.

Ogasawara, W., Yamaguchi, R., Shida, Y., Ishii, Y., Wakayama, & Imada, M. 2013. *Variant of cellulase-producing fungus, cellulase manufacturing method, and cello-oligosaccharide manufacturing method.* International Application Published under the Patent WO2014142325A1 Cooperation Treaty (PCT).

Orlandelli, R. C., Specian, V., Felber, A. C., & Pamphile, J. A. 2012. Enzimas de interesse industrial: Produção por fungos e aplicações. *Revista Saúde e Bioilogia* 7(3): 97–109.

Pan, S., Ding, N., Ren, J., Gu, Z., Li, C., & Hong, Y. 2017. Maltooligosaccharide-forming amylase: Characteristics, preparation, and application. *Biotechnology Advances* 35: 619–632.

Pasin, T. M., Benassi, V. M., Heinen, P. R., *et al.* 2017. Purification and functional properties of a novel glucoamylase activated by manganese and lead produced by *Aspergillus japonicus*. *International Journal of Biological Macromolecules* 102: 779–788.

Passos, D. F., Pereira Jr., N., & Castro, A. M.. 2018. A comparative review of recent advances in cellulases production by *Aspergillus*, *Penicillium* and *Trichoderma* strains and their use for lignocellulose deconstruction. *Current Opinion in Green and Sustainable Chemistry* 14: 60–66.

Phukon, L. C., Chourasia, R., Padhi, S. *et al.* 2022. Cold-adaptive traits identified by comparative genomic analysis of a lipase-producing *Pseudomonas* sp. HS6 isolated from snow-covered soil of Sikkim Himalaya and molecular simulation of lipase for wide substrate specificity. *Current Genetics* 68: 375–391.

Potprommanee, L. *et al.* 2017. Characterization of a thermophilic cellulase from *Geobacillus* sp. HTA426, an efficient cellulase-producer on alkali pretreated of lignocellulosic biomass. *PLOS ONE*: 1–16.

Prakash, O. & Jaiswal, N. 2010. α-Amylase: An ideal representative of thermostable enzymes. *Applied Biochemistry and Biotechnology* 160(8): 2401–2414.

Pyo, Y. H. & Lee, T. C. 2007. The potential antioxidant capacity and angiotensin I-converting enzyme inhibitory activity of Monascus-fermented soybean extracts: Evaluation of Monascus-fermented soybean extracts as multifunctional food additives. *Journal of Food Science* 72: 218–223. https://doi.org/10.1111/j.1750-3841.2007.00312.x

Rana, N., Walia, A., & Gaur, A. 2013. α-Amylases from microbial sources and its potential applications in various industries. *National Academy Science Letters* 36(1): 9–17.

Rani, K., Rana, R., & Datt, S. 2012. Review on latest overview of proteases. *International Journal of Current Life Sciences* 2(1): 12–18.

Ravindran, R., Sharma, S., & Jaiswal, A. 2016. Enzymes in processing of functional food and nutraceuticals. In D. M. Martirosyan (Ed), *Functional Foods for Chronic Diseases* (1st ed., pp. 360–385). Technological University Dublin.

Reddy, N. S., Nimmagadda, A., Rao, K. R. S. S. 2003. An overview of the microbial α -amylase family. *African Journal of Biotechnology* 2: 645–648.

Reque, P. M., Pinilla, C. M. B., Gautério, G. V., Kalil, S. J., & Brandelli, A. 2019. Xylooligosaccharides production from wheat middlings bioprocessed with *Bacillus subtilis*. *Food Research Internationa* 126: 108673.

Reyna-Beltrán, E., Iranzo, M., Calderón-González, K. G., *et al.* 2018. The *Candida albicans* ENO1 gene encodes a transglutaminase involved in growth, cell division, morphogenesis, and osmotic protection. *Journal of Biological Chemistry* 293(12): 4304–4323.

Romeih, E., & Walker, G. 2017. Recent advances on microbial transglutaminase and dairy application. *Trends in Food Science and Technology* 62: 133–140.

Ruiz, H. A., Rodríguez-Jasso, R. M., Rodríguez, R., Contreras-Esquivel, J. C., & Aguilar, C. N. 2012. Pectinase production from lemon peel pomace as support and carbon source in solid-state fermentation column-tray bioreactor. *Biochemical Engineering Journal* 65: 90–95.

Sadhu, S. & Maiti, T. K. 2013. Cellulase production by bacteria : A review. British *Microbiology Research Journal* 3(3): 235–258.

Sajith, S., Priji, P., Sreedevi, S., & Benjamin, S. 2016. An overview on fungal cellulases with an industrial perspective. *Journal of Nutrition & Food Sciences* 6(1): 1–13.

Sandri, I. G. & Moura, M. 2018. Production and application of pectinases from *Aspergillus niger* obtained in solid state cultivation. *Beverages* 4(3): 48.

Santhosh Kumar, D. 2015. Fungal lipase production by solid state fermentation-An overview. *Journal of Analytical & Bioanalytical Techniques* 6(1): 1–10.

Sarethy, I. P., Saxena, Y., & Kapoor, A. 2011. Alkaliphilic bacteria: Applications in industrial biotechnology. *Journal of Industrial Microbiology & Biotechnology* 38(7): 769–790.

Schneider, W. H. D. *et al.* 2014. Morphogenesis and production of enzymes by *Penicillium echinulatum* in response to different carbon sources. *BioMed Research International*. doi: 10.1155/2014/254863

Sharma, S. & Yazdani, S. S. 2016. Diversity of microbial cellulase system. In V. Gupta (Ed), *New and Future Developments in Microbial Biotechnology and Bioengineering* (1st ed., pp. 49–64). Elsevier B.V. https://doi.org/10.1016/B978-0-444-63507-5/00006-X.

Shinde, V. K. & Vamkudoth, K. R. 2021. Maltooligosaccharide forming amylases and their applications in food and pharma industry. *Journal of Food Science and Technology*: 1–12. https://doi.org/10.1007/s13197-021-05262-7

Sindhu, R., Binod, P., & Pandey, A. 2016. Biological pretreatment of lignocellulosic biomass – An overview. *Bioresource Technology* 199: 76–82.

Singh, A. *et al.* 2011. Enhanced saccharification of rice straw and hull by microwave – alkali pretreatment and lignocellulolytic enzyme production. *Bioresource Technology* 102(2): 1773–1782. doi: 10.1016/j.biortech.2010.08.113

Singh, A., Bajar, S., Devi, A., & Pant, D. 2021. An overview on the recent developments in fungal cellulase production and their industrial applications. *Bioresource Technology Reports* 14: 100652.

Singh, S., Ghosh, A., & Goyal, A. 2018. Manno-oligosaccharides as prebiotic-valued products from agro-waste. In S. J. Varjani, B. Parameswaran, S. Kumar, & S. K. Khare *(Eds), Biosynthetic Technology and Environmental Challenges* (pp. 205–221). Springer.

Sivaramakrishnan, S., Gangadharan, D., Nampoothiri, K. M., & Soccol, C. R. 2007. Alpha amylase production by Aspergillus oryzae employing solid-state fermentation. *Journal of Scientific & Industrial Research* 66: 621–626.

Souza, P. M., Assis Bittencourt, M. L., Caprara, C. C., *et al.* 2015. A biotechnology perspective of fungal proteases. *Brazilian Journal of Microbiology* 46(2): 337–346.

Srivastava, N., Srivastava, M., Mishra, P. K., & Singh, P. 2015. Application of cellulases in biofuels industries: an overview. *Journal of Biofuels and Bioenergy* 1: 55–63.

Sundarram, A., Pandurangappa, T., & Murthy, K. 2014. α-amylase production and applications: A review. *Journal of Applied & Environmental Microbiology* 2(4): 166–175.

Suriya, J., Bharathiraja, S., Krishnan, M., Manivasagan, P., & Kim, S. K. 2016. Marine microbial amylases: Properties and applications. In S.-K. Kim & F. Toldrá (Eds), *Advances in Food and Nutrition Research* (1st ed., Vol. 79, pp. 161–177). Elsevier Inc. https://doi.org/10.1016/bs.afnr.2016.07.001.

Swapnil, J. S. & Arpana, J. H. 2019. Applications of Lipases. *Research Journal of Biotechnology* 14(11): 130–138.

Szymczak, T., Cybulska, J., Podlesny, M., & Frac, M. 2021. Various perspectives on microbial lipase production using agri-food waste and renewable products. *Agriculture-Basel* 11(6): 540.

Takahashi, J. A., Barbosa, B. V. R., Martins, B. de A., Guirlanda, C. P., & Moura, M. A. F. 2020. Use of the versatility of fungal metabolism to meet modern demands for healthy aging, functional foods, and sustainability. *Journal of Fungi* 6(4): 223.

Tanyildizi, M. S., Ozer, D., & Elibol, M. 2005. Optimization of α-amylase production by *Bacillus* sp. using response surface methodology. *Process Biochemistry* 40: 2291–2296.

Timón, M. L., Andrés, A. I., Otte, J., Petrón, M. J. 2019. Antioxidant peptides (<3 kDa) identified on hard cow milk cheese with rennet from different origin. *Food Research International* 120: 643–649. https://doi.org/10.1016/j.foodres.2018.11.019

Toushik, S. H., Lee, K.-T., Lee, J.-S., & Kim, K.-S. 2017. Functional applications of lignocellulolytic enzymes in the fruit and vegetable processing industries. *Journal of Food Science* 82(3): 585–593.

Ünal, A., Subaşı, A. S., Malkoç, S., *et al.* 2022. Potential of fungal thermostable alpha amylase enzyme isolated from Hot springs of Central Anatolia (Turkey) in wheat bread quality. *Food Bioscience* 45: 101492.

UNEP 2021. *Food waste index report 2021*. UN Environment Programme. https://www.unep.org/resources/report/unep-food-waste-index-report-2021.

Vaidyanathan, S., Macaloney, G., Vaughan, J., *et al.* 1999. Monitoring of Submerged Bioprocesses. *Critical Reviews in Biotechnology* 19(4): 277–316.

Vaishnav, N., Singh, A., Adsul, M., *et al.* 2018. *Penicillium*: The next emerging champion for cellulase production. *Bioresource Technology Reports* 2: 131–140.

Vandenberghe, L. P. S., Karp, S. G., Pagnoncelli, M. G. B., Rodrigues, C., Medeiros, A. B. P., & Soccol, C. R. 2019. Digestive enzymes: Industrial applications in food products. In B. Parameswaran, S. Varjani, & S. Raveendran (Eds), *Green Bio-Processes* (pp. 267–291). Springer.

Vandenberghe, L. P. S., Karp, S. G., Pagnoncelli, M. G. B., Tavares, M. L., Linbardi Junior, N., Diestra, K. V., Viesser, J. A., Soccol, C. R. 2020. Classification of enzymes and catalytic properties. In: *Advances in Enzyme Catalysis and Techonologies* (pp. 11–30). Elsevier Inc.

Vandenberghe, L. P. S., Pandey, A., Carvalho, J. C., *et al.* 2021. Solid-state fermentation technology and innovation for the production of agricultural and animal feed bioproducts. *Systems Microbiology and Biomanufacturing* 1(2): 142–165.

Wang, D., Wang, L. J., Zhu, F. X., *et al.* 2008. In vitro and in vivo studies on the antioxidant activities of the aqueous extracts of Douchi (a traditional Chinese salt-fermented soybean food). *Food Chemistry* 107(4): 1421–1428.

Wang, L., Yu, B., Wang, R., & Xie, J. 2018. Biotechnological routes for transglutaminase production: Recent achievements, perspectives and limits. *Trends in Food Science and Technology* 81: 116–120.

Wang, Z., Wang, C., Zhang, C., & Li, W. 2017. Ultrasound-assisted enzyme catalyzed hydrolysis of olive waste and recovery of antioxidant phenolic compounds. *Technologies* 44: 224–234.

Ward, O. P., Rao, M. B., & Kulkarni, A. 2009. Proteases, production. In M. Schaechter (Ed) *Encyclopedia of Microbiology* (pp. 495–511). Academic Press. https://doi.org/10.1016/B978-012373944-5.00172-3.

WIPO. 2022. *International Patent Classification.* https://www.wipo.int/classifications/ipc/en/.

Wongkaew, M., Tinpovong, B., Sringarm, K., *et al.* 2021. Crude pectic oligosaccharide recovery from Thai Chok Anan. *Foods* 10(3) 1–16.

Yahya, S., Muhammad, F., Sohail, M., & Khan, S. A. 2021. Amylase production and growth pattern of two indigenously isolated Aspergilli under submerged fermentation: Influence of physico-chemical parameters. *Pakistan Journal of Botany* 53(3) 1147–1155.

Yang, W. & Huang, G. 2021. Extraction methods and activities of natural glucans. *Trends in Food Science and Technology* 112: 50–57.

Yang, X. & Zhang, Y. 2019. Expression of recombinant transglutaminase gene in *Pichia pastoris* and its uses in restructured meat products. *Food Chemistry* 291: 245–252.

Yang, Y., Yang, J., Liu, J., Wang, R., Liu, L., Wang, F., & Yuan, H. 2018. The composition of accessory enzymes of *Penicillium chrysogenum* P33 revealed by secretome and synergistic effects with commercial cellulase on lignocellulose hydrolysis. *Bioresource Technology* 257: 54–61.

Yokoyama, K., Nio, N., & Kikuchi, Y. 2004. Properties and applications of microbial transglutaminase. *Applied Microbiology and Biotechnology* 64(4): 447–454.

Zhang, J. H., Tatsumi, E., Ding, C. H., Li, L. Te. 2006. Angiotensin I-converting enzyme inhibitory peptides in douchi, a Chinese traditional fermented soybean product. *Food Chemistry* 98: 551–557. https://doi.org/10.1016/j.foodchem.2005.06.024

Zhang, Q., Han, Y., & Xiao, H. 2017. Microbial α-amylase: A biomolecular overview. *Process Biochemistry* 53: 88–101.

Zhu, F., Du, B., & Xu, B. 2016. A critical review on production and industrial applications of beta-glucans. *Food Hydrocolloids* 52: 275–288.

4 Enzymes from *Bacillus* spp. for Nutraceutical Production

Luiz Alberto Junior Letti, Leonardo Wedderhoff Herrmann, Rafaela de Oliveira Penha, Ariane Fátima Murawski de Mello, Susan Grace Karp, and Carlos Ricardo Soccol
Department do Bioprocess Engineering and Biotechnology, Federal University of Paraná, Curitiba, Paraná, Brazil

CONTENTS

4.1 INTRODUCTION

As described in previous chapters, enzymes are biological catalysts which present high specificity over their substrates. They can be found in virtually all animal, plant, and microbial species, and play an important role in a wide range of applications, such as biofuels, textile, pulp and paper, chemistry, and of course, for feed, food, and the growing nutraceutical industries, among others. The industrial production of enzymes comprises several aspects, such as economic, logistic, and even social ones. In the technical field, the most usual challenges are finding proper hosts to produce the desired enzyme in high amounts (high productivity), with the desired characteristics for each application (high selectivity), and in a form they can act efficiently (high activity). In this context, in addition to yeasts and fungi, bacteria play a very important role, especially the genus *Bacillus*.

The next section details the characteristics of the *Bacillus* genus and provides explanation for the main reasons which make it one of the most important hosts for enzyme production. The subsequent section describes the most frequent and value-added enzymes produced by the genus, focusing on food, feed, and nutraceutical applications. The importance of molecular biology tools is also discussed in the chapter, adding many recent technological developments. Finally, the chapter presents an overview of patented processes regarding the production of enzymes by *Bacillus* spp., with some examples related to nutraceutical products.

4.2 *BACILLUS* SPP. AS A FACTORY FOR ENZYMES PRODUCTION

One of the most well-known genera among the microorganisms is the *Bacillus*, mainly represented by the *Bacillus subtilis* species, as remarkable as *Escherichia coli* and *Saccharomyces cerevisiae.*

DOI: 10.1201/9781003311164-6

The *Bacillus* genus consists in rod-shaped bacteria, classified as gram-positive, positive catalase, and aerobic. They naturally occur in a great variety of soil, water, and air environments, and are intensively applied in industrial processes, especially for the production of enzymes. The first species was identified by Ferdinand Cohn in 1872, and was *Bacillus subtilis* itself. Currently, the genus has more than 250 recognized species, and a large number of them are used as biological factories for the production of proteins, antibiotics, surfactants, biofertilizers, chemicals, biopolymers, pharmaceuticals, nutraceuticals, and, of course, enzymes (Harwood, 1989; Berkeley et al., 2002; Graumann, 2017; Gu et al., 2018; Park et al., 2021).

Bacillus bacteria have interesting features, which make them attractive for industrial purposes. One of them is the capacity to produce spores, also called endospores. The spores are resistant structures that protect the genetic material inside five protecting layers, allowing the bacteria to survive extreme pH, high temperatures, salinity, and presence of organic solvents (Pandey et al., 2013; Sella et al., 2014; Abhyankar et al., 2016). These layers can be used as a support for enzymes and proteins anchorage in order to improve stability or efficacy of the heterologous ones (Zhang, Al-Dossary et al., 2020). *Bacillus* are also capable of producing a biofilm formed mainly by exopolysaccharides and proteins, which gather the colonies over the growth substrate (Branda et al., 2006; Beauregard et al., 2013).

Both endospore and biofilm formation capacities, as well as other features, enable the genus to develop in a large variety of substrates. The bacteria are able to grow in both submerged fermentation (SmF) and solid-state fermentation (SSF), and several agroindustrial or domestic wastes can be used as a nutrient source, which minimizes industrial costs and allows a better destination for residues. For instance, wheat bran, wheat straw, corncobs, banana peels, mandarin peels (Khardziani, Kachlishvili et al., 2017; Khardziani, Sokhadze et al., 2017), corn steep liquor, soybean flour (Chen et al., 2010), wheat middlings (Reque et al., 2019), and even feather waste (Cedrola et al., 2012) can be used as substrate to grow *Bacillus*. In addition, both SmF and SSF strategies present advantages on their own, such as better mass and heat homogeneity in submerged conditions, or higher yields and less expensive sterilization in solid state (Pandey, 2003; Vandenberghe et al., 2021). Depending on the desired characteristic for the enzyme and the final product, as well as the purification processes, each technique presents unique characteristics for the production of biological compounds and have already been tested with promising results (Terlabie et al., 2006; Posada-Uribe et al., 2015), even combining both strategies (Reque et al., 2019).

Bacillus can also count on its secretory capacity with an exceptional advantage for the production of enzymes. The genus has at least four different secretory pathways and naturally exports more than 300 native proteins. As gram-positive bacteria, the absence of an outer membrane facilitates the excretion, with secretory capacity superior to *E. coli*. The most common pathways include the general secretion (Sec), the twin-arginine translocation (Tat), pseudopilin export, and the ATP-binding cassette (Sarvas et al., 2004; Zweers et al., 2008; Zhang, Su et al., 2020;). Each of them fold and translocate the proteins in different stages before sending them to the extracellular environment, conferring proteolytic degradation resistance, especially interesting for heterologous protein expression and for their production on an industrial scale (Gu et al., 2018; Zhang, Su et al., 2020; Park et al., 2021).

Several enzymes produced by the genus show particular characteristics according to the growth environment profile, as the *Bacillus* produces several extracellular hydrolytic enzymes to survive in complex nutrient sources and extreme conditions. For instance, plenty of their enzymes present alkaliphilic and thermostable characteristics, which indicates that the optimum pH for the enzyme activity varies between 7.0 and 10.0, and the ideal temperature as well as the protein limit of stability can be kept between 35 and 70°C (Berkeley et al., 2002; Eggert et al., 2002; Al-Quadan et al., 2011; Barros et al., 2013; Kannan & Kanagaraj, 2019). This comes from the *Bacillus* ability to grow in protein-rich substrates liberating ammonia and turning the environment alkaline, in addition to the ability of the endospore to resist high temperatures (Mazas et al., 1997; Steinkraus, 2004; Parkouda et al., 2009; Sella et al., 2014).

Some of the genus species, such as *B. subtilis*, *Bacillus licheniformis* and *Bacillus clausii*, are classified as "generally recognized as safe" (GRAS) and "qualified presumption of safety" (QPS) by the United States and European Union respectively (Andreoletti et al., 2008; Gu et al., 2018). This indicates that the enzymes and by-products derived from them can easily be added in food, healthcare, and medicinal products. Besides, when the bacteria are added directly to the food, much of the texture and sensorial properties come from the degradation provided by the enzymes (Elshaghabee et al., 2017; Kimura & Yokoyama, 2019). All of these advantages indicate that the *Bacillus* is an extremely promising and suitable genus to be used as a factory for the production of nutraceutical enzymes, both in terms of industrial scale production and safe application in health products. More details about each *Bacillus* enzyme produced and applied for improving the host health can be seen in the following sections of this chapter.

4.3 ENZYMES FROM *BACILLUS* SPP. AND INDUSTRIAL APPLICATIONS

4.3.1 AMYLASES

Amylases are among the most important enzymes in the biotechnology field, singly responsible for approximately 25% of the enzyme market, as they can be applied in several industries, such as food, textile, paper, and detergent (Souza & Magalhães, 2010). Regarding the nutraceutical and food industry specifically, amylases are extensively employed in baking, brewing, preparation of digestive aids, production of cakes, fruit juices, and starch syrups (Afzaljavan & Mobini-Dehkordi, 2012; Kumari et al., 2021).

Amylase-type enzymes are able to hydrolyse starch molecules to produce progressive smaller polymers composed of glucose (Naidu & Saranraj, 2013). They can be divided into three main classes: α-amylases, β-amylases, and γ-amylases. Alpha-amylases can be found in humans, animals, plants, and microbes; β-amylases are present in microbes and plants; and γ-amylases are found in animals and plants (Azzopardi et al., 2016). More specifically, α-amylases (EC 3.2.1.1) are the most important form of industrial amylases and are able catalyse the hydrolysis of the α-1,4-glucosidic bond in starch, acting over amylose and amylopectin (Naidu & Saranraj, 2013). The α-amylases are multidomain proteins and can have the domains A (the catalytic (β/α)8-barrel), B, C, D, and E (Janeček et al., 2003).

Alpha-amylases can be produced by different species of microorganisms; however, *Bacillus* α-amylases are the main ones used for commercial purposes. Among the *Bacillus* species, *Bacillus licheniformis*, *Bacillus stearothermophilus*, and *Bacillus amyloliquefaciens* are some of the producers whose amylases are applied in industry, such as in the food, fermentation, textiles, and paper industries (Souza & Magalhães, 2010). Generally, *Bacillus* amylases have their optimal pH between 6.0 and 7.0 and optimal temperature from 50–80°C (Farias et al., 2021).

Bacillus bacteria are well known to produce α-amylases in starch medium (Rajagopalan & Krishnan, 2008). A number of reports also describe the use of by-products to produce the enzyme such as sugarcane bagasse hydrolysate (Rajagopalan & Krishnan, 2008), wheat bran (Haq et al., 2003), potato starchy waste (Abd-Elhalem et al., 2015), corn pericarp (Salim et al., 2017), mustard oil seed cake (Saxena & Singh, 2011), oil palm empty fruit bunch, and rice straw fibres (Hassan & Karim, 2012). Table 4.1 shows different studies on the production of α-amylases by *Bacillus* species.

Amylase production by *Bacillus* can be performed both in SSF and SmF. One of the main advantages of SSF is the use of agro-industrial wastes and by-products as substrates (Salim et al., 2017). SSF also presents low capital investment and can produce enzymes with better physicochemical properties (Saxena & Singh, 2011). On the other hand, SmF processes present as advantages an easier system sterilization and an overall better process control, e.g., pH, temperature, aeration, and oxygen transfer (Vidyalakshmi et al., 2009; Souza & Magalhães, 2010). For this reasons SmF has been traditionally applied for the production of industrially important enzymes, such as amylases (Souza & Magalhães, 2010). Enzymatic liquefication and saccharification of starch is performed

TABLE 4.1
Different Processes for Production of α-Amylases by *Bacillus* Species

Species	Fermentation Method	Substrate	Enzyme Activity	Reference
B. licheniformis GCBU-8	SmF[a]	Wheat bran	240 U/mL/min	(Haq et al., 2003)
Bacillus sp. PN5	SmF	Starch	65 U/mL	(Saxena et al., 2007)
B. subtilis KCC103	SmF	Sugarcane bagasse hydrolysate	144 U/mL	(Rajagopalan & Krishnan, 2008)
Bacillus sp.	SSF[b]	Mustard oil seed cake	5,400 U/g	(Saxena & Singh, 2011)
B. licheniformis MTCC 2618	SmF	Starch	3.6 U/mL/min	(Divakaran et al., 2011)
		Rice powder	2.9 U/ml/min	
		Wheat powder	2.7 U/mL/min	
		Ragi powder	2.4 U/mL/min	
B. subtilis	SSF	Rice straw	276 U/g	(Hassan & Karim, 2012)
		Oil palm empty fruit bunch	127 U/g	
Bacillus spp.	SmF	Starch	11 U/mL	(Vidyalakshmi et al., 2009)
B. amyloliquefaciens	SmF	Potato starchy waste	155 U/mL	(Abd-Elhalem et al., 2015)
Bacillus sp. TMF-1	SSF	Corn pericarp	51 U/g	(Salim et al., 2017)

[a] Submerged fermentation.
[b] Solid state fermentation.

at high temperatures (100–110°C). For this reason, thermostable and thermoactive amylases are of great industrial interest for the production of several products such as glucose, maltose, maltodextrins, crystalline dextrose, and dextrose syrup (Gomes et al., 2003). *Bacillus megaterium* and *B. licheniformis* are examples that produce α-amylases with optimum temperature between 100 and 110°C (Farias et al., 2021)

Thermostable amylases used in the industry are mainly produced by *B. stearothermophilus* or *B. licheniformis* (Gomes et al., 2003). *B. licheniformis* α-amylase is rather interesting since the bacterium is mesophilic and its amylase is highly thermostable (Declerck et al., 2000). Declerck et al. (2000) were able to replace several important amino acid residues for thermostability to create *B. licheniformis* amylase variants. The researchers concluded that a triadic Ca-Na-Ca metal-binding site located in domain B and at its interface with central domain A was crucial for maintaining the amylase structure at high temperatures. Furthermore, other studies showed that modifications of amino acids to form other Ca^2+ binding sites in *B. licheniformis* amylase resulted in an increase in enzyme stability (Farias et al., 2021). The thermostability of *B. stearothermophilus* amylases was also linked to calcium binding sites, in this case, located at the interface of domains A and C (Xie et al., 2020; Farias et al., 2021). In addition, different studies on thermostable amylases from other *Bacillus* species are being produced frequently, e.g., studies using *Bacillus mojavensis* (Ozdemir et al., 2018), *Bacillus cereus* (Vaikundamoorthy et al., 2018; Priyadarshini et al., 2020), *B. amyloliquefaciens* (Du et al., 2018; Devaraj et al., 2019), and *B. megaterium* (Abootalebi et al., 2020).

In terms of specific applications of *Bacillus* α-amylases in the food industry, the use of *B. subtilis* US586 amylases in wheat flour improved its quality by decreasing the ratio between elasticity and extensibility, making the flour more suitable for baking (Trabelsi et al., 2019). The addition of *B. subtilis* M13 amylase to bread preparation resulted in maximum leavening activity (Rana et al., 2017). The use of *B. amyloliquefaciens* α-amylase in instant noodles made its structure more porous, improving palatability of the food (Li, Jiao et al., 2018). *Bacillus* α-amylases can also be used in the clarification of beer and fruit juices, avoiding precipitation in the beverage and reducing cloudiness and viscosity (Farias et al., 2021). The use of these amylases can also improve juice production yield (Rana et al., 2017).

Beta-amylases hydrolyse the α-1,4 glycosidic bonds of starch from the nonreducing end. They act in amylose, producing 90% of maltose and 10% of glucose and maltotriose, and its hydrolysis of amylopectin is incomplete. *Bacillus* β-amylases can be used in malt supplementation in addition to α-amylase and β-glucanase. With the use of the enzyme, higher brewing yields as well as lower malt consumption and process time can be achieved (Farias et al., 2021).

4.3.2 Xylanases

Xylan is a complex polymer, which requires several enzymes for its complete breakdown. The xylanolytic enzyme group includes β-1,4-endoxylanase, β-xylosidase, α-glucuronidase, α-L-arabinofuranosidase, acetyl xylan esterase, and phenolic acid (ferulic and p-coumaric acid) esterase. Among them, endoxylanase and β-xylosidase are the most important ones (Burlacu et al., 2016; Phukon et al., 2020).

Xylanases are of great industrial interest since they have potential for several biotechnological applications, such as bioconversion of lignocellulosic biomass into simple sugars for the production of various industrial products, clarification of juices and wines, pulp bleaching, animal feed processing, bread-quality improvement, biobleaching, and the processing of fabrics, silage production, and waste treatment (Bajaj & Manhas, 2012). This industrial interest is reflected on the xylanase global market value, which will be worth over US$35 million in 2024 (Sharma et al., 2019). In the nutraceutical industry, xylanases can be applied to produce xylooligosaccharides. These non-digestible oligosaccharides have the potential to lower the cholesterol levels, improve the biological availability of calcium and have antioxidant, anti-inflammatory, immunomodulatory, antidiabetic, and anti-cancer activities (Jagtap et al., 2017).

Xylanases can be produced by several microorganisms, wherein *Bacillus* spp. are among the main producers of endoxylanase and β-xylosidase (Burlacu et al., 2016). Due to the tolerance to high temperatures and broad pH ranges of activity, some *Bacillus* species such as *B. stearothermophilus*, *B. amyloliquefaciens*, *Bacillus pumilus*, *Bacillus halodurans*, *B. subtilis*, *Bacillus coagulans*, and *Bacillus circulans* are considered important sources for the enzyme (Rashid & Sohail, 2021). The thermostability traits of *Bacillus* xylanases are especially interesting since, although fungi are also able to produce this type of enzyme, the ones produced by bacteria generally have better half-lives at temperatures of 80°C or higher (Dahlberg et al., 1993).

Purified and commercial xylose can be used for the production of xylanase by *Bacillus* (Khasin et al., 1993); however, a number of substrates containing xylan (a hemicellulose) can be used for the production of xylanases, e.g., sugarcane bagasse (Irfan et al., 2019), wheat bran (Archana & Satyanarayana, 1997; Gessesse & Mamo, 1999; Kamble & Jadhav, 2012), rice bran (Bajaj & Manhas, 2012), and oat bran (Sepahy et al., 2011).

Bacillus xylanases can be produced by either SSF (Gessesse & Mamo, 1999) or SmF (Bataillon et al., 2000). The SSF processes can produce enzymes with higher specific activities. However, most xylanases industries use SmF processes, since they provide better control of the conditions during fermentation (Motta et al., 2013). Table 4.2 shows different studies using SFF and SmF for different *Bacillus* species and substrates.

4.3.3 Pullulanases

Pullulanases (EC 3.2.1.41) are called debranching enzymes, as they are able to hydrolyse the α-1,6-glycosidic bonds present in polysaccharides. They can cleave pullulan and also hydrolyse amylopectin α-1,6-bonds. They are usually preferred when longer chains of amylopectin must be cleaved, while other debranching enzymes are favoured when shorter chains are the target (Farias et al., 2021).

Pullulanses are often used synergistically with other enzymes (such as isoamylases, α-amylase, β-amylase, glucoamylase, and α-glucosidase) for complete hydrolysis and depolymerization of starch (Xia et al., 2021). Their use allows the reduction in the dosage of other enzymes, and also

TABLE 4.2
Different Processes for Production of Xylanases by *Bacillus* Species

Species	Fermentation Method	Substrate	Enzyme Activity	Reference
B. stearothermophilus T-6	SmF	d-xylan	288 U/mg	(Khasin et al., 1993)
B. licheniformis A99	SSF	Wheat bran	19 U/g	(Archana & Satyanarayana, 1997)
Bacillus sp. AR-009	SSF	Wheat bran	720 U/g	(Gessesse & Mamo, 1999)
B. pumilus B20	SmF	Wheat bran	313 U/mL	(Geetha & Gunasekaran, 2010)
B. mojavensis AG137	SmF	Oat bran	249 U/mL	(Sepahy et al., 2011)
Bacillus arseniciselenatis DSM 15340	SSF	Wheat Bran	910 U/g	(Kamble & Jadhav, 2012)
B. licheniformis P11(C)	SmF	Xylan	30 U/mL	(Bajaj & Manhas, 2012)
		Wheat bran	28 U/mL	
		Rice bran	26 U/mL	
B. subtilis BS04	SmF	Sugarcane bagasse	52 U/mL	(Irfan et al., 2019)
B. megaterium BM07			46 U/mL	

enhances efficiency and reduces process time (Zhang, Guo et al., 2020). In 2020 the pullulanase market size was around US$182 million, with Novozymes, Genencor, Amano Enzyme, and Longda responsible for 70% of worldwide production. Europe leads the market share with around 45%, followed by the USA and Japan (MarketWatch, 2022).

Many starchy substrates can be used for pullulanase production, such as corn, wheat rice, potato, wheat, and their residues. However, just a few studies focus on the use of agroindustrial wastes as substrates for its production; most of the research efforts during the last decades have been dedicated to finding thermoactive and/or thermostable pullulanases (Singh & Singh, 2019). This trend is explained by the condition in which starch saccharification takes place: around 60°C in most cases. However, more recent technologies aiming at lowering the saccharification temperature (in order to save energy consumption) are also opening room for development of "cold" active pullulanases (Lu et al., 2018; Zhang, Guo et al., 2020; Thakur et al., 2021).

Several genera of microorganisms are used for pullulanase production, and among them, *Bacillus* sp. play a remarkable role. Although there are several types of pullulanases, the main classes produced by the *Bacillus* genus are GH13 type I, which, when attacking starchy materials, are able to cleave just α-1,6-glycosidic bonds in amylopectin, and GH13 type II, which, in the same context, cleave both α-1,6-glycosidic bonds in amylopectin and also α-1,4-glycosidic bonds in amylose and amylopectin (Li, Wang et al., 2018). Table 4.3 shows some examples of *Bacillus* sp. able to produce pullulanases, their optimum conditions of application, and their specific activity (adapted from Xia et al., 2021).

The debranched starch, obtained from the action of pullulanase over native starch has shown very interesting alternatives and potential to be used in the nutraceutical industry, such as production of slowly digestible starch, fat and/or protein substitute in food products, tableting excipient for controlled drug release, and also as emulsion stabilizer (Liu et al., 2017). With the application of pullulanases, there is also a great potential in using modified rice, potato, maize, and oat starches as components for functional food; the main desired effects for industrial application are the lowering of starch digestibility and lowering of starch viscoelasticity (Li, Wang et al., 2018). Furthermore, pullulanases have also been investigated for use in the production of cyclodextrins and biodegradable packing materials

TABLE 4.3
Pullulanase Produced by *Bacillus* Strains, Optimum Conditions, and Specific Activities

Strain	Pullulanase Type	Optimum Temperature (°C) and pH	Specific Activity (U/mg)	Reference
Bacillus acidopullulyticus	GH13 type I	60 and 6.0	1,096	(Chen et al., 2014)
B. cereus	GH13 type I	60 and 6.0	45	(Wei et al., 2014)
Bacillus deramificans	GH13 type I	55 and 4.5	470	(Zou et al., 2014)
Bacillus flavocaldarius	GH13 type I	80 and 7.0	54	(Kashiwabara et al., 1999)
B. megaterium	GH13 type I	55 and 6.5	83	(Yang et al., 2017)
Bacillus methanolicus	GH13 type I	50 and 5.5	292	(Zhang, Guo et al., 2020)
Bacillus naganoensis	GH13 type I	62 and 5.0	700	(Mu et al., 2015)
Bacillus pseudofirmus	GH13 type I	45 and 7.0	270	(Lu et al., 2018)
B. stearothermophilus	GH13 type I	60 and 7.0	50	(Kambourova & Emanuilova, 1992)
B. subtilis	GH13 type I	70 and 9.5	26	(Asoodeh & Lagzian, 2012)
B. circulans	GH13 type II	50 and 7.0	1400	(Kim & Kim, 1995)
B. megaterium	GH13 type II	45 and 6.5	not informed	(Liu et al., 2020)
Geobacillus stearothermophilus	GH13 type II	65 and 5.5	967	(Zareian et al., 2010)
Geobacillus thermoleovorans	GH13 type II	60 and 7.0	795	(Nisha & Satyanarayana, 2015)

4.3.4 Other Enzymes

In the previous sections, the most important enzymes produced by *Bacillus* regarding interest in food and/or nutraceutical industry were discussed. In the same context, other relevant enzymes are briefly shown in the sequence. Nattokinase (EC 3.4.21.62) is a serine protease discovered in 1987 in the traditional fermented food called natto. It is produced during the natural fermentation of natto by *B. subtilis* and acts as a relevant fibrinolytic and thrombolytic agent (Pagnoncelli et al., 2017). In the last decades, many studies have aimed to develop and optimize the controlled production, recovery, and purification of this enzyme, as well as the activity and stability improvement and process cost reduction (Deepak et al., 2008; Hsieh et al., 2009; Cho et al., 2010; Garg & Thorat, 2014; Weng et al., 2015). However, nattokinase is sensitive to degradation in the human gastrointestinal tract, turning recent studies to focus on the development of strategies for its delivery, mainly due to advances in microencapsulation techniques (Zhou et al., 2021). As an example, Kou et al. (2020) have developed a delivery drug system based on lyposomes for delivery of nattokinase-polysialic acid complexes, which proved to be effective in the treatment of advanced tumours (due to thrombolytic activity of nattokinase). Other recent studies have also shown other potential applications of nattokinase in nutraceutical industry. Zhang et al. (2021) have demonstrated that *B. subtilis* JNFE0126 is able to produce both nattokinase and a milk clotting enzyme during milk fermentation. The final fermented milk presented both thrombolytic and clotting activities, and poses as an interesting candidate as functional food to prevent cardiovascular diseases. Zhou et al. (2020) has shown nattokinase as a potential candidate to be used in treatment of inflammatory bowel diseases, more specifically against colitis. They have demonstrated nattokinase was able to attenuate symptoms and also to alleviate damages caused by induced colitis. Li et al. (2021) have developed a fusion polypeptide using nattokinase and a glucagon-like peptide. The resulting polypeptide, besides presenting thrombolytic action, was able to reduce glucose to almost normal levels in mice when administered orally.

Glucoamylases (EC 3.2.1.3) act mainly in the α-1,4-glycosidic links of the non-reducing end of dextrin, amylose, and amylopectin. Therefore, they have application for starch and/or other glucose-rich polymer hydrolysis and generation of glucose-rich syrups (for beverages and baking) (Farias et al., 2021). They are also applied in brewing, for the production of fructose-rich syrups (Negi & Vibha, 2017). Filamentous fungi are the main producers of glucoamylases, but some research has been made using *Bacillus* sp. with the aim of obtaining thermostable enzymes and to reducing fermentation time (Ghani et al., 2019; Ilyas et al., 2020).

Isoamlyases (EC 3.2.1.41) hydrolyse α-1,6-glycosidic bonds present in amylopectin, so they are also referred as debranching enzymes, like pullulanases. While pullulanases act mainly in longer amylopectin chains, isoamylases are preferred for medium and shorter sized chains, as for dextrins. They usually act together with other enzymes for complete hydrolysis, and are important in the generation of monossaccharide-rich syrups (Farias et al., 2021). The production of isoamylases can be carried in several microorganisms, such as *Rhizopus oryzae* and many bacteria, among them *Bacillus* genus, such as *B. circulans* and *Bacillus lentus* (Castro et al., 1992; Li et al., 2013).

Pectinases (EC 3.2.1.99 and 3.1.1.11) constitute an essential class of enzymes for the food and beverage sectors, sharing around 25% of the global market of enzymes in this niche. They act in the depolymerization of pectin, which is composed of subunits of d-galacturonic acid linked by α-1,4-bonds (John et al., 2020). There are some reports of the applications of bacterial pectinases, e.g., the production of pectate lyase from *B. pumilus* to bioscoure cotton (Klug-Santner et al., 2006); the production of exopolygalacturanase from *B. subtilis* for wastewater treatment and degumming (Gupta et al., 2008); and the production of an extracellular pectinase from *B. subtilis* for clarification of fruit juices (Takci & Turkmen, 2016). However, they are mostly produced by fungal species (John et al., 2020), so they are not discussed in detail in the present chapter.

Lipases (EC 3.1.1.3) constitute one of the most important enzymes in the food sector (market value projected to be approximately US$590 million in 2023), due to their ability to hydrolyse triacylglicerols and release glycerol, monoglycerides, diglycerides, and fatty acids. They can act in several biochemical reactions, such as transesterification, interesterification, acidolysis, and aminolysis; hence, they have several biotechnological applications in oil, fat, dairy, meat, bakery, and flavour and aroma among other food industries (Phukon et al., 2022; Salgado et al., 2022). Most of industrial applications use fungi or yeast to produce lipases, but there are some interesting reports of their production from *Bacillus* genus, as follows: *Bacillus tequilensis* lipase was able to reduce the maturation time of Swiss cheese (Rani & Jagtap, 2019); *G. stearothermophilus* lipase was successfully employed for interesterification of sunflower oil and hydrogenated soybean oil (Samoylova et al., 2017); and *Bacillus aerius* lipase was isolated and produced isoamyl acetate (which is an important aroma and flavouring agent) from acetic acid and isoamyl alcohol (Narwal et al., 2016).

Mannanases represent a class of enzymes with several subtypes depending on the kind of mannan they are able to hydrolyse. The most important class for the food industry are the endo β-mannanases (3.2.1.78) which act in the β-1,4-links of β-mannan, mainly present in hemicellulose. Oligo- and polysaccharides derived from hemicellulose can present interesting prebiotic activities, despite therapeutic and immunological benefits, consisting in potential nutraceuticals. Hence, the production and/or extraction of β-mannanases by several kinds of microorganisms is of great economic importance. Many microorganisms produce β-mannanases, being the bacterial ones usually overexpressed in *E. coli* and the fungal ones usually overexpressed in *Pichia pastoris* (Srivastava & Kapoor, 2017). Nonetheless, it is possible to find some successful examples employing *Bacillus* genus: *B. subtilis* CAe24 was able to produce manno-oligosaccharides from copra meal with potential to be used as food ingredient (prebiotic) (Rungrassamee et al., 2014); while *Bacillus* sp. CFR1601 have shown the same potential from guar gum and locust bean gum (Srivastava et al., 2014); and *B. pumilus* was used to clarify apple juice due its ability to produce and excrete appropriate β-mannanases (Adiguzel et al., 2015).

Other enzymes, such as cellulases, laccases, and proteases, are classically produced by *Bacillus* spp., but are not discussed in this chapter as their main applications are in other fields, such as biofuel, textile, pulp and paper, among other industries.

4.4 GENETIC ENGINEERING IN *BACILLUS* FOR ENHANCED PRODUCTION OF ENZYMES

The *Bacillus* genus is recognized as one of the most reliable genera to perform genetic engineering among microorganisms, especially the *B. subtilis* species, which had its complete genome sequenced in 1997, second only to *E. coli* (Harwood, 1989; Zweers et al., 2008; Gu et al., 2018). With a well-known and well-characterized genetic material, coupled with the efficient capacity of secreting proteins to extracellular space, the genus is an attractive host for metabolic engineering and for the production of recombinant proteins (Zhang et al., 2020; Park et al., 2021). Through recombinant DNA techniques, numerous products have been induced in *Bacillus* cells, including peptides, biofuels, organic acids, biopolymers, oligosaccharides, and, of course, enzymes (Berkeley et al., 2002; Schallmey et al., 2004; Park et al., 2021).

As the transcriptome, proteome, secretome, and metabolome are well determined within the genus, several procedures have already been applied to modify the bacteria chromosome or plasmid. The techniques go from physical or chemical mutagenic agents to favour gene insertion, deletion or mutation, to the modern sequence-specific genome editing through clustered, regularly interspaced, short palindromic repeats coupled with the endonuclease Cas9 (CRISPR-cas9) (Gu et al., 2018; Hong et al., 2018).

The current most frequent strategy is the use of plasmid insertion. The plasmid consists in a small circular piece of DNA comprising some genes, markers, promoters, and a specific region to insert the gene of interest. The plasmid is inserted into a competent cell and, according to its structure and composition, expresses the desired protein (Schumann, 2007; Liu et al., 2013;). There are some desired characteristics for a plasmid that will industrially produce enzymes. For instance, several plasmids use isopropyl β-D-1-thiogalactopyranoside (IPTG) as induction factor, such as pMT1, pHT100, and pLus-Hyb; however, IPTG is expensive, toxic, and unaffordable for large scales (Liu et al., 2013). With that in mind, alternatives for IPTG, e.g., maltose (Ming et al., 2010) or mannitol (Heravi et al., 2011), are being developed. Another direction for the research is the development of less labour- and less time-consuming markers, reducing the number of markers to one or zero. For instance, the *upp* market-free gene, which encodes for uracil phosphoribosyl-transferase, acting as a counter-selectable marker and identifying the bacteria that has the plasmid in a single step (Fabret et al., 2002; Zhang & Zhang, 2011). It is interesting to mention that temperature-sensitive markers in the plasmid can easily be applied in the genus due to its heat resistance as well (Widner et al., 2000; Liu et al., 2013). Several examples using plasmid techniques to increase enzymes production can be seen in Table 4.4.

Besides the plasmid vector insertion and expression, *Bacillus* cells are susceptible to chromosome modification through genome integration methods, such as the counter-selectable markers and site-specific recombination systems, enabling a long-lasting better performance alteration (Wang et al., 2012; Jeong et al., 2015; Gu et al., 2018). An interesting approach is to reduce the total genome size by removing specific regions, which allows the discovery of dispensable regions and direct the cell mechanisms to produce heterologous proteins and enzymes. This method has been applied in the strain *Bacillus subtilis* MGB874, enhancing the production of aconitase and glutamate dehydrogenase (e.g. Morimoto et al., 2008; Manabe et al., 2011).

The CRISPR-cas9 technique, which allows site-specific mutation or gene-specific knockdown, is also being tested in *Bacillus*, leading to different protein conformations and expressions (Schallmey et al., 2004; Altenbuchner, 2016; Hong et al., 2018). The system is a natural bacteria defence mechanism that uses a RNA-guided DNA endonuclease enzyme to break specific regions according to the

TABLE 4.4
Example of Enzymes Production Increased by Plasmids in *Bacillus*

Host Strain	Enhanced Enzyme	Genetic Engineering Strategy	Enhancement	Reference
B. subtilis	α-amylase	Plasmid pC194Amy insertion by electroporation	2.9–3.0 μM/mg (thermostable)	(Özcan & Özcan, 2008)
B. subtilis	α-amylase	Site-directed mutagenized by polymerase chain. Insertion of vector pWB980 with P_{43} promoter	836 to 917 U/mg (acid-resistant)	(Liu et al., 2008)
B. subtilis	α-amylase	Transformed with vector pMA5	5.6–441 U/mL (alkaline enzyme)	(Yang et al., 2011)
B. subtilis	α-amylase	Transformed with vector pWB980	187–723 U/mL	(Liu et al., 2010)
B. subtilis	Xylanase	Transformed with vector pET28a	93–380 U/mg	(Huang et al., 2015)
B. subtilis	Xylanase	Transformed with vector pDG364	24 U/mL (extracellular and extreme thermophile)	(Zhang et al., 2010)
B. subtilis	Xylanase	Transformed with vector pWH1520	40–119 U/mL	(Verma & Satyanarayana, 2013)
B. subtilis	Pullulanase	Transformed with vector pMA0911, with the promoter P_{HpaII} replaced by a stronger constitutive promoter P43	3.9–25 U/mL	(Song et al., 2016)
B. subtilis	Pullulanase	The plasmid pDL was employed as a universal integration vector to integrate the fusion with the *amyE* promoter by double homologous recombination	25–61 U/mL	(Wang et al., 2019)
B. subtilis	Pullulanase	Transformed with vector pCBS3 with the combinations of the promoters *sodA*, *fusA*, and *amyE*	71–1,555 U/mL	(Meng et al., 2018)
B. megaterium	Levansucrase	Transformed with vector p3STOP1623hp by protoplast transformation	0.55–55 U/mL	(Korneli et al., 2013)
B. subtilis	Pectate lyase	Transformed with vector pWB980	15–870 U/mL	(Liu et al., 2012)
B. subtilis	Lipase	Transformed with vector pBSR2	0.86–8.6 mg protein/g of wet weight of cell mass	(Ma et al., 2006)
B. subtilis	Lipase	Transformed with vector pUC19 by electroporation	0.5–8 U/mL	(Kim et al., 2002)
B. licheniformis	Nattokinase	Transformed with vector pHY300PLK	15–34 U/mL	(Wei et al., 2015)
B. subtilis	Nattokinase	Transformed with vector pMA5-BSAP with an arrangement of promoters containing *HpaII-HpaII-P43*	111–817 U/mL	(Guan et al., 2016; Liu et al., 2019)

CRISPR sequence provided (Deltcheva et al., 2011). This strategy has been tested successfully in *B. subtilis* (Altenbuchner, 2016); however, the mechanisms are being adapted from other microorganisms, and the efficiency is yet to be optimized, as in the example of *B. subtilis* ATCC 6051a reaching 33–55% gene disruption efficiency with *Streptococcus pyogenes* CRISPR/Cas9 based system (Zhang et al., 2016; Hong et al., 2018).

Beyond alterations targeting genes, plasmids, and chromosomes, other molecular tools can be applied in order to modify secretion pathways, transporters, and different metabolic modulators for enhanced enzyme production (Liu et al., 2013; Gu et al., 2018). For this purpose, it is interesting to highlight the importance of the databases focused on proteins, metabolic pathways, regulatory pathways, protein–protein interaction, enzymes, and even gene knockdown or deletion effects, especially when updated for *B. subtilis*, for instance the SubiWiki database (Michna et al., 2016; Gu et al., 2018).

As the cell naturally lacks specific secretory pathways and protein-conducting channels for the new heterologous proteins, altering the existing secretory pathways turns into a vital approach to increase the bacteria excretion capacity. In this case, the Tat pathway and the Sec-SRP pathway are the most investigated excretion ways (Liu et al., 2013; Gu et al., 2018; Zhang et al., 2020). For instance, Zhu et al. (2008) verified the secretion of 24 different peptides to mediate the secretion in Tat, reaching four peptides that enhanced the excretion of Penicillin G Acylase. Goosens et al. (2013) also discovered a new protein that serves as a substrate for the same pathway, the QcrA membrane protein. In the Sec-SRP pathway, the deletion of 61 amino acids of the translocator motor SecA increased 2.2 times the expression of heterologous human interferon α (Kakeshtia et al., 2010).

Another approach to intensify the production of enzymes is altering expression levels through different molecular tools. Yang et al. (2016), for instance, constructed an expression system based on the type II toxin–antitoxin natural system of *B. subtilis*, by expressing both toxin EndoA and antitoxin EndoB. The systems can be applied in food producing fermentations with stability during 100 generations, eliminating the need for antibiotics of traditional techniques (Yang et al., 2016). Radeck et al. (2013) developed the *Bacillus* BioBrick Box, a genetic tool that comprises several compatible integration plasmid vectors with different strength magnitude promoters. The tool has cassettes that enable β-galactosidase and luciferase reporter assays, four constitutive and two inducible promoters, and five widely used epitope tags to facilitate the expression of any desired protein (Radeck et al., 2013).

Alternative strategies to produce enzymes using genetic engineering are being developed every day as the research gains more and more control of the proteomic and metabolomic of *Bacillus*. Gu et al. (2017) slightly modified the phosphotransferase system (PTS), which is responsible for the transportation of glucose, for the production of N-acetylglucosamine. The technique successfully altered the stoichiometry balance, inducing the generation of by-products such as acetate, lactate, acetoin, and 2,3-butanediol, with a twofold increase in the yield of the product (Gu et al., 2017). Liu et al. (2016), also targeting the N-acetylglucosamine production increase, deeply investigated the metabolomics of *B. subtilis* and detected a cycle between N-acetylglucosamine-6-phosphate and N-acetylglucosamine that was energy-dissipating and futile. Deleting a gene encoding for an enzyme of the cycle, glucokinase, also led to a twofold enhancement in the production (Liu et al. 2016).

Most of the improvements desired for the enzymes through genetic engineering are focused on increasing yield or adapting the enzyme to different conditions. The greater the amount of protein or enzyme produced and the greater the activity of the enzyme, while maintaining stability in altered pH, temperature, or presence of solvents or oxidizing agents, the better. Also, qualities that are especially attractive for nutraceutical enzymes are an enhanced substrate specificity, greater metal binding capacity, non-production of toxic compounds, and the required safety level to be applied in food, hygiene, and medicinal products (Schallmey et al., 2004; Liu et al., 2013;).

For α-amylases, for instance, Yang et al. (2020) performed slight changes in the signal peptide residues sequence and in the promoter sequence to improve the *B. subtilis* 168 enzyme yield. The signal peptide, responsible for facilitating the intracellular transport to the membrane, was YwbN' with an arginine between residues 5 and 6, and the promoter was P_{HpaII} with a deletion on the 27th nucleotide. Both alterations combined resulted in a 250-fold increase of the original alkaline α-amylase activity (Yang et al., 2020). Wang et al. (2018) utilized three different mutations – R172K, A270K and N271H – to adapt *B. subtilis* Ca-independent α-amylase to acidic pH values. With these mutations focused on reducing the pKa, double mutants showed a reduction of 2 units in the optimum pH and an enhancement of 3.9-fold efficiency (Wang et al., 2018).

For xylanases, Bai et al. (2016) performed several error-prone polymerase chain reaction (PCR) and reached a mutation named E135V. This mutation alone was capable of changing the optimum *Bacillus* sp. SN5 xylanase pH from 7.5 to values between 8.0 and 10.0, even improving its efficiency 57% (Bai et al., 2016). Zhang et al. (2016) tested two different promoters and 138 different signal peptides to reach a higher yield of alkaline xylanase from *B. pumilus* BYG. The enzyme was expressed in *B. subtilis* to evaluate both Sec and Tat pathways, reaching an increased activity of 193.7 U/mL through the Sec pathway (Zhang et al., 2016).

Bi et al. (2020) targeted the enhancement of the thermostability of pullulanase from *Bacillus thermoleovorans*. By predicting and confirming 6 mutants, the G692M mutant increased 3.8°C in optimum temperature of activity and presented a 2.1-fold longer half-life at 70°C (Bi et al., 2020). Cai et al. (2016) tested different signal peptides to optimize the production of *B. licheniformis* nattokinase. They obtained a 4.7-fold improvement by using the AprE signal peptide coupled with a mutation and overesxpression of sipV (Cai et al., 2016). Tang et al. (2015) modulated the hydrophobicity of *Bacillus thermocatenulatus* lipase lid domains. Within the three mutants developed, Y225F+S232A showed an increase of 35% in activity catalysing heterogeneous hydrolytic reactions (Tang et al., 2015).

New approaches and technologies in the area of molecular biology will definitely reach higher efficiencies as research advances, making the industrial production of enzymes more feasible and economically attractive while using the *Bacillus* genus as a host.

4.5 PATENT OVERVIEW

The rising awareness of food safety and quality around the world leads to a higher demand for bio-based food additives and enzymes in the industry. The food enzymes market was valued at US$1.69 billion in 2019 and is projected to reach US$2.39 billion in 2027, representing a compound annual growth rate (CAGR) of 4.70% (FortuneBusinessInsights, 2021). *Bacillus* spp. derived enzymes can be applied for ingredient processing and manufacturing of functional and nutraceutical foods. Research and development strategies involving enzymes generally focus on new production processes with reduced costs, enabling product competitivity, which can be achieved through new methods for fabricating the enzymes with specific conditions or alternative substrates, new methods for purifying the molecules, new microorganisms with specific characteristics that can enhance enzymatic activity or production (usually obtained with genetic engineering), new formulations containing various enzymes in a blend for specific purposes, or even new formulations with higher quality and specificity. Furthermore, new processes for enzyme application are constantly being developed as well, either in the application for the development of new commercial products or specific industrial solutions.

These technological novelties can be protected through the patent system, which can guarantee exclusive exploitation of technology to inventors and applicants. In order to reveal innovation tendencies regarding the production and application of enzymes from *Bacillus* spp., a patent search and analysis was conducted using the Derwent Innovation Index (DII) advanced search tool. Terms related to enzymes (amylase, cellulase, xylanase, pullulanase, pectinase, levansucrase, galactosidases, mannanases), their production (process, fabrication, manufacture, among others) and their source (*Bacillus*) were combined with Bollean language and truncated when necessary. The selected time period for analysis was defined from 2011 to January 2022 and 616 documents were retrieved. After revision of title and abstracts, 266 documents were selected for analysis.

Over the last few years, research, development, and patenting of new technologies involving the production and application of enzymes derived from *Bacillus* spp. are in an incipient phase, with nonconstant growth and with few countries dominating the market (Figure 4.1a and b). Moreover, universities and research centres are the main patent applicants with almost 54% of the documents, followed by enterprises with almost 37% (Figure 4.1c), which may mean that technology is still being developed in laboratory and pilot scales. The majority of commercial enzymes are derived

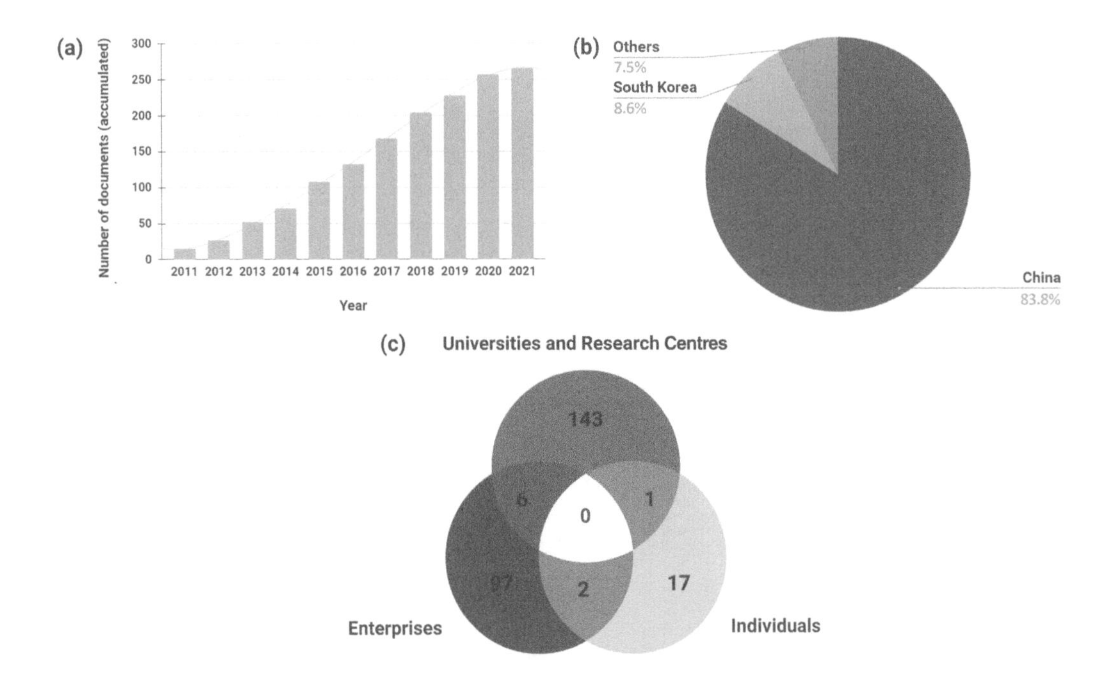

FIGURE 4.1 Number of registers regarding the search for patents of the production and application of enzymes derived from *Bacillus* spp. (a) Accumulated documents over the years, (b) Profile of applying countries, and (c) Profile of appliers.

from fungi and molecules from this source dominate industrial applications (Liu & Kokare, 2017; Obeng et al., 2017). However, bacterial sources can present an advantage when compared to fungi due to easier manipulation, faster fermentation processes and growth, and, therefore, reduced production costs. Furthermore, genetic engineering techniques for the production of recombinant enzymes are easier to perform in bacteria, which can be an opportunity for technology development (Gopinath et al., 2017). Population growth, rise of food demand, and increased awareness and search for healthier foods and habits – boosted by the Sars-Cov2 pandemic (Monteloeder, 2021) – are also opportunities that may stimulate the development of new enzyme formulations and, therefore, increase the number of applied documents in the years to come.

Regarding International Patent Classification (IPCs), the most used codes were derivations of C12N 01, C12R 01, and C12N 09, which refer generally to microorganisms (bacteria) and the enzymes produced by them (Figure 4.2a). These codes are directly related to the main object of protection (Figure 4.2b). Genetically engineered or newly isolated microorganisms were found to be the main interest of research (38% of applied documents), followed by new production processes (31%) and new enzyme formulations (9%). The main microorganisms reported in the documents were *B. subtilis* (IPC C12R-001/125 in 27% of the applied documents) and *B. licheniformis* (IPC C12R-001/10 in 6%). The main application areas were food, feed, and medicine, through enhancing functional properties of the ingredients. In the majority of the documents analysed, the produced or applied enzymes were contained in a blend or complex (Figure 4.2c). This can be advantageous for intricate substrates and in animal feeding, in which a variety of enzymes need to be applied for proper processing. Beyond the conventional enzymes applied in food industries (amylases, xylanases, pectinases, among others), records protecting the production or application of thrombolytic (or fibrinolytic) enzymes were reported. This enzyme is a type of protease able to degrade fibrin clots and, therefore, prevent and/or treat thrombosis, one of the main causes of deaths in the 21st century (Altaf et al., 2021). Nattokinase is a type of thrombolytic enzyme and can be found in Asian fermented foods such as Japanese natto and Chinese douchi and doufuru (variations of soybean

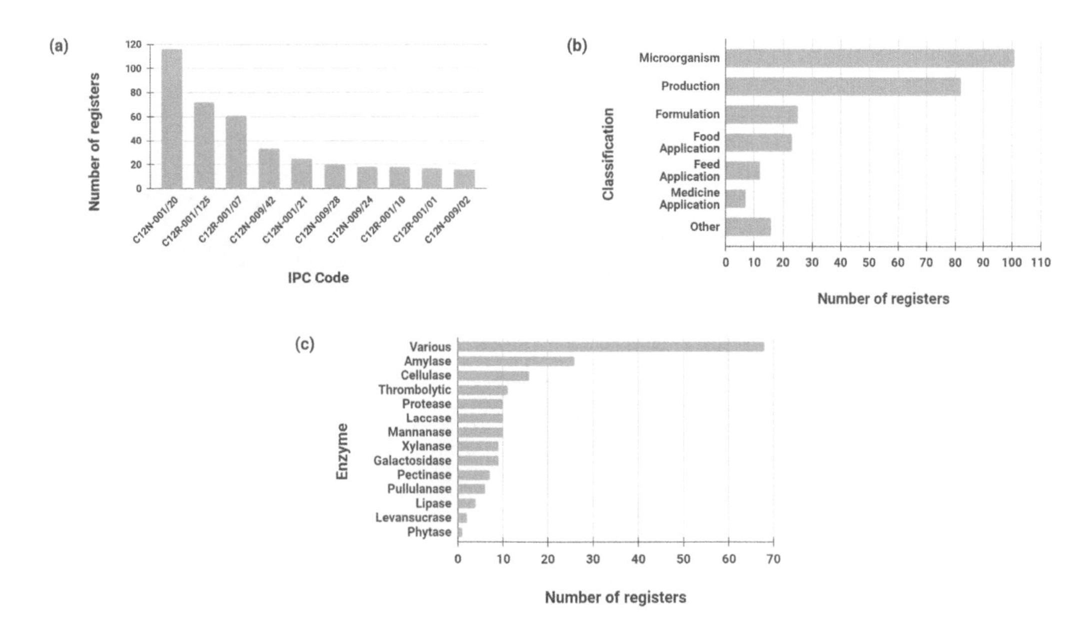

FIGURE 4.2 Profile of classifications and protected objects regarding the search for patents of production and application of enzymes derived from *Bacillus* spp. (a) IPC codes, (b) Protected object or process, and (c) Enzyme produced or applied.

fermented with *Bacillus* spp. strains) (Mine et al., 2005; Altaf et al., 2021). These enzymes can be either isolated from fermented foods or produced by various *Bacillus* strains and further applied for nutraceutical objectives.

Relevant patent documents describing technologies related to the application of *Bacillus* spp. or *Bacillus* enzymes in nutraceutical products or related processes are presented as follows. In the segment of feed products, the patent document VN46412-A (Dao & Duong, 2015) describes a process to produce a multi-enzyme probiotic composition by the cultivation of *B. amyloliquefaciens* plantarum SP1901 in SSF. The composition contains at least three enzymes synthetized by the strain, namely phytase, protease, and xylase, and is useful as a feed supplement.

In the area of food products, two patent documents are highlighted. The first, KR2163993-B1 (Shin et al., 2020), presents a composition used as medicine and food for preventing or treating metabolic syndromes including obesity, diabetes, arteriosclerosis, hypertension, and hyperlipidemia. The composition comprises *B. coagulans* enzymes produced by fermentation of mixed grains including brown rice, barley, soybean, corn, wheat, and adlay, and is presented in the form of dry powder. Some claimed benefits include weight loss, fatty liver inhibitory effect, hyperlipidemia improving effect, and blood sugar lowering effect. The second, CN103876155-A/CN103876155-B (Tian, 2014), describes a method for preparing a composition containing probiotics and plant enzymes, e.g., lipase, amylase, protease, cellulase, beta-glucanase, xylanase, and lactase, useful in food and chemical applications. The composition is prepared by fermenting fruit, vegetables, grains, beans, medicinal and edible plants, and fungus as substrates using *Dekkera* yeast, *Streptococcus thermophilus* and *B. amyloliquefaciens* as inoculum.

Some patent documents protect specific *Bacillus* strains that synthesize novel enzyme preparations, or genetic engineering methods to improve the enzymatic activity obtained from *Bacillus* strains. The document KR2020126739-A/CN111961605-A (Choi et al., 2019) describes the new *B. subtilis* natto GFNK-01 KCTC 18770P strain, isolated from cheonggukjang, for producing the thrombolytic enzyme nattokinase. Independent claims are included for a fermentation starter for fermentation, a microbial preparation, fermented food, e.g., meju, Korean miso, and red pepper paste, and a method for producing thrombolytic enzyme. The document CN110144320-A (Wu et al., 2019) reports a new genetically engineered strain of *B. subtilis* hosting pHY300PLK as an expression vector, useful for the preparation of maltotetraose amylase or maltotetraose amylase containing products. The enzyme activity produced by the genetically engineered strain can reach 1900 U/mL, and the preferred fermentation medium contains 240–260 g/L glucose, 170–190 g/L industrial peptone, 65–75 g/L soybean meal, and a trace element solution. The document CN106754605-A/CN106754605-B (Jiao et al., 2017) describes a method for improving α-amylase activity in *B. subtilis* fermentation broth, which involves performing genetic modification of the genome of *B. subtilis* strain 168. The primer is designed for PCR cloning amplification of sdpBC partial sequence DNA, and the genomic DNA of *B. subtilis* strain 168 is used as a template for PCR amplification. Lastly, the document CN105112348-A/CN105112348-B (Duan et al., 2015) describes the preparation of recombinant bacteria used in fermenting and producing pullulanase for preparing a food product, through the expression of the recombinant plasmid pulA/pSVEB in a *Bacillus* host. The preparation of the recombinant plasmid comprises taking the missing promoter and signal peptide pWB980, the pullulanase expression element (from *B. pumilus* outer protein promoter P2, outer signal peptide and *Bacillus* pullulanase mutant gene), and the missing signal peptide pET20b (+) and in-fusion ligating to get the recombinant plasmid pulA/pSVEB. Recombinant bacteria are obtained by electrically transferring the pulA/pSVEB plasmid to *Brevibacillus brevis* (CCTCC NO: AB94025).

Finally, some patent documents describe methods or media compositions to improve the production process of *Bacillus* enzymes. The document IN202131002187-A (Sathiyamoorthy et al., 2021) specifies a low-cost medium composition containing agricultural by-products to produce alkaline α-amylase from *B. licheniformis*. The medium comprises a cheap source of nitrogen (e.g., corn steep liquor), 0.5% soybean, 2% mustard seed, and 3% cotton seed. The document CN112029750-A (Xia & Liu, 2020) describes a fed-batch fermentation method based on an oxygen uptake rate (OUR) useful for producing mesophilic alpha-amylase from *B. subtilis*. The method comprises the online detection of the change trend of OUR in the exhaust gas, to feed the substrate in a way that keeps the OUR in a steady or slow decline state. This method regulates fermentation parameters to promote the production of enzyme and increases the enzyme activity output.

4.6 CONCLUSIONS

This chapter has shown the importance of the genus *Bacillus* for the production of enzymes with commercial application, focusing on the growing nutraceutical industry. In the past decades many developments have been achieved, especially in the molecular biology field, which has allowed the use of potent cell-factory hosts to produce enzymes. In addition, advances in other fields (i.e., new techniques of microencapsulation) have allowed the enhancement of known processes, the investigation of new applications, and the proposal of novel strategies to overcome unresolved problems. Considering the trend of recent scientific advances, as well as current research around the world, the promise is that the nutraceutical industry will keep growing at a pace in the years to come. In that direction, many of its new products and solutions will keep coming from enzymes produced by *Bacillus* spp., revealing their importance in the biotechnology field.

REFERENCES

Abd-Elhalem, B. T., El-Sawy, M., Gamal, R. F., & Abou-Taleb, K. A. 2015. Production of amylases from Bacillus amyloliquefaciens under submerged fermentation using some agro-industrial by-products. *Annals of Agricultural Sciences* 60(2): 193–202. https://doi.org/10.1016/J.AOAS.2015.06.001.

Abhyankar, W. R., Kamphorst, K., Swarge, B. N. et al. 2016. The influence of sporulation conditions on the spore coat protein composition of *Bacillus subtilis* spores. *Frontiers in Microbiology* 7: 1636. https://doi.org/10.3389/fmicb.2016.01636.

Abootalebi, S. N., Saeed, A., Gholami, A. et al. 2020. Screening, characterization and production of thermostable alpha-amylase produced by a novel thermophilic *Bacillus megaterium* isolated from pediatric intensive care unit. *Journal of Environmental Treatment Techniques* 8(3): 952–960.

Adiguzel, A., Nadaroglu, H., & Adiguzel, G. 2015. Purification and characterization of b-mannanase from *Bacillus pumilus* (M27) and its applications in some fruit juices. *Journal of Food Science and Technology* 52(8): 5292.

Afzaljavan, F. & Mobini-Dehkordi, M. 2012. Application of alpha-amylase in biotechnology. *Journal of Biology and Today's World* 1(1): 39–50.

Al-Quadan, F., Akel, H., & Natshi, R. 2011. Characteristics of a novel, highly acid-and thermo-stable amylase from thermophilic *Bacillus* strain HUTBS62 under different environmental conditions. *Annals of Microbiology* 61(4): 887–892.

Altaf, F., Wu, S., & Kasim, V. 2021. Role of fibrinolytic enzymes in anti-thrombosis therapy. *Frontiers in Molecular Biosciences* 8: 680397.

Altenbuchner, J. 2016. Editing of the *Bacillus subtilis* genome by the CRISPR-Cas9 system. *Applied and Environmental Microbiology* 82(17): 5421–5427.

Andreoletti, O., Budka, H., Buncic, S., Colin, P., Collins, J. D., & Koeijer, A. De. 2008. The maintenance of the list of QPS microorganisms intentionally added to food or feed - Scientific Opinion of the Panel on Biological Hazards. *EFSA Journal* 6(12): 1–48. doi: 10.2903/J.EFSA.2008.923.

Archana, A., & Satyanarayana, T. 1997. Xylanase production by thermophilic *Bacillus licheniformis* A99 in solid-state fermentation. *Enzyme and Microbial Technology* 21(1): 12–17. https://doi.org/10.1016/S0141-0229(96)00207-4.

Asoodeh, A. & Lagzian, M. 2012. Purification and characterization of a new glucoamylopullulanase from thermotolerant alkaliphilic *Bacillus subtilis* DR8806 of a hot mineral spring. *Process Biochemistry* 47(5): 806–815. doi:10.1016/j.procbio.2012.02.018.

Azzopardi, E., Lloyd, C., Teixeira, S. R., Conlan, R. S., & Whitaker, I. S. 2016. Clinical applications of amylase: Novel perspectives. *Surgery* 160(1): 26–37.

Bai, W., Cao, Y., Liu, J., Wang, Q., & Jia, Z. 2016. Improvement of alkalophilicity of an alkaline xylanase Xyn11A-LC from *Bacillus* sp. SN5 by random mutation and Glu135 saturation mutagenesis. *BMC Biotechnology* 16(1): 1–9.

Bajaj, B. K. & Manhas, K. 2012. Production and characterization of xylanase from *Bacillus licheniformis* P11(C) with potential for fruit juice and bakery industry. *Biocatalysis and Agricultural Biotechnology* 1(4): 330–337.

Barros, F. F. C., Simiqueli, A. P. R., de Andrade, C. J., & Pastore, G. M. 2013. Production of enzymes from agroindustrial wastes by biosurfactant-producing strains of *Bacillus subtilis*. *Biotechnology Research International* 2013: 1–9. doi: 10.1155/2013/103960.

Bataillon, M., Nunes Cardinali, A. P., Castillon, N., & Duchiron, F. 2000. Purification and characterization of a moderately thermostable xylanase from *Bacillus* sp. strain SPS-0. *Enzyme and Microbial Technology* 26(2–4): 187–192.

Beauregard, P. B., Chai, Y., Vlamakis, H., Losick, R., & Kolter, R. 2013. *Bacillus subtilis* biofilm induction by plant polysaccharides. *Proceedings of the National Academy of Sciences of the United States of America* 110(17): 1621–1630.

Berkeley, R., Heyndrickx, M., Logan, N., & Vos, P. De. 2002. *Applications and Systematics of Bacillus and Relatives* (R. Berkeley, M. Heyndrickx, N. Logan, & P. De Vos (Eds.), 1st ed.) Blackwell Publishing Ltd. https://doi.org/10.1002/9780470696743.

Bi, J., Chen, S., Zhao, X., Nie, Y., & Xu, Y. 2020. Computation-aided engineering of starch-debranching pullulanase from *Bacillus thermoleovorans* for enhanced thermostability. *Applied Microbiology and Biotechnology* 104(17): 7551–7562.

Branda, S. S., Chu, F., Kearns, D. B., Losick, R., & Kolter, R. 2006. A major protein component of the *Bacillus subtilis* biofilm matrix. *Molecular Microbiology* 59(4): 1229–1238.

Burlacu, A., Cornea, C. P., & Israel-Roming, F. 2016. Microbial xylanase: A review. *Scientific Bulletin. Series F. Biotechnologies.*

Cai, D., Wei, X., Qiu, Y., Chen, Y., Chen, J., Wen, Z., & Chen, S. 2016. High-level expression of nattokinase in *Bacillus licheniformis* by manipulating signal peptide and signal peptidase. *Journal of Applied Microbiology* 121(3): 704–712.

Castro, G. R., Garcia, G. F., & Siñeriz, F. 1992. Extracellular isoamylase produced by *Bacillus circulans* MIR-137. *Journal of Applied Microbiology* 73: 520–523.

Cedrola, S. M. L., de Melo, A. C. N., Mazotto, A. M. et al. 2012. Keratinases and sulfide from *Bacillus subtilis* SLC to recycle feather waste. *World Journal of Microbiology and Biotechnology* 28(3): 1259–1269.

Chen, A., Li, Y., Liu, X., Long, Q., Yang, Y., & Bai, Z. 2014. Soluble expression of pullulanase from Bacillus acidopullulyticus in *Escherichia coli* by tightly controlling basal expression. *Journal of Industrial Microbiology and Biotechnology* 41(12): 1803–1810. doi:10.1007/s10295-014-1523-3.

Chen, Z. M., Li, Q., Liu, H. M. et al. 2010. Greater enhancement of *Bacillus subtilis* spore yields in submerged cultures by optimization of medium composition through statistical experimental designs. *Applied Microbiology and Biotechnology* 85(5): 1353–1360.

Cho, Y. H., Song, J. Y., Kim, K. M. et al. 2010. Production of nattokinase by batch and fed-batch culture of *Bacillus subtilis*. *New Biotechnology* 27(4): 341–346.

Choi, J. Y., Yang, T. H., Hyeon, K. J. et al. 2019. *New Bacillus subtilis natto GFNK-01 KCTC 18770P strain, for producing thrombolytic enzyme (e.g. nattokinase), fermentation starter for fermentation, microbial preparation, and fermented food, e.g. meju, Korean miso, and red pepper paste* (Patent No. KR2020126739-A; CN111961605-A).

Dahlberg, L., Holst, O., & Kristjansson, J. K. 1993. Thermostable xylanolytic enzymes from Rhodothermus marinus grown on xylan. *Applied Microbiology and Biotechnology* 40(1): 63–68.

Dao, T. L., & Duong, V. H. 2015. *Producing multi-enzyme probiotic composition by using cellular biomass of Bacillus amyloliquefaciens plantarum SP1901 and enzymes of phytase, protease and xylase generated from such strain, useful as a supplement in feed production industry* (Patent No. VN46412-A).

Declerck, N., Machius, M., Wiegand, G., Huber, R., & Gaillardin, C. 2000. Probing structural determinants specifying high thermostability in *Bacillus licheniformis* α-amylase. *Journal of Molecular Biology* 301(4): 1041–1057.

Deepak, V., Kalishwaralal, K., Ramkumarpandian, S., Babu, S. V., Senthilkumar, S. R., & Sangiliyandi, G. 2008. Optimization of media composition for Nattokinase production by *Bacillus subtilis* using response surface methodology. *Bioresource Technology* 99(17): 8170–8174.

Deltcheva, E., Chylinski, K., Sharma, C. M. et al. 2011. CRISPR RNA maturation by trans-encoded small RNA and host factor RNase III. *Nature* 471(7340): 602–607.

Devaraj, K., Aathika, S., Periyasamy, K., Manickam Periyaraman, P., Palaniyandi, S., & Subramanian, S. 2019. Production of thermostable multiple enzymes from Bacillus amyloliquefaciens KUB29. *Natural Product Research* 33(11): 1674–1677.

Divakaran, D., Chandran, A., & Chandran, R. P. 2011. Comparative study on production of α-amylase from *Bacillus licheniformis* strains. *Brazilian Journal of Microbiology* 42(4): 1397–1404. doi:10.1590/S1517-83822011000400022.

Du, R., Song, Q., Zhang, Q., Zhao, F. et al. 2018. Purification and characterization of novel thermostable and Ca-independent α-amylase produced by Bacillus amyloliquefaciens BH072. *International Journal of Biological Macromolecules* 115: 1151–1156.

Duan, X., Wu, J., & Zou, C. 2015. *Preparation of recombinant bacteria used in fermenting and producing pullulanase for preparing food product comprises taking Bacillus as host and expressing with recombinant plasmid pulA/pSVEB* (Patent No. CN105112348-A; CN105112348-B).

Eggert, T., Van Pouderoyen, G., Pencreac'h, G. et al. 2002. Biochemical properties and three-dimensional structures of two extracellular lipolytic enzymes from *Bacillus subtilis*. *Colloids and Surfaces B: Biointerfaces* 26(1–2): 37–46.

Elshaghabee, F. M. F., Rokana, N., Gulhane, R. D., Sharma, C., & Panwar, H. 2017. Bacillus as potential probiotics: Status, concerns, and future perspectives. *Frontiers in Microbiology* 8(1490): 1–15.

Fabret, C., Ehrlich, S. D., & Noirot, P. 2002. A new mutation delivery system for genome-scale approaches in *Bacillus subtilis*. *Molecular Microbiology* 46(1): 25–36.

Farias, T. C., Kawaguti, H. Y., & Bello Koblitz, M. G. 2021. Microbial amylolytic enzymes in foods: Technological importance of the Bacillus genus. *Biocatalysis and Agricultural Biotechnology* 35: 102054.

FortuneBusinessInsights. (2021). *Food Enzymes Market Size, Share – Global Industry Trends:* 2027.

Garg, R. & Thorat, B. N. 2014. Nattokinase purification by three phase partitioning and impact of t-butanol on freeze drying. *Separation and Purification Technology* 131: 19–26. https://doi.org/10.1016/j.seppur.2014.04.011.

Geetha, K. & Gunasekaran, P. 2010. Optimization of nutrient medium containing agricultural waste for xylanase production by *Bacillus pumilus* B20. *Biotechnology and Bioprocess Engineering* 15(5): 882–889. doi:10.1007/S12257-009-3094-0.

Gessesse, A. & Mamo, G. 1999. High-level xylanase production by an alkaliphilic Bacillus sp. by using solid-state fermentation. *Enzyme and Microbial Technology* 25(1–2): 68–72.

Ghani, M., Ansari, A., Haider, M. S. et al. 2019. Purification and characterization of a thermostable starch-saccharifying Alpha-1,4-Glucan-Glucohydrolase produced by *Bacillus licheniformis*. *Starch/Staerke* 71(11–12): 1–10.

Gomes, I., Gomes, J., & Steiner, W. 2003. Highly thermostable amylase and pullulanase of the extreme thermophilic eubacterium Rhodothermus marinus: Production and partial characterization. *Bioresource Technology* 90(2): 207–214.

Goosens, V. J., Otto, A., Glasner, C. et al. 2013. Novel twin-arginine translocation pathway-dependent phenotypes of *Bacillus subtilis* unveiled by quantitative proteomics. *Journal of Proteome Research* 12(2): 796–807.

Gopinath, S. C. B., Anbu, P., Arshad, M. K. M. et al. 2017. Biotechnological processes in microbial amylase production. *BioMed Research International* 2017: 1–9. doi: 10.1155/2017/1272193.

Graumann, P. 2017. *Bacillus: Cellular and Molecular Biology.* Caister Academic Press. https://doi.org/10.21775/9781910190753.

Gu, Y., Deng, J., Liu, Y. et al. 2017. Rewiring the glucose transportation and central metabolic pathways for overproduction of N-acetylglucosamine in *Bacillus subtilis. Biotechnology Journal* 12(10).

Gu, Y., Xu, X., Wu, Y. et al. 2018. Advances and prospects of *Bacillus subtilis* cellular factories: From rational design to industrial applications. *Metabolic Engineering* 50: 109–121.

Guan, C., Cui, W., Cheng, J., Liu, R., Liu, Z., Zhou, L., & Zhou, Z. 2016. Construction of a highly active secretory expression system via an engineered dual promoter and a highly efficient signal peptide in *Bacillus subtilis. New Biotechnology* 33(3): 372–379. doi:10.1016/J.NBT.2016.01.005.

Gupta, S., Kapoor, M., Sharma, K. K., Nair, L. M., & Kuhad, R. C. 2008. Production and recovery of an alkaline exo-polygalacturonase from *Bacillus subtilis* RCK under solid-state fermentation using statistical approach. *Bioresource Technology* 99(5): 937–945.

Haq, I. U., Ashraf, H., Iqbal, J., & Qadeer, M. A. 2003. Production of alpha amylase by *Bacillus licheniformis* using an economical medium. *Bioresource Technology* 87(1): 57–61.

Harwood, C. R. 1989. *Bacillus* (1st ed.). Springer Science+Business Media. https://doi.org/10.1007/978-1-4899-3502-1.

Hassan, H. & Karim, K. A. 2012. Utilization of agricultural by-products for alpha-amylase production under solid state fermentation by *Bacillus subtilis. Engineering Journal* 16(5): 177–186.

Heravi, K. M., Wenzel, M., & Altenbuchner, J. 2011. Regulation of mtl operon promoter of *Bacillus subtilis*: Requirements of its use in expression vectors. *Microbial Cell Factories* 10(1): 1–19.

Hong, K. Q., Liu, D. Y., Chen, T., & Wang, Z. W. 2018. Recent advances in CRISPR/Cas9 mediated genome editing in *Bacillus subtilis. World Journal of Microbiology and Biotechnology* 34(10): 1–9.

Hsieh, C. W., Lu, W. C., Hsieh, W. C., Huang, Y. P., Lai, C. H., & Ko, W. C. 2009. Improvement of the stability of nattokinase using γ-polyglutamic acid as a coating material for microencapsulation. *LWT – Food Science and Technology* 42(1): 144–149.

Huang, X., Li, Z., Du, C., Wang, J., & Li, S. 2015. Improved expression and characterization of a multidomain xylanase from thermoanaerobacterium aotearoense SCUT27 in *Bacillus subtilis. Journal of Agricultural and Food Chemistry* 63(28): 6430–6439. doi:10.1021/ACS.JAFC.5B01259.

Ilyas, R., Ahmed, A., Sohail, M., & Syed, M. N. 2020. Glucoamylase from a thermophilic strain of *Bacillus licheniformis* RT-17: Production and characterization. *Pakistan Journal of Botany* 52(1): 329–333.

Irfan, M., Asghar, U., Nadeem, M., Nelofer, R., & Syed, Q. 2019. Optimization of process parameters for xylanase production by Bacillus sp. in submerged fermentation. *New Pub: Elsevier* 9(2): 139–147.

Jagtap, S., Deshmukh, R. A., Menon, S., & Das, S. 2017. Xylooligosaccharides production by crude microbial enzymes from agricultural waste without prior treatment and their potential application as nutraceuticals. *Bioresource Technology* 245: 283–288.

Janeček, Š., Svensson, B., & MacGregor, E. A. 2003. Relation between domain evolution, specificity, and taxonomy of the α-amylase family members containing a C-terminal starch-binding domain. *European Journal of Biochemistry* 270(4): 635–645.

Jeong, D. E., Park, S. H., Pan, J. G., Kim, E. J., & Choi, S. K. 2015. Genome engineering using a synthetic gene circuit in *Bacillus subtilis. Nucleic Acids Research* 43(6): e42–e42.

Jiao, G., Sun, L., Qiu, L., Gao, Y., Huang, T., Tian, F., & Liu, Z. 2017. *Improving alpha-amylase activity in Bacillus subtilis fermentation broth involves performing genetic modification of genome of Bacillus subtilis strain 168, designing primer and carrying out PCR amplification to obtain PCR product* (Patent No. CN106754605-A; CN106754605-B).

John, J., Kaimal, K. K. S., Smith, M. L., Rahman, P. K. S. M., & Chellam, P. V. 2020. Advances in upstream and downstream strategies of pectinase bioprocessing: A review. *International Journal of Biological Macromolecules* 162: 1086–1099.

Kakeshtia, H., Kageyama, Y., Ara, K., Ozaki, K., & Nakamura, K. 2010. Enhanced extracellular production of heterologous proteins in *Bacillus subtilis* by deleting the C-terminal region of the SecA secretory machinery. *Molecular Biotechnology* 46(3): 250–257.

Kamble, R. D., & Jadhav, A. R. 2012. Isolation, purification, and characterization of xylanase produced by a new species of *Bacillus* in solid state fermentation. *International Journal of Microbiology* 2012: 683193.

Kambourova, M. S. & Emanuilova, E. I. 1992. Purification and general biochemical properties of thermostable pullulanase from *Bacillus stearothermophilus* G-82. *Applied Biochemistry and Biotechnology* 33(3): 193–203. doi:10.1007/BF02921835.

Kannan, T. R. & Kanagaraj, C. 2019. Molecular characteristic of α-amylase enzymes producing from Bacillus lichenformis (JQ946317) using solid state fermentation. *Biocatalysis and Agricultural Biotechnology* 20: 101240.

Kashiwabara, S.-I., Ogawa, S., Miyoshi, N., Oda, M., & Suzuki, Y. 1999. Three domains comprised in thermostable molecular weight 54,000 pullulanase of type I from *Bacillus flavocaldarius* KP1228. *Biosciences, Biotechnology and Biochemistry* 63(10): 1736–1748.

Khardziani, T., Kachlishvili, E., Sokhadze, K. et al. 2017. Elucidation of *Bacillus subtilis* KATMIRA 1933 potential for spore production in submerged fermentation of plant raw materials. *Probiotics and Antimicrobial Proteins* 9(4): 435–443.

Khardziani, T., Sokhadze, K., Kachlishvili, E., Chistyakov, V., & Elisashvili, V. 2017. Optimization of enhanced probiotic spores production in submerged cultivation of Bacillus amyloliquefaciens B-1895. *Journal of Microbiology, Biotechnology and Food Sciences* 7(2): 132–136.

Khasin, A., Alchanati, I., & Shoham, Y. 1993. Purification and characterization of a thermostable xylanase from *Bacillus stearothermophilus* T-6. *Applied and Environmental Microbiology* 59(6): 1725–1730.

Kim, C. H. & Kim, Y. S. 1995. Substrate specificity and detailed characterization of a bifunctional amylase-pullulanase enzyme from *Bacillus circulans* F-2 having two different active sites on one polypeptide. *European Journal of Biochemistry* 227(3): 687–693. doi:10.1111/J.1432-1033.1995.TB20189.X.

Kim, H. K., Choi, H. J., Kim, M. H., Sohn, C. B., & Oh, T. K. 2002. Expression and characterization of Ca^{2+}-independent lipase from *Bacillus pumilus* B26. *Biochimica et Biophysica Acta (BBA) - Molecular and Cell Biology of Lipids* 1583(2): 205–212. doi:10.1016/S1388-1981(02)00214-7.

Kimura, K. & Yokoyama, S. 2019. Trends in the application of Bacillus in fermented foods. *Current Opinion in Biotechnology* 56(1): 36–42.

Klug-Santner, B. G., Schnitzhofer, W., Vršanská, M. et al. 2006. Purification and characterization of a new bioscouring pectate lyase from *Bacillus pumilus* BK2. *Journal of Biotechnology* 121(3): 390–401.

Korneli, C., Biedendieck, R., David, F., Jahn, D., & Wittmann, C. 2013. High yield production of extracellular recombinant levansucrase by *Bacillus megaterium*. *Applied Microbiology and Biotechnology* 97(8): 3343–3353. doi:10.1007/S00253-012-4567-1/TABLES/2.

Kou, Y., Feng, R., Chen, J., et al. 2020. Development of a nattokinase–polysialic acid complex for advanced tumor treatment. *European Journal of Pharmaceutical Sciences* 145: 105241.

Kumari, M., Padhi, S., Sharma, S., Phukon, L. C., Singh, S. P., Rai, A. K. 2021. Biotechnological potential of psychrophilic microorganisms as the source of cold-active enzymes in food processing applications. *3 Biotech* 11: 1–18.

Li, J., Jiao, A., Deng, L., Rashed, M. M. A., & Jin, Z. 2018. Porous-structured extruded instant noodles induced by the medium temperature α-amylase and its effect on selected cooking properties and sensory characteristics. *International Journal of Food Science & Technology* 53(10): 2265–2272.

Li, X., Wang, Y., Lee, B. H., & Li, D. 2018. Reducing digestibility and viscoelasticity of oat starch after hydrolysis by pullulanase from Bacillus acidopullulyticus. *Food Hydrocolloids* 75: 88–94.

Li, X., Yang, X., Umar, M. et al. 2021. Expression of a novel dual-functional polypeptide and its pharmacological action research. *Life Sciences* 267: 118890.

Li, Y., Zhang, L., Ding, Z., & Shi, G. 2013. Constitutive expression of a novel isoamylase from Bacillus lentus in Pichia pastoris for starch processing. *Process Biochemistry* 48(9): 1303–1310.

Liu, G., Gu, Z., Hong, Y., Cheng, L., & Li, C. 2017. Structure, functionality and applications of debranched starch: A review. *Trends in Food Science and Technology* 63: 70–79. doi:10.1016/j.tifs.2017.03.004.

Liu, L., Liu, Y., Shin, H. D. et al. 2013. Developing *Bacillus* Spp. as a cell factory for production of microbial enzymes and industrially important biochemicals in the context of systems and synthetic biology. *Applied Microbiology and Biotechnology* 97(14): 6113–6127. doi:10.1007/s00253-013-4960-4.

Liu, X., Chen, H., Tao, H. Y., Chen, Z., Liang, X. B., Han, P., & Tao, J. H. 2020. Cloning and characterization of a novel amylopullulanase from *Bacillus megaterium* Y103 with transglycosylation activity. *Biotechnology Letters* 42(9): 1719–1726. doi:10.1007/s10529-020-02891-4.

Liu, X. & Kokare, C. 2017. Microbial Enzymes of Use in Industry. In G. Brahmachari, A. Demain, & J. Adrio (Eds), *Biotechnology of Microbial Enzymes: Production, Biocatalysis and Industrial Applications* (pp. 267–298). Academic Press.

Liu, Y., Chen, G., Wang, J., Hao, Y., Li, M., Li, Y., Hu, B., & Lu, F. 2012. Efficient expression of an alkaline pectate lyase gene from *Bacillus subtilis* and the characterization of the recombinant protein. *Biotechnology Letters* 34(1): 109–115. doi:10.1007/S10529-011-0734-1/FIGURES/5.

Liu, Y., Link, H., Liu, L., Du, G., Chen, J., & Sauer, U. 2016. A dynamic pathway analysis approach reveals a limiting futile cycle in N-acetylglucosamine overproducing *Bacillus subtilis*. *Nature Communications* 7(1): 1–9. doi:10.1038/ncomms11933.

Liu, Y. H., Lu, F. P., Li, Y. Yin, X. B., Wang, Y., & Gao, C. 2008. Characterisation of mutagenised acid-resistant alpha-amylase expressed in *Bacillus subtilis* WB600. *Applied Microbiology and Biotechnology* 78(1): 85–94. doi:10.1007/S00253-007-1287-Z/FIGURES/7.

Lu, Z., Hu, X., Shen, P. et al. 2018. A pH-stable, detergent and chelator resistant type I pullulanase from Bacillus pseudofirmus 703 with high catalytic efficiency. *International Journal of Biological Macromolecules* 109: 1302–1310.

Liu, Y., Lu, F., Chen, G., Snyder, C. L., Sun, J., Li, Y., Wang, J., & Xiao, J. 2010. High-level expression, purification and characterization of a recombinant medium-temperature α-amylase from *Bacillus subtilis*. *Biotechnology Letters* 32(1). Springer Netherlands: 119–124. doi:10.1007/S10529-009-0112-4/FIGURES/4.

Liu, Z., Zheng, W., Ge, C., Cui, W., Zhou, L., & Zhou, Z. 2019. High-level extracellular production of recombinant nattokinase in *Bacillus subtilis* WB800 by multiple tandem promoters. *BMC Microbiology* 19(1): 1–14. doi:10.1186/S12866-019-1461-3/TABLES/3.

Ma, J., Zhang, Z., Wang, B., Kong, X., Wang, Y., Cao, S., & Feng, Y. 2006. Overexpression and characterization of a lipase from *Bacillus subtilis*. *Protein Expression and Purification* 45(1): 22–29. doi:10.1016/J.PEP.2005.06.004.

Manabe, K., Kageyama, Y., Morimoto, T. et al. 2011. Combined effect of improved cell yield and increased specific productivity enhances recombinant enzyme production in genome-reduced *Bacillus subtilis* strain MGB874. *Applied and Environmental Microbiology* 77(23): 8370.

MarketWatch. 2022. *Pullulanase Market In 2022: -1.1% CAGR with Top Countries Data, Which aspirants are pulling the development of the Pullulanase Industry?*. Press Release.

Mazas, M., López, M., González, I., Bernardo, A., & Martín, R. 1997. Effects of sporulation pH on the heat resistance and the sporulation of *Bacillus cereus*. *Letters in Applied Microbiology* 25(5): 331–334.

Meng, F., Zhu, X., Nie, T., Lu, F., Bie, X., Lu, Y., Trouth, F., & Lu, Z. 2018. Enhanced expression of pullulanase in *Bacillus subtilis* by new strong promoters mined from transcriptome data, both alone and in combination. *Frontiers in Microbiology* 9: 2635. doi:10.3389/FMICB.2018.02635/BIBTEX.

Michna, R. H., Zhu, B., Mäder, U., & Stülke, J. 2016. SubtiWiki 2.0 – An integrated database for the model organism *Bacillus subtilis*. *Nucleic Acids Research* 44(D1): D654–D662.

Mine, Y., Kwan Wong, A. H., & Jiang, B. 2005. Fibrinolytic enzymes in Asian traditional fermented foods. *Food Research International* 38(3): 243–250.

Ming, Y. M., Wei, Z. W., Lin, C. Y., & Sheng, G. Y. 2010. Development of a *Bacillus subtilis* expression system using the improved Pglv promoter. *Microbial Cell Factories* 9: 55. doi: 10.1186/1475-2859-9-55.

Monteloeder. 2021. *Chinese health functional food market extension due to COVID*. URL: https://monteloeder.com/chinese-health-functional-food-market-extension-due-to-covid-19/.

Morimoto, T., Kadoya, R., Endo, K. et al. 2008. Enhanced recombinant protein productivity by genome reduction in *Bacillus subtilis*. *DNA Research: An International Journal for Rapid Publication of Reports on Genes and Genomes* 15(2): 73.

Motta, L. F., Andrade, C. C. P., & Santana, M. H. A. 2013. A Review of Xylanase Production by the Fermentation of Xylan: Classification, Characterization and Applications. In K. C. Anuj & S. S. Silvio (Eds), *Sustainable Degradation of Lignocellulosic Biomass – Techniques, Applications and Commercialization* (pp. 251–266). InTech.

Mu, G. C., Nie, Y., Mu, X. Q., Xu, Y., & Xiao, R. 2015. Single amino acid substitution in the pullulanase of Klebsiella variicola for enhancing thermostability and catalytic efficiency. *Applied Biochemistry and Biotechnology* 176(6): 1736–1745. doi:10.1007/s12010-015-1675-2.

Naidu, M. A. & Saranraj, P. 2013. Bacterial amylase: A review. *International Journal of Pharmaceutical & Biological Archives* 4(2): 274–287.

Narwal, S. K., Saun, N. K., Dogra, P., & Gupta, R. 2016. Green synthesis of isoamyl acetate via silica immobilized novel thermophilic lipase from Bacillus aerius. *Russian Journal of Bioorganic Chemistry* 42(1): 69–73.

Negi, S. & Vibha, K. 2017. Amylolytic Enzymes: Glucoamylases. In A. Pandey, S. Negi, & C. R. Soccol (Eds), *Current Developments in Biotechnology and Bioengineering - Production, Isolation and Purification of Industrial Products* (pp. 25–46). Elsevier B.V.

Obeng, E. M., Adam, S. N. N., Budiman, C., Ongkudon, C. M., Maas, R., & Jose, J. 2017. Lignocellulases: A review of emerging and developing enzymes, systems, and practices. *Bioresources and Bioprocessing* 4(1): 1–22.

Özcan, B. D. & Özcan, N. 2008. Expression of thermostable α-amylase gene from *Bacillus stearothermophilus* in various *Bacillus subtilis* strains. *Annals of Microbiology* 58(2): 265–268. doi:10.1007/BF03175327.

Ozdemir, S., Fincan, S. A., Karakaya, A., & Enez, B. 2018. A novel raw starch hydrolyzing thermostable α-Amylase produced by newly isolated Bacillus mojavensis SO-10: Purification, characterization and usage in starch industries. *Brazilian Archives of Biology and Technology* 61: 18160399.

Pagnoncelli, M. G. B., Fernandes, M. J., Rodrigues, C., & Soccol, C. R. 2017. Nattokinases. In A. Pandey, S. Negi, & C. R. Soccol (Eds), *Current Developments in Biotechnology and Bioengineering - Production, Isolation and Purification of Industrial Products* (pp. 509–526). Elsevier B.V.

Pandey, A. 2003. Solid-state fermentation. *Biochemical Engineering Journal* 13(2–3): 81–84.

Pandey, R., Ter Beek, A., Vischer, N. O. E., Smelt, J. P. P. M., Brul, S., & Manders, E. M. M. 2013. Live cell imaging of germination and outgrowth of individual *Bacillus subtilis* spores; the effect of heat stress quantitatively analyzed with SporeTracker. *PLOS ONE* 8(3): e58972.

Park, S. A., Bhatia, S. K., Park, H. A. et al. 2021. *Bacillus subtilis* as a robust host for biochemical production utilizing biomass. *Critical Reviews in Biotechnology* 41(6): 827–848.

Parkouda, C., Nielsen, D. S., Azokpota, P. et al. 2009. The microbiology of alkaline-fermentation of indigenous seeds used as food condiments in Africa and Asia. *Critical Reviews in Microbiology* 35(2): 139–156.

Phukon, L. C., Abedin, M. M., Chourasia, R., Sahoo, D., Parameswaran, B., & Rai, A. K. 2020. Production of xylanase under submerged fermentation from Bacillus firmus HS11 isolated from Sikkim Himalayan region. *Indian Journal of Experimental Biology* 58: 563–570.

Phukon, L.C., Chourasia, R., Padhi, S. et al. 2022. Cold-adaptive traits identified by comparative genomic analysis of a lipase-producing Pseudomonas sp. HS6 isolated from snow-covered soil of Sikkim Himalaya and molecular simulation of lipase for wide substrate specificity. *Current Genetics*: 1–17.

Posada-Uribe, L. F., Romero-Tabarez, M., & Villegas-Escobar, V. 2015. Effect of medium components and culture conditions in *Bacillus subtilis* EA-CB0575 spore production. *Bioprocess and Biosystems Engineering* 38(10): 1879–1888.

Priyadarshini, S., Pradhan, S. K., & Ray, P. 2020. Production, characterization and application of thermostable, alkaline α-amylase (AA11) from *Bacillus cereus* strain SP-CH11 isolated from Chilika Lake. *International Journal of Biological Macromolecules* 145: 804–812.

Radeck, J., Kraft, K., Bartels, J. et al. 2013. The Bacillus BioBrick Box: Generation and evaluation of essential genetic building blocks for standardized work with *Bacillus subtilis. Journal of Biological Engineering* 7(1): 1–17.

Rajagopalan, G. & Krishnan, C. 2008. α-Amylase production from catabolite derepressed *Bacillus subtilis* KCC103 utilizing sugarcane bagasse hydrolysate. *Bioresource Technology* 99(8): 3044–3050.

Rana, N., Verma, N., Vaidya, D., Parmar, Y. S., Dipta, B., & Bhawna Dipta, C. 2017. Application of bacterial amylase in clarification of juices and bun making. *Journal of Pharmacognosy and Phytochemistry* 6(5): 859–864.

Rani, S., & Jagtap, S. 2019. Acceleration of Swiss cheese ripening by microbial lipase without affecting its quality characteristics. *Journal of Food Science and Technology* 56(1): 497.

Rashid, R., & Sohail, M. 2021. Xylanolytic Bacillus species for xylooligosaccharides production: A critical review. *Bioresources and Bioprocessing* 8(1): 1–14.

Reque, P. M., Pinilla, C. M. B., Gautério, G. V., Kalil, S. J., & Brandelli, A. 2019. Xylooligosaccharides production from wheat middlings bioprocessed with *Bacillus subtilis. Food Research International* 126: 108673.

Rungrassamee, W., Kingcha, Y., Srimarut, Y., Maibunkaew, S., Karoonuthaisiri, N., & Visessanguan, W. 2014. Mannooligosaccharides from copra meal improves survival of the Pacific white shrimp (Litopenaeus vannamei) after exposure to Vibrio harveyi. *Aquaculture* 434: 403–410.

Salgado, C. A., dos Santos, C. I. A., & Vanetti, M. C. D. 2022. Microbial lipases: Propitious biocatalysts for the food industry. *Food Bioscience* 45: 101509.

Salim, A. A., Grbavčić, S., Šekuljica, N. et al. 2017. Production of enzymes by a newly isolated Bacillus sp. TMF-1 in solid state fermentation on agricultural by-products: The evaluation of substrate pretreatment methods. *Bioresource Technology* 228: 193–200.

Samoylova, Y. V., Piligaev, A. V., Sorokina, K. N., & Parmon, V. N. 2017. Enzymatic interesterification of sunflower oil and hydrogenated soybean oil with the immobilized bacterial recombinant lipase from *Geobacillus stearothermophilus* G3. *Biocatalysis* 9(1): 62–70.

Sarvas, M., Harwood, C. R., Bron, S., & Van Dijl, J. M. 2004. Post-translocational folding of secretory proteins in Gram-positive bacteria. *Biochimica et Biophysica Acta (BBA) – Molecular Cell Research* 1694(1–3): 311–327.

Sathiyamoorthy, M., Debaprasad, D., & Sharma, D. K. 2021. *Media composition from agricultural byproducts for production of alkaline alpha-amylase enzyme from Bacillus licheniformis comprises cheap source of nitrogen (e.g. corn steep liquor), soyabean, mustard seed and cotton seed* (Patent No. IN202131002187-A).

Saxena, R. & Singh, R. 2011. Amylase production by solid-state fermentation of agro-industrial wastes using Bacillus sp. *Brazilian Journal of Microbiology* 42(4): 1334–1342.

Saxena, R. K., Dutt, K., Agarwal, L., & Nayyar, P. 2007. A highly thermostable and alkaline amylase from a *Bacillus* sp. PN5. *Bioresource Technology* 98(2): 260–265. doi:10.1016/J.BIORTECH.2006.01.016.

Schallmey, M., Singh, A., & Ward, O. P. 2004. Developments in the use of Bacillus species for industrial production. *Canadian Journal of Microbiology* 50(1): 1–17.

Schumann, W. 2007. Production of Recombinant Proteins in *Bacillus Subtilis.* In A. I. Laskin, S. Sariaslani, & G. M. Gadd (Eds), *Advances in Applied Microbiology* (1st ed., pp. 137–191). Elsevier.

Sella, S. R. B. R., Vandenberghe, L. P. de S., & Soccol, C. R. 2014. Life cycle and spore resistance of spore-forming Bacillus atrophaeus. *Microbiological Research* 169(12): 931–939.

Sepahy, A. A., Ghazi, S., & Sepahy, M. A. 2011. Cost-effective production and optimization of alkaline xylanase by indigenous Bacillus mojavensis AG137 fermented on agricultural waste. *Enzyme Research* 2011(1) doi: 10.4061/2011/593624.

Sharma, D., Sharma, G., & Mahajan, R. 2019. Development of strategy for simultaneous enhanced production of alkaline xylanase-pectinase enzymes by a bacterial isolate in short submerged fermentation cycle. *Enzyme and Microbial Technology* 122: 90–100.

Shin, Y. C., Park, C., Shin, D. K. et al. 2020. *Composition used as medicine and food composition for preventing or treating metabolic syndrome including i.e. obesity, diabetes, arteriosclerosis, hypertension, and hyperlipidemia, comprises Bacillus coagulans fermentation enzyme product* (Patent No. KR2163993-B1).

Singh, R. S. & Singh, T. 2019. Inulinase and Pullulanase Production from Agro-industrial Residues. In *Industrial Biotechnology.* https://doi.org/10.1515/9783110563337-001.

Song, W., Nie, Y., Mu, X. Q., & Xu, Y. 2016. Enhancement of extracellular expression of *Bacillus naganoensis* pullulanase from recombinant *Bacillus subtilis*: Effects of promoter and host. *Protein Expression and Purification* 124(August): 23–31. doi:10.1016/j.pep.2016.04.008.

Souza, P. M. & Magalhães, P. O. 2010. Application of microbial α-amylase in industry – A review. *Brazilian Journal of Microbiology* 41: 850–861.

Srivastava, P., Appu Rao, A. R., & Kapoor, M. 2014. Structural insights into the thermal stability of endo-mannanase belonging to family 26 from Bacillus sp. CFR1601 (580.2). *The FASEB Journal* 28(S1). doi: 10.1096/FASEBJ.28.1_SUPPLEMENT.580.2.

Srivastava, P. K., & Kapoor, M. 2017. Production, properties, and applications of endo-β-mannanases. *Biotechnology Advances* 35(1): 1–19.

Steinkraus, K. H. 2004. *Industrialization of Indigenous Fermented Foods* (O. R. Fennema, Y. H. Rui, M. Karel, P. Walstra, & J. R. Whitaker (Eds), 2nd ed.). Marcel Dekker, Inc.

Takci, H. A. M. & Turkmen, F. U. 2016. Extracellular pectinase production and purification from a newly isolated *Bacillus subtilis* strain. *International Journal of Food Properties* 19(11): 2443–2450.

Tang, T., Yuan, C., Hwang, H. T., Zhao, X., Ramkrishna, D., Liu, D., & Varma, A. 2015. Engineering surface hydrophobicity improves activity of Bacillus thermocatenulatus lipase 2 enzyme. *Biotechnology Journal* 10(11): 1762–1769.

Terlabie, N. N., Sakyi-Dawson, E., & Amoa-Awua, W. K. 2006. The comparative ability of four isolates of *Bacillus subtilis* to ferment soybeans into dawadawa. *International Journal of Food Microbiology* 106(2): 145–152.

Thakur, M., Sharma, N., Rai, A. K., & Singh, S. P. 2021. A novel cold-active type I pullulanase from a hot-spring metagenome for effective debranching and production of resistant starch. *Bioresource Technology* 320: 124288.

Tian, L. 2014. *Preparing composition containing probiotics and plant enzymes e.g. lipase, by fermenting inoculum comprising e.g. Streptococcus thermophilus and Bacillus amyloliquefaciens and fermenting obtained liquid in substrate comprising e.g. fruit* (Patent No. CN103876155-A; CN103876155-B).

Trabelsi, S., Ben Mabrouk, S., Kriaa, M. et al. 2019. The optimized production, purification, characterization, and application in the bread making industry of three acid-stable alpha-amylases isoforms from a new isolated *Bacillus subtilis* strain US586. *Journal of Food Biochemistry* 43(5): e12826.

Vaikundamoorthy, R., Rajendran, R., Selvaraju, A., Moorthy, K., & Perumal, S. 2018. Development of thermostable amylase enzyme from *Bacillus cereus* for potential antibiofilm activity. *Bioorganic Chemistry* 77: 494–506.

Vandenberghe, L. P. S., Pandey, A., Carvalho, J. C. et al. 2021. Solid-state fermentation technology and innovation for the production of agricultural and animal feed bioproducts. *Systems Microbiology and Biomanufacturing* 1(2): 142–165.

Verma, D. & Satyanarayana, T. 2013. Production of cellulase-free xylanase by the recombinant *Bacillus subtilis* and its applicability in paper pulp bleaching. *Biotechnology Progress* 29(6): 1441–1447. doi:10.1002/BTPR.1826.

Vidyalakshmi, R., Paranthaman, R., & Indhumathi, J. 2009. Amylase production on submerged fermentation by Bacillus spp. *World Journal of Chemistry* 4(1): 89–91.

Wang, C. H., Liu, X. L., Huang, R. B., He, B. F., & Zhao, M. M. 2018. Enhanced acidic adaptation of *Bacillus subtilis* Ca-independent alpha-amylase by rational engineering of pKa values. *Biochemical Engineering Journal* 139: 146–153.

Wang, Y., Chen, S., Zhao, X., Zhang, Y., Wang, X., Nie, Y., & Xu, Y. 2019. Enhancement of the production of *Bacillus naganoensis* pullulanase in recombinant *Bacillus subtilis* by integrative expression. *Protein Expression and Purification* 159(July): 42–48. doi:10.1016/J.PEP.2019.03.006.

Wang, Y., Weng, J., Waseem, R., Yin, X., Zhang, R., & Shen, Q. 2012. *Bacillus subtilis* genome editing using ssDNA with short homology regions. *Nucleic Acids Research* 40(12): e91–e91.

Wei, W., Ma, J., Guo, S., & Wei, D. Z. 2014. A type I pullulanase of *Bacillus cereus* Nws-Bc5 screening from stinky tofu brine: Functional expression in *Escherichia coli* and *Bacillus subtilis* and enzyme characterization. *Process Biochemistry* 49(11). Elsevier Ltd: 1893–1902. doi:10.1016/j.procbio.2014.07.008.

Wei, X., Zhou, Y., Chen, J., Cai, D., Wang, D., Qi, G., & Chen, S. 2015. Efficient expression of nattokinase in *Bacillus licheniformis*: Host strain construction and signal peptide optimization. *Journal of Industrial Microbiology and Biotechnology* 42(2): 287–295. doi:10.1007/S10295-014-1559-4.

Weng, M., Deng, X., Bao, W. et al. 2015. Improving the activity of the subtilisin nattokinase by site-directed mutagenesis and molecular dynamics simulation. *Biochemical and Biophysical Research Communications* 465(3): 580–586.

Widner, B., Thomas, M., Sternberg, D., Lammon, D., Behr, R., & Sloma, A. 2000. Development of marker-free strains of *Bacillus subtilis* capable of secreting high levels of industrial enzymes. *Journal of Industrial Microbiology and Biotechnology* 25(4): 204–212.

Wu, J., Su, L., & Yang, Y. 2019. *New genetically engineered strain useful for preparation of maltotetraose amylase or maltotetraose amylase containing products comprises Bacillus subtilis as host and pHY-300PLK as an expression vector* (Patent No. CN110144320-A).

Xia, J., & Liu, S. 2020. *OUR-based fed-batch fermentation method useful for producing mesophilic alpha-amylase from Bacillus subtilis, comprises e.g. detecting trend of changes in OUR in exhaust gas online in batch culture fermentation of Bacillus subtilis* (Patent No. CN112029750-A).

Xia, W., Zhang, K., Su, L., & Wu, J. 2021. Microbial starch debranching enzymes: Developments and applications. *Biotechnology Advances* 50: 107786.

Xie, X., Ban, X., Gu, Z., Li, C., Hong, Y., Cheng, L., & Li, Z. 2020. Insights into the thermostability and product specificity of a maltooligosaccharide-forming amylase from *Bacillus stearothermophilus* STB04. *Biotechnology Letters* 42(2): 295–303.

Yang, H., Liu, L., Li, J., Du, G., & Chen, J. 2011. Heterologous expression, biochemical characterization, and overproduction of alkaline α-amylase from *Bacillus alcalophilus* in *Bacillus subtilis*. *Microbial Cell Factories* 10(1): 77. doi:10.1186/1475-2859-10-77/FIGURES/6.

Yang, H., Ma, Y., Zhao, Y., Shen, W., & Chen, X. 2020. Systematic engineering of transport and transcription to boost alkaline α-amylase production in *Bacillus subtilis*. *Applied Microbiology and Biotechnology* 104(7): 2973–2985.

Yang, S., Kang, Z., Cao, W., Du, G., & Chen, J. 2016. Construction of a novel, stable, food-grade expression system by engineering the endogenous toxin-antitoxin system in *Bacillus subtilis*. *Journal of Biotechnology* 219: 40–47.

Yang, S., Yan, Q., Bao, Q., Liu, J., & Jiang, Z. 2017. Expression and biochemical characterization of a novel type I pullulanase from *Bacillus megaterium*. *Biotechnology Letters* 39(3): 397–405. doi:10.1007/s10529-016-2255-4.

Zareian, S., Khajeh, K., Ranjbar, B., Dabirmanesh, B., Ghollasi, M., & Mollania, N. 2010. Purification and characterization of a novel amylopullulanase that converts pullulan to glucose, maltose, and maltotriose and starch to glucose and maltose. *Enzyme and Microbial Technology* 46(2): 57–63. doi:10.1016/j.enzmictec.2009.09.012.

Zhang, K., Duan, X., & Wu, J. 2016. Multigene disruption in undomesticated *Bacillus subtilis* ATCC 6051a using the CRISPR/Cas9 system. *Scientific Reports* 6. doi: 10.1038/SREP27943.

Zhang, K., Su, L., & Wu, J. 2020. Recent advances in recombinant protein production by *Bacillus subtilis*. *Annual Review of Food Science and Technology* 11: 295–318.

Zhang, S. Y., Guo, Z. W., Wu, X. L., Ou, X. Y., Zong, M. H., & Lou, W. Y. 2020. Recombinant expression and characterization of a novel cold-adapted type I pullulanase for efficient amylopectin hydrolysis. *Journal of Biotechnology* 313: 39–47.

Zhang, W., Lou, K., & Li, G. 2010. Expression and characterization of the dictyoglomus thermophilum Rt46B.1 xylanase gene (XynB) in *Bacillus subtilis*. *Applied Biochemistry and Biotechnology* 160(5): 1484–1495. doi:10.1007/S12010-009-8634-8/TABLES/2.

Zhang, W., Yang, M., Yang, Y., Zhan, J., Zhou, Y., & Zhao, X. 2016. Optimal secretion of alkali-tolerant xylanase in *Bacillus subtilis* by signal peptide screening. *Applied Microbiology and Biotechnology* 100(20): 8745–8756.

Zhang, X., Al-Dossary, A., Hussain, M., Setlow, P., & Li, J. 2020. Applications of *Bacillus subtilis* spores in biotechnology and advanced materials. *Applied and Environmental Microbiology* 86(17).

Zhang, X., Tong, Y., Wang, J., Lyu, X., & Yang, R. 2021. Screening of a *Bacillus subtilis* strain producing both nattokinase and milk-clotting enzyme and its application in fermented milk with thrombolytic activity. *Journal of Dairy Science* 104(9): 9437–9449.

Zhang, X. Z. & Zhang, Y. H. P. 2011. Simple, fast and high-efficiency transformation system for directed evolution of cellulase in *Bacillus subtilis*. *Microbial Biotechnology* 4(1): 98–105.

Zhou, L., Hao, N., Li, X. et al. 2020. Nattokinase mitigated dextran sulfate sodium-induced chronic colitis by regulating microbiota and suppressing tryptophan metabolism via inhibiting IDO-1. *Journal of Functional Foods* 75: 104251.

Zhou, X., Liu, L., & Zeng, X. 2021. Research progress on the utilisation of embedding technology and suitable delivery systems for improving the bioavailability of nattokinase: A review. *Food Structure* 30: 100219.

Zhu, F. M., Ji, S. Y., Zhang, W. W., Li, W., Cao, B. Y., & Yang, M. M. 2008. Development and application of a novel signal peptide probe vector with PGA as reporter in *Bacillus subtilis* WB700: Twenty-four tat pathway signal peptides from *Bacillus subtilis* were monitored. *Molecular Biotechnology* 39(3): 225–230.

Zou, C., Duan, X., & Wu, J. 2014. Enhanced extracellular production of recombinant *Bacillus deramificans* pullulanase in *Escherichia coli* through induction mode optimization and a glycine feeding strategy. *Bioresource Technology* 172: 174–179. doi:10.1016/j.biortech.2014.09.035.

Zweers, J. C., Barák, I., Becher, D. et al. 2008. Towards the development of *Bacillus subtilis* as a cell factory for membrane proteins and protein complexes. *Microbial Cell Factories* 7(1): 1–20.

Section III

Specific Microbial Enzymes and Their Application in Functional Foods Production

5 Microbial Production and Application of Pullulanases

Krishna Gautam[1], *Poonam Sharma*[2], *Pallavi Gupta*[3], and *Vivek Kumar Gaur*[1,4]

[1] Centre for Energy and Environmental Sustainability, Lucknow, Uttar Pradesh, India

[2] Department of Bioengineering, Integral University, Lucknow, Uttar Pradesh, India

[3] Bioscience and Biotechnology Department, Banasthali University, Jaipur, Rajasthan, India

[4] School of Energy and Chemical Engineering, UNIST, Ulsan, Republic of Korea

CONTENTS

5.1 INTRODUCTION

Enzymatic reactions, in which enzymes function as biocatalysts, involve a major portion of biotechnology. The novel enzyme-based approaches are being developed to enhance the production and quality of value-added compounds on a commercial scale (Kumari et al., 2021). Currently, microbial synthesis of enzymes is being explored more because of the convenience, cost-effectiveness, short duration, consistency, range of catalytic properties, and higher yield of the product (Razzaq et al., 2019). For thousands of years, the kinetics of enzymes derived from microbes have been studied in various industries such as cheese production, wine production, and beer processing.

DOI: 10.1201/9781003311164-8

Pullulanase is a debranching enzyme for starch belonging to the α-amylase family. Pullulan, glycogen, amylopectin, and β-limited dextrin are a few examples of starch-type polysaccharides which specifically cleave α-1,6-glycosidic bonds (Xu et al., 2021). The complex polysaccharides polymeric starch was extracted from food waste (potato peels), followed by enzymatic debranching using PulM, a type 1 pullulanase, and it was discovered that type III resistant starch (RS3) was produced by the retrogradation of debranched starch The debranching capability of pullulanase may aid in producing starch-lipid complexes by increasing the unbound amylose content of the starch. As a response, the fraction of resistant starch increased, as did the overall resistance to digestive enzyme activity (Tu et al., 2021). Pullulanases have two terminal domains: the N-terminal domain (catalytic) and the C-terminal domain. Carbohydrate-binding module (CBM) proteins such as CBM68, CBM48, and CBM41 are found near the N-terminus of pullulanases. One of the roles of these CBM domains is to bind enzymes to polysaccharide substrates and perform enzyme-based hydrolysis (Janeček et al., 2017). In addition, specific carbohydrate-binding modules may also serve as a framework for the catalytic core, allowing enzymes to remain in their active state (Armenta et al., 2017). Therefore, sugar-based syrup, resistant starch, and beer result in higher product quality, productivity, and production costs (Xu et al., 2021). However, just a few examples fulfil the needs of industrial applications (Meng et al., 2021).

Pullulanase has recently been identified from *Bacillus acidopullulyticus, Thermotoganeapolitana, Thermococcus Marinus, Geobacillusthermocatenulatus,* and *Klebsiella variicola*, among others (Wang et al., 2019a). The most common strategy for increasing pullulanase productivity is to create recombinant microbes with high expression efficiency. However, pullulanases from the wild-type strain has poor yield, unstable enzymatic activities, and difficult growing conditions. These make it unsuitable for large-scale commercialization. Consequently, adopting recombinant technology to improve the extracellular production of pullulanase is required (Meng et al., 2021). The genes responsible for pullulanase production are cloned and expressed using the cloning approach from various microorganisms. These include bacterial strains such as *Klebsiella variicola, Raoultella planticola, Bacillus deramificans, Bacillus naganoensis, Bacillus flavothermus*, and other *Bacillus sp.* (Chen et al., 2013; Duan et al., 2013; Hii et al., 2009; Li et al., 2012; Shankar et al., 2014; Song et al., 2016). *Bacillus subtilis, Escherichia coli,* and *Pichia pastoris* have all been used to express pullulanases during recombination (Xu et al., 2021). Among them, *B. subtilis* has been identified as a safe strain with strong protein secretion abilities.

It is commonly recognized that various microbial (bacterial, viral, and fungal) strains can produce enzymes. On the other hand, researchers are interested in recombinant biocatalysts with high productivity and thermostability, all of which affect their potential for industrial applications. Furthermore, a more mechanistic understanding of enzyme characterization is necessary to design a de novo biocatalyst with more predicted features. As a result, efforts should be made to employ low-cost substrates for enzyme manufacturing, which would improve bioprocess economics. The production methodologies from wild-type and recombinant strains and the primary factors affecting the activity of pullulanases enzyme are comprehensively covered in this chapter. We also discuss how waste might be used as a substrate for pullulanase synthesis and large-scale production applications in industries.

5.2 MICROBIAL PRODUCTION OF PULLULANASES

Pullulanase was first isolated from mesophilic bacterium *Klebsiella pneumonia* by Bender and Wallenfels (1961). Pullalanases are widely distributed among different genera of bacteria but less common in fungi. Till date, pullulanases have been discovered among wide ranges of bacterial species such as *Geobacillus thermoleovorans* (Ayadi et al., 2008), *Raoultella planticola* (Hii et al., 2012a), *Thermotoga neapolitana* (Kang et al., 2011), *Clostridium thermosulfurogenes* (Ramesh et al., 2001), *Anaerobran cagottschakii* (Bertoldo et al., 2004) (Table 5.1) and even from the

TABLE 5.1
Pullulanases Derived from Different Microorganisms and Their Activities

Microorganisms	Type of Pullulanases	Substrate	Fermentation Strategy	Enzyme Activity	Reference
Anoxybacillus sp. LM18-11	Type I	Pullulan	-	750 U/mg	Xu et al. (2014)
E. coli BL21	Type I	Pullulan	Fed-batch fermentation	1567.9 U/ml	Zou et al. (2014)
Klebsilla aerogenes NCIM 2239	Type II	Starch as carbon source	Submerged fermentation	92.76 U/ml	Prabhu et al. (2018)
Klebsilla aerogenes NCIM 2239	Type I	Rice Bran + Corn Bran	Solid substrate fermentation	151.54 U/ml	Prabhu. et al. (2018)
Bacillus licheniformis (Bs18)	Type I	Rice Bran + Corn Bran	Solid substrate fermentation	24 U/mg	Khalaf and Aldeen (2013)
Bacillus licheniformis (Bs18)	Type II	Starch+ Pullulan	Liquid-state fermentation	10.5 U/mg	Khalaf and Aldeen (2013)
Bacillus acidopullulyticus	Type I	Pullulan	-	1096 U/mg	Chen et al. (2015)
Bacillus methanolicus PB1	Type I	Pullulan	-	292 U/mg	Zhang et al. (2020a)
Anoxybacillus sp. SK3-4	Type I	Pullulan	-	10 U/mg	Kahar et al. (2016)
Geobacillusthermocatenulatus NP33	Type II	Starch + Pullulan	-	851 U/mg	Nisha and Satyanarayana (2013)
Aspergillus flavus	Type II	Wheat bran	Solid-state fermentation	648.0 U/g	Naik et al. (2022)
Aspergillus sp.	Type II	Wheat bran	Solid-state fermentation	67 U/g	Naik et al. (2019)
Bacillus subtilis BS001	Type II	Corn starch and Corn steep liquor	Batch fermentation	1743 U/ml	Meng et al. (2021)
B. subtilis BS001	Type II	Corn starch and Corn steep liquor	Fed-batch fermentation	896 U/ml,	Meng et al. (2021)
Bacillus subtilis WB600	Type I	Pullulan	Shake flask fermentation	60.85 U/ml	Wang et al. (2019b)
Bacillus subtilis WB800	Type I	Pullulan	Batch fermentation	84.54 U/ml	Zhang et al. (2020b)
Bacillus subtilis WB800	Type I	Pullulan	Fed-batch fermentation	102.75 U/ml	Zhang et al. (2020b)

uncultured environment (Nie et al., 2013). Thereafter, pullulanase production was discovered in and characterized from various extremophilic microorganisms such as mesophilic *Bacillus cereus*, psychrotrophic *Exiguobacterium sp.* SH3, hyperthermophilic *Pyrococcusyayanosii* CH1, and thermophilic *Geobacillus thermocatenulatus* (Pang et al., 2020). However, the exploitation of these strains could not meet the demand of pullulanase production at an industrial scale due to insufficient cognizance of bioprocess. The development in fermentation technology was therefore aimed specifically at producing pullulanase using selected producers. Pullulanase enzyme was successfully purified from the *K. aerogenes* NCIM2239 strain and found stable at a higher range of temperatures and pH levels. Optimization of various carbon/nitrogen sources and comparative investigation between solid-state fermentation (SSF) and submerged fermentation (SmF) concluded that in the presence of starch as a carbon source and peptone as a nitrogen source under SSF, a higher yield of pullulanase by *K. aerogenes* NCIM2239 resulted (Prabhu et al., 2018).

Pullulanase has applications in the detergent industry as effective additives under alkaline conditions. Owing to its increased industrial application, a highly effective alkaline pullulanase was isolated and characterized from *Bacillus halodurans*. The enzyme showed highest activity at pH 10.0 in the presence of 1.5% pullulan at 37°C thereby suggesting its application in the detergent industry and other biotechnological functions (Asha et al., 2013). Similarly, thermostable extracellular pullulanase was also produced by a thermophilic *Bacillus sp.* strain AN-7. The isolated enzymes were evaluated for their stability at high temperatures and it was found that the thermophile-derived enzymes had the highest activity at 90°C and rapidly reduced activity at temperatures >105°C (Kunamneni and Singh, 2006). Additionally, a gene named pulN, encoded a cold-adapted pullulanase, was obtained from the strain *Paenibacillus polymyxa* Nws-pp2 suggesting this strain as a prominent candidate for food industries (Wei et al., 2015). Furthermore, newer approaches like genetic engineering have involved the construction of recombinant strains to enhance the expression of the pullulanase enzyme in engineered strains thereby strengthening the production on a large scale.

Pullulanase production from a recombinant strain of *Bacillus subtilis* were systemically investigated in batch and fed-batch fermentation. After optimization of feeding ingredients, feeding methods, and feeding concentration, it was found that production of pullulanase was enhanced and the enzymatic activity of 102.75 U ml^{-1} was recorded, which has provided the prerequisite for the strategy of constructing engineered strain (Zhang et al., 2020b). Pullulanase production from *Bacillus flavothermus* KWF-1 in batch and fed-batch culture were investigated in the presence of sago starch and, when compared, it was found that an exponential feeding system yielded higher amounts of enzyme than the batch system (Shankar et al., 2014). Due to their involvement in the food industry, they have been isolated from amylolytic lactic acid bacteria (ALAB). *Lactococcus lactis* IBB500, an amylolytic strain, was employed for the production of extracellularly produced pullulanase type I, which suggested its exploitation as a producer of an effective de-branching agent (Waśko et al., 2011). Only few studies have been reported for the production of pullulanase with native strains of fungi because of low productivity (Naik et al., 2019). With the rising demand for pullulanase, the search for high yielding microbes has become important. For example, *Aureobasidium pullulans* was found to produce pullulanase on the third day of incubation. Statistical analysis revealed that the optimum parameters for enhanced production of pullulanase enzyme were with 100 g/L sucrose as carbon source and 2 g/L of sodium nitrate as nitrogen source at 25°C for 5 days (Moubasher et al., 2013). A fungus, *Hypocreajecorina* QM9414, produced extracellular pullulanase in shake flask culture. Pullulanase activity was shown to be higher in fermentation media containing 0.1% amylopectin at 30°C and a pH of 6.5 (Orhan et al., 2014). Further, *Aerobasidium pullulans* was found to produce pullulanase in the presence of *Eichhornia crassipes* (water hyacinth) and *Pistia stratiotes* (water cabbage) as substrates. Agricultural waste such as rice bran, banana peel, sugarcane bagasse, wheat bran, mosambi peel, and orange peel were used to screen *Aspergillus sp.* for the synthesis of pullulanase. It was found that the fungus was able to produce 65.33±2.08 U/gds of enzyme when wheat bran was used as a substrate (Naik et al., 2019).

5.3 ENGINEERING APPROACHES FOR PULLULANASE PRODUCTION

Pullulanases are the enzymes which aid in reducing the reaction time in the starch conversion process and in the use of glucoamylase. Due to the industrial importance of these enzymes it is imperative to search for better production abilities with enhanced enzyme activity. Therefore, molecular biology techniques and protein engineering approaches including truncation and mutagenesis had served as an important tool for improving the substrate specificity, catalytic activity, and thermostability (Figure 5.1) (Nisha and Satyanarayana, 2016). This section highlights the recent biotechnological advances in the production of more effective and efficient pullulanase.

5.3.1 Recombinant Microorganism

The limited productivity of wild-type strains and the development of acidophilic/thermostable pullulanases have been limitations for the fulfilment of commercial applications such as starch

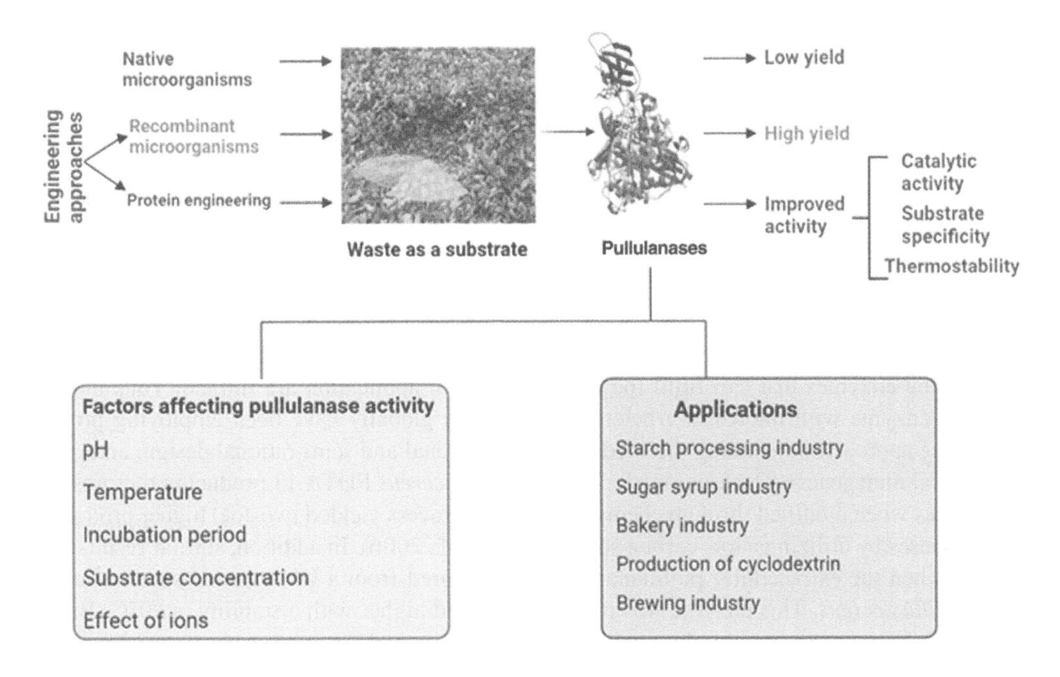

FIGURE 5.1 Effect of engineering approaches on pullulanase production and its application.

saccharification, sugar syrup, and beer production in large-scale synthesis of pullulanases. Researchers have preferred the development of recombinant strains with high expression efficiency to produce pullulanases. Presently, microorganisms such as *Pichia pastoris*, *Bacillus subtilis*, and *Escherichia coli* have all been used as hosts in heterologously producing pullulanases (Xu et al., 2021). The *E. coli* expression system has been widely employed among these species. This is due to its ability to produce high levels of recombinant protein production and a clear genetic background. Another key point to remember is that the majority of pullulanases are bacterial in origin, which enables *E. coli* to become a desirable host.

Thermus thermophilus HB27 derived pullulanase gene was cloned and expressed in *Escherichia coli* BL21 (DE3) by using two different expression systems. The activity of pullulanase was recorded to be 13.8±0.4 U/ml when expressed in pHsh vector. The activity was recorded to be 3.6-fold higher when compared with the expression in pET-28a. Furthermore, maximal activity of the isolated enzyme was found to be optimum at 70°C and 6.5 pH. By attacking pullulans α-1,4 and α-1,6 glycosidic linkages, the isolated enzyme produces a combination of glucose, maltose, and maltotriose and is advised for use in the starch saccharification process (Wu et al., 2014). Meanwhile, in recombinant *E. coli* strains, process parameters are also important in the production of the pullulanase enzyme. In 2014, Zou and colleagues optimized process parameters for the production of pullulanases from *Bacillus deramificans* in *E. coli*. The concentration of inducer (lactose) and glycine was optimized. This strategy has led to a 1.2-fold increase in the total pullulanases activity 2523.5 U/ml and also the extracellular pullulanases activity increases to 22.6-fold, i.e., 1567.9 U/ml in a 3L fermenter (Zou et al., 2014). Similarly, the pullulanase derived from *Geobacillus castophilus* species was found to be significant or overexpressed in *Bacillus subtilis*, with an extracellular yield of 0.08 mg/ml and activity of 64.8 U/mg that was higher than the wild-type strain (Xu et al., 2021). In another study the *P. pastoris* was engineered to express the amylopullulanases from thermophilic *Geobacillus thermoleovorans*. They used two different promoter systems and it was found that Gap promoter showed high expression as compared to the AOX1 promoter. Interestingly the recombinant amylopullulanase expressed in *P. pastoris* showed thermostability, higher starch saccharification efficiency, and substrate specificity as compared to the ones expressed in *E. coli* (Nisha and Satyanarayana, 2016).

These studies have led to the suggestion that few factors are crucial in terms of getting high pullulanase activity and production from recombinant strains. These may include the selection of specific robust enzymes from the source organism, the expression system, the promoters, and the conditions under which the protein is expressed (Nisha and Satyanarayana, 2016; Xu et al., 2021). However, overcoming these barriers, the recombinant strains have proven to be more efficient as compared to the wild-type strains for the production of pullulanases with better activity, stability, and productivity.

5.3.2 Protein Engineering

The protein database contains a diverse list of enzymes that have been characterized over time. However, the enzymes that can fulfil the desired practical application are limited. Thus in order to get the enzyme with improved efficiency, researchers globally have been employing protein-engineering approaches including directed-evolution, rational and semi-rational design, and physical/chemical mutagenesis (Tarafdar et al., 2021). *Bacillus cereus* FDTA 13 producing thermostable pullulanases when modified through chemical mutation process yielded two-fold higher production of pullulanases by utilizing a low carbon source (Nair et al., 2006). In addition, similar results were obtained when the extracellular pullulanases were measured from a UV-mutated strain of mesophilic *Bacillus cereus.* This enzyme was produced twofold higher with a stability at 70°C (Bakshi et al., 1992). In a similar context, directed evolution can be one of the approaches to make variants. This method has one disadvantage of generating a huge mutant library that makes it laborious to screen and identify the mutant of interest. In pullulanases, this challenge is overcome by the use of a chromogenic substrate that serves as a high-throughput screening method.

To date pullulanase engineering has been centred on rational/semi-rational design employing structure or sequence information through domain truncation or site-directed mutagenesis. The structure-guided consensus approach was utilized to study the site-directed mutagenesis in specific regions of *Bacillus deramificans* pullulanases. The D503F, D437H, and D503Y mutant and wild protein had the same optimal pH of 4.5 and a temperature of 55°C. However, at 60°C, mutants' half-lives were twice as long as those of the wild-type enzyme. Interestingly the double mutant showed a 4.3-fold increase in the half-life at 60°C. The mutation has led to a 10–100% increase in the K_{cat}/K_m values. It was found that compared to the wild-type enzyme, the mutant enzyme has better glucose yield during starch hydrolysis and improved reaction rate (Duan et al., 2013). Pullulanase (PulB) from *Bacillus naganoensis* with high enzymatic activity has improved thermostability when using rational protein design. The three single-site mutants when screened for their thermostability were found to have improved compared to the wild-type PulB. The T_m values for the mutants were 0.9°C, 0.4°C, and 1.8°C, i.e., higher than the control. Moreover, the mutant showed better K_{cat} and K_{cat}/K_m in addition to improved catalytic efficiency and thermostability (Chang et al., 2016). *B. naganoensis* pullulanase (BnPul) structural analysis revealed that its catalytic pocket has two sites that are critical for enzymatic activity. Mutant generated by employing triple-code saturation mutagenesis revealed a 1.8-fold and 1.7-fold increase in the K_{cat} and K_{cat}/K_m values as compared to the wild type. This mutant was generated by modifying the hydrogen bond affecting the loop region which in turn increased the acid resistance and thermostability of the mutant enzyme (Wang et al., 2018).

The pullulanase enzymes contain a conserved carbohydrate-binding module at N-terminal and a C-terminus whose function is not clear and is non-conserved (Xu et al., 2021). Therefore, it was suggested that these domains may serve as better target for pullulanases engineering approaches to obtain better catalytic properties. In lieu of this, N1 deletion construct (gt-apuΔN) was developed and expressed in *E. coli* to better understand the role of the N1 domain (1–257 aa) in the amylopullulanase (gt-apu) of the thermophilic bacteria *Geobacillusthermoleovorans* NP33. The mutant has a similar optimum temperature and pH, whereas the thermostability is improved as compared to the wild enzyme. Interestingly, the mutant has the potential to hydrolyse maltotetraose (Nisha and

Satyanarayana, 2016). Furthermore, it was also found that the truncation in N-terminal of pullulanase (PulPB1) negatively affected the specific activity and the deletion made the enzyme prone to inactivation at 50°C (Zhang et al., 2020a). These studies on pullulanase engineering suggest that the effect of the individual domain on the overall pullulanase structure must be taken into consideration. The protein-based engineering approach of pullulanase has been proven as an efficient tool for improving the activity, stability, and robustness of the enzyme. This strategy does not affect the productivity of the enzyme through the wild or even the recombinant strains.

5.4 USE OF WASTE AS SUBSTRATE FOR PULLULANASE PRODUCTION BY MICROORGANISMS

With rapid increase in industrialization of different sectors, generation of waste and their improper disposal represents a notable environmental menace. Waste from agricultural and food industries have been receiving much attention, as they are rich in minerals, proteins and carbohydrates which may be converted and assimilated by microorganisms (Awasthi et al., 2022; Gaur et al., 2022; Sharma et al., 2022). An attractive practice of valorization of these wastes is their exploitation as substrate in bio-refineries for microbial sustainability to produce innumerable products such as industrial enzymes. Enzymes such as pullulanases are comparatively high-priced commodities and around 30–40% of their production cost is accounted for in the substrate used in the fermentation, which can be significantly minimized by using low-cost waste residues (Ravindran et al., 2018). Pullulanase production from a variety of bacterial (*Bacillus, Klebsiella, Anoxybacillus, Paenibacillus, Clostridium*) and fungal (*Aureobasidium and Aspergillus*) species has been reported, although the optimal concentration of waste used as a nutritional source has been little investigated.

Wheat bran was found to be suitable among the easily available complex organic substrates to optimize the effect of a substrate on pullulanase production by thermophilic *Clostridium thermosulforegenes* SVM17. It was found to be suitable among the easily available complex organic substrates such as de-oiled ground nut and coconut cakes, coarse types of brans of pulses such as green gramme, Bengal gramme, black gramme, red gramme, fine types of wheat bran, and rice (Mrudula et al., 2011). Apart from classical strategy, statistical strategy is generally preferred because of its advantages in short-listing substrates and their correlation with the yield of enzymes (Srinivas et al., 1994). Designs such as a Plackett-Burman for the production of thermostable pullulanase enzyme using *Clostridium thermosulfurogenes* SV2 in SSF was studied and it was found that among 15 nutrient sources screened, 14% pearl millet, 2.5% corn steep liquor, 16.5% potato starch, and 0.015% ferrous sulphate was most optimal for enzyme production. It has been noted that, when *Clostridium thermosulfurogenes* SV2 was supplied with only these four nutrients, it produced 10% more pullulanase via optimised SSF (Rama Mohan Reddy et al., 1999). Moreover, pullulanase production using *Bacillus licheniformis* strain (BS18) was studied by exploiting organic waste as substrate in SSF and it was observed that the enzyme gave the highest activity of 29 U/mg in a media supplemented with a mixed ratio of rice bran and corn bran, whereas the media containing a mixture of rice bran and soya bran produced the lowest enzymatic activity of 7 U/mg (Khalaf and Aldeen, 2013). However, as well as bacterial species, endophytic fungi have also contributed to the successful production of the pullulanase enzyme. *Aspergillus sp.* BHU-46 was screened for pullulanase production using agricultural residue as substrate during fermentation. Wheat bran, sugarcane bagasse, orange peel, banana peel, and mosambi peel were employed along with optimized parameters to promote and stimulate the production in SSF and SmF. For example, wheat bran was shown to be the most effective substrate for pullulanase production, yielding 65.33±2.08 U/gds during SSF and 39.33±0.571 U/ml during SmF, whereas rice bran gave 44.66±1.527 U/gds and 41.33±1.154 U/ml during SSF and SmF, respectively. The lowest enzyme yield was obtained when orange peel, mosambi peel, banana peel, and sugarcane bagasse were added to the fermentation medium (Naik et al., 2019). Another statistical technique, Box-Behnken Design (BBD), was used to optimize the concentration of nutrients for *Aspergillus flavus* pullulanase production. Production of pullulanase was carried out on a large scale (flask size 500–5000 ml of capacity containing 10–100 g dry substrate) with optimized parameters in which nutrients were supplemented with wheat bran. The overall yield was found to be 1.63-fold higher thereby suggesting that the pullulanase production can be scaled up using agricultural waste in SSF (Naik et al., 2022). Notably, the cost of producing pullulanase may be decreased by using different categories of waste as substrate in fermentation processes, thus positively impacting the economics of this enzyme across the world, where it has a booming presence in the industries related to its exploitation.

5.5 FACTOR AFFECTING THE ACTIVITY OF PULLULANASE ENZYME

Microorganisms produce a variety of extracellular enzymes that have been commercially employed in the medicine, soap/detergent, fabric, nutrition, food, and beverage sectors (Adrio and Demain, 2014; Beygmoradi and Homaei, 2017; Bilal and Iqbal, 2020; Gurung et al., 2013). The pullulanase enzyme has been used in various applications as a biocatalyst. This enzyme is used to produce several types of products such as sugar syrups, glucose, low-calorie beer, maltose, bio-ethanol, and maltodextrins (Farias et al., 2021; Lu et al., 2018). However, minimal research has been carried out to explore the potential of the pullulanase enzyme. Nowadays, scientists have reported various optimal processes and parameters on a laboratory as well as industrial scale to enhance the production and enzymatic properties of potential enzymes such as pullulanase (Farias et al., 2021; Zeng et al., 2019; Zhang et al., 2018; Zheng et al., 2020). It is well known that microorganisms release ideal enzymes for their growth. Therefore, these enzymes may further be characterized and optimized for their extensive production at a commercial scale. Multiple parameters, including pH, temperature, incubation length, substrate concentration, and ions, have been discovered to alter the activity of the pullulanase enzyme by reducing the production rate and development of microorganisms, despite the fact that they are not well optimized. As a result, the factors mentioned above must be optimized to obtain a proper and enhanced yield of the pullulanase enzyme with good enzymatic potential for large-scale processing in food, beverage, and pharmaceutical industries (Akassou and Groleau, 2019).

5.5.1 Effect of pH, Temperature and Incubation Period on the Activity of Pullulanase Enzyme

pH is the most important physical factor affecting microbial growth and enzyme extracellular secretion via microorganisms (HusnainGondal et al., 2021). Several studies have reported the production of pullulanase using microbial species such as *Bacillus acidopullulyticus, Klebsiella pneumoniae, Micrococcus sp., Thermococcus sp.,* and *Rhodotehrmus marinus* at ideal pH (4.0–7.0) and temperature (50–60°C) for maltotriose syrup formation (Bender and Wallenfels, 1961; Gantelet et al., 1998; Gomes et al., 2003; Jensen, 1984; Kimura and Horikoshi, 1990; Stefanova et al., 1999; Singh et al., 2010b). The pullulanase enzyme is reported to be active between pH 5.0 and 6.0, with 5.0 being just the optimum pH, and it was completely stabilized (100%) at 50°C after 6 h incubation. Between pH 5.0 and 6.5, the enzyme maintained over 89% activity, but had over 75% activity at pH 4.5. Pullulanase activity was also found to be 92.51% and 82.52% at 55°C and 60°C, respectively, with a dramatic reduction at 70°C. For example, Chen et al. (2015) reported that the highest specific activity of pulA produced from mesophilic *Bacillus acidopullulyticus* against pullulan was 1096 U/mg at 60°C (optimal), with a half-life of 34.6 min at this temperature. Similarly, after a 9 h incubation at 60°C (the optimal reaction temperature) the pullulanasePulASK isolated from *Anoxybacillus sp.* (SK3-4) sustained 50% of its original activity (Kahar et al., 2016). A mesophilic bacteria, *Bacillus pseudofirmus* 703, was reported to clone a new pullulanase gene called pul703 (Lu et al., 2018). Pul703 was discovered as a type I pullulanase with the highest activity at 45°C temperature with good consistency. During 72 h of incubation at a temperature between 25 and 35°C, more than 70% activity was retained. Although, after a 12 h incubation period, the Pul703 demonstrated maximal activity at pH between 7.0 and 8.0 and was generally active and stable at a pH between 5.5 and 9.5, retaining about 80% of its activity. Overall, there was a significant and robust link between pH, temperature, and incubation duration, which affects pullulanase production and activity.

5.5.2 Effect of Substrate Specificity/Concentration on the Activity of Pullulanase Enzyme

Pullulanase is found in various organisms such as plants, animals, and microorganisms. In each scenario, the detailed enzymatic pathways of substrate degradation and the end products differ. Basically, pullulanase is split into two classes based on substrate specificity: (1) Type I – pullulanase cleave the α- 1,6-glycosidic connections in amylopectin, polysaccharides, and pullulan; and (2) Type II – pullulanase hydrolyse both the 1,4 and 1,6 glycosidic links in starch (Xu et al., 2021). The carbohydrate-binding module domain (CMD) is located at the N-terminus of pullulanase. It is essential in interacting with enzymes to the substrate (polysaccharide) and to the hydrolysis process (Janeček et al., 2017). It has been found that relative activity on dextran and soluble starch was 23.26% and 59.45%, respectively, compared to pullulan, which was considered 100% (Singh et al., 2010a). The results showed that the enzyme had 61.93% activity on soluble starch compared to 100% activity on pullulan. Similarly, another study from the same research group (Singh et al., 2010b), showed the affinities of pure pullulanase towards pullulan, soluble starch, and dextran were determined by incubating the enzyme at various concentrations between 0.25 to 1.75%. Furthermore, the putative type I pullulanase gene (Pul_M) enzyme's substrate specificity was determined for various substrates. These substrates, namely pullulan, dextran, amylose, soluble starch, rice flour, and amylopectin, were used at 5 mg/ml concentration (Thakur et al., 2021).

5.5.3 Effect of Ions on the Activity of Pullulanase Enzyme

Enzymes have long been recognized as stabilized, activated, or inhibited by metal ions. They function as co-enzymes and can speed up or slow down reactions in various ways, including general catalysis, covalent catalysis, and acid-base catalysis. The influence of possible metal ions and enzyme inhibitors on pullulanase activity was examined at various concentrations. Pullulanase activity was tested in the presence of several metal ions (Ca^{2+}, Zn^{2+}, Cu^{2+}, Ni^{2+}, Fe^{2+}, Cd^{2+}, Mg^{2+}, Ba^{2+}, Co^{2+}, Mn^{2+}, K^+, Na^+, and Hg^{2+}) at various doses (2–10 mM) to investigate the effect of metal ions (Singh et al., 2010b). The addition of Ca^{2+} and Mn^{2+} boosted pullulanase activity by 2.1 and 1.84 times, respectively, compared to a control with no metal ions, implying that ions are required as a catalyst. On the other hand, Cu^{2+}, Hg^{2+}, and Ni^{2+} were discovered to be effective inhibitors. Even at a lesser dosage of 4 mM, $Cu^{2+}$and Hg^{2+} completely inhibited the action. Waśko et al. (2011) reported a similar observation of suppression of activity via Hg^{2+}. However, no apparent effect of Ca+ ion on the activity of pullulanase isolated from *Bacillus licheniformis* was detected (Khan et al., 2021). It was also shown that adding $MgCl_2$ to the culture medium significantly increased the activity of *B. Deramificans* pullulanase produced by *Brevibacillus choshinensis* (Zou et al., 2016). The pullulanase activity produced from the $MgCl_2$-supplemented medium was likewise greater than control (without ion). These studies suggest the importance of metal ions for the efficient activity of pullulanases.

5.6 APPLICATION OF PULLULANASES

Enzymes make a remarkable contribution to industries, performing a significant role in bioconversion, synthesis, extraction efficacy improvement, viscosity reduction, flavour modification, colour, texture, and appearance modification etc. (Hii et al., 2012b). Pullulanases are an amylolytic pullulan degrading glucanase, which have attained maximum attention on an industrial scale due to their large-scale production convenience, ease in characterization, and low production cost. This section discusses the application of pullulanases in starch processing industries, baking applications, and detergent industries.

5.6.1 Starch Processing

In starch processing industries, pullulanase is rewarded as an imperative debranching enzyme, capable of simplifying the branched polysaccharides into small fermentable sugars during the saccharification process via the gearing up of the hydrolyse of α-1,6 glucosidic bonds in amylopectin, starch, pullulan, and other oligosaccharides. At an industrial level, glucose manufacturing involves the liquefaction and gelatinization processes successively (Hii et al., 2012b). Starch is composed of two polymeric units: (1) linear chain-amylose, and (2) branched polymer-amylopectin. To perform hydrolysis of starch, debranching enzymes (DBEs) are required to cleave α-1,6 linkages to release branched polysaccharides from the matrix (Hii et al., 2012b). DBEs possess the distinctive ability to limit dextrin production 'while processing amylopectin. It also increases the conversion rate and utilization of starch as raw material (Xie et al., 2021). According to their substrate selectivity, these enzymes can be grouped into three primary groups: (1) microbial pullulanases and plant R-enzymes (pullulan-6-glucanohydrolases), (2) isoamylases (glycogen-6-glucanohydrolases), and (3) microbial amylo1,6-glucosidases (dextrin 6-D-glucosidases) (Hii et al., 2012b).

As per its industrial applications, pullulanase can be utilized in five forms: (1) pullulan hydrolase type I or neopullulanase, which cleaves α -1, 4 glycosidic bond in pullulan, and forms panose; (2) pullulan hydrolase type II or isopullulanase, which cleaves α -1, 4 glycosidic bonds in pullulan, and produces isopanose; (3) pullulan hydrolase type III, which attacks both α -1, 6 and α -1, 4 glycosidic linkages in pullulan to produce a mixture of maltose, panose, and maltotriose; (4) pullulanase type I, which specifically hydrolyses α -1, 6 glycosidic linkages in pullulan or branched oligosaccharides to produce maltotriose or linear oligomers; and (5) pullulanase type II, or amylopullulanases, which are prominently used in starch processing industries for their specific capacity of hydrolysing α-1, 6 glycosidic linkages in pullulan and other branched polysaccharides to produce maltotriose which subsequently cleaves α -1, 4 glycosidic bonds of other linear polysaccharides like α-amaylase to yield panose, maltose, and glucose as final products (Hii et al., 2012b; Liu et al., 2020; Thakur et al., 2019; Xia et al., 2021).

In the sugar syrup industry, pullulanase is used to complete the starch hydrolysis initiated by α-amylases. The use of pullulanase in conjunction with other amylolytic enzymes improves the quality of sugar syrups. The efficiency of a saccharification reaction is boosted when starch is treated with amylase and pullulanase at the same time (Ramdas Malakar and Malviya, 2010). Pullulanases are advantageous in industrial applications to produce higher yields of a desired end product from starch. Pullulanase is employed in conjugation with a glucoamylase enzyme like β-amylase in the saccharification process to synthesize maltose or glucose syrup. Pullulanase is employed in the food industry as a processing aid, such as in the debranching of corn starch to manufacture corn sweeteners and in retrograde starch production from starchy foods (Ramdas Malakar and Malviya, 2010; Zhang et al., 2013)

At an industrial scale, the conversion of starch into value-added products involves liquefaction by α-amylase at 110°C followed by saccharification and debranching by pullulanase and glucoamylase at 65°C (Pang et al., 2019; Shankar et al., 2014). The optimum efficacy of pullulanase was observed at pH 6.0–6.5 and 55°C in absence of Ca^{2+} ions; therefore, thermostable pullulanase is warranted for industrial processing (Ramdas Malakar and Malviya, 2010). Thermostable pullulanases can concurrently accomplish liquefaction and saccharification reaction at 95°C along with sufficient reduction in energy consumption, and reaction time with enhanced productivity. Thermostable pullulanases can be extracted from wild microbes like *Thermotoganeapolitana, Bacillus flavothermus, Bacillus thermoleovorans, Shewanellaarctica Bacillus acidopullulyticus, Exiguobacterium sp.* and *Anoxybacillus sp.*, etc. but their lower yield and low productivity are hurdles in industrial processing (Liu et al., 2016). Thermostable pullulanase recovered from genetically modified strains is highly valued in the food processing industries for production of health-promoting sugars like maltotriose and panose (Shankar et al., 2014).

5.6.2 Bakery Industry

In the bakery industry staling is a major challenge, causing moisture loss from the crust, loss of crispiness of the crust, loss of firmness in the crumb, loss of bread flavour, and quality deterioration during storage. Treatment of flour with starch-modifying enzymes like pullulanase is highly acceptable, and is greatly helpful in solving this problem (Ramdas Malakar and Malviya, 2010). Pullulanase acts as an anti-staling agent which retards the process of retrogradation in amylopectin fraction of starch. This shelf life extension is achieved by shortening the chain length of amylopectin (Prakash et al., 2012). The use of huge quantities of α-amylase in bread dough treatment produces branched maltodextrin, which provides a sticky texture to the bread. Therefore, a combination of debranching enzyme, amylopullulanases, and α-amylase is preferential in bread preparation. This combination is highly effective in hydrolysing the branched maltodextrins produced due to α-amylase treatment thereby reducing the gumminess of the bread. It is also efficient in reducing the chance of staling in baked goods (Nisha and Satyanarayana, 2016; Prakash et al., 2012; Singh and Singh, 2019).

5.6.3 Production of Cyclodextrin

Cyclodextrins (CDs) belong to the cyclic oligosaccharides family, made up of five or more α-D-glucopyranoside units arranged in a ring-like structure via α-1, 4-glycosidic linkages. CDs are of two types: linear and branched. Branched CDs such as glucosylCDs and maltosyl-CDs are homogeneous, cyclic oligosaccharides composed of glucose moieties. The manufacture of CDs is a straightforward process which utilizes a series of starch-modifying enzymes to treat regular starch. Commonly, α-amylase, along with cyclodextrin glycosyltransferase (CGTase) is used for starch modification. However, CGTase is insufficient to cleave α-1, 6-glycosidic linkage of amylopectin, so a huge part of starch remains untreated. The pullulanase is effective in the dissociation of these bonds, thereby boosting the yield percentage of cyclodextrin synthesis. Pullulanase is also used to produce CDs and branched CDs, such as maltosyl-CDs and glucosyl-CDs (Ramdas Malakar and Malviya, 2010). CDs and branched CDs are frequently employed as odour maskers, labile material stabilisers, and solvents for insoluble or poorly soluble pharmaceutical drugs. These are also utilized in the food processing industries to provide cholesterol-free goods (Prakash et al., 2012). CDs have been widely used in a variety of applications including complexing materials, antioxidants, emulsifiers, and stabilizing agents in foods, agricultural, and pharmaceuticals products (Hii et al., 2012b).

Pullulanase has the ability to produce branched CDs in a reverse hydrolysis process by utilizing a high concentration of oligosaccharides. The interaction affinity of β-CD with pullulanase is strongest as compared to α-CD and γ-CD due to confined cavity geometric dimensions; therefore, β-CD have inhibitory effect on pullulanase activity of starch conversion to CDs. Mutations or genetic modifications of pullulanase-producing microbes were successful to overcome the persisting challenge. A mutant pullulanase D465E was capable of reducing the β-CD's inhibition effect and retained the activity. The β-CGTase and D465E shows higher yield of β-CD and starch substrate utilization in a synchronous bioconversion process (Li et al., 2020a,b).

5.6.4 Brewing Industry

Enzymes have been used in the brewing industries for the conversion of cereals into beer. Use of enzymes with low activity will lead to undesired ramifications such as prolonged wort separation, slow fermentation process, inferior stability and flavour of the product, and low extraction yield (Blanco et al., 2014). Versatility of the pullulanase enzyme has provided some advancement to the brewing industries. They have significantly replaced the use of α-amylases and β-amylases that are commonly used in starch liquefaction and further saccharification respectively (Prakash et al., 2012). Pullulanases can efficiently combine these two into a one-step process, and thus could qualify techno-economic aspects for growing industries. It has a similar effect in limiting dextrinase, which helps in debranching, thereby promoting the fermentation of beers. The addition of pullulanases spontaneously reduces the amount of dextrin in the fermented medium thus resulting in the improved quality of beer with low calorific content. Pullulanases also can reduce the amount of amylopectin in order to produce dry beer. During saccharification, pullulanases can improve the yield of glucose by reducing the amount of amylopectin and thus can further reduce the supply of glucoamylase. It is well studied that, in the preparation of maltose syrup, the use of pullulanases could improve the yield of maltose with content greater than 70%. Moreover, rich maltotriose syrup has been used in brewing industries with many excellent properties. While maltotriose-producing amylases have been used for making the syrup, due to the poor yield and impurity, an alternative way to producing high maltotriose syrup was investigated using pullulanases. Pullulanases derived from *Aurebasidium pullulan- SB*-1 were able to produce maltotriose by hydrolysing pullulan. The total amount of maltotriose in the product produced, as well as the yield of maltotriose, were determined to be 93.5% and 87.3% (w/w), respectively, indicating that it may be used in the brewing business (Mishra et al., 2016). Therefore, the applications of pullulanase will increase manifold and thus new possibilities for industrial processes shall emerge.

5.7 CHALLENGES ASSOCIATED WITH THE PRODUCTION OF PULLULANASES

Considering the diversity of pullulanases, there are many complications which limit the production at an industrial scale. Microorganisms provide an excellent array of valuable commodities, yet only in an amount they require for their own good. Microorganisms tend not to up-regulate the machinery and overproduce the enzymes. Therefore, exploiting the wild strain cannot meet the assigned yield due to the low productivity of enzymes. However, genetic manipulations could offer recombinant microbial hosts for production of pullulanases. The mutant strains of *B. subtilis* ATCC6051Δ10 was constructed using a knockout technique which displayed only 1.48 times more pullulanase production than the wild-type strain *B. subtilis* ATCC 6051 (Liu et al., 2018).

A lack of engineered microorganisms with highly efficient pullulanase production is considered to be a restraining factor for its market and commercialization. Notably, the application of this enzyme has been repressed due to its high cost of production (Velhaal et al., 2014). Expensive substrates for feeding the inoculum in fermentation media has led to the high-cost production of pullulanases. Although many organic waste residues are abundant in the environment, they have no commercial value. The use of cheap organic substrate as a carbon or nitrogen source for microorganisms in pullulanase production has not yet replaced the expensive raw materials, and thus affects the techno-economic evaluation at a commercial scale. Non-optimized parameters such as pH, temperature, and substrate concentration for the scaling-up of this enzyme has also proved to be a shortcoming for the production on a large scale, thereby positively influencing overall productivity. Moreover, previous studies have exhibited that pullulanases are widely distributed in mesophilic microorganisms with similar types of enzymatic property. Consequently, finding novel thermo-tolerant and cold adaptive type of pullulanases from thermophilic and psychrophilic microorganisms is still a challenge. A scarcity of temperature-resistant pullulanases has made the industries discontented, thus restricting its applications. To increase the reusability of pullulanases, the immobilization approach using a novel carrier or materials is another challenge for the pullulanase production industry. Owing to their practical application, the hunt for novel carriers is still the focus of researchers.

5.8 CONCLUSIONS

The diverse application of pullulanase has led to the exploration of its superior activity. These are of great importance for exploiting these enzymes in different industrial applications. With the continuous revolution in industrial production, pullulanases have shown great value in the preparation of several products, but there remains a need to overcome the challenges that are faced during their production. Several microbial species have been reported to produce pullulanases with/without the use of wastes as a substrate. Yet owing to the low yield, recombinant strains have been employed to overexpress the pullulanases and protein engineering for increasing the robustness of the enzyme itself. The productivity and activity also depend on several factors that need to be optimized to better fulfil the industrial need. Pullulanases have been very efficient in starch processing and in the baking and brewery industries. Therefore, more research into strengthening the activity and productivity is warranted.

REFERENCES

Adrio, J. L., and A. L. Demain. 2014. Microbial enzymes: Tools for biotechnological processes. *Biomolecules*, 4(1):117–139.

Akassou, M., and D. Groleau. 2019. Advances and challenges in the production of extracellular thermoduric pullulanases by wild-type and recombinant microorganisms: A review. *Critical Reviews in Biotechnology*, 39(3):337–350.

Armenta, S., S. Moreno-Mendieta, Z. Sánchez-Cuapio, S. Sánchez, and R. Rodríguez-Sanoja. 2017. Advances in molecular engineering of carbohydrate-binding modules. *Proteins: Structure, Function and Bioinformatics*, 85(9):1602–1617.

Asha, R., F. N. Niyonzima, and S. M. Sunil. 2013. Purification and properties of pullulanase from Bacillus halodurans. *International Research Journal of Biological Sciences*, 2(3):35–43.

Awasthi, M. K., A. Tarafdar, V. K. Gaur, K. Amulya, V. Narisetty, D. K. Yadav, R. Sindhu, P. Binod, T. Negi, A. Pandey, Z. Zhang, and R. Sirohi. 2022. Emerging trends of microbial technology for the production of oligosaccharides from biowaste and their potential application as prebiotic. *International Journal of Food Microbiology*, 368:109610.

Ayadi, DZ., M. Ben Ali, S. Jemli, S. Ben Mabrouk, M. Mezghani, E. Ben Messaoud, and S. Bejar. 2008. Heterologous expression, secretion and characterization of the Geobacillusthermoleovorans US105 type I pullulanase. *Applied Microbiology and Biotechnology*, 78:473–481.

Bakshi, A., P. R. Patnaik, and J. K. Gupta. 1992. Thermostable pullulanase from a mesophilic bacillus cereus isolate and its mutant UVT.4. *Biotechnology Letters*, 14(8):689–694.

Bender, H., and K. Wallenfels. 1961. Investigations on pullulan. II. Specific degradation by means of a bacterial enzyme. *Biochemische Zeitschrift*, 334:79–95.

Bertoldo, C., M. Armbrecht, F. Becker, T. Schäfer, G. Antranikian, and W. Liebl. 2004. Cloning, sequencing, and characterization of a heat- and alkali-stable type I pullulanase from anaerobrancagottschalkii. *Applied and Environmental Microbiology*, 70(6):3407–3416.

Beygmoradi, A., and A. Homaei. 2017. Marine microbes as a valuable resource for brand new industrial biocatalysts. *Biocatalysis and Agricultural Biotechnology*, 11:131–152.

Bilal, M., and H. M. N. Iqbal. 2020. State-of-the-art strategies and applied perspectives of enzyme biocatalysis in food sector — current status and future trends. *Critical Reviews in Food Science and Nutrition*, 60(12):2052–2066.

Blanco, C. A., I. Caballero, R. Barrios, and A. Rojas. 2014. Innovations in the brewing industry: Light beer. *International Journal of Food Sciences and Nutrition*, 65(6):655–660.

Chang, M., X. Chu, J. Lv, Q. Li, J. Tian, and N. Wu. 2016. Improving the thermostability of acidic pullulanase from bacillus naganoensis by rational design. *PlOS ONE*, 11(10):e0165006.

Chen, A., Y. Li, J. Nie, B. McNeil, L. Jeffrey, Y. Yang, and Z. Bai. 2015. Protein engineering of Bacillus acidopullulyticuspullulanase for enhanced thermostability using in silico data driven rational design methods. *Enzyme and Microbial Technology*, 78:74–83.

Chen, W. B., Y. Nie, and Y. Xu. 2013. Signal peptide-independent secretory expression and characterization of pullulanase from a newly isolated klebsiella variicola shn-1 in escherichia coli. *Applied Biochemistry and Biotechnology*, 169:41–54.

Duan, X., J. Chen, and J. Wu. 2013. Improving the thermostability and catalytic efficiency of Bacillus deramificanspullulanase by site-directed mutagenesis. *Applied and Environmental Microbiology* 79(13):4072–4077.

Farias, T. C., H. Y. Kawaguti, and M. G. Bello Koblitz. 2021. Microbial amylolytic enzymes in foods: Technological importance of the Bacillus genus. *Biocatalysis and Agricultural Biotechnology*, 35:102054.

Gantelet, H., C. Ladrat, A. Godfroy, G. Barbier, and F. Duchiron. 1998. Characteristics of pullulanases from extremely thermophilic archaea isolated from deep-sea hydrothermal vents. *Biotechnology Letters*, 20:819–823.

Gaur, V. K., P. Sharma, R. Sirohi, S. Varjani, M. J. Taherzadeh, J.-S. Chang, H. Y. Ng, J. W. C. Wong, and S.-H. Kim. 2022. Production of biosurfactants from agro-industrial waste and waste cooking oil in a circular bioeconomy: An overview. *Bioresource Technology*, 343:126059.

Gomes, I., J. Gomes, and W. Steiner. 2003. Highly thermostable amylase and pullulanase of the extreme thermophilic eubacterium Rhodothermus marinus: Production and partial characterization. *Bioresource Technology*, 90(2):207–214.

Gurung, N., S. Ray, S. Bose, and V. Rai. 2013. A broader view: Microbial enzymes and their relevance in industries, medicine, and beyond. *BioMed Research International*, 2013:1–18.

Hii, L. S., M. Rosfarizan, T. C. Ling, and A. B. Ariff. 2012a. Statistical optimization of pullulanase production by Raoultellaplanticola DSMZ 4617 using sago starch as carbon and peptone as nitrogen sources. *Food and Bioprocess Technology*, 5:729–737.

Hii, S. L., T. C. Ling, R. Mohamad, and A. B. Ariff. 2009. Enhancement of extracellular pullulanase production by Raoultellaplanticola DSMZ 4617 using optimized medium based on sago starch. *The Open Biotechnology Journal*, 3:1–8.

Hii, S. L., J. S. Tan, T. C. Ling, and A. Bin Ariff. 2012b. Pullulanase: Role in starch hydrolysis and potential industrial applications. *Enzyme Research*, 2012:1–14.

HusnainGondal, A., Q. Farooq, S. Sohail, S. Shasang Kumar, M. Danish Toor, A. Zafar, and B. Rehman. 2021. Adaptability of soil pH through innovative microbial approach. *Current Research in Agricultural Sciences*, 8(2):71–79.

Janeček, Š., K. Majzlová, B. Svensson, and E. A. MacGregor. 2017. The starch-binding domain family CBM41—An in silico analysis of evolutionary relationships. *Proteins: Structure, Function and Bioinformatics*, 85(8):1480–1492.

Jensen, B. F. 1984. Bacillus acidopullulyticuspullulanase: Application and regulatory aspects for use in the food industry. *Process Biochemistry* 19:129–134.

Kahar, U. M., C. L. Ng, K. G. Chan, and K. M. Goh. 2016. Characterization of a type I pullulanase from Anoxybacillus sp. SK3-4 reveals an unusual substrate hydrolysis. *Applied Microbiology and Biotechnology*, 100:6291–6307.

Kang, J., K. M. Park, K. H. Choi, C. S. Park, G. E. Kim, D. Kim, and J. Cha. 2011. Molecular cloning and biochemical characterization of a heat-stable type I pullulanase from Thermotoganeapolitana. *Enzyme and Microbial Technology*, 48(3):260–266.

Khalaf, A. K., and S. B. Aldeen. 2013. Optimum condition of pullulanase production by liquid state and solid state fermentation (SSF) method from Bacillus licheniforms (BS18). *Iraqi Journal of Science*, 54:35–49.

Khan, A., M. Irfan, U. Rahman, F. Azhar, A. A. Shah, M. Badshah, F. Hasan, F. U. Rehman, Z. A. Malik, and S. Khan. 2021. A ca+2 independent pullulanase from bacillus licheniformis and its application in the synthesis of resistant starch. *Pakistan Journal of Agricultural Sciences*, 58(2):699–709.

Kimura, T., and K. Horikoshi. 1990. Characterization of pullulan-hydrolysing enzyme from an alkalopsychrotrophic Micrococcus sp. *Applied Microbiology and Biotechnology*, 34(1):52–56.

Kumari, M., Padhi, S., Sharma, S. et al. 2021. Biotechnological potential of psychrophilic microorganisms as the source of cold-active enzymes in food processing applications. *3 Biotech* 11:479.

Kunamneni, A., and S. Singh. 2006. Improved high thermal stability of pullulanase from a newly isolated thermophilic Bacillus sp. AN-7. *Enzyme and Microbial Technology*, 39(7):1399–1404.

Li, X., Y. Bai, H. Ji, Y. Wang, and Z. Jin. 2020a. Phenylalanine476 mutation of pullulanase from Bacillus subtilis str. 168 improves the starch substrate utilization by weakening the product β-cyclodextrin inhibition. *International Journal of Biological Macromolecules*, 155:490–497.

Li, X., Y. Bai, H. Ji, Y. Wang, Z. Jin, Y. Bai, and Z. Jin. 2020b. Development of pullulanase mutants to enhance starch substrate utilization for efficient production of β-CD. *International Journal of Biological Macromolecules*, 155:490–497.

Li, Y., L. Zhang, D. Niu, Z. Wang, and G. Shi. 2012. Cloning, expression, characterization, and biocatalytic investigation of a novel Bacilli thermostable type I pullulanase from Bacillus sp. CICIM 263. *Journal of Agricultural and Food Chemistry*, 60(44):11164–11172.

Liu, J., Y. Liu, F. Yan, Z. Jiang, S. Yang, and Q. Yan. 2016. Gene cloning, functional expression and characterisation of a novel type I pullulanase from Paenibacillusbarengoltzii and its application in resistant starch production. *Protein Expression and Purification*, 121:22–30.

Liu, P., W. Gao, X. Zhang, Z. Wu, B. Yu, and B. Cui. 2020. Physicochemical properties of pea starch-lauric acid complex modified by maltogenic amylase and pullulanase. *Carbohydrate Polymers*, 242:116332.

Liu, X., H. Wang, B. Wang, and L. Pan. 2018. Efficient production of extracellular pullulanase in Bacillus subtilis ATCC6051 using the host strain construction and promoter optimization expression system. *Microbial Cell Factories*, 17(1):1–12.

Lu, Z., X. Hu, P. Shen, Q. Wang, Y. Zhou, G. Zhang, and Y. Ma. 2018. A pH-stable, detergent and chelator resistant type I pullulanase from Bacillus pseudofirmus 703 with high catalytic efficiency. *International Journal of Biological Macromolecules*, 109:1302–1310.

Meng, F., X. Zhu, H. Zhao, F. Lu, Y. Lu, and Z. Lu. 2021. Improve production of pullulanase of bacillus subtilis in batch and fed-batch cultures. *Applied Biochemistry and Biotechnology*, 193(1):296–306.

Mishra, B., A. Manikanta, and D. Zamare. 2016. Preparation of maltotriose syrup from microbial pullulan by using pullulanase enzyme. *Biosciences Biotechnology Research Asia*, 13(1):481–485.

Moubasher, H., S. S. Wahsh, and N. A. El-Kassem. 2013. Isolation of Aureobasidium pullulans and the effect of different conditions for pullulanase and pullulan production. *Microbiology (Russian Federation)*, 82:155–161.

Mrudula, S., Reddy, G., and Seenayya, G. 2011. Effect of substrate and culture conditions on the production of amylase and pullulanase by thermophilic Clostridium thermosulforegenes SVM17 in solid state fermentation. *Malaysian Journal of Microbiology*, 7(1):19–25.

Naik, B., S. K. Goyal, A. D. Tripathi, and V. Kumar. 2019. Screening of agro-industrial waste and physical factors for the optimum production of pullulanase in solid-state fermentation from endophytic Aspergillus sp. *Biocatalysis and Agricultural Biotechnology*, 22:101423.

Naik, B., S. K. Goyal, A. D. Tripathi, and V. Kumar. 2022. Optimization of pullulanase production by Aspergillus flavus under solid-state fermentation. *Bioresource Technology Reports*, 100963.

Nair, S. U., R. S. Singhal, and M. Y. Kamat. 2006. Enhanced production of thermostable pullulanase type 1 using Bacillus cereus FDTA 13 and Its mutant. *Food Technology and Biotechnology* 44(2):275–282.

Nie, Y., W. Yan, Y. Xu, W. B. Chen, X. Q. Mu, X. Wang, and R. Xiao. 2013. High-level expression of Bacillus naganoensispullulanase from recombinant Escherichia coli with auto-induction: Effect of lac operator. *PLOS ONE*, 8(10):e78416.

Nisha, M., and T. Satyanarayana. 2016. Characteristics, protein engineering and applications of microbial thermostable pullulanases and pullulan hydrolases. *Applied Microbiology and Biotechnology*, 100(13):5661–5679.

Orhan, N., N. AltasKiymaz, and A. Peksel. 2014. A novel pullulanase from a fungus Hypocreajecorina QM9414: Production and biochemical characterization. *Indian Journal of Biochemistry and Biophysics*, 51(2):149–55.

Pang, B., L. Zhou, W. Cui, Z. Liu, S. Zhou, J. Xu, and Z. Zhou. 2019. A hyperthermostable type II pullulanase from a deep-sea microorganism Pyrococcusyayanosii CH1. *Journal of Agricultural and Food Chemistry*, 67(34):9611–9617.

Pang, B., L. Zhou, W. Cui, Z. Liu, and Z. Zhou. 2020. Production of a thermostable pullulanase in Bacillus subtilis by optimization of the expression elements. *Starch/Staerke*, 72(11-12):2000018.

Prabhu, N., Maheswari U. R., Singh, M. V. P., Karunakaran, S., Kaliappan, C., and Gajendran, T. 2018. Production and purification of extracellular pullulanase by Klebsilla aerogenes NCIM 2239. *African Journal of Biotechnology*, 17(14):486–494.

Prakash, N., S. Gupta, M. Ansari, Z. A. Khan, and V. Suneetha. 2012. Production of economically important products by the use of pullulanase enzyme. *International Journal of Science Innovations and Discoveries*, 2(2):266–273.

Rama Mohan Reddy, P., G. Reddy, and G. Seenayya. 1999. Production of thermostable pullulanase by Clostridium thermosulfurogenes SV2 in solid-state fermentation: Optimization of nutrients levels using response surface methodology. *Bioprocess Engineering*, 21:497–503.

Ramdas Malakar, D., and S. N. Malviya. 2010. Pullulanase: A potential enzyme for industrial application. *International Journal of Biomedical Research*, 1(2):10–20.

Ramesh, B., P. R. M. Reddy, G. Seenayya, and G. Reddy. 2001. Effect of various flours on the production of thermostable β-amylase and pullulanase by Clostridium thermosulfurogenes SV2. *Bioresource Technology*, 76(2):169–171.

Ravindran, R., S. S. Hassan, G. A. Williams, and A. K. Jaiswal. 2018. A review on bioconversion of agro-industrial wastes to industrially important enzymes. *Bioengineering*, 5(4):1–20.

Razzaq, A., S. Shamsi, A. Ali, Q. Ali, M. Sajjad, A. Malik, and M. Ashraf. 2019. Microbial proteases applications. *Frontiers in Bioengineering and Biotechnology*, 7:1–10.

Shankar, R., M. S. Madihah, E. M. Shaza, N. A. KO, A. A. Suraini, and K. Kamarulzaman. 2014. Application of different feeding strategies in fed batch culture for pullulanase production using sago starch. *Carbohydrate Polymers*, 102:962–969.

Sharma, P., V. K. Gaur, S. Gupta, S. Varjani, A. Pandey, E. Gnansounou, S. You, H. H. Ngo, and J. W. C. Wong. 2022. Trends in mitigation of industrial waste: Global health hazards, environmental implications and waste derived economy for environmental sustainability. *Science of The Total Environment*, 811:152357.

Singh, R. S., G. K. Saini, and J. F. Kennedy. 2010a. Covalent immobilization and thermodynamic characterization of pullulanase for the hydrolysis of pullulan in batch system. *Carbohydrate Polymers*, 81(2):252–259.

Singh, R. S., G. K. Saini, and J. F. Kennedy. 2010b. Maltotriose syrup preparation from pullulan using pullulanase. *Carbohydrate Polymers*, 80(2):401–407.

Singh, R. S., and T. Singh. 2019. Microbial Inulinases and Pullulanases in the Food Industry. Nova Science Publishers, Inc, New York, NY, USA, 23–52.

Song, W., Y. Nie, X. Q. Mu, and Y. Xu. 2016. Enhancement of extracellular expression of Bacillus naganoensispullulanase from recombinant Bacillus subtilis: Effects of promoter and host. *Protein Expression and Purification*, 124:23–31.

Srinivas, M. R. S., N. Chand, and B. K. Lonsane. 1994. Use of Plackett–Burman design for rapid screening of several nitrogen sources, growth/product promoters, minerals and enzyme inducers for the production of alpha-galactosidase by Aspergillus niger MRSS 234 in solid state fermentation system. *Bioprocess Engineering*, 10:139–144.

Stefanova, M. E., R. Schwerdtfeger, G. Antranikian, and R. Scandurra. 1999. Heat-stable pullulanase from Bacillus acidopullulyticus: Characterization and refolding after guanidinium chloride-induced unfolding. *Extremophiles* 3(2):147–152.

Tarafdar, A., R. Sirohi, V. K. Gaur, S. Kumar, P. Sharma, S. Varjani, H. O. Pandey, R. Sindhu, A. Madhavan, R. Rajasekharan, and S. J. Sim. 2021. Engineering interventions in enzyme production: Lab to industrial scale. *Bioresource Technology*, 326:124771.

Thakur, M., N. Sharma, A. K. Rai, and S. P. Singh. 2021. A novel cold-active type I pullulanase from a hot-spring metagenome for effective debranching and production of resistant starch. *Bioresource Technology*, 320(A): 124288.

Thakur, S., J. Chaudhary, V. Kumar, and V. K. Thakur. 2019. Progress in pectin based hydrogels for water purification: Trends and challenges. *Journal of Environmental Management*, 238:210–223.

Tu, D., Y. Ou, Y. Zheng, Y. Zhang, B. Zheng, and H. Zeng. 2021. Effects of freeze-thaw treatment and pullulanase debranching on the structural properties and digestibility of lotus seed starch-glycerinmonostearin complexes. *International Journal of Biological Macromolecules*, 177:447–454.

Wang, X., Y. Nie, and Y. Xu. 2018. Improvement of the activity and stability of starch-debranching pullulanase from Bacillus naganoensis via tailoring of the active sites lining the catalytic pocket. *Journal of Agricultural and Food Chemistry*, 66(50):13236–13242.

Wang, X., Y. Nie, and Y. Xu. 2019b. Industrially produced pullulanases with thermostability: Discovery, engineering, and heterologous expression. *Bioresource Technology*, 278:360–371.

Wang, Y., S. Chen, X. Zhao, Y. Zhang, X. Wang, Y. Nie, and Y. Xu. 2019a. Enhancement of the production of Bacillus naganoensispullulanase in recombinant Bacillus subtilis by integrative expression. *Protein Expression and Purification*, 159:42–48.

Waśko, A., M. Polak-Berecka, and Z. Targoński. 2011. Purification and characterization of pullulanase from Lactococcus lactis. *Preparative Biochemistry and Biotechnology*, 41(3):252–261.

Wei, W., J. Ma, S. Q. Chen, X. H. Cai, and D. Z. Wei. 2015. A novel cold-adapted type I pullulanase of Paenibacilluspolymyxa Nws-pp2: In vivo functional expression and biochemical characterization of glucans hydrolyzates analysis. *BMC Biotechnology*, 15:1–13.

Wu, H., X. Yu, L. Chen, and G. Wu. 2014. Cloning, overexpression and characterization of a thermostable pullulanase from Thermus thermophilus HB27. *Protein Expression and Purification*, 95:22–27.

Xia, W., K. Zhang, L. Su, and J. Wu. 2021. Microbial starch debranching enzymes: Developments and applications. *Biotechnology Advances*, 50:107786.

Xie, F., B.-J. Gu, S. R. Saunders, and G. M. Ganjyal. 2021. High methoxyl pectin enhances the expansion characteristics of the cornstarch relative to the low methoxyl pectin. *Food Hydrocolloids*, 110:106131.

Xu, J., F. Ren, C. H. Huang, Y. Zheng, J. Zhen, H. Sun, T. P. Ko, M. He, C. C. Chen, H. C. Chan, R. T. Guo, H. Song, and Y. Ma. 2014. Functional and structural studies of pullulanase from Anoxybacillus sp. LM18-11. *Proteins: Structure, Function and Bioinformatics*, 82(9):1685–1693.

Xu, P., S.-Y. Zhang, Z.-G. Luo, M.-H. Zong, X.-X. Li, and W.-Y. Lou. 2021. Biotechnology and bioengineering of pullulanase: State of the art and perspectives. *World Journal of Microbiology and Biotechnology*, 37(3):1–10.

Zeng, Y., J. Xu, X. Fu, M. Tan, F. Liu, H. Zheng, and H. Song. 2019. Effects of different carbohydrate-binding modules on the enzymatic properties of pullulanase. *International Journal of Biological Macromolecules*, 137:973–981.

Zhang, H., Y. Tian, Y. Bai, X. Xu, and Z. Jin. 2013. Structure and properties of maize starch processed with a combination of α-amylase and pullulanase. *International Journal of Biological Macromolecules*, 52:38–44.

Zhang, K., L. Su, and J. Wu. 2018. Enhanced extracellular pullulanase production in Bacillus subtilis using protease-deficient strains and optimal feeding. *Applied Microbiology and Biotechnology*, 102:5089–5103.

Zhang, S. Y., Z. W. Guo, X. L. Wu, X. Y. Ou, M. H. Zong, and W. Y. Lou. 2020a. Recombinant expression and characterization of a novel cold-adapted type I pullulanase for efficient amylopectin hydrolysis. *Journal of Biotechnology*, 313:39–47.

Zhang, Y., Y. Nie, X. Zhou, J. Bi, and Y. Xu. 2020b. Enhancement of pullulanase production from recombinant Bacillus subtilis by optimization of feeding strategy and fermentation conditions. *AMB Express*, 10:1–9.

Zheng, Y., Y. Ou, Y. Zhang, B. Zheng, S. Zeng, and H. Zeng. 2020. Effects of pullulanasepretreatment on the structural properties and digestibility of lotus seed starch-glycerinmonostearin complexes. *Carbohydrate Polymers*, 240:116324.

Zou, C., X. Duan, and J. Wu. 2014. Enhanced extracellular production of recombinant Bacillus deramificans-pullulanase in Escherichia coli through induction mode optimization and a glycine feeding strategy. *Bioresource Technology*, 172:174–179.

Zou, C., X. Duan, and J. Wu. 2016. Magnesium ions increase the activity of Bacillus deramificanspullulanase expressed by Brevibacilluschoshinensis. *Applied Microbiology and Biotechnology*, 100:7115–7123.

6 Microbial Proteases for Production of Bioactive Peptides

Divyang Solanki[1,*], *Reena Kumari*[2,*], *Sangeeta Prakash*[1], *Amit Kumar Rai*[2], *and Subrota Hati*[3]
[1]School of Agriculture and Food Sciences, The University of Queensland, Brisbane, Australia
[2]Institute of Bioresources and Sustainable Development, Regional Centre, Gangtok, Sikkim, India
[3]Dairy Microbiology Department, SMC College of Dairy Science, Kamdhenu University, Anand, Gujarat, India

CONTENTS

[*] Authors contributed equally.

DOI: 10.1201/9781003311164-9

6.1 INTRODUCTION

Proteins are a vital part of the human diet that not only provide nutrition but have also evolved as a component having functional properties beyond nutrition. Nowadays, food proteins are gaining interest because of their techno-functional properties and health-promoting activities in humans. Food protein/hydrolysates have been reported to accelerate certain "hormone-like" responses *in vivo* and/or *in vitro* (Korhonen, 2009). In short, protein-based bioactive components such as bioactive peptides have properties that reduce or prevent diseases (non-communicable diseases) like diabetes, and hypertension (Chourasia et al., 2022a; Singh et al., 2022). A bioactive peptide is defined as a sequence of 2–20 amino acids, with a molecular weight of less than 3 KDa, encoded within a parental protein that, when released, presents a specific activity with a beneficial impact on human health. Release of amino acids in the form of bioactive peptides is due to the protein hydrolysis which imparts the bioactivities upon application to the human body where, protein hydrolysis, the release of selected bioactive peptide(s), and their respective nutritional benefits are investigated through *in vitro* or *in vivo*, *in silico* analysis.

Different bioactivities have been determined and reported in the literature such as anticarcinogenic, antioxidant, antimicrobial, antithrombotic, anti-inflammatory, antihypertensive, mineral-binding activities, antidiabetic, dipeptidyl–peptidase IV-inhibitory (DPPIV), antimicrobial, opioid activities and immunomodulatory (Jia et al., 2021; Kumari et al., 2022; Padhi et al., 2022). Mellander (1950) and Oshima (1979) reported bioactive peptides from foods. Different sources of bioactive peptides have been determined in recent years. They can be derived from milk proteins, animal protein, seafoods, plant sources, or other sources like microalgae and edible insects (Devi et al., 2022; Kumari et al., 2022). Various bioactive peptides are also reported with multifunctional activities depending on composition and sequence of amino acids (Abedin et al., 2022; Padhi et al., 2021). These peptides are proved to have more than one bioactivity in a single sequence of amino acids (Lammi et al., 2019). Peptides can be generated using enzymes, microbial fermentation, or during food processing. Online available databases like PIR, AHTPDB, and BIOPEP are also useful to determine the homology match with the proteins of interest and bioactivity.

Now a days, molecular docking is becoming a popular approach to study the affinity of peptides at the targeted site in the body (Ashokbhai et al., 2022; Chourasia et al., 2022a; Sanjukta et al., 2021). *In silico* approaches are revolutionizing the field of bioactive peptides with the help of bioinformatics (Padhi et al., 2021). Validating the bio-functional activity with the required animal or *in vivo* studies is also necessary before recommending for human health care purposes. Overall, safety analysis and bioavailability at the required specific site is also needed to recommend a sequence as a valid bioactive peptide having health promoting activity. This chapter provides overview of bioactive peptides, sources, methods of generation of peptides, various bio-functional activities with mechanisms, techniques used for the isolation and characterization of bioactive peptides, and future applications of bioactive peptides.

6.2 APPROACHES FOR PRODUCTION OF BIOACTIVE PEPTIDES FROM DIFFERENT FOOD PROTEINS

Bioactive peptides can be produced from a variety of food proteins. Peptides are produced from plant and animal sources including rice, wheat, soy, broccoli, sweet potato, garlic, pulses, cheese, milk, beef, fish, pork, eggs, and so on (Jia et al., 2021). Various sources of bioactive peptides with determined bio-functional activities are presented in the Table 6.1. There are three distinct methods for generating bioactive peptides including (1) enzymatic hydrolysis, (2) fermentation through microorganisms, and (3) food processing (Danquah & Agyei, 2012).

TABLE 6.1
Bioactive Peptides, Functional Properties and Their Method of Production

Source	Bio-Functional Properties	Bioactive Peptide	Method of Generation	Reference
Milk Proteins				
Caseins	Antioxidation, antibacterial, hypocholesterolemic,anti-hypertension, opioid agonist, and mineral binding	PEL and MKP	Fermentation (*Bacilluscereus* and pepsin) Hydrolysis (subtilisin,bacillolysin, trypsin)	Ouertani et al. (2018)
Cheese	Antihypertensive activity	RPKHPIKHQ	Fermentation	Saito et al. (2000)
Whey protein	Anti-hypertension, anti-inflammatory, antioxidation, Immunomodulatory and hypocholesterolemic	LQKW	Hydrolysis (thermolysine)	Gluvic & Ulrih (2019).
Fermented milk (Camel milk, & Goat milk)	Angiotensin-I converting enzyme (ACE)-inhibitory activity	MQTDIMIFTIGPA, AFPEHK	Fermentation	Parmar et al. (2017); Solanki & Hati (2018)
Yoghurt	Antihypertensive, antithrombotic, anti-cholesterol micellar solubility and antioxidant activities	N/A	Fermentation	Abd El-Fattah et al. (2018)
Lassi	ACE-inhibitory activity	KVLPVPQK	Fermentation	Padghan et al. (2017)
Buttermilk	ACE-inhibitory activity	N/A	Fermentation	Parekh et al. (2017)
Dahi	ACE-inhibitory activity	Tyr-Gly-Gly-Phe-Leu	Fermentarian	Dabarera et al. (2015)
Plant Proteins				
Soy	ACE-inhibitory, Antiamnestic,Antioxidative, Antithrombotic, Dipeptidyl peptidase IV inhibitor	IRHFNEGDVLVIPPGVPY	Enzymatic hydrolysis	Coscueta et al. (2019)
Wheat	ACE-inhibitory	Ile-Ala-Pro	Enzymatic hydrolysis (acid protease)	Motoi & Kodama (2003)
Rice	ACE-inhibitory, antioxidant	FGGSGGPGG, FGGGGAGAGG	Enzymatic hydrolysis	Selamassakul et al. (2020)
Flaxseed	Antioxidant activity	GFPGRLDHWCASE	Enzymatic hydrolysis (Alcalase)	eSilva et al. (2017)
Chia seeds	Antimicrobial activity	GDVIAIR	Enzymatic hydrolysis	Aguilar-Toala et al. (2020)
Pea	Anti-hypertension andhypocholesterolemic	PPI and PPH	Enzymatic hydrolysis (thermolysin)	Bougle & Bouhallab (2017)
Oat bran	Antioxidant activity	YRISRQEARNLKNNRGQE	Enzymatic hydrolysis (pepsin)	Vanvi & Tsopmo (2016)
Hemp seed	Antioxidant, antihypertensive properties	WVYY, PSLPA	Enzymatic hydrolysis	Girgih et al. (2014)

(Continued)

TABLE 6.1 *(Continued)*
Bioactive Peptides, Functional Properties and Their Method of Production

Source	Bio-Functional Properties	Bioactive Peptide	Method of Generation	Reference
Animal Proteins				
Egg	ACE-inhibitory	RADHPFL and YAEERYPIL	Pepsin hydrolysate	Miguel et al. (2005)
Ham	ACE-inhibitory	KAAAAP, AAPLAP, KPVAAP, IAGRP, and KAAAATP	N/A	Escudero et al. (2014)
Oyster	Anticancer activity	Leu-Ala-Asn-Ala-Lys	Enzymatic hydrolysis	Umayaparvathi et al. (2014)
Fish	Antioxidative activity	KAGFAWTANQQLS	Enzymatic hydrolysis	Je et al. (2007)
Insect Proteins				
Edible cricket	ACE-inhibitory	YKPRP, PHGAP, and VGPPQ	Enzymatic hydrolysis (Alcalase)	Hall et al. (2020)
Hermetiaillucens larvae	Antioxidative activity	GYGFGGGAGCLSMDTGAHLNR	Enzymatic hydrolysis	Lu et al. (2022)
Alphitobius diaperinus	ACE-inhibitory activity, antimicrobial activity	Not identified	Enzymatic hydrolysis	Sousa et al. (2020)
Algal Proteins				
Chlorella vulgaris	Antioxidant activity	VECYGPNRPQF	Enzymatic hydrolysis	Sheih et al. (2009)
Navicula incerta	Antioxidant activity	PGWNQWFL,VEVLPPAEL	Enzymatic hydrolysis	Kang et al. (2012)
Chlorella ellipsoidea	Antioxidant activity	LNGDVW	Enzymatic hydrolysis	Ko et al. (2012)
Chlorella vulgaris	Anticancer activity	VECYGPNRPQF	Enzymatic hydrolysis (pepsin)	Sheih et al. (2010)
Spirulina platensis	Anticancer activity	Polypeptide Y2	Enzymatic hydrolysis (pepsin)	Zhang & Zhang (2013)
Undaria pinnatifida	ACE-inhibitory activity	YH, KY, FY, IY	Hot water extraction	Suetsuna et al. (2004)
Palmariapalmata	Renin-inhibitory activity	IRLIIVLMPILMA	Enzymatic hydrolysis	Fitzgerald et al. (2012)
Palmariapalmata	Antiatherosclerosis peptides	NIGK	Enzymatic hydrolysis	Fitzgerald et al. (2013)

6.2.1 Hydrolysis through Proteolytic Enzymes

Many bioactive peptides have been reported to be released on protein hydrolysis using microbial enzymes (Gobbetti et al., 2007). Commercially produced proteolytic enzymes like thermolysin, alcalase, subtilisin, neutrase®, actinase E®, and flavourzyme® are used to produce bioactive peptides from a wide range of protein sources (Hathwar et al., 2011; Korhonen, 2009; Mora, 2014). Many commercial neutral proteases like validase from *Aspergillus oryzae* and alcalase from *Bacillus licheniformis* are employed to produce soybean protein hydrolysates. Novel hydrolytic enzymes (extracellular proteases) from keratinolytic bacterium *Chryseobacterium species* kr6 have also been found to produce hydrolytic enzymes (de Oliveira et al., 2015). Among different types of proteases, alcalase is one of the commonly used proteases for the production of bioactive peptides.

6.2.2 Microbial Fermentation

Microbial fermentation is one of the methods used to produce bioactive peptides where the peptides are produced using proteolytic starter culture (Chourasia et al., 2022b; Sanjukta et al., 2021). A wide range of fermented foods are reported for the production of bioactive peptides including fermented milk, fermented legumes (including soybean), and fermented meat and fish products (Kumari et al., 2022). Proteins in fermented dairy foods are reported for production of various peptides due to the proteolytic activity of starter cultures and non-starter lactic cultures. Peptidases and proteinases present in lactic acid bacteria can generate bioactive peptides (Christensen, 1999; Williams et al., 2002). In their study, Solanki and Hati (2018) produced, isolated, and characterized the bioactive peptides from fermented camel milk using proteolytic lactic culture. Kathiriya (2014) studied Angiotensin-I converting enzyme (ACE)-inhibitory activity in milk fermented by different isolates and production of milk peptides. Hati et al. (2015) used milk fermented with *Lactobacilli* isolates to study ACE-inhibitory and antimicrobial activity. There are possibilities of production of bioactive peptides in novel fermented foods with specific health benefits.

6.2.3 Food Processing

It was discovered that bioactive peptides can be produced during food processing and storage. The plasmin present in milk can contribute to hydrolysing the milk proteins to produce bioactive peptides (Dalsgaard, 2008). Endogenous muscle peptidases produced peptides after several months of ripening in meat and meat products, such as dry cured ham (Mora et al., 2014). Several bioactive peptides are also produced during storage and the ripening of milk product (Rai et al., 2017). Food components having active proteases can result in proteolysis and production of bioactive peptides. The different approaches of bioactive peptide production are given in Table 6.2.

TABLE 6.2
Methods to Produce Bioactive Peptides

Method	Advantages	Disadvantages	Reference
Enzymatic hydrolysis	Controllable reaction, easy to repeat, highly safe, low cost	Lower efficiency, more time for enzymatic reaction	Jia et al. (2021)
Microbial fermentation	Maximum yield and lower cost	More time of fermentation and safety aspects	
Chemical extraction	Extensive application and advanced technologies	More costly, environment pollution, risk of health	
Recombination of DNA	Bulk production and lower cost	Limited production of short peptides and more time in research analysis	

6.3 BIOACTIVE PEPTIDES: HEALTH BENEFITS

6.3.1 Antihypertensive Peptides

Hypertension is a health problem affecting people worldwide and can lead to serious health issues including the development of cardiovascular diseases and sudden death. Caseins, whey proteins, meat, and other foods proteins are applied as a source to generate bioactive peptides using enzymatic/bacterial hydrolysis (Murray & Fitzgerald, 2007). The renin-angiotensin system (RAS) and kinin-nitric oxide system (KNOS), include different metabolic pathways related to hypertension (Martinez-Maqueda et al., 2012). The effect of bioactive peptides in reducing the chances of hypertension by different mechanisms is explained in the following sections.

6.3.1.1 Angiotensin-I Converting Enzyme (ACE) Inhibitory Peptides

In the RAS, angiotensinogen is converted to angiotensin II (vasoconstrictor peptide), which is catalysed by ACE and renin during intermediate steps. Reduction in the activity of ACE can lead to lower blood pressure. ACE-inhibitory peptides are reported from different sources including animals, plants, and microbial sources (Keska & Stadnik, 2022; Lu et al., 2022; Wang et al., 2022; Xue et al., 2021). In kinin-nitric oxide system, ACE inhibits the vasodilatory peptides kallidin and bradykinin. Bradykininbinds to β-receptors and through stimulation of nitric oxide synthase (NOS) produces vasodilation (Fitzgerald et al., 2004).

6.3.1.2 Renin-Inhibitory Peptides

Renin inhibition prevents the formation of potent vasoconstrictor (Ang- II) (Staessen et al., 2006). Renin-inhibitory peptides are more specific and reported from hydrolysis of several food proteins (Staessen et al., 2006; Udenigwe et al., 2012). Pea protein-derived peptides have been reported to possess renin-inhibitory activity (Li & Aluko, 2010). Formation of these peptides depends on the protein source and the specificity of the microbial enzyme applied for hydrolysis.

6.3.1.3 Calcium Channel Blocking Effects

Calcium channel blockers interact with voltage-gated calcium channels (VGCCS) in cardiac muscle and blood vessel walls to reduce intracellular calcium and, as a result, vasoconstriction. Several studies have demonstrated that peptides can act as calcium channel blockers (Depuydt et al., 2021; Pringos et al., 2011; Sousa et al., 2013). His-Arg-Trp peptides have been defined to have vaso-relaxative properties in the phenylephrine-contracted thoracic aorta (Tanaka et al., 2009).

6.3.1.4 Opioid-Induced Blood Pressure Regulation

Opioid peptides are defined for having the ability to bind to opioid receptors and produce morphine-like effects. These receptors may be involved in a variety of regulation processes in the body, such as blood circulation regulation, which can affect blood pressure. Casein, wheat gluten, and rapeseed have all been found to contain opioid peptides (Jauhiainen & Korpela,2007). Oral administration of α-lactorphin (Tyr-Gly-Leu-Phe) had an antihypertensive effect in spontaneously hypertensive rats (SHR) and normotensive Wistar Kyoto rats (WKY) (Nurminen et al., 2000).

6.3.1.5 Endothelin-Converting Enzyme (ECE) Inhibitory Peptides

Endothelin-converting enzyme (ECE) releases the vasoconstrictory peptide endothelin-1 (ET-1) from big endothelin-1 (big ET- 1) (ECE). ET-1 regulates vasoconstriction through two receptors, *eta* and *etb* (Fitzgerald et al., 2004). The antihypertensive effect of the egg protein-derived peptide ovokinin f (2-7) is mediated by endothelial-dependent NOS release (Matoba et al., 1999). Ovokinin (2-7) modulates cardiovascular activity by interacting with B2 bradykinin receptors (Scruggs et al., 2004). Okitsu et al. (1995) found ECE-inhibitory peptides from beef and bonito pyrolic appendix. Bovine β-lactoglobulin-based peptide Ala-Leu-Pro-Met-His-Ile-Arg inhibited the release of ET-1 in cultured porcine aortic endothelial cells (PAECs) (Maes et al., 2004).

6.3.2 Antimicrobial Peptides

These peptides are recognized as a critical "first line of defense against many pathogens" (Mulero et al., 2008). Antimicrobial peptides (Amps) are small peptides (from 12 to 80 amino acids), with low molecular weight (from 1 to 5 KDa) and mostly found as cationic and amphipathic (Brogden et al., 2003). Amps play an important function in inflammation and its modulation in the innate immune system (Cuesta et al., 2008). Amps can regulate the adaptive response (Oppenheim et al., 2003), act as chemokines to recruit other effectors cells (Chertov et al., 1996), and are promising candidates as potential therapeutic molecules (Bridle et al., 2011). Most Amps have a well-defined mechanism of action, which involves either better penetration and interrupting microbial membrane integrity or translocating across the membrane and acting on internal targets (Steinstraesser et al., 2011). Many Amps isolated from milk proteins (Demers-Mathieu et al., 2013; Shin et al., 1998) and soy proteins (Vasconcellos et al., 2014) and egg protein (Zambrowicz et al., 2014) are used in day to day lives through food sources and therefore is always a focus of many researchers for a healthy functional foods with most potent Amps.

6.3.3 Antidiabetic Peptides

Being overweight places extra stress on the body, including difficulty maintaining blood sugar levels, which leads to diabetes. A hike in blood glucose level (hyperglycemia) is a prime concern in the context of diabetes mellitus caused by unhealthy food habits. The human digestive system includes glycosidic cleaving enzymes like α-amylase from the pancreas and saliva, and α-glucosidase from the intestine for carbohydrate metabolism. The need for naturally available peptides is rising due to the side effects (diarrhoea, abdominal pain etc.) caused by synthetic α-amylase and α-glucosidases inhibitors that are available commercially (Arise et al., 2019). Aglycin is a natural antidiabetic peptide from soybean (Dun et al., 2007) and shows antidiabetic effects in diabetic mice (Lu et al., 2012). Similarly, a study on Nigerian seed protein hydrolysate (alcalase) from sponge guard (*Luffa cylindrica*) showed that it expresses effective α-amylase inhibitory activity with IC_{50} value of 1.02 ± 0.19 mg/mL (Arise et al., 2019). Another study on walnut alcalase-derived protein hydrolysate hasreported to show the DPP-IV inhibitory peptides IVVTRGRAT, QEDDNRR, RAPRMRWI, REEEQQR, QEERQEQR, AGGEPRDGQSGQ, MRPDEDEQEGQ, DDEENPRDPRE and GNPDDEFRPQ (Kong et al., 2021). Among the animal sources, naturally occurring peptides AAATP and KA from Spanish dry cured ham are proven to be important DPP-IV inhibitors with IC_{50} value of 6.47mM and 6.27mM, respectively (Gallego et al., 2014). In the recent past, the salmon skin collagen hydrolysate has been claimed to owe DPP-IV inhibitory hexapeptide LDKVFR with an IC_{50} value of 128.71μM, effective for preventing or managing type 2 diabetes (Jin et al., 2020). The limited but continuous exploration of antidiabetic peptides from natural sources is a global process among researchers in the hope of combatingpoor dietary habits leading to diabetes and other disorders.

6.3.4 Antioxidative Peptides

Oxidative stresses advertently trigger certain cardiovascular diseases (CVD) as well as inflammatory diseases (Ramalingam & Kim, 2012). Bioactive peptides show antioxidative activity through various mechanisms (Aluko, 2012; Sanjukta et al., 2021). Peptides exhibit antioxidant properties due to the presence of specific hydrophobic amino acid residues such as methionine, cysteine, and histidine; and aromatic amino acids like phenylalanine, tryptophan, and tyrosine (Wong et al., 2020). Further, bioactive peptides have been reported with the ability to inhibit specific enzymes responsible for oxidative stress (Sanjukta et al., 2021). Antioxidant peptides have already been isolated in hydrolysate from soybean (Park et al., 2010), sweet potato (Zhang et al., 2012), milk proteins (Rival et al., 2001), eggs (Aluko, 2012), and meat proteins (Mora et al., 2014). Future research on finding naturally occurring antioxidant peptides from different protein sources using microbial enzymes has always captivated researchers' interest to explore for more.

6.3.5 Anticancer Peptides

Cancer is a globally occurring health issue, which can be controlled by nutritional supplements and diet (Uster et al., 2013). Food proteins/peptides can play a significant role in the prevention of cancer in different stages, such as initiation, promotion, and progression (de Mejia & Dia, 2010). Anticancer peptides are isolated from milk proteins (caseins), soy proteins (Lunasin) and many marine organisms (Kim et al., 2000, 2012; Otani & Suzuki, 2003; Roy et al., 1999). Bioactive peptides may exert the protective effect through the mechanisms supporting cell viability and immune cell functions (Meisel, 2005). There are possibilities of applying enzymatic hydrolysis procedures for the production of novel anticancer peptides.

6.3.6 Opioid Peptides

Opioid peptides possess some form of affinity for opiate receptors and can exert effects on the nervous system and gastrointestinal functions. Endorphins and enkephalins are naturally occurring opioid peptides in the body (usually in the brain and pituitary gland), whereas exorphins are opioid peptides obtained from the enzymatic digestion of food proteins. Opioid peptides are categorized as opioid agonists and antagonists with different functions (Aluko, 2012). Many milk casein-based opioid agonist peptides were reported such as, β-Casomorphins (Rokka, 1997), and peptides derived from whey protein, such as α-lactorphin and β-lactorphin (Pihlanto-Leppala, 2000). Opioid antagonists (casoxins) are encoded in the bovine κ-casein sequences (Yoshikawa, 1994).

6.3.7 Antithrombotic Peptides

Thrombosis is a local coagulation of blood in the circulatory system that can lead to CVD (Grundy et al., 1998). Antithrombotic peptides are substances that prevent blood clotting and may be used to treat thrombosis. Bovine κ-casein and human lactoferrin have been shown to inhibit thrombosis (Fosset & Tome, 2000). Foods with anti-thrombotic peptides have been progressively revealed in the past two–three decades. Peptides from fishes *viz.* blue mussel (Qiao et al., 2018), yellowfin sole, goby mussel, and mackerel skin have been reported for their anti-thrombotic activities (Cheng et al., 2019). Similarly, daily dietary foods like oyster (Chen et al., 2019), granulated ark clam shells and echiuroid worms have also been reported to carry anti-thrombotic peptides. Certain plant sources likeoaks, peanuts, highland barley, amaranth, buckwheat and even sea weeds – are also claimed to possess anti-thrombotic peptides rich in platelet aggregation and interact with thrombin to stop blood coagulation. A study on casein-derived peptide also shows anti-thrombotic properties (Cheng et al., 2019). There are still possibilities of finding novel antithrombic peptides on enzymatic hydrolysis of food proteins.

6.3.8 Hypocholesterolemic Peptides

Hypercholesterolemia is positively linked with the likelihood of developing CVD (Grundy et al., 1998). Many proteins/peptides have been shown to lower cholesterol levels (e.g., soy glycinin fragment, soy protein, pork protein hydrolysate, milk β-lactoglobulin hydrolysate) with soybean being one of the most well-known sources of hypocholesterolemic peptides. The ability of soybean protein/peptides to lower cholesterol appears to be related to their bile-acid-binding capacity (Makino et al., 1988). By influencing intestinal emulsification and the nature of the resulting micelles, whey proteins/peptides influence cholesterol absorption and serum cholesterol levels (Nagaoka et al., 2001). Bioactive peptides from other protein sources can be explored for their hypocholesterolemic property.

6.3.9 Immunomodulatory and Cytomodulatory Peptides

Immunomodulatory peptides have the capacity to boost the activity of immune cells (Horiguchi et al., 2005). Immunomodulatory peptides have the potential to reduce allergies and boost mucosal immunity in the gastrointestinal tract (Korhonen & Pihlanto, 2003). Immunomodulatory peptides also activate the body's nonspecific defence system (Kitts & Weiler, 2003). These peptides enhance the response by the immune system through proliferating lymphocytes, cytokine regulation, and natural killer (NK) cell activity. Immunomdulatory peptides like clavanin-MO, clavanin-A, and defensin are known for destroying biofilms, influencing the inflammatory response, cytokine modulation, and influencing innate and adaptive immune system in various ways (Kang et al., 2019). Peptides from yak (*Bos grunniens*) bone (collagen) hydrolysates have been found to be a potential immunomodulatory agent (Gao et al., 2019). Many plant sources like *Amaranthus* species, *Linum* species, *Helianthus* species, *Glycine max*, *Clitoria ternatea*, *Hordium vulgare*, *Juglans regia*, *Oryza sativa*, and *Pinus* species have been reported to hold pro and anti-inflammatory peptides for human defence system (Pavlicevica et al., 2022). Similarly, oligopeptides from fermented cottonseed meal have also been found with immunomodulatory activities in Balb/c immunosuppressed mice (Liu et al., 2018). Food-based bioactive peptides influence the viability (proliferation, differentiation, and apoptosis) of various cell types. Apoptosis has been shown to be triggered by some milk-derived peptides. These cytomodulatory and immunomodulatory properties may work together to protect against tumour formation (Meisel & Fitzgerald, 2003).

6.3.10 Mineral-Binding Peptides

Mineral-binding peptides are derived from milk casein and egg having anticariogenic-like activity. These peptides are composed of serine-phosphate residues. Casein-derived phosphorylated peptides can form soluble organophosphate salt and lead to enhanced mineral uptake, and increase recalcification of enamel (Meisel, 1998). Structural features of each peptide vary in specific amino acid sequence, based on which the peptides show different biological activity (Erdmann et al., 2008). An overview of the production of bioactive peptides and different health benefits is given in Figure 6.1.

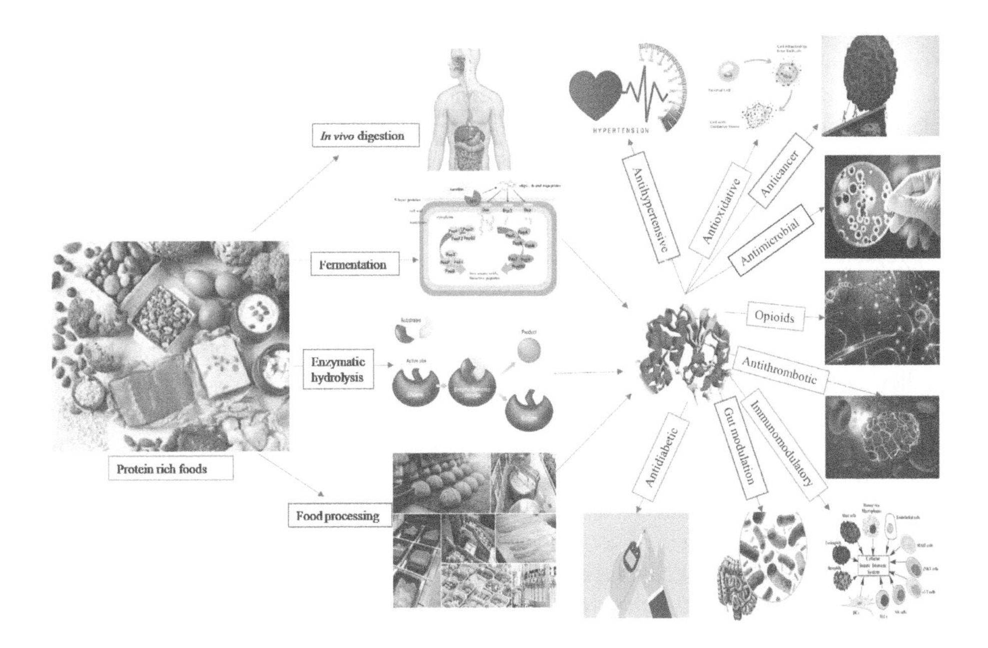

FIGURE 6.1 Health benefits of bioactive peptides produced using different approaches.

6.4 CLINICAL STUDIES: APPLICATION OF BIOACTIVE PEPTIDES

A large proportion of human studies have been carried out for the validation of antihypertensive activities as compared to other activities like antioxidant and anticancer (Agyei & Danquah, 2012; Rai et al., 2017). Before proposing the health benefits, it is compulsory to find out dosage amount and correct states. Health functions are studied following many screening methods and series of experiments (Hartmann & Meisel, 2007). Many clinical trials have been performed to study the specific bio-functionality of bioactive peptides (Aihara et al., 2005; Mizuno et al., 2005; Mizushima et al., 2004; Walker et al., 2009).

6.5 LIMITATIONS: ALLERGENICITY, STABILITY, AND BIOAVAILABILITY

Bioactive peptides have been reported to be less allergenic than their native proteins (Host & Halken, 2004). Bioactive peptides resist enzymatic attack as they are more stable and are more bioavailable due to the presence of a specific sequence of amino acids. Despite the richness in availability and consumption of bioactive peptides from naturally occurring food sources and commercials, there are a few barriers which can still limit their use. The prime concern with these bioactive peptides is low bioavailability and shorter half-lives when ingested. Many bioactive peptides are also sometimes unequally distributed due to changes in confirmations, lack of receptor selectivity, and activation, gradually resulting in side effects. Peptides are also found to be adsorbed or aggregated, or become denatured while passing through different environments of the gastrointestinal tract. Additionally, bioactive peptides synthesized commercially are expensive due to high production costs (Haggag et al., 2018). These limitations again challenge the researchers worldwide to look out for effective, natural, and potent bioactive peptides with cost effectiveness for the general public.

6.6 MICROBES FROM FERMENTED FOODS IN THE GENERATION OF BIOACTIVE PEPTIDES

Microbes are capable of biosynthesizing certain metabolites (including peptides with bioactive properties) while utilizing the substrates during fermentation. This is the most economical and the easiest process in peptide generation. Foods rich in protein sources (e.g.,milk, fish, and soybean) serve as substrates to many microbes during fermentation (Kumari et al., 2022). The fragmentation of larger protein molecules through natural fermentation of foods rich in proteins demands a combined action of different enzymes which are released by microbes (Sharma et al., 2020). The association of certain microflora with "generally recognized as safe" (GRAS) status not only is economical and helps to improve the organoleptic properties, it also assists in the generation of different peptides. Consequently, due to their numerous advantageous roles, microflora are widely applied in food industries in the generation of bioactive peptides with specific bioactivity (Pessione & Cirrincione, 2016; Sharma et al., 2020; Venegas-Ortega et al., 2019). Some yeasts, *Bacillus* sp., lactic acid bacteria (LAB), and moulds are eminent producers of health-promoting bioactive peptides through their complex yet only partially understood enzymatic systems (Fernandez et al., 2015).

The amalgamation of different enzymes like proteinases, proteases, and amylases released during fermentation is helpful in the generation of different bioactive peptides. These enzymes are released by proteolytic microorganisms (e.g. LAB, *Bacillus*, yeasts, and fungal species) during the process of fermentation (Fernandez et al., 2015; Panchal et al., 2021; Phelan et al., 2009; Rai et al., 2016; Sharma et al., 2020). However, certain species of *Staphylococcus* are also active participants in releasing bioactive peptides (Phelan et al., 2009; Toldra et al., 2017). For example, *Staphylococcus carnosus* and *Staphylococcus xylosus* are found to release antioxidant peptides which also prevent food spoilage (Gallego et al., 2018). These microorganisms are a potential source of proteolytic enzymes that can be applied to the production of bioactive peptides.

6.6.1 Types of Microbial Proteases and Their Role in the Generation of Bioactive Peptides

Microbial proteases account for two-thirds of the available commercial proteases in the enzyme industries and are produced by a diverse group of microbes either by submerged or solid-state fermentation process. For their growth and survival, microbes utilize proteins as substrate and release certain enzymes through their proteolytic system to generate peptides, many of which hold bioactive properties also. The breakdown of proteins from their conjugated forms is usually originated by proteinases, secretory proteases, and peptidases. These can be related to either intra- or extra-cellular processes. The intra-cellular proteases are useful for cell sustenance and carrying out the metabolic processes through cell differentiation and sporulation, whereas the extra-cellular proteases hydrolyse the protein molecules for utilization as substrate in their simplest forms. Depending on the proteolytic system in different microbial species, proteases are generated for a wide range of applications in different industries (Martinez-Medina et al., 2019; Sony & Potty, 2016). Based on their cleavage site, microbial proteases are classified into four classes: (1) exopeptidases – release free amino acids by acting on terminal peptide bonds, either at N or C terminal; (2) endopeptidases – release short oligo-peptides while acting on internal peptide bonds of proteins; (3) aminopeptidases – release di-, tri- peptides or free amino acids by acting only on N terminal of the polypeptide chain; and (4) carboxypeptidases – act only on C terminal to release peptides (Jisha et al., 2013). However, proteases, on the basis of their functional groups at the active site, are again classified into four important groups:(1) serine proteases – having serine group at their active site; (2) cysteine proteases – consisting of cysteine and histidine at their active site; (3) aspartic proteases – consisting of aspartic acid residue at their active site; and (4) metaloproteases – require divalent metal ions for their action. All these group of proteases can either be acidic (pH 2–6), alkaline (pH8–13), or neutral (pH ~7) based on the environment of their release and action (Martinez-Medina et al., 2019).

Depending on their source of isolation, microbial proteases can be bacterial or fungal (Martinez-Medina et al., 2019). Bacterial proteases are generally released by species of *Bacillus*, LAB, some *Staphylococci, Streptococci,* and *Enterococci* (Cheng et al., 2017; Choeisoongnern et al., 2021; Gallego et al., 2018; Rai et al., 2017). Similarly, certain yeasts and fungi (e.g. species of *Aspergillus, Mucor, Rhizopus, Saccharomyces*) are widely reported for their proteases so as to form bioactive peptides (Akbarian et al., 2022; Jisha et al., 2013; Martinez-Medina et al., 2019; Rai et al., 2016,2019). The role of proteases or proteinases released by microbes (specifically LAB) in bioactive peptide formation is discussed in the next section.

6.6.2 Proteolytic Systems of Lactic Acid Bacteria and Its Role in the Generation of Bioactive Peptides

Putrefaction and proteolysis of different proteins in food products through fermentation by LAB species, and generation of health-benefitting bioactive peptides through natural fermentation using microbes is one of nature's gifts to human beings. Several species of LAB are capable of fermenting protein-rich foods (specifically caseins or other milk proteins where nitrogen is abundantly available as a substrate) so as to release oligopeptides and free amines with bioactivities in an efficient though very complicated way (Chourasia et al.,2022a; Tagliazucchi et al., 2019). The metabolic pathway in LAB is efficient enough in acidifying the environment during fermentation which leads to the formation of lactic, formic, and acetic acid (in hetero fermenters) and lactic acid (in homo fermenters). The proteolysis and peptide generation in LAB varies from species to species. For example, the genome of *Lactobacillus* species encodes a variety of amino acid permeases, proteases, and peptidases, and oligopeptide and di-tripeptide transport system (ATP-binding cassette transport system) in comparison to many *Lactococci* (Venegas-Ortega et al., 2019). Majority of oligopeptides with bioactivity are generally released in the cell via cell autolysis (e.g., in *Lactococcus lactis*). The cell envelope proteinases in LAB play a major role in the formation of oligopeptides by hydrolysing the extracellular proteins (Pessione & Cirrincione,2016; Tagliazucchi et al., 2019). There are four different cell envelope proteinases in LAB *viz.* (1) Prt B in *L. delbrueckii* subsp. *Bulgaricus*, (2) Prthin *L. helveticus*, (3) Prtpin *L. casei* and *L. paracasei*, and (4) Prtrin *L. rhamnosus* and *L. plantarum* (Sharma et al., 2020). Post formation of the oligopeptides, they are then transported to the cytoplasm through an oligopeptide transport system for further fragmentation into short peptides, free amines, or precursor molecules of bioactive peptides. The oligopeptide transport system in LAB belongs to an ATP-binding cassettes transporter system, a highly conserved superfamily in cell transport systems. Some strains of *Lactobacillus delbrueckii* subsp. *Lactis* can generate antihypertensive and metal-chelating peptides directly from caseins (Pessione & Cirrincione, 2016).

In dairy sources, casein proteins are classified into four major types namely αs1, αs2, β, and κ-casein the ratio 38:11:38:13. The species of LAB (e.g., *Lactococcus*) are also found to possess the cell envelope proteinases (serine proteases, intracellular peptidases, and some carboxypeptidases) classified into different types and subtypes in accordance with the utilization of their substrate (Hajirostamloo et al., 2010; Tagliazucchi et al., 2019; Toldra et al., 2017). Species of *Lactobacillus* and *Lactococcus* are able to hydrolyse the milk and other proteins in accordance with their substrate utilization strategy. When hydrolysed, the LAB proteolytic system leads to the formation of bioactive peptides. For example, the breakdown of casein proteins generates immunomodulatory, mineral binding, antimicrobial, antihypertensive, antithrombotic, and opioid or anti-opioid activities. Similarly, hypocholesterolemic, antioxidant, and immunomodulating peptides are released when whey proteins like α-lacto albumin, β-lacto globulin, immunoglobulins, and lactoferrins are broken down. Proteins of vegetable origin like β-conglycinin, when broken down, generate opioid and antihypertensive peptides (Sanjukta & Rai, 2016). Fragmentation of bovine meat proteins (e.g., collagen, serum albumin, haemoglobin, elastin, and fibrinogen) leads to generation of antihypertensive peptides (Kumari et al., 2022). Opioid peptides, some of which also have analgesic effects on the central nervous system (CNS), have been detected from wheat gluten (gluten exorphins) breakdown (Pessione & Cirrincione, 2016). All the peptides generated during fermentation are dependent on some specific factors (like acidic environment). These factors aid in proteasection, production of other important enzymes, expression of proteolytic system and release of required enzymes in order to maintain a stationary phase (Venegas-Ortega et al., 2019). Concerning to the production of effective bioactive peptides, many researchers are still in the process of discovering an ideal environment through co-culture fermentation techniques.

6.7 BIOACTIVE PEPTIDE PRODUCTION BY ENZYMATIC HYDROLYSIS AND MICROBIAL FERMENTATION

There have been some developments in methods of peptide formation by chemical synthesis, autolysis, and acid hydrolysis *in vitro*. An omnipresence of enzymatic hydrolysis and microbial fermentation methods are still observed as classic methods for breaking down the proteins so as to generate different peptides with bioactivities. Enzymatic hydrolysis is popular as it is considered to be a cost-effective process in laboratories and industries. A set of proteases like alcalase^{tm}, flavourzyme^{tm}, and protamex^{tm} are frequently used in the production of hydrolysates, either individually or sequentially in a set, depending upon the source of protein and type of peptide targeted for relevant bioactivity (Boukil et al., 2020; Marciniak et al., 2018; Rai et al., 2016; Shahidi & Zhong, 2008; Sun, 2011). A major advantage of enzymatic hydrolysis is that it does not leave behind any toxic or organic solvent as by-product. Further, the degree of hydrolysis can be controlled by parameters like time, temperature, pH, and enzyme-substrate ratio. In other words, these are the four major factors for the controlled production of the unique peptide (Marciniak et al., 2018; Shahidi & Zhong, 2008).

Similarly, another common and economical approach for peptide production is microbial fermentation. This method requires use of GRAS-status microorganisms. Species of *Lactobacillus* (Chourasia et al., 2022b; Mishra et al., 2018), some selected species of *Bacillus* (Sanjukta et al., 2015, 2021), *Bifidobacterium* (Marciniak et al., 2018), and yeast (Rai et al., 2016), are found extensively in use for protein hydrolysis and production of free amino acids and bioactive peptides (Kumari et al., 2022). Due to the minimal media required for culture growth and production to ferment the substrates on a large scale, it is easy for both researchers and industrialists to produce bioactive-rich peptides via microbial fermentation (Marciniak et al., 2018). Many researchers prefer microbial fermentation as it increases the chances of production of novel peptides with relevant bioactivity by the action of microbes, either in group or independently in the natural environment. A consortium of microbes can even be used for production of peptide(s) with more than one bioactivity (Mishra et al., 2019; Patel & Hati, 2017; Sanjukta et al., 2015). Many LAB species produce proteases at the cell membrane, which again are easy to harvest. Nevertheless, the process of microbial fermentation can be unserviceable if not carried out appropriately. For example, degree of fermentation, pH, incubation time and temperature, maintenance of sterile conditions, use of fresh inoculums, etc. are the factors which need to be optimized (Marciniak et al., 2018). In order to produce efficient bioactive peptides, a combined method of both enzymatic hydrolysis followed by microbial fermentation are also used by researchers (El-Salam & El-Shibiny, 2019; Hong LE et al., 2021).

6.8 SAFETY ASPECTS OF PEPTIDES GENERATION USING MICROBES

Production of bioactive peptides and free amino acids through microbial fermentation can also result in formation of other allergic and toxic by-products. While microbes during fermentation aid to aroma, texture, flavour, shelf-life, and bioactivity of the resultant product, they also bring concomitant release of unwanted metabolites like biogenic amines, cyanogenic glycosides, and toxins like bacterial toxins (enterotoxins, shiga toxin, neurotoxins etc.) and mycotoxins (e.g., Aflatoxins, ochratoxins). The over-production of these toxic metabolites can bring health-related issues. Also, failure to check for pathogenic microbes (e.g., species of *Staphylococcus*, *Streptococcus*, *Listeria*, *Brucella*, *Salmonella* and pathogenic strains of *E.coli* like STECO157:H7), specifically during milk, soybean, and fish fermentation can have serious consequences (Beermann & Hartung, 2012; Fernandez et al., 2015; Sivamaruthi et al., 2018). Aiming to avoid these obstacles, basic microbiology techniques like examining the pathogens by isolation in selected media and molecular methods (e.g., PCR, ELISA) for specific species and toxin identification can serve the purpose (Alahi & Mukhopadhyay, 2017; Rohm et al., 2010). The excessive production of toxins (mycotoxins) can also be reduced by use of LAB (*Lactobacillus*, *Streptococcus* and *Lactococcus*) and yeast (*Saccharomyces*) species.

Species of *Bacillus* (*Bacillus subtilis* HJ18-4 and *B. subtilis* RD7-7) have been reported to be useful in reducing food contamination (Kumari et al., 2022). Some *Aspergillus* strains (*Aspergillus oryzae*MAO103 and *A. oryzae* MAO104) have also been found to reduce the content of aflatoxin B1 upto 90%. For a qualitative fermented product with reduced contaminants, the use of extracts of some medicinal plants (e.g., *Syzygiumaromaticum*, *Ginkgo biloba,* and *Nelumbo nucifera*) is also popular (Sivamaruthi et al., 2018). The researchers are continually finding novel ideas to increase safety, aroma, flavour, bioactivity, shelf-life, texture, and reduced cooking time with decreased toxicity, cytotoxicity, and allergenicity in different protein-rich fermented food products.

6.9 EXTRACTION AND PURIFICATION OF PEPTIDES FROM FERMENTED MEDIA

Bioactive peptides are usually released in the non-proliferative broth media from where they are extracted and purified by downstream processing. The extraction and purification of peptides become essential for an appropriate demonstration of the selected bioactivity (Raveschot et al., 2018). The classical method of peptide segregation involves the membrane separation techniques (chiefly for milk proteins) based on the difference in molecular weight. Smaller peptides from residues with high molecular mass can be easily extracted using cut-off membranes with low molecular mass (below 2000 Da) by the stepwise ultra-filtration technique. Selective ultrafiltration membranes

(30 to 1KDa) are also used for production and enrichment of opioid and ACE-inhibitory peptides (Korhonen & Philanto, 2006). Another standard and cost-effective method of peptide separation is precipitation method followed by centrifugation. The method involves the addition of ethanol or ammonium sulphate $(NH_4)_2SO_4$. This method is easy to scaleup and selectively segregate the hydrolysed proteins quickly (Raveschot et al., 2018). Trichloroacetic acid (TCA) precipitation methods have also been applied to study the hydrolysis of several *Lactobacillus* fermented milk (Aguilar-Toala, 2017; Gandhi & Shah, 2014; Solieri et al., 2015). Peptide separation without added chemicals includes membrane ultra- and nano-filtration techniques. Through these techniques, the hydrolysis of proteins and separation of peptides can be carried on simultaneously in a membrane reactor (Raveschot et al., 2018).

Post extraction of peptides, the best applied techniques for high selectivity and separation with the highest resolutions are chromatographic methods like ion-exchange and size-exclusion chromatography (Hernandez-Ledesma et al., 2014; Wu et al., 2013). This method is advantageous for the purification of a specific peptide(s) selected for its related bioactivity. However, the aforementioned technique is not very cost effective and involves the dilution of peptides and accumulation of solvent waste, hence is preferrable for research rather than in industrial applications (Raveschot et al., 2018). In order to separate the peptides of acceptable standards with better efficiency, electro-membrane filtration (EMF) and selective membrane chromatography (MC) techniques are applied. The EMF involves the combined principles of membrane filtration and electrophoresis for separating the charged molecules, specifically neutral bioactive peptides (Agyei et al., 2016). This method is beneficial in improving the protein solution's permeable flux and subsequently controlling the concentration polarization and membrane fouling (Bargeman et al., 2002). Likewise, MC techniques are utilized for peptide separation from different sources vialigand- protein/peptide interactions (Agyei et al., 2016; Saxena et al., 2009).

6.10 CONCLUSIONS

Microbial enzymes are produced by LAB, *Bacillus* spp., yeasts and fungi groups and have been shown to have potential application in the production of bioactive peptides. Health-benefitting effects like antioxidative, antihypertensive, immunomodulatory, antimicrobial, and hypocholesterolemic are achieved from bioactive peptides formed through different enzymatic processes. The enzymatic hydrolysis of complex protein molecules and microbial fermentation methods for peptide generation, are important for application to many food industries so as to bring out effective, and health benefiting outgrowth(s). These methods,however, produce an ample number and variety of peptides in different food products, but also need extra attention in process development. The purpose of safe and effective bioactive peptide generation, their extraction, and purification can be more efficient with additional recommendations to the guidelines and inclusions of advance formulae. Future studies on exploring novel microbial strains need to be carried out for production of efficient enzymes with novel catalytic ability. These enzymes can be applied for production of novel bioactive peptides with multifunctional properties.

REFERENCES

Abd El-Fattah, A., Sakr, S., El-Dieb, S. and Elkashef, H. 2018. Developing functional yogurt rich in bioactive peptides and gamma-aminobutyric acid related to cardiovascular health. *LWT Food Science and Technology* 98: 390–397.

Abedin, M.M., Chourasia, R., Phukon, C. L., Singh, S.P. and Rai, A. K. 2022. Characterization of ACE inhibitory and antioxidant peptides in yak and cow milk hard chhurpi cheese of the Sikkim Himalayan region. *Food Chemistry: X* 13: 100231.

Aguilar-Toala, J.E., Deering, A.J. and Liceaga, A.M. 2020. New insights into the antimicrobial properties of hydrolysates and peptide fractions derived from chia seed (*Salvia hispanica L.*). *Probiotics and Antimicrobial Proteins* 12(4): 1571–1581.

Aguilar-Toala, J.E., Santiago-Lopez, L., Peres, C.M., Peres, C., Garcia, H.S., Vallejo-Cordoba, et al. 2017. Assessment of multifunctional activity of bioactive peptides derived from fermented milk by specific *Lactobacillus plantarum* strains. *Journal of Dairy Science* 100(1): 65–75.

Agyei, D. and Danquah, M.K. 2012. Industrial-scale manufacturing of pharmaceutical-grade bioactive peptides. *Biotechnology Advances* 29(3): 272–277.

Agyei, D., Ongkudon, C.M., Wei, C.Y., Chan, A.S. and Danquah, M.K. 2016. Bioprocess Challenges to the Isolation and Purification of Bioactive Peptides. *Food and Bioproducts Processing* 98: 244–256.

Aihara, K., Kajimoto, O., Hirata, H., Takahashi, R. and Nakamura, Y. 2005. Effect of powdered fermented milk with *Lactobacillus helveticus* on subjects with high-normal blood pressure or mild hypertension. *Journal of the American College of Nutrition* 24(4): 257–265.

Akbarian, M., Khani, A., Eghbalpour, S. and Uversky, V.N. 2022. Bioactive peptides: Synthesis, sources, applications, and proposed mechanisms of action. *InternationalJournal of Molecular Sciences* 23(3): 1445.

Alahi, Md E.E. and Mukhopadhyay, S.C. 2017. Detection methodologies for pathogen and toxins: A review. *Sensors* 17(8): 1885.

Aluko, R. E.2012. Functional Foods and Nutraceuticals. New York, NY, USA: Springer. (pp. 37–61).

Arise, R.O., Idi, J.J., Mic-Braimoh, I.M., Korode, E., Ahmed, R.N. and Osemwegie, O. 2019. *In vitro* Angiotesin-1-converting enzyme, α-amylase and α-glucosidase inhibitory and antioxidant activities of *Luffa cylindrical* (L.) M. Roem seed protein hydrolysate. *Heliyon* 5(5): e01634.

Ashokbhai, J.K., Basaiawmoit, B., Das, S., et al. 2022. Antioxidant, antimicrobial and anti-inflammatory activities and release of ultrafiltered antioxidative and antimicrobial peptides during fermentation of sheep milk: Invitro, insilico and molecular interaction studies. *Food Bioscience* 47: 101666.

Bargeman, G., Houwing, J., Recio, I., Koops, G.H. and van der Horst, C. 2002. Electro-Membrane Filtration for the Selective Isolation of Bioactive Peptides from an α_{s2}-Casein Hydrolysate. *Biotechnology and Bioengineering* 80(6): 599–609.

Beermann, C. and Hartung, J. 2012. Physiological properties of milk ingredients released by fermentation. *Food & Function* 4(2): 185–99.

Bougle, D. and Bouhallab, S. 2017. Dietary bioactive peptides: Human studies. *Critical Reviews in Food Science and Nutrition* 57(2): 335–343.

Boukil, A., Perreault V., Chamberland, J., Mezdour, S., Pouliot Y. and Doyen A. 2020. High hydrostatic pressure-assisted enzymatic hydrolysis affect mealworm allergenic proteins. *Molecules* 25(11): 2685.

Bridle, A., Nosworthy, E., Polinski, M. and Nowak, B. 2011. Evidence of an antimicrobial-immunomodulatory role of Atlantic salmon cathelicidins during infection with Yersinia ruckeri. *Plos One* 6(8): e23417.

Brogden, K.A., Ackermann, M., McCray Jr, P.B. and Tack, B.F. 2003. Antimicrobial peptides in animals and their role in host defences. *International Journal of Antimicrobial Agents* 22(5): 465–478.

Chen, H., Cheng, S., Fan, F., Tu, M., Xu, Z. and Du, M. 2019. Identification and molecular mechanism of antithrombotic peptides from oyster proteins released in simulated gastro-intestinal digestion. *Food & Function* 10: 5426–5435.

Cheng, A.C., Lin, H.L., Shiu, Y.L., Tyan, Y.C. and Liu, C.H. 2017. Isolation and characterization of antimicrobial peptides derived from Bacillus subtilis E20-fermented soybean meal and its use for preventing Vibrio infection in shrimp aquaculture. *Fish and Shellfish Immunology* 67:270–279.

Cheng, S., Tu, M., Liu, H., Zhao, G. and Du, M. 2019. Food-derived antithrombotic peptides: Preparation, identification, and interactions with thrombin. *Critical Reviews in Food Science and Nutrition* 59: 1–15.

Chertov, O., Michiel, D.F., Xu, L.L., Wang, J.M., Tani, K., Murphy, W.J., Longo, D.L., Taub, D.D. and Oppenheim, J.J. 1996. Identification of defensin-1, defensin-2, and CAP37/azurocidin as T-cell chemoattractant proteins released from interleukin-8- stimulated neutrophils. *Journal of Biological Chemistry* 271: 2935–2940.

Choeisoongnern, T., Sirilun, S., Waditee-Sirisattha, R., Pintha, K., Peerajan, S. and Chaiyasut, C. 2021. Potential Probiotic Enterococcus faecium OV3-6 and Its Bioactive Peptide as Alternative Bio-Preservation. *Foods* 10(10): 2264.

Chourasia, R., Kumari, R., Singh S.P., Sahoo, D. and Rai A.K. 2022b. Characterization of native lactic acid bacteria from traditionally fermented chhurpi of Sikkim Himalayan region for the production of chhurpi cheese with enhanced antioxidant effect. *LWT Food Science and Technology* 154: 112801.

Chourasia, R., Phukon, L. Chiring., Minhajul Abedin, M., Sahoo, D. and Rai, A.K. 2022a. Production and characterization of bioactive peptides in novel functional soybean chhurpi produced using Lactobacillus delbrueckii WS4. *Food Chemistry* 387: 132889.

Christensen, J.E., Dudley, E.G., Pederson, J.A. and Steele, J.L. 1999. Peptidases and amino acid catabolism in lactic acid bacteria. *Antonie van Leeuwenhoek* 76(1): 217–246.

Coscueta, E.R., Campos, D.A., Osorio, H., Nerli, B.B. and Pintado, M. 2019. Enzymatic soy protein hydrolysis: A tool for biofunctional food ingredient production. *Food chemistry: X* 1: 100006.

Cuesta, A., Meseguer, J. and Esteban M.A. 2008. The antimicrobial peptide hepcidin exerts an important role in the innate immunity against bacteria in the bony fish gilthead seabream. *Molecular Immunology* 45: 2333–2342.

Dabarera, M.C., Athiththan, L.V. and Perera, R.P. 2015. Antihypertensive peptides from curd. *An International Quarterly Journal of Research in Ayurveda* 36(2): 214–219. Https://doi.org/10.4103/0974-8520.175534.

Dalsgaard, T.K., Heegaard, C.W. and Larsen, L.B. 2008. Plasmin digestion of photooxidized milk proteins. *Journal of Dairy Science* 91: 2175–2183.

Danquah, M.K. and Agyei, D. 2012. Pharmaceutical applications of bioactive peptides. *OA Biotechnology* 1(2): 1–7.

de Mejia, E.G. and Dia, V.P. 2010. The role of nutraceutical proteins and peptides in apoptosis, angiogenesis, and metastasis of cancer cells. *Cancer and Metastasis Reviews* 29(3): 511–528.

de Oliveira, C.F., Correa, A.P.F., Coletto, D., Daroit, D.J., Cladera-Olivera, F. and Brandelli, A. 2015. Soy protein hydrolysis with microbial protease to improve antioxidant and functional properties. *Journal of Food Science and Technology* 52(5): 2668–2678.

Demers-Mathieu, V., Gauthier, S.F., Britten, M., Fliss, I., Robitaille, G. and Jean, J. 2013. Antibacterial activity of peptides extracted from tryptic hydrolyzate of whey protein by nanofiltration. *International Dairy Journal* 28(2): 94–101.

Depuydt, A., Rihon, J., Cheneval, O., Vanmeert, M., Schroeder, C.I., Craik, D.J., et al. 2021. Cyclic Peptides as T-Type Calcium Channel Blockers: Characterization and Molecular Mapping of the Binding Site. *ACS Pharmacology and Translational Science* 4(4): 1379–1389.

Devi, W.D., Bonysana, R., Kapesa, K. Rai, A.K., Mukherjee, P.M. and Rajshekhar, Y. 2022. Potential of edible insects as source of functional foods: Biotechnological approaches for improving functionality.*System Microbiology and Biomanufacturing*. https://doi.org/10.1007/s43393-022-00089-5

Dun, X.P., Wang, J.H., Chen, L., Lu, J., Li, F.F., Zhao, Y.Y., et al. 2007. Activity of the plant peptide aglycin in mammalian systems. *The FEBS Journal* 274(3):751–759.

e Silva, F.G.D., Hernandez-Ledesma, B., Amigo, L., Netto, F.M. and Miralles, B. 2017. Identification of peptides released from flaxseed (*Linumusitatissimum*) protein by Alcalase® hydrolysis: Antioxidant activity. *LWT-Food Science and Technology* 76: 140–146.

El-Salam, M.H.A. and El-Shibiny, S. 2019. Reduction of Milk protein antigenicity by enzymatic hydrolysis and fermentation. a review. *Food Reviews International* 37(3): 1–20.

Erdmann, K., Cheung, B.W. and Schroder, H. 2008. The possible roles of food-derived bioactive peptides in reducing the risk of cardiovascular disease. *The Journal of Nutritional Biochemistry* 19(10): 643–654.

Escudero, E., Mora, L. and Toldra, F. 2014. Stability of ACE inhibitory ham peptides against heat treatment and in vitro digestion. *Food Chemistry* 161: 305–311.

Fernandez, M., Hudson, J.A., Korpela, R. and Reyes-Gavilan, C.G. de los. 2015. Role of microorganisms present in dairy fermented products in health and disease. *Biomed Research International* 412714.

Fitzgerald, C., Gallagher, E., O'Connor, P., Prieto, J., Mora-Soler, L., Grealy, M. and Hayes, M. 2013. Development of a seaweed derived platelet activating factor acetylhydrolase (PAF-AH) inhibitory hydrolysate, synthesis of inhibitory peptides and assessment of their toxicity using the Zebrafish larvae assay. *Peptides* 50: 119–124.

Fitzgerald, C., Mora-Soler, L., Gallagher, E., O'Connor, P., Prieto, J., Soler-Vila, A. and Hayes, M. 2012. Isolation and characterization of bioactive pro-peptides with *in vitro* renin inhibitory activities from the macroalga *Palmariapalmata.Journal of Agricultural and Food Chemistry* 60(30): 7421–7427.

Fitzgerald, R.J., Murray, B.A. and Walsh, D.J. 2004. Hypotensive peptides from milk proteins. *The Journal of Nutrition* 134(4): 980S–988S.

Fosset, S. and Tome, D. 2000. Dietary protein-derived peptides with antithrombotic activity. *Bulletin of the International Dairy Federation* (353): 65–68.

Gallego, M., Aristoy, M.C. and Toldra, F. 2014. Dipeptidyl peptidase IV inhibitory peptides generated in Spanish dry-cured ham. *Meat Science* 96(2): 757–761.

Gallego, M., Mora, L., Escudero, E. and Toldra F. 2018. Bioactive peptides and free amino acids profiles in different types of European dry-fermented sausages. *International Journal of Food Microbiology* 2(276): 71–78.

Gandhi, A. and Shah, N.P. 2014. Cell growth and proteolytic activity of *Lactobacillusacidophilus, Lactobacillus helveticus, Lactobacillus delbrueckii* ssp. *bulgaricus*, and *Streptococcus thermophilus* in milk as affected by supplementation with peptide fractions. *International Journal of Food Science and Nutrition* 65(8): 937–941.

Gao, S., Hong, H.Zhang, C., Wang, K., Zhang, B., Han, Q., et al. 2019. Immunomodulatory effects of collagen hydrolysates from yak (*Bos grunniens*) bone on cyclophosphamide-induced immunosuppression in BALB/c mice. *Journal of Functional Foods* 60: 103420.

Girgih, A.T., He, R., Malomo, S., Offengenden, M., Wu, J. and Aluko, R.E. 2014. Structural and functional characterization of hemp seed (*Cannabis sativa* L.) Protein-derived antioxidant and antihypertensive peptides. *Journal of Functional Foods* 6: 384–394.

Gluvic, A. and Ulrih, N.P. 2019. Peptides derived from food sources: Antioxidative activities and interactions with model lipid membranes. *Food Chemistry* 287: 324–332.

Gobbetti, M., Minervini, F. and Rizzello, C.G. 2007. Bioactive peptides in dairy products. *Handbook of Food Products Manufacturing* 2: 489–517.

Grundy, S.M., Balady, G.J., Criqui, M.H., Fletcher, G., Greenland, P., Hiratzka, L.F., et al. 1998. Primary prevention of coronary heart disease: Guidance from Framingham: A statement for healthcare professionals from the AHA Task Force on Risk Reduction. *Circulation* 97(18): 1876–1887.

Haggag, Y.A., Donia, A.A., Osman, M.A. and El-Gizawy, S.A. 2018. Peptides as drug candidates: Limitations and recent development perspectives. *Biomedical Journal of Scientific & Technical Research* 8(4): 6659–6662.

Hajirostamloo, B. 2010. Bioactive component in milk and dairy product. *World Academy of Science, Engineering and Technology* 72: 162–166.

Hall, F., Reddivari, L. and Liceaga, A.M. 2020. Identification and characterization of edible cricket peptides on hypertensive and glycemic in vitro inhibition and their anti-inflammatory activity on RAW 264.7 macrophage cells. *Nutrients* 12(11): 3588.

Hartmann, R. and Meisel, H. 2007. Food-derived peptides with biological activity: From research to food applications. *Current Opinion in Biotechnology* 18(2): 163–169.

Hathwar, S.C., Bijinu, B., Rai, A.K. and Bhaskar, N. 2011. Simultaneous recovery of lipids and proteins by enzymatic hydrolysis of fish industry waste using different commercial proteases.*Applied Biochemistry and Biotechnology* 164: 115–124.

Hati, S., Dabhi, B., Sakure, A. and Prajapati, J. B. 2015. Significance of Proteolytic *Lactobacilli* on ACE-inhibitory activity and release of bioactive peptides during fermentation of milk, National seminar & 11th Alumni Convection on Indian Dairy Industry: Opportunities and Challenges, AAU, Anand, pp. 91.

Hernandez-Ledesma, B., Garcia-Nebot, M.J., Fernandez-Tome, S., Amigo, L. and Recio I. 2014. Dairy protein hydrolysates: Peptides for health benefits. *International Dairy Journal* 38(2): 82–100.

Hong LE, P., Parmentier, N., Trung LE, T. and Raes, K. 2021. Evaluation of using a combination of enzymatic hydrolysis and lactic acid fermentation for γ-aminobutyric acid production from soymilk. *LWT - Food Science and Technology* 142: 111044.

Horiguchi, N., Horiguchi, H. and Suzuki, Y. 2005. Effect of wheat gluten hydrolysate on the immune system in healthy human subjects. *Bioscience, Biotechnology, and Biochemistry* 69(12): 2445–2449.

Host, A. and Halken, S. 2004. Hypoallergenic formulas–when, to whom and how long: After more than 15 years we know the right indication! *Allergy* 59: 45–52.

Jauhiainen, T. and Korpela, R. 2007. Milk peptides and blood pressure. *The Journal of Nutrition* 137(3): 825S–829S.

Je, J.Y., Qian, Z.J., Byun, H.G. and Kim, S.K. 2007. Purification and characterization of an antioxidant peptide obtained from tuna backbone protein by enzymatic hydrolysis. *Process Biochemistry* 42(5): 840–846.

Jia, L., Wang, L., Liu, C., Liang, Y. and Lin, Q. 2021. Bioactive peptides from foods: Production, function, and application. *Food & Function* 12: 7108–7125.

Jin, R., Teng, X., Shang, J., Wang, D. and Liu, N. 2020. Identification of novel DPP–IV inhibitory peptides from Atlantic salmon (Salmo salar) skin. *Food Research International* 133: 109161.

Jisha, V.N., Smitha, R.B., Pradeep, S., Sreedevi, S., Unni, K.N., Sajith, S., et al. 2013. Versatility of microbial proteases. *Advances in Enzyme Research* 1(3): 39–51.

Kang, H.K., Lee, H.H., Seo, C.H. and Park, Y. 2019. Antimicrobial and immunomodulatory properties and applications of marine-derived proteins and peptides. *Marine Drugs* 17(6): 350.

Kang, K.H., Qian, Z.J., Ryu, B., Karadeniz, F., Kim, D. and Kim, S.K. 2012. Antioxidant peptides from protein hydrolysate of microalgae *Navicula incerta* and their protective effects in hepg2/CYP2E1 cells induced by ethanol. *Phytotherapy Research* 26(10): 1555–1563.

Kathiriya, M. 2014. Study on functional and probiotic potential of lactic acid bacteria. A thesis submitted at Anand Agricultural University, Anand, Gujarat, India.

Keska, P. and Stadnik, J. 2022. Dipeptidyl Peptidase IV Inhibitory Peptides Generated in Dry-Cured Pork Loin during Aging and Gastrointestinal Digestion. *Nutrients* 14(4):770.

Kim, E.K., Joung, H.J., Kim, Y.S., Hwang, J.W., Ahn, C.B., Jeon, Y.J., Moon, S.H., Song, B.C. and Park, P.J. 2012. Purification of a novel anticancer peptide from enzymatic hydrolysate of *Mytilus coruscus*. *Journal of Microbiology and Biotechnology* 22(10):1381–1387.

Kim, S.E., Kim, H.H., Kim, J.Y., Im Kang, Y., Woo, H.J. and Lee, H.J. 2000. Anticancer activity of hydrophobic peptides from soy proteins. *Biofactors* 12(1–4): 151–155.

Kitts, D.D. and Weiler, K. 2003. Bioactive proteins and peptides from food sources. Applications of bioprocesses used in isolation and recovery. *Current Pharmaceutical Design* 9(16): 1309–1323.

Ko, S.C., Kim, D. and Jeon, Y.J. 2012. Protective effect of a novel antioxidative peptide purified from a marine Chlorella ellipsoidea protein against free radical-induced oxidative stress. *Food and Chemical Toxicology* 50(7): 2294–2302.

Kong, X., Zhang, L., Song, W., Zhang, C., Hua, Y., Chen, Y. et al. 2021. Separation, identification and molecular binding mechanism of dipeptidyl peptidase IV inhibitory peptides derived from walnut (*Juglans regia* L.) protein. *Food Chemistry* 347: 129062.

Korhonen, H. 2009. Milk-derived bioactive peptides: From science to applications. *Journal of Functional Foods* 1(2): 177–187.

Korhonen, H. and Philanto, A. 2003. Food-derived bioactive peptides-opportunities for designing future foods. *Current Pharmaceutical Design* 9(16): 1297–1308.

Korhonen, H. and Philanto, A. 2006. Bioactive peptides: Production and functionality. *International Dairy Journal* 16(9): 945–960.

Kumari, R., Sanjukta, S., Sahoo D. and Rai, A.K. 2022. Functional peptides in Asian protein rich fermented foods: Production and health benefits. *Systems Microbiology and Biomanufacturing* 2: 1–13.

Lammi, C., Aiello, G., Boschin, G. and Arnoldi, A. 2019. Multifunctional peptides for the prevention of cardiovascular disease: A new concept in the area of bioactive food-derived peptides. *Journal of Functional Foods* 55: 135–145.

Li, H. and Aluko, R.E. 2010. Identification and inhibitory properties of multifunctional peptides from pea protein hydrolysate. *Journal of Agricultural and Food Chemistry* 58(21): 11471–11476.

Liu, J., Sun, H., Nie, C., Ge, W., Wang, Y. and Zhang, W. 2018. Oligopeptide derived from solid-state fermented cottonseed meal significantly affect the immunomodulatory in BALB/c mice treated with cyclophosphamide. *Food Science and Biotechnology* 27(6):1791–1799.

Lu, J., Guo, Y., Muhmood, A., Zeng, B., Qiu, Y., Wang, P., et al. 2022. Probing the antioxidant activity of functional proteins and bioactive peptides in *Hermetiaillucens* larvae fed with food wastes. *Scientific Reports* 12(1): 1–10.

Lu, J., Zeng, Y., Hou, W., Zhang, S., Li, L., Luo, X., et al. 2012. The soybean peptide aglycin regulates glucose homeostasis in type 2 diabetic mice via IR/IRS1 pathway. *The Journal of Nutritional Biochemistry* 1;23(11):1449–1457.

Lu, Y., Wu, Y., Hou, X., Lu, Y., Meng, H., Pei, S., et al. 2022. Separation and identification of ACE inhibitory peptides from lizard fish proteins hydrolysates by metal affinity-immobilized magnetic liposome. *Protein Expression and Purification* (191): 106027.

Maes, W., Van Camp, J., Vermeirssen, V., Hemeryck, M., Ketelslegers, J.M., Schrezenmeir, J., et al. 2004. Influence of the lactokinin Ala-Leu-Pro-Met-His-Ile-Arg (ALPMHIR) on the release of endothelin-1 by endothelial cells. *Regulatory Peptides* 118(1–2):105–109.

Makino, S., Nakashima, H., Minami, K., Moriyama, R. and Takao, S. 1988. Bile acid-binding protein from soybean seed: Isolation, partial characterization and insulin-stimulating activity. *Agricultural and Biological Chemistry* 52(3): 803–809.

Marciniak, A., Suwal, S., Naderi, N., Pouliot, Y. and Doyen, A. 2018. Enhancing enzymatic hydrolysis of food proteins and- production of bioactive peptides using high hydrostatic pressure technology. *Trends in Food Science & Technology* 80:187–198.

Martinez-Maqueda, D., Miralles, B., Recio, I. and Hernandez-Ledesma, B. 2012. Antihypertensive peptides from food proteins: A review. *Food & Function* 3(4): 350–361.

Martinez-Medina, G.A., Barragan, A.P., Ruiz, H.A., Ilyina, A., MartinezHernandez, J.L. Rodriguez-Jasso, R.M., et al. 2019. Fungal proteases and production of bioactive peptides for the food industry. *Enzymes in Food Biotechnology: Production, Applications, and Future Prospects*, pp. 221–246. Oxford: Academia Press.

Matoba, N., Usui, H., Fujita, H. and Yoshikawa, M. 1999. A novel anti-hypertensive peptide derived from ovalbumin induces nitric oxide-mediated vasorelaxation in an isolated SHR mesenteric artery. *FEBS Letters* 452(3): 181–184.

Meisel, H. 1998. Overview on milk protein-derived peptides. *International Dairy Journal* 8(5–6): 363–373.

Meisel, H. 2005. Biochemical properties of peptides encrypted in bovine milk proteins. *Current Medicinal Chemistry* 12(16): 1905–1919.

Meisel, H. and Fitzgerald, R.J. 2003. Biofunctional peptides from milk proteins: Mineral binding and cytomodulatory effects. *Current Pharmaceutical Design* 9(16): 1289–1296.

Mellander, O.1950. The physiological importance of the casein phosphopeptide calcium salts II: Peroral calcium dosage in infants. *Acta Societatis Medicorum Upsaliensis, Uppsala* 55: 247–255.

Miguel, M., Lopez-Fandino, R., Ramos, M. and Aleixandre, A. 2005. Short-term effect of eggwhite hydrolysate products on the arterial blood pressure of hypertensive rats. *British Journal of Nutrition* 94: 731–737.

Mishra, B.K., Hati, S., Das, S. and Kumari, R. 2018. Evaluation of probiotic potentials of *Lactobacillus* isolated from traditional fermented foods of Garo Hills, Meghalaya, India. *Reviews in Medical Microbiology* 29(3): 120–128.

Mishra, B.K., Hati, S., Das, S. and Prajapati, J.B. 2019. Biofunctional attributes and storage study of soy milk fermented by *Lactobacillus rhamnosus* and *Lactobacillus helveticus. Food Technology and Biotechnology* 57(3): 399–407.

Mizuno, S., Matsuura, K., Gotou, T., Nishimura, S., Kajimoto, O., Yabune, M., et al. 2005. Antihypertensive effect of casein hydrolysate in a placebo-controlled study in subjects with high-normal blood pressure and mild hypertension. *British Journal of Nutrition* 94(1):84–91.

Mizushima, S., Ohshige, K., Watanabe, J., Kimura, M., Kadowaki, T., Nakamura, Y., et al. 2004. Randomized controlled trial of sour milk on blood pressure in borderline hypertensive men. *American Journal of Hypertension* 17(8):701–706.

Mora, L., Reig, M. and Toldra, F. 2014. Bioactive peptides generated from meat industry by-products. *Food Research International* 65: 344–349.

Motoi, H. and Kodama, T. 2003. Isolation and characterization of angiotensin I-converting enzyme inhibitory peptides from wheat gliadin hydrolysate. *Food/Nahrung* 47(5): 354–358.

Mulero, I., Noga, E.J., Meseguer, J., Garcia-Ayala, A. and Mulero, V. 2008. The antimicrobial peptides piscidins are stored in the granules of professional phagocytic granulocytes of fish and are delivered to the bacteria-containing phagosome upon phagocytosis. *Developmental & Comparative Immunology* 32(12): 1531–1538.

Murray, B.A. and Fitzgerald, R.J. 2007. Angiotensin converting enzyme inhibitory peptides derived from food proteins: Biochemistry, bioactivity and production. *Current Pharmaceutical Design* 13(8): 773–791.

Nagaoka, S., Futamura, Y., Miwa, K., Awano, T., Yamauchi, K., Kanamaru, Y., et al. 2001. Identification of novel hypocholesterolemic peptides derived from bovine milk β-lactoglobulin. *Biochemical and Biophysical Research Communications* 281(1):11–7.

Nurminen, M.L., Sipola, M., Kaarto, H., Pihlanto-Leppala, A., Piilola, K., Korpela, R., et al. 2000. A-Lactorphin lowers blood pressure measured by radiotelemetry in normotensive and spontaneously hypertensive rats. *Life Sciences* 66(16): 1535–1543.

Okitsu, M., Morita, A., Kakitani, M., Okada, M. and Yokogoshi, H. 1995. Inhibition of the endothelin-converting enzyme by pepsin digests of food proteins. *Bioscience, Biotechnology, and Biochemistry* 59(2): 325–326.

Oppenheim, J.J., Biragyn, A., Kwak, L.W. and Yang, D. 2003. Roles of antimicrobial peptides such as defensins in innate and adaptive immunity. *Annals of the Rheumatic Diseases* 62(suppl 2): ii17–ii21.

Oshima, G., Shimabukuro, H., Nagasawa, K. 1979. Peptide inhibitors of angiotensin I-converting enzyme in digests of gelatin by bacterial collagenase. *Biochimica et Biophysica Acta* 566:128–137.

Otani, H., and Suzuki, H. 2003. Isolation and characterization of cytotoxic small peptides, α-casecidins, from bovine α_{s1}-casein digested with bovine trypsin. *Animal Science Journal* 74: 427–435.

Ouertani, A., Chaabouni, I., Mosbah, A., Long, J., Barakat, M., Mansuelle, P., et al. 2018. Two new secreted proteases generate a casein-derived antimicrobial peptide in *Bacillus cereus* food born isolate leading to bacterial competition in milk. *Frontiers in Microbiology* 9:1148.

Padghan, P. V., Mann, B., Sharma, R., Bajaj, R. and Saini, P. 2017. Production of angiotensin-I-converting-enzyme-inhibitory peptides in fermented milks (Lassi) fermented by *Lactobacillus acidophillus* with consideration of incubation period and simmering treatment. *International Journal of Peptide Research and Therapeutics* 23(1): 69–79.

Padhi, S., Chourasia, R., Kumari, M., Singh, S.P. and Rai, A.K. 2022. Production and characterization of bioactive peptides from rice beans using Bacillus subtilis. *Bioresource Technology* 351: 126932.

Padhi, S., Sanjukta, S., Chourasia, R., Labala, R.K., Singh, S.P. and Rai, A.K. 2021. A multifunctional peptide from *Bacillus* fermented soybean for effective inhibition of SARS-CoV-2 S1 receptor binding domain and modulation of toll like receptor 4: A molecular docking study. *Frontiers in Molecular Bioscience* 31: 636647.

Panchal, G., Sakure, A. and Hati, S. 2021. Peptidomic profiling of fermented goat milk: Considering the fermentation-time dependent proteolysis by Lactobacillus and characterization of novel peptides with Antioxidative activity. *Journal of Food Science and Technology*, DOI: 10.1007/s13197-021-05243-w.

Parekh, S.L., Balakrishnan, S., Hati, S. and Aparnathi, K. D. 2017. Bio-functional properties of cultured buttermilk prepared by incorporation of fermented paneer whey. *International Journal of Current Microbiology and Applied Sciences* 6(2): 933–945.

Park, S.Y., Lee, J. S., Baek, H.H. and Lee, H.G. 2010. Purification and characterization of antioxidant peptides from soy protein hydrolysate. *Journal of Food Biochemistry* 34: 120–132.

Parmar, H., Hati, S. and Sakure, A. 2017. *In Vitro* and *In Silico* analysis of novel ACE-inhibitory bioactive peptides derived from fermented goat milk. *International Journal of Peptide Research and Therapeutics* 24: 441–453. https://doi.org/10.1007/s10989-017-9630-4.

Patel, R. and Hati, S. 2017. Production of antihypertensive (angiotensin I-converting enzyme inhibitory) peptides derived from fermented milk supplemented with WPC70 and Calcium caseinate by *Lactobacillus* cultures. *Reviews in Medical Microbiology* 29(1): 30–40.

Pavlicevica, M., Marmirolib, N. and Maestri, E. 2022. Immunomodulatory peptides—A promising source for novel functional food production and drug discovery. *Peptides* 148:170696.

Pessione, E. and Cirrincione, S. 2016. Bioactive molecules released in food by lactic acid bacteria: Encrypted peptides and biogenic amines. *Frontiers in Microbiology* 7:876.

Phelan, M., Aherne, A., Fitzgerald, R.J. and O'Brien, N.M. 2009. Casein-derived bioactive peptides: Biological effects, industrial uses, safety aspects and regulatory status. *International Dairy Journal* 19(11): 643–654.

Pihlanto-Leppala, A. 2000. Bioactive peptides derived from bovine whey proteins: Opioid and ace-inhibitory peptides. *Trends in Food Science & Technology* 11(9–10): 347–356.

Pringos, E., Vignes, M., Martinez, J. and Rolland, V. 2011. Peptide neurotoxins that affect voltage-gated calcium channels: A close-up on ω-Agatoxins. *Toxins* 3(1): 17–42.

Qiao, M., Tu, M., Wang, Z., Mao, F., Chen, H., Qin, L., et al. 2018. Identification and antithrombotic activity of peptides from blue mussel (*Mytilus edulis*) protein. *International Journal of Molecular Sciences* 19(1):138.

Rai, A.K., Kumari, R., Sanjukta, S. and Sahoo, D. 2016. Production of bioactive protein hydrolysate using the yeasts isolated from soft chhurpi. *Bioresource Technology* 219: 239–245.

Rai, A.K., Pandey, A. and Sahoo, D. 2019. Biotechnological potential of yeasts in functional food industry. *Trends in Food Science & Technology* 83:129–137.

Rai, A.K., Sanjukta, S., Chourasia, R., Bhat, I., Bhardwaj, P.K. and Sahoo, D. 2017. Production of bioactive hydrolysate using protease, β-glucosidase and α-amylase of Bacillus spp. Isolated from kinema. *Bioresource Technology* 235: 358–365.

Ramalingam, M. and Kim, S.J. 2012. Reactive oxygen/nitrogen species and their functional correlations in neurodegenerative diseases. *Journal of Neural Transmission* 119(8): 891–910.

Raveschot, C., Cudennec, B., Coutte, F., Flahaut, C., Fremont, M., Drider, D. et al. 2018. Production of bioactive peptides by *lactobacillus* species: From gene to application. *Frontiers in Microbiology* 9: 2354.

Rival, S.G., Boeriu, C.G. and Wichers, H.J. 2001. Caseins and casein hydrolysates. 2. Antioxidative properties and relevance to lipoxygenase inhibition. *Journal of Agricultural and Food chemistry* 49(1): 295–302.

Rohm, B., Scherlach, K., Mobius, N., Partida-Martinez, L.P. and Hertweck, C. 2010. Toxin production by bacterial endosymbionts of a Rhizopus microsporus strain used for tempe/sufu processing. *International Journal of Food Microbiology* 136(3): 368–371.

Rokka, T., Syvaoja, E.L., Tuominen, J. and Korhonen, H.J. 1997. Release of bioactive peptides by enzymatic proteolysis of *Lactobacillus* GG fermented UHT-milk. *Series Milchwissenschaft - Milk Science International* 52: 675–678.

Roy, M.K., Watanabe, Y. and Tamai, Y. 1999. Induction of apoptosis in HL-60 cells by skimmed milk digested with a proteolytic enzyme from the yeast *Saccharomyces cerevisiae. Journal of Bioscience and Bioengineering* 88(4): 426–432.

Saito, T., Nakamura, T., Kitazawa, H., Kawai, Y. and Itoh, T. 2000. Isolation and structural analysis of antihypertensive peptides that exist naturally in Gouda cheese. *Journal of Dairy Science* 83(7): 1434–1440.

Sanjukta, S., Padhi, S., Sarkar, P., Singh, S.P., Sahoo, D. and Rai, A.K. 2021. Production, characterization and molecular docking of antioxidant peptides from peptidome of kinema fermented with proteolytic Bacillus spp. *Food Research International* 141:110161.

Sanjukta, S. and Rai, K. A. 2016. Production of bioactive peptides during soybean fermentation and their potential health benefits. *Trends in Food Science & Technology* 50:1–10.

Sanjukta, S., Rai, A.K., Muhammed, A., Jeyaram K. and Talukdar N.C. 2015. Enhancement of antioxidant properties of two soybean varieties of Sikkim Himalayan region by proteolytic Bacillus subtilis fermentation. *Journal of Functional Foods* 14: 650–658

Saxena, A., Tripathi, B.P., Kumar, M. and Shahi, V.K. 2009. Membrane-based techniques for the separation and purification of proteins: An overview. *Advances in Colloid and Interface Science* 145(1–2): 1–22.

Scruggs, P., Filipeanu, C.M., Yang, J., Chang, J.K. and Dun, N.J. 2004. Interaction of ovokinin (2–7) with vascular bradykinin 2 receptors. *Regulatory Peptides* 120(1–3): 85–91.

Selamassakul, O., Laohakunjit, N., Kerdchoechuen, O., Yang, L. and Maier, C.S. 2020. Bioactive peptides from brown rice protein hydrolyzed by bromelain: Relationship between biofunctional activities and flavor characteristics. *Journal of Food Science* 85(3): 707–717.

Shahidi, F. and Zhong, Y. 2008. Bioactive Peptides. *Journal of AOAC International* 91(4): 914–931.

Sharma, R., Garg, P., Kumar, P., Bhatia, S.K. and Kulshrestha, S. 2020. Microbial fermentation and its role in quality improvement of fermented foods. *Fermentation* 6(4): 106.

Sheih, I.C., Fang, T.J., Wu, T.K. and Lin, P.H. 2010. Anticancer and antioxidant activities of the peptide fraction from algae protein waste. *Journal of Agricultural and Food Chemistry* 58(2): 1202–1207.

Sheih, I.C., Wu, T.K. and Fang, T.J. 2009. Antioxidant properties of a new antioxidative peptide from algae protein waste hydrolysate in different oxidation systems. *Bioresource Technology* 100(13): 3419–3425.

Shin, K., Yamauchi, K., Teraguchi, S., Hayasawa, H., Tomita, M., Otsuka, Y. et al. 1998. Antibacterial activity of bovine lactoferrin and its peptides against enterohaemorrhagic *Escherichia coli* O157: H7. *Letters in Applied Microbiology* 26(6): 407–411.

Singh, B.P., Aluko, R.E., Hati, S. and Solanki, D. 2022. Bioactive peptides in the management of lifestyle-related diseases: Current trends and future perspectives. *Critical Reviews in Food Science and Nutrition* 62: 1–14.

Sivamaruthi, B.S., Kesika, P. and Chaiyasut, C. 2018. Toxins in fermented foods: Prevalence and preventions—A mini review. *Toxins* 11(1): 1–16.

Solanki, D. and Hati, S. 2018. Considering the potential of *Lactobacillus rhamnosus* for producing Angiotensin I-Converting Enzyme (ACE) inhibitory peptides in fermented camel milk (Indian breed). *Food Bioscience* 23: 16–22.

Solieri, L., Rutella, G.S. and Tagliazucchi, D. 2015. Impact of non-starter lactobacilli on release of peptides with angiotensin-converting enzyme inhibitory and antioxidant activities during bovine milk fermentation. *Food Microbiology* 51: 108–116.

Sony, I.S. and Potty, V.P. 2016. Quantitative estimation of protease produced by bacterial isolates from food processing industries. *International Journal of Engineering Research & Technology* 5(10): 238–244.

Sousa, P., Borges, S. and Pintado, M. 2020. Enzymatic hydrolysis of insect *Alphitobiusdiaperinus* towards the development of bioactive peptide hydrolysates. *Food & Function* 11(4): 3539–3548.

Sousa, S.R., Vetter I. and Lewis, R.J. 2013. Venom peptides as a rich source of Cav2.2 channel blockers. *Toxin* 5: 286–314.

Staessen, J.A., Li, Y. and Richart, T. 2006. Oral renin inhibitors. *The Lancet* 368(9545): 1449–1456.

Steinstraesser, L., Kraneburg, U., Jacobsen, F. and Al-Benna, S. 2011. Host defense peptides and their antimicrobial-immunomodulatory duality. *Immunobiology* 216(3): 322–333.

Suetsuna, K., Maekawa, K. and Chen, J.R. 2004. Antihypertensive effects of *Undaria pinnatifida* (wakame) peptide on blood pressure in spontaneously hypertensive rats. *TheJournal of Nutritional Biochemistry* 15(5): 267–272.

Sun, X.D. 2011. Enzymatic hydrolysis of soy proteins and the hydrolysates utilization. *International Journal of Food Science and Technology* 46: 2447–2459.

Tagliazucchi, D., Martini, S. and Solieri, L. 2019. Bioprospecting for Bioactive Peptide Production by Lactic Acid Bacteria Isolated from Fermented Dairy Food. *Fermentation* 5(4): 96.

Tanaka, M., Watanabe, S., Wang, Z., Matsumoto, K. and Matsui, T. 2009. His-Arg-Trp potently attenuates contracted tension of thoracic aorta of Sprague-Dawley rats through the suppression of extracellular Ca2+ influx. *Peptides* 30(8): 1502–1507.

Toldra, F., Reig, M., Aristoy, M.C. and Mora, L. 2017. Generation of bioactive peptides during food processing. *Food Chemistry* 267: 395–404.

Udenigwe, C. C., Li, H. and Aluko, R.E. 2012. Quantitative structure–activity relationship modeling of renin-inhibiting dipeptides. *Amino Acids* 42(4): 1379–1386.

Umayaparvathi, S., Meenakshi, S., Vimalraj, V., Arumugam, M., Sivagami, G. and Balasubramanian, T. 2014. Antioxidant activity and anticancer effect of bioactive peptide from enzymatic hydrolysate of oyster (*Saccostrea cucullata*). *Biomedicine & Preventive Nutrition* 4(3): 343–353.

Uster, A., Ruefenacht, U., Ruehlin, M., Pless, M., Siano, M., Haefner, M., et al. 2013. Influence of a nutritional intervention on dietary intake and quality of life in cancer patients: A randomized controlled trial. *Nutrition* 29(11–12):1342–1349.

Vanvi, A. and Tsopmo, A. 2016. Pepsin digested oat bran proteins: Separation, antioxidant activity, and identification of new peptides. *Journal of Chemistry* Article ID 8216378.

Vasconcellos, F.C.S., Woiciechowski, A.L., Soccol, V.T., Mantovani, D. and Soccol, C.R. 2014. Antimicrobial and antioxidant properties of-conglycinin and glycinin from soy protein isolate. *International Journal of Current Microbiology and Applied Science* 3: 144–157.

Venegas-Ortega, M.G., Flores-Gallegos, A.C., Martınez-Hernandez, J.L., Aguilar, C.N. and Nevarez-Moorillon, G.V. 2019. Production of bioactive peptides from lactic acid bacteria: A sustainable approach for healthier foods. *Comprehensive Reviews in Food Science and Food Safety* 18(4): 1039–1051.

Walker, G.D., Cai, F., Shen, P., Bailey, D.L., Yuan, Y., Cochrane, N.J. and Reynolds, E.C. 2009. Consumption of milk with added casein phosphopeptide-amorphous calcium phosphate remineralizes enamel subsurface lesions in situ. *Australian Dental Journal* 54(3): 245–249.

Wang, Z., Shu, G., Chen, L., Dai, C., Li, Y., Niu, J., et al. 2022. Directed-Vat-Set starter producing ACE-inhibitory peptides: Optimization and evaluation of stability. *Journal of Food processing and Preservation* 46(4): e16475.

Williams, A.G., Noble, J., Tammam, J., Lloyd, D. and Banks, J.M. 2002. Factors affecting the activity of enzymes involved in peptide and amino acid catabolism in non-starter lactic acid bacteria isolated from Cheddar cheese. *International Dairy Journal* 12(10): 841–852.

Wong, F.C., Xiao, J., Wang, S., Ee, K.Y. and Chai, T.T. 2020. Advances on the antioxidant peptides from edible plant sources. *Trends in Food Science & Technology* 99: 44–57.

Wu, S., Qi, W., Li, T., Lu, D., Su, R., et al. 2013. Simultaneous production of multi-functional peptides by pancreatic hydrolysis of bovine casein in an enzymatic membrane reactor via combinational chromatography. *Food Chemistry* 141(3): 2944–2951.

Xue, L., Yin, R., Howell, K. and Zhang, P. 2021. Activity and bioavailability of food protein-derived angiotensin-I-converting enzyme–inhibitory peptides. *Comprehensive Reviews in Food Science and Food Safety* 20(2): 1541–4337.

Yoshikawa, M., Tani, F., Shiota, H., Suganuma, H., Usui, H., Kurahashi, K., et al. 1994. Casoxin D, an opioid antagonist ileum-contracting/vasorelaxing peptide derived from human α_{s1}-casein. In: Brantl, V. and Teschemacher, H. (eds.). B-Casomorphins and Related Peptides: Recent Developments, pp. 43–48. Germany: VCH-Weinheim.

Zambrowicz, A., Dąbrowska, A., Bobak, L. and Szołtysik, M. 2014. Egg yolk proteins and peptides with biological activity. *Advanced Hygiene* 68: 1524–1529.

Zhang, B. and Zhang, X. 2013. Separation and nanoencapsulation of antitumor polypeptide from Spirulina platensis. *Biotechnology Progress* 29(5): 1230–1238.

Zhang, M., Mu, T.H. and Sun, M.J. 2012. Sweet potato protein hydrolysates: Antioxidant activity and protective effects on oxidative DNA damage. *International Journal of Food Science & Technology* 47(11): 2304–2310.

7 Microbial Enzymes for the Production of Xylooligosaccharides

Cristina Álvarez[1], Elia Tomás-Pejó[2], Cristina González-Fernández[2,3,4], and María José Negro[1]
[1] Advanced Biofuels and Bioproducts Unit, Department of Energy, CIEMAT, Madrid, Spain
[2] Biotechnological Processes Unit, IMDEA Energy, Móstoles, Spain
[3] Department of Chemical Engineering and Environmental Technology, School of Industrial Engineering, Valladolid University, Valladolid, Spain
[4] Institute of Sustainable Processes, Valladolid, Spain

CONTENTS

7.1 INTRODUCTION

During the last three decades, prebiotic compounds have raised great interest in various fields such as nutrition, biomedicine, and the food industry. Different definitions have been proposed, and the debate persists on whether they reflect all the properties that prebiotics can present. Gibson and Roberfroid (1995) first defined the term prebiotic as a "non-digestible food ingredient that beneficially affects the host by selectively stimulating the growth and/or activity of one or a limited number of bacteria in the colon, thus improving the health of the host."

The term is constantly under review without losing its original sense. The International Scientific Association for Probiotics and Prebiotics (ISAPP) defined prebiotics in 2017 as "a substrate that is

DOI: 10.1201/9781003311164-10

selectively used by host microorganisms, conferring a health benefit" (Gibson et al., 2017). Thus, for an ingredient to be considered a prebiotic, it must meet a series of requirements: (1) show resistance to gastric acidity, digestive enzymes, and absorption in the gastrointestinal tract; (2) be selectively fermented by the beneficial bacteria of the microbiota, and (3) induce physiological effects beneficial for the health of the host (Gibson and Roberfroid, 1995).

Among all food ingredients, non-digestible carbohydrates (oligo- and polysaccharides) are the most important candidates for prebiotics. The concept of non-digestible oligosaccharides refers to the fact that digestive enzymes do not hydrolyse the anomeric carbon of the monosaccharide. It should be highlighted that the anomeric carbon is the carbon that is part of the carbonyl group in the straight-chain structure when the sugar is its cyclic form. Therefore, with the anomeric carbon not hydrolysed, this food ingredient should reach the colon in intact form. Carbohydrates with a degree of polymerization (DP) between 3 and 10 units are considered oligosaccharides (IUPAC, 1982). However, some authors consider that oligosaccharides are formed by several units ranging from 2 to 20 (Manning et al., 2004).

Among the compounds reported to possess prebiotic properties are fructooligosaccharides (FOS), galactooligosaccharides, inulin, and lactulose. However, xylooligosaccharides (XOS), mannooligosaccharides, and isomaltooligosaccharides are emerging prebiotic compounds. Most of the beneficial effects of prebiotics are related to their fermentation products. The main derived products resulting from the intestinal bacterial fermentation of prebiotics are short-chain fatty acids (SCFAs) like acetate, propionate, and butyrate. SCFA accumulation causes a drop in pH favouring the absorption of minerals like Ca and Mg, which are crucial elements in avoiding osteoporosis (Gibson et al., 2017). Furthermore, SCFAs play a critical role in lipid, glucose, and cholesterol metabolism in various tissues and organs such as the liver or pancreas (Palaniappan et al., 2021) (Figure 7.1).

Acidification of the medium also positively affects the intestine by hindering the proliferation of some pathogenic microorganisms. Thus, SCFAs affect the immune system by changing the balance of the gut microbiota (Santibañez et al., 2021).

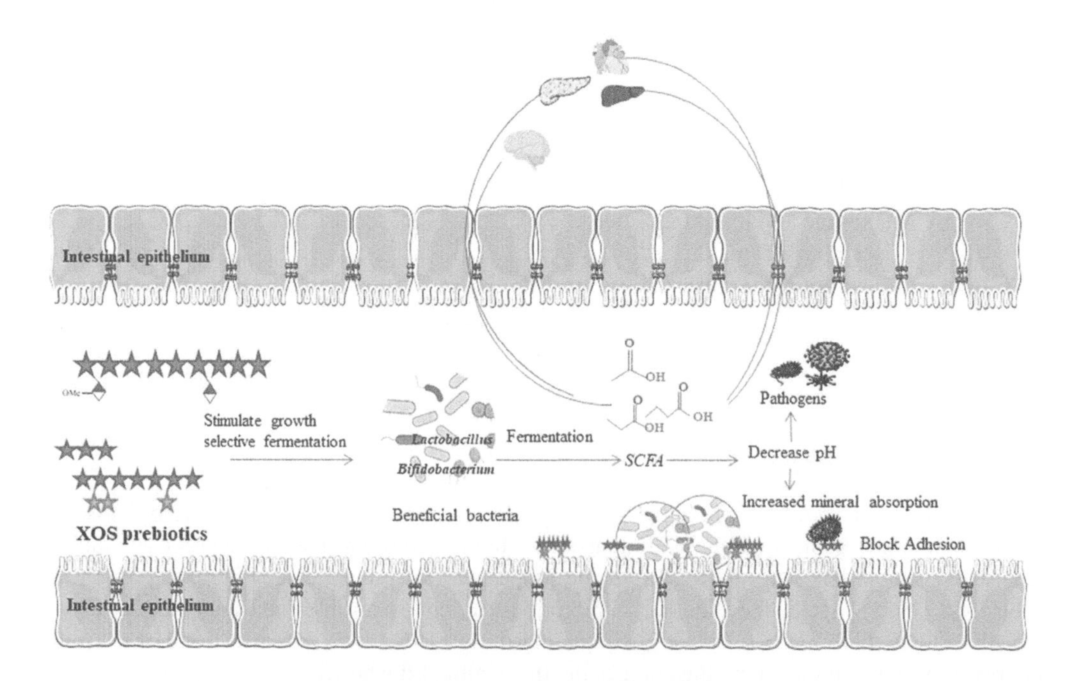

FIGURE 7.1 Potential health benefits of prebiotics. (Graphical illustrations were created using items from Servier Medical Art by Servier.)

7.1.1 XOS as Emerging Prebiotic: Beneficial Effects

XOS are made up of xylose units linked by β-(1→4) bonds with a low degree of polymerization (DP2-DP6) (Santibañez et al., 2021). Lignocellulosic biomass (LB) has a valuable fraction containing hemicellulose, which can be transformed into XOS. Because of that, LB is regarded as an important source for prebiotics' cost-effective and sustainable production.

Some studies have shown the impact of XOS fermentation on the regulation of lipids and blood glucose (Yang et al., 2015), the digestive health and microbiota (Finegold et al., 2014), and immune markers (Childs et al., 2014). Studies conducted *in vivo* and *in vitro* have shown that XOS significantly enriched *Bifidobacteria* populations, mitigating inflammatory diseases. Besides, human and animal studies demonstrated that beneficial XOS dosages (1.4 g/d) were much lower than the required levels for FOS (5 g/d) or galactooligosaccharides (8 g/d). In human trials, XOS led to a significant increase in the *Bifidobacterium* population, enhancing the total number of anaerobic bacteria and *Bacteroides fragilis* (Finegold et al., 2014).

XOS obtained from rice husk showed significant antidiabetic potential in a high-fat diet and streptozotocin-induced Type 2 diabetic rat model. In this case, XOS improved insulin resistance and achieved reasonable glycaemic control in diabetes, modulated gut microbiota, increased beneficial bacteria, *Lactobacillus* and *Bifidobacterium* spp., and increased SCFA production (Khat-Udomkiri et al., 2020). Yang et al. (2015) found that the ingestion de XOS increased the quantity of positively associated microflorae species, such as *Blautia hydrogenotrophica,* in healthy and prediabetic subjects. Li et al. (2021a) used *in vitro* fermentation models to assess the effect of XOS on the intestinal microbiota with ulcerative colitis in clinical remission. Results suggested that XOS could potentially alleviate dysbiosis (imbalance in the intestinal flora); however, the authors highlighted the need for *in vivo* studies to verify the real effects of XOS.

Dietary XOS supplementation also induced changes in gut microbial composition. Long et al. (2019) reported a reduced obese-related microbiota ratio (Firmicutes/Bacteroidetes =obesity marker) in a mouse model that was using XOS supplementation. This fact resulted in a reduction of visceral fat depots, as it could be concluded from the decreased gene expression of adipogenesis markers and fat synthesis (Long et al., 2019). Xylobiose has also demonstrated therapeutic potential for obesity. The effect of xylobiose supplementation on mice fed with a high-fat diet reduced lipogenesis and adipogenesis in mesenteric adipose tissues, leading to suppressed lipolysis and inflammatory response (Lim et al., 2018).

7.1.2 Prebiotic Market

There is a wide variety of commercial prebiotic products used by the pharmaceutical and food industries. Depending on the region of the world where these prebiotic products are to be marketed, different regulatory laws are implemented. Foods for Specified Health Use was created in Japan, which led it to becoming the first country to regulate functional foods in the 1990s. Japan is also the leading producer and consumer of XOS globally (Santibañez et al., 2021). XOS are globally commercialized by companies such as Longlive, Kangwei, HFsugar, Henan, Shengtai, YIBIN YATAI, HBTX, YuHua, and ShunTian.

In Europe, the responsible organization for authorizing food ingredients is the European Food Safety Authority (EFSA), created in 2002. EFSA scientifically advises and communicates the risks associated with the food chain. In 2018, EFSA's Panel on Dietetic Products reported that enzymatically produced XOS from corncobs (novel food) were a safe product to be used by the general population and determined the dosage levels (Turck et al., 2018). Consumers currently demand healthy products with low fat and salt content, but also ingredients such as prebiotics and probiotics are nowadays of particular importance. For this reason, the market demand for functional food has significantly increased in the last few years. In 2018, the prebiotic market was valued at USD 3.4 billion, and it is projected to reach a value of USD 8.34 billion by 2026, with a compound annual growth rate (CAGR) of 10.1% (Watson, 2019).

In particular, the XOS global market was USD 61 million in 2019, and it is expected to reach USD 81 million in 2026 at a CAGR of 4.1% (Capetti et al., 2021). In this sense, due to their price competitiveness compared to other prebiotics, XOS present a remarkable potential as a food ingredient offering new opportunities for research, development, and commercialization of prebiotics (Palaniappan et al., 2021).

7.2 RAW MATERIALS

The global annual production of LB is estimated at 182 billion tonnes (Dahmen et al., 2019), being the major renewable material on the Earth. Besides its wide distribution, it is considered a low-cost carbohydrate-rich feedstock readily and locally available. These facts make LB an excellent raw material to provide value-added compounds, including biofuels and other chemicals. Indeed, 1.2 billion tonnes of lignocellulosic residues are currently valorized for different purposes (Dahmen et al., 2019).

LB is composed of two carbohydrate fractions (cellulose and hemicellulose), lignin, and small amounts of pectin, proteins, extractives, and ash. Cellulose is a linear polymer of glucose linked with β-1,4-glucosidic bonds. The polymerization degree of cellulose ranges from 10,000 to 15,000 glucose moieties (Agbor et al., 2011) and it can account for the 30–50% (w w^{-1}) of LB dry weight (Agbor et al., 2011). Hemicellulose consists of polymers made up of more than one type of sugar, including D-xylose, L-arabinose, D-mannose, D-glucose, D-galactose, etc., and by various uronic acids, such as glucuronic and galacturonic acid. Hemicellulose primary function is to provide the linkage between cellulose and lignin. Lignin is the third major component in LB. It is, therefore, the third most abundant natural polymer in nature after cellulose and hemicellulose. Lignin is an amorphous three-dimensional polymer formed by the basic monomers ρ-coumaryl alcohol, coniferyl alcohol, and sinapyl alcohol. These three aromatic alcohols give rise to ρ-hydroxyphenyl units, guaiacil units, and syringyl units, respectively, the proportion of which varies in hardwood, softwood, and herbaceous biomass.

Both sugar-rich fractions, cellulose, and hemicellulose, can be used for the production of sugars as platform molecules via enzymatic hydrolysis. Most microorganisms are able to metabolize hexoses, like D-glucose, but not pentoses, like D-xylose and L-arabinose. In this sense, natural C5-utilizing organisms that produce targeted products are very unusual. The utilization of cellulosic glucose from different LB sources has been studied for years, especially for bioethanol production. However, the valorization of the hemicellulosic fraction is still a big challenge due to the absence of microorganisms that can efficiently use C5 sugars as a carbon source. In recent years, attention to the use of hemicellulose has significantly grown due to its attractive properties and unexplored applications. Because of that, the production of XOS from hemicellulose is gaining huge attention among the scientific community and food and health sectors.

7.2.1 Hemicellulose as Source of Xilooligosaccharides

The composition and structure of hemicelluloses is biomass specific. Depending on the nature of the monomers present in the backbone of hemicellulose, this polymer can be classified into xylans, xyloglucans, mannans, and β-(1→3, 1→4)-glucans (Scheller and Ulvskov, 2010). Xylan backbone is formed by D-xylose units (90%) linked by β-(1→4) bonds with side branches of other pentoses, hexoses, and acetyl groups. Xylan is mainly present in hardwoods (25–35% dry matter) and agricultural residues such as cereal straw, sugar cane bagasse and sorghum bagasse (Mikkonen and Tenkanen, 2012).

Heteroxylans can be substituted at C2 and/or C3 positions with α-L-arabinofuranose, 4-O-methyl-α-D-glucuronic acid, α-D-glucuronic acid, and other neutral sugar units such as D-xylose, and D/L-galactose (Vuong and Master, 2022). Heteroxylans rich in L-arabinofuranose and 4-O-methyl-α-D-glucuronic acid are grouped into three main types: glucuronoxylans, arabinoxylans, and

FIGURE 7.2 Scheme of three main types of heteroxylans: (a) glucoronoxylan, (b) arabinoxylan, and (c) arabinoglucuronoxylans.

glucuronoarabinoxylans or arabinoglucuronoxylans (Figure 7.2). Xyloglucans' backbone is formed by glucose units joined by β-(1→4) bonds, with 50–75% of these substituted with D-xyloses by α-(1→6) bonds. Xylose can be further replaced by fucose, galactose, and occasionally by arabinosethrough (1—2) linkage (Kim et al., 2020).

Mannans comprise of a linear backbone of β-(1→4) mannose residues and, depending on the substitutions in the backbone, they can be classified as galactomannans, glucomannans, or galactoglucomannans (Singh et al., 2018). Glucans are polymers of β-glucoses linked by a β (1→3;1→4) bond. Glucans differ from cellulose because they show a lower degree of polymerization and crystallinity. Glucans have mainly structural function and their covalent linkages to proteins make them insoluble. Having all these chemical compositions and structures in mind, it should be highlighted that xylan-rich biomass such as hardwood, grass, and cereal biomass is more appropriate when targeting at XOS production when it comes to prebiotics. When starting with xylan-rich feedstock, it can be inferred that pretreatment and specific enzyme activities are crucial to attain these desired target compounds (XOS).

7.3 PRETREATMENT STEP AND HEMICELLULOSE SOLUBILIZATION

To utilize the carbohydrate fraction of LB, sugars have to be firstly released from the cellulosic or hemicellulosic fractions. Enzymes are considered among the most adequate catalysis to release sugars from LB. However, due to the natural recalcitrance of LB, a first pretreatment to increase enzyme accessibility and favour sugar release in the subsequent enzymatic hydrolysis is required. Different pretreatment technologies have been studied on a wide variety of lignocellulosic feedstocks during the last decades. These pretreatment methods can be divided into physical, chemical, physicochemical, and biological processes and their use depends on the further application of the lignocellulosic sugars. As a matter of fact, in the case of XOS production, a pretreatment technology enabling high hemicellulosic sugars solubilization is required, with both chemical and physico-chemical pretreatment among the most suitable.

Steam explosion is a physico-chemical process in which LB is subjected to high temperatures and pressure in the presence of saturated steam for a certain period of time, followed by a sudden depressurization (Moreno et al., 2019). These conditions allow the release of the acetyl groups that are part of the hemicellulose and result in the formation of acetic acid. Acetic acid accumulation provokes a pH drop producing an acid autohydrolysis process. This autohydrolysis affects the glycosidic bonds of the hemicelluloses and the ester bonds linking hemicellulose with the lignin. Thus, the hemicellulose heteropolymer is solubilized and different oligomers, including XOS, can be obtained. Moreover, the autohydrolysis reaction also generates lignin redistribution and partial solubilisation.

Steam explosion is one of the most commonly used methods to pretreat lignocellulose. It has been successfully applied in bioethanol production from a wide range of feedstocks such as agricultural residues (Álvarez et al., 2021, Montipó et al., 2018) and hardwoods (del Rio et al., 2022). However, due to the low content of acetyl groups in some materials like softwoods, acid catalysts are required to conduct the steam pretreatment (Pielhop et al., 2017). The high enzymatic hydrolysis yields, the complete sugar recovery, the lower environmental impact, and its applicability at industrial scale are among the attractive features of steam explosion. However, due to the harsh conditions in steam explosion, sugar and lignin degradation can occur resulting in the formation of inhibitory compounds such as furfural, hydroxymethylfurfural, formic and acetic acids, as well as phenolic compounds like vanillin and syringaldehyde (Moreno et al., 2019).

Liquid hot water is also a hydrothermal pretreatment that applies high pressures to maintain water in the liquid phase at elevated temperatures (160–240°C). In these conditions, alteration in the structure of lignocellulose occurs. In this case, the solubilization of the hemicelluloses is produced, making the cellulose polymer more accessible. To avoid the formation of inhibitors, the process pH is maintained between 4 and 7, since at these conditions hemicellulosic sugars remain in oligomeric form (Moreno et al., 2019). Liquid hot water has been used to produce XOS from chestnut shells (180°C) and peanut shells (210°C) achieving 57 mg XOS/g and 93 mg XOS/g biomass, respectively (Gullón et al., 2018; Rico et al., 2018).

Chemical pretreatment of LB with acid catalysts gives rise to a rupture of the ester bonds of the hemicellulose side chains. In acid pretreatment, a high number of monosaccharides and degradation compounds (such as furfural, hydroxymethylfurfural, and acetic acid) are produced. The most common acids used as catalysts are H_2SO_4, HCl, HNO_3, and H_3PO_4 (Hu and Ragauskas, 2012). Concentrations of monomeric sugars and degradation products depend on the severity of the process. Indeed, the use of acids presents many disadvantages like corrosivity, necessity of special reactors, and requirement of neutralizers (Brodeur et al., 2011). Mild acid hydrolysis conditions (e.g., 0.7M H_2SO_4; 90°C and 45 min) allowed obtaining XOS from commercial birch xylan, with an average structure formed by 6 xylose units and a single acid unit 4-O-methyl-D-glucuronic acid (Chemin et al., 2015). Alkalis can also be used as catalysts. In this case, lower temperatures than with acids are required but reaction times increase. For instance, NaOH (4-40 g/g substrate) at room temperature was capable of solubilizing 46.5% of the hemicellulose present in soybean straw (Wan et al., 2011).

Alternatively, the reagents can also be based on ionic liquids. This is the case of 1-ethyl-3-methylimidazolium acetate that was applied as a pretreatment method for XOS production from sugarcane reporting 115 mg XOS/g biomass using this ionic liquid at 100°C for 30 min (Ávila et al., 2020). As discussed later in this chapter, after pretreatment, enzymes allow the production of oligomers with a certain degree of polymerization, whereby avoiding monomer generation, which is crucial when aiming at XOS production.

7.4 ENZYMATIC HYDROLYSIS AND KEY ENZYMES

Once the hemicellulose is solubilized, it can be hydrolysed to oligomers or even monosaccharides with enzymes. These enzymes can be obtained from fungi, yeast, or bacteria, among others (Li et al., 2022). The hydrolytic activity of the enzymes is bond-specific. The use of particular enzymatic activities can result in greater depolymerization of the xylan and the production of low-degree polymerization XOS. The reaction conditions in enzymatic hydrolysis are milder than that with chemical agents and produce lower degradation products that can be potentially inhibitory. However, enzymes have narrow pH and temperature ranges and can be inhibited by both the product and the substrate.

The enzymes that hydrolyse hemicellulose can be divided into enzymes that cleave the backbone and those that affect the side branches. In some cases, these substituents can hinder enzymes' mediated cleavage. The enzymes that act on the main chain of xylan belong to the endoxylase family, and their action results in the release of XOS with a lower degree of polymerization and monomeric xylose. Due to the different substituents in the hemicellulose main chain, different types of enzyme with diverse activities must work synergistically to effectively hydrolyse hemicellulose (Biely et al., 2016; Li et al., 2022).

7.4.1 Xylanolytic Enzymes

The xylanolytic enzyme complex includes a series of enzymes that act on different groups that are part of the hemicellulose chain. The linkages are chosen depending on a chain length, degree of polymerization, and the presence of substituents. Xylanases system is a group of: endo-β-(1,4)-D xylanases (EC 3.2.1.8), β-D-xylosidase (EC3.2.1.37), α-glucuronidase (EC3.2.1.139), acetylxylan esterase (EC 3.1.1.72), α-L-arabinofuranosidase (EC 3.2.1.55), *p*-coumaric esterase (3.1.1.B10), and ferulic acid esterase (EC3.1.1.73) (Bhardwaj et al., 2019).

7.4.1.1 Main Xylan-Degrading Enzymes

7.4.1.1.1 Endo-β-(1-4)-D Xylanases

The endo-β-(1-4)-D xylanase enzymes are necessary to break the β-(1-4) bond between the xylose units in the hemicellulose backbone. Depending on the bond they hydrolyse, they can be classified as endo-xylanases or exo-xylanases. Endo-β-(1-4)-D xylanases catalyse the cleavage of the β-(1→4) bond of the xylan chain, resulting in a reduction in the degree of polymerization of the xylan. Enzymes belonging to groups EC 3.2.1.8 and EC 3.2.1.32 are classified in this category.

The endo-xylanases, which are the most frequently used enzymes with hemicellulase activity, are classified into different families of glycosyl hydrolases (GH). This classification depends on their 3D structure and their similarity in the amino acid sequence and hydrophobicity. Endo-β-(1,4)-D xylanases belong to the GH5, GH7, GH8, GH10, GH11, and GH43 families. Besides, the majority belongs to GH10 and GH11 families. The enzymatic hydrolysis of xylan occurs via inversion or retention pathway mechanisms, where the anomeric carbon is inverted or retained in its stereochemical configuration, respectively (Lombard et al., 2014). In the cyclic form of a sugar, the anomeric carbon is the carbon that was part of the carbonyl group in the straight-chain structure. Figure 7.3 shows these two mentioned mechanisms. The first mechanism is followed by GH43 and GH8, whereas GH10, GH11, GH30, and GH51 are retaining enzymes.

In the inversion mechanism (Figure 7.3a), the hydrolysis reaction occurs via a single displacement involving two enzymatic residues. One of the residues acts as an acid and the other one as a base. Firstly, the carboxylic groups of the first enzyme act as an acid activating water molecule. The water acts as nucleophilic and attacks the anomeric carbon providing the inversion of the configuration (Bhardwaj et al., 2019). In the retention mechanism, a double displacement occurs (Figure 7.3b). One of the catalytic residues works as a general acid catalyst by protonating the substrate, while the second one

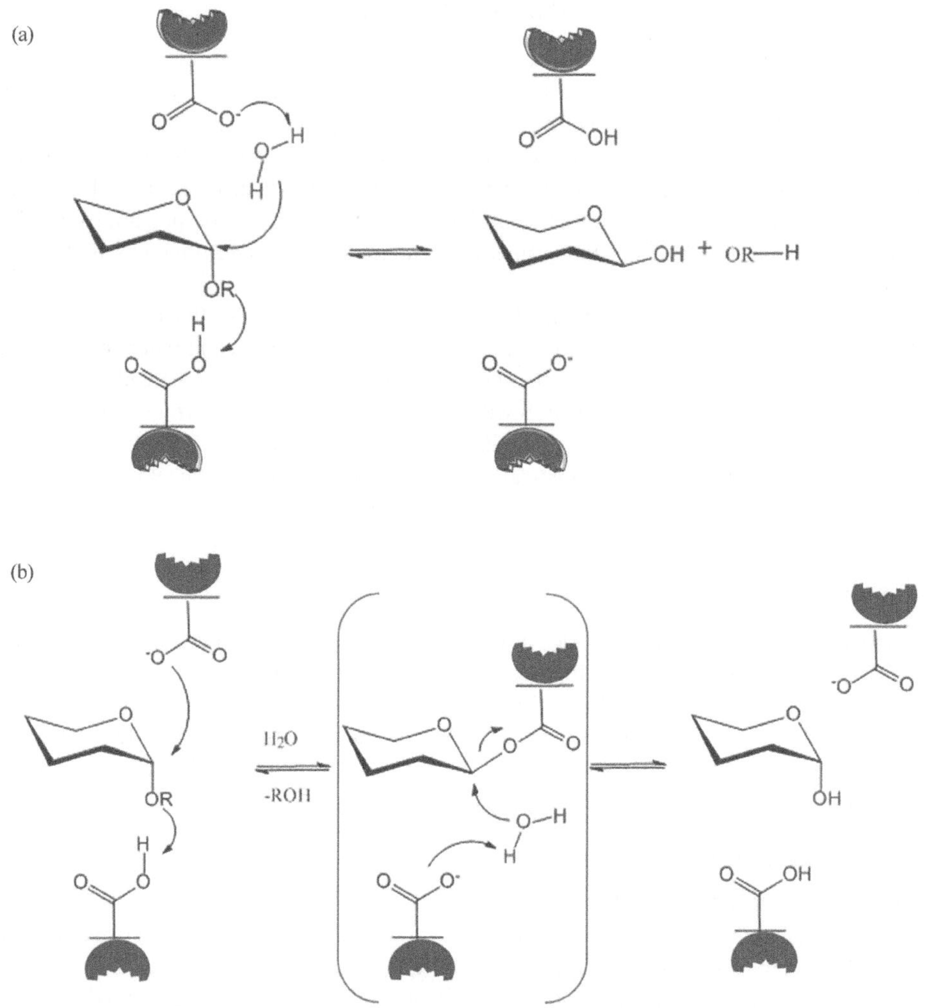

FIGURE 7.3 Mechanisms of inverting (a) and retaining (b) β-glycosidase.

performs a nucleophilic attack to the anomeric carbon generating a covalent glycosyl-enzyme intermediate. In the second step, the first carboxylate group functions as a general base, subtracting a proton from an incoming water molecule that attacks the anomeric carbon. The overall result of this reaction mechanism is a retention of the configuration at the anomeric centre (Bhardwaj et al., 2019; Motta et al., 2013).

Endo-xylanases belonging to GH10 family have high molecular weight and low isoelectric points (Bhardwaj et al., 2019). They usually show the catalytic domains comprising a barrel $(\alpha/\beta)_8$ fold (typical of Clan A), consisting of alternating α-helices and β-strands (Puchart et al., 2021). Most of the enzymes belonging to this family are EC 3.2.1.8 and some exceptions are EC 3.2.1.32. The endo-β-(1-4)-D xylanases of this family hydrolyse both linear and branched substrates because they can also accommodate on side chains attached to xylan. These endo-β-(1-4)-D xylanases require the presence of two unsubstituted xyloses to hydrolyse the main chain towards the reducing end of a single or double xylose substitution (Nordberg Karlsson et al., 2018). For this reason, the endo-β-(1-4)-D xylanases of the GH10 family release oligosaccharides with a low degree of polymerization (Motta et al., 2013), being the main enzymatic products xylobiose and xylotriose (Capetti et al., 2021).

The endo β-(1-4)-D xylanases belonging to the GH11 family are the most used enzymes, given that they act at wide range of temperature and pH. They have small molecular mass (<30 KDa) with high isoelectric points and their catalytic domain has a β-jelly roll structure (Capetti et al., 2021). This family has denominated "true xylanases" because they are exclusively active on xylose containing substrates. The GH11 family show high substrate selectivity and are substrate-specific. These enzymes require three consecutive unsubstituted xylose units to act, making them somewhat inefficient in substrates with many substituents (Biely et al., 2016; Nordberg Karlsson et al., 2018). Generally, GH11 are most active on long-chain XOS because they preferentially cleave the unsubstituted regions of the xylan backbone (Motta et al., 2013). In comparison, GH10 show greater catalytic versatility and lower substrate specificity than GH11 enzymes (Bhardwaj et al., 2019).

Other families of endo-β-(1-4)-D xylanases are GH30.With these enzymes, the cleavage of the main chain takes place primarily at the second glycosidic linkage from the branch toward the reducing end of the polysaccharides. The mode of cleaving GH30 from glucuronoxylan is determined by acid substituents (Biely et al., 2016). Since the 1990s, the scientific community has focused on new xylanases of GH30. These enzymes were previously classified as xylanases from family GH5 (St John et al., 2006). However, in 2010, these xylanases were transferred to family GH30 based on a bioinformatic and structural study of the catalytic domains of both members (St John et al., 2010). This analysis was based on the organization of a secondary structure, around the catalytic domains comprising a $(\beta/\alpha)_8$-fold central barrel formed by 8 parallel β-strands.The assembly of a 9-stranded aligned β-sheet structure tightly tethered to the $(\beta/\alpha)_8$ barrel through a dual linker is a different structural component used as a differentiating element for GH5. This assembly is not identical inside GH30; therefore, two groups have been distinguished. In turn, these two groups have been divided into 10 subfamilies (January 2022, www.cazy.org) (Puchart et al., 2021). This GH30 show a wide diversity including β-glucosidases (3.2.1.21), β-glucuronosidases (EC 3.2.1.31), β-xylosidases (3.2.1.37), β-fucosidases (EC 3.2.1.38), glucosylceramidases (EC 3.2.1.45), β-1,6-glucanases (EC 3.2.1.75), glucuronoarabinoxylan endo-β-1,4-xylanase (EC 3.2.1.136), and endo-β-1,6-galactanase (EC 3.2.1.164).

In general, these enzymes show a high selectivity for glucuronoxylan and XOS substituted with glucuronic acids or methyl glucuronic acid via a 1-2 linkage. In detail, GH30_7 and GH30_8 require substitutions with uronic acid to be able to develop their activity (Šuchová et al., 2020). This property makes them different from GH10 and GH11 (Linares-Pasten et al., 2018). Endoxylanases from bacterial and fungal sources are used for producing different types of XOS. Bacterial endoxylanases are frequently recombinantly produced from Gram-positive bacteria classified under the phyla *Firmicutes*, including *Bacilli* and Actinobacteria (derived from Streptomyces) (Linares-Pasten et al., 2018). Regarding fungal enzymes, the ones derived from strains of *Aspergillus* and *Trichoderma* are regarded to be the best producers (Santibañez et al., 2021). Also, a combined enzyme production system with recombinant and native organisms has been used to obtain endoxylanases from *Paecilomyces thermophila, Trichoderma reesei, Trichoderma longibranchiatum*or *Scheffersomycesstipites* (Linares-Pasten et al., 2018). Figure 7.4 shows the scheme of enzymatic breakdown of xylan by endoxylanases GH10, GH11, and GH30.

7.4.1.1.2 Exoxylanases and β-Xylosidases

Exoxylanases are active on natural xylan substances, hydrolysing long-chain xylo-oligomers from the reducing end to produce short-chain xylooligomers and xylose. This enzyme belongs to GH family 8, according to CAZY. Exoxylanase activity must be avoided or reduced to achieve a high concentration of XOS (Santibañez et al., 2021). However, β-xylosidase (EC 3.2.1.37) is required to catalyse the non-reducing ends of the XOS and β-xylobiose. These enzymes hydrolyse XOS of low degree of polymerization accumulated in the medium that, indeed, can inhibit endo-β-(1,4)-D-xylanases. However, the use of β-xylosidases hampers XOS accumulation and favours xylose production which in turn is unfavourable for XOS production (Santibañez et al., 2021).

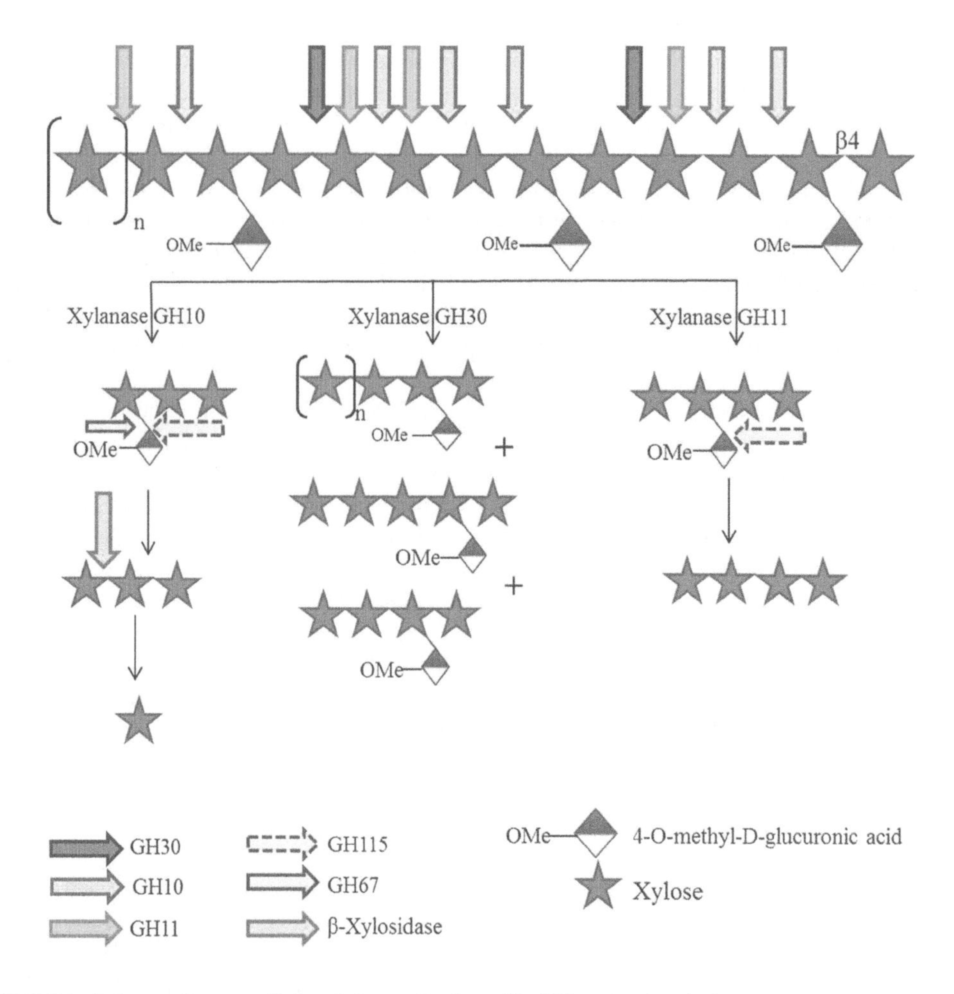

FIGURE 7.4 Scheme of enzymatic breakdown of xylan with different xylanolytic enzymes.

7.4.1.2 Accessory Enzymes

The xylan from hemicellulose is usually substituted with different residues as described in previous sections. These side chains can hamper the action of endoxylanases by steric hindrance. Consequently, the removal of these ramifications can improve enzymatic hydrolysis and XOS production. Accessory activities including α-L-arabinofuranosidase, α-glucuronidase, and esterase are frequently used. The action of these enzymes produces a synergistic effect with endoxylanases. The mode of the action of these enzymes can be simultaneous or sequential. Previous reports have shown synergisms between xylanolytic enzymes (Malgas et al., 2019). This synergism can be (1) between endoxylanases, mainly GH10 and GH11or (2) between endoxylanases and side-chain-cleaving enzymes.

7.4.1.2.1 α-L-Arabinofuranosidases

These enzymes (EC 3.2.1.55) catalyse the cleavage of the bonds between the arabinoses that are part of the main chain as substituents, releasing α-L-arabinoses. They typically belong to the GH43, 51, 54, and 62 families (Borsenberger et al., 2014). These enzymes are classified into three groups: (1) enzymes that act on monosubstituted arabinoses in positions (1→2) and (1→3) (AXH-m), (2) Enzymes that release the substituted arabinose from disubstituted xyloses (AXH-d), and (3) α-L-arabinofuranosidases that act on both monosubstituted and terminal double-substituted

D-xylanopyranosyl substituted residues (AXH-m,d). The α-L-arabinofuranosidases belonging to the AXH-d category are exclusive of the GH43 family (Biely et al., 2016). Numbers can often express substrate specificity, thus AXH-d3 indicates the release of α-L-arabinofuranosyl linked (1→3) to doubly substituted xylanopyranol residues (Borsenberger et al., 2014).

7.4.1.2.2 Esterases

These enzymes hydrolyse the ester bond formed between the xylan chain and various acetyl, ferulic, or coumaric groups.

Acetyl esterase: This type of enzyme (E.C.3.1.1.6) acts on the bond formed between the acetyl groups and the xylan chain, releasing the acetyl groups attached to XOS. These enzymes play an essential role in the hydrolysis of xylan since the acetyl side groups can interfere with the accessibility of the enzymes cleaving the backbone due to steric hindrance. Thus, the removal of acetyl groups facilitates the action of endoxylanases.

Acetylxylanesterases (AXE): These cleave the ester bond between the acetyl groups and the xylan chain in mono- or di-O-acetylated polysaccharides. They are classified as carbohydrate esterase (CE) families 1, 4, 5 and 6. Enzymes belonging to the CE4 family do not attack doubly acetylated positions and none of them are able to deacetylate the 3-position of the residue substituted with 4-O-methyl-D-glucuronic acid (Neumüller et al., 2015).

-Feruloyl esterase: These enzymes release phenolic acids (ferulic acid and *p*-coumaric acid) attached to the 5-position of arabinose residues.

7.4.1.2.3 α-Glucuronidases

The α(1→2) linkage of the 4-O-methyl-D-glucuronic acid to the main chain is one of the most stable glycosidic acid linkages in plant cells. α-Glucuronidases are necessary to hydrolyse this bond. The α-glucuronidases (EC 3.2.1.131/139) have been classified into the GH67 and GH115 families. The α-glucuronidases of the GH67 family only liberate 4-O-methyl-D-glucuronic acid to the non-reducing ends of the main chain, being unable to act on groups that are internally attached to the chain (Biely et al., 2016) (Figure 7.5). For the release of this group, it is necessary to use an α-glucuronidase belonging to the GH115 family (Wang et al., 2016). It should be highlighted that the presence of acetyl groups in nearby positions can hinder the action of α-glucuronidase.

7.5 XYLOOLIGOSACCHARIDES PROCESS PRODUCTION BY MICROBIAL ENZYMES

As previously mentioned, autohydrolysis, consisting of steam produced from water at high temperatures and pressure, can produce XOS by depolymerizing the hemicellulose with hydronium ions (Moreno et al., 2019). In addition to the high-energy requirement needed for these thermal pretreatments, unwanted by-products might be released as well during hydrolysis. While chemical processes might be cheaper when compared with thermal ones, chemical ones might also produce not only XOS but also some side-products or even end up with degradation of products. For this reason, LB physico-chemical or chemical pretreatments are usually combined with substrate-specific enzymatic treatments. With enzymes, only the targeted product is hydrolysed, which indirectly brings benefits in the step of XOS purification. In the case of enzymatic hydrolysis, the drawback associated with this treatment is the economic costs of enzymatic cocktails. Because enzymatic hydrolysis generates fewer residues and avoids equipment corrosion, enzymatic processes are the most widely used treatment to produce XOS.

As previously introduced, enzymes responsible for the hydrolysis of xylan belong to the GH10 and GH11 families (Poletto et al., 2020). Enzymatic cocktails for XOS production are mainly

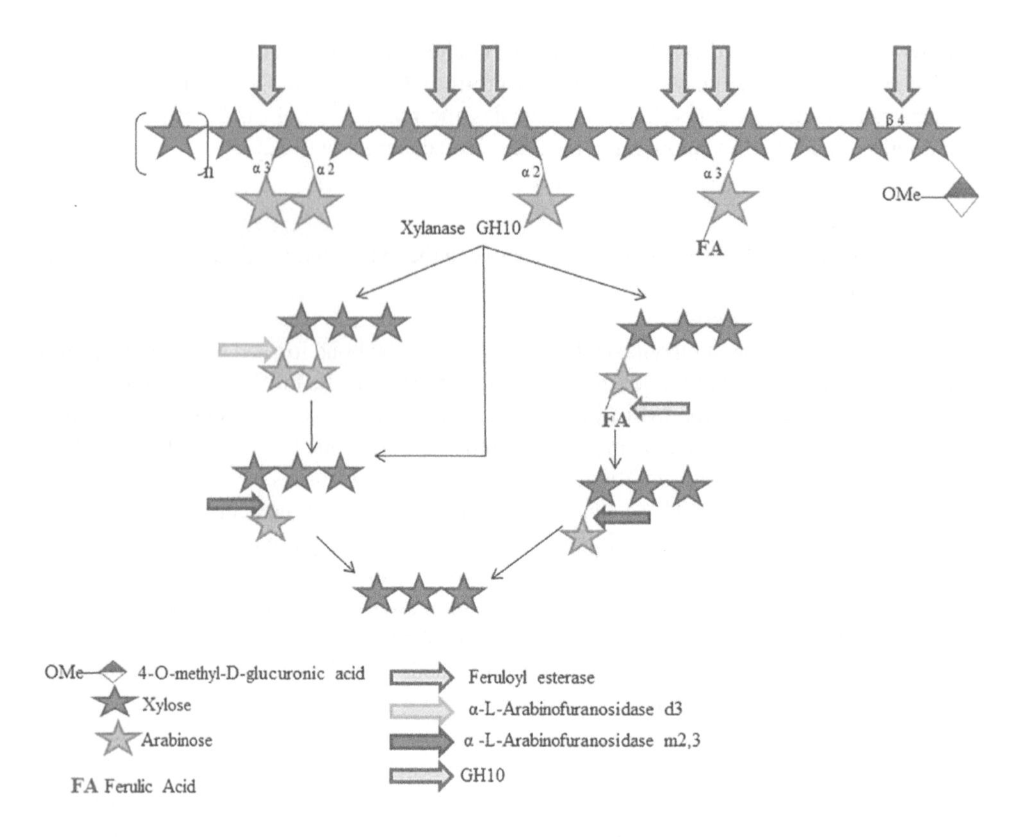

FIGURE 7.5 Schematic representation of arabinoxylan cleavage by GH10, feruloyl esterase, and α-L-arabinofuranosidase.

composed of endo-xylanases and should exhibit as low β-xylosidases activity as possible. The use of these two enzymes should be accomplished in such a way that the activity of the latter one remains low while the activity of endo-xylanases prevails. The underlying reason for this interplay of activities relies on the fact that exoxylanase might lead to xylose, which has been reported to inhibit endo-xylanase activity (Santibañez et al., 2021). Endoxylanases with low β-xylosidase activity favour XOS production over xylose (Bouiche et al., 2020), which is a desirable feature when aiming at prebiotics. Concerning xylanase production, a compilation of conditions employed, producing microorganisms, substrate employed, and enzymatic activities can be found in Corim Marim and Gabardo (2021). In terms of endoxylanase-producing microorganisms, it should be highlighted that filamentous fungi are by far the most widely used sources.

Together with these enzymes, some other accessory enzymes might also be used when dealing with XOS production. Heretogenousxylan might exhibit some ramification that can interfere with xylan hydrolysis and thereby XOS production. Endo-xylanase activity could interfere with the stearic hindrance of those ramifications that do not allow optimum binding of endo-xylanases. Some of those xylan ramifications can be, in fact, removed by the use of accessory enzymes. α-L-arabinofuranosidades, α-glucuronidases, and acetyl-esterase are the most reported accessory enzymes (as deeply discussed in Section 7.4.1.). While α-L-arabinofuranosidades remove arabinose, α-glucuronidases are mostly used in materials with greater glucuronoxylan presence, such is the case of woody material (Santibañez et al., 2021). The negative effect of acetyl side-groups, most commonly found in grasses and cereals-based feedstocks, can also be avoided by the use of acetyl-xylan esterase. A recent investigation evidenced that the hydrolytic activity of xylanase could be increased by 40% when using those later accessory enzymes (Hettiarachchi et al., 2019).

Indeed, the different combinations of pre-treatments that open up the recalcitrant nature of LB and different enzymes present in hydrolytic cocktails might result in XOS with different degrees of depolymerization (Pinales-Márquez et al., 2021). Some other parameters like the feedstock employed and the enzymatic conditions might also affect XOS production yields (Corim Marim and Gabardo, 2021). Table 7.1 summarizes several examples of different combinations of endoxylanases and accessory enzymes used to produce XOS from the literature. Table 7.1 shows in detail the biomass source, enzyme, operational conditions, and mg XOS/g xylan.

As it can be seen in Table 7.1 several enzyme cocktails are being studied to provide better XOS yields. Goldbeck et al. (2016) studied the effect of a mixture of recombinant endo-xylanase and a feruloylesterase using sugar cane bagasse as raw material. The obtained yield was 356 mg of XOS/g xylan. Liu et al. (2018) evaluated XOS production in corn cobs using *Paenibacillusbarengoltzii* endoxylanase (GH10). These authors achieved 750 mg XOS/g xylan. On the other hand, the use of endo-β-(1,4)-D-xylanase and α-L-arabinofuranosidase (GH51) mixture in a sugarcane bagasse produced 115 mg of XOS/g sugarcane straw (Ávila et al., 2020). Several feedstocks such as corncob, rice straw, and almond shells were also used for producing XOS producing 180 and 110 mg of XOS/g of biomass (Han et al., 2020; Le and Yang, 2019; Singh et al., 2019).

TABLE 7.1
Examples of Different Combinations of Endoxylanases and Accessory Enzymes Used to Produce XOS

Raw Material	Pretreatment	Conditions of Enzymatic Hydrolysis	XOS Yield mg XOS/g Xylan	Reference
Rice husk	Alkaline pretreatment NaOH 18% (w/v) 120°C 45 min	Endo-1,4-xylanase (Pentopan™) pH 6, 50°C	347	Khat-Udomkiri et al. (2018)
Almond shell	Liquid hot Water 200°C 15 min	Endo-1,4-xylanase (*Thermomyces lanuginosus* expressed in *Aspergillus oryzae*) pH 5.5, 50°C	545	Singh et al. (2019)
Corncobs	Steam explosion (acidic electrolyzed) 165°C 35 min	Endo-1,4-xylanase (GH10) pH 6.5, 60°C	750	Liu et al. (2018)
Wheat straw	Steam explosion 200°C 4 min	Endo-1,4-xylanase (*NS50030*) + β-glucosidase (Novozym 188) pH 4.8, 50°C	350	Álvarez et al. (2017)
Brewer's spent grain	-	Endo-1,4-xylanase (GH11) pH 4.5 40° C	444	Amorim et al. (2019)
Barley straw	Steam explosion 180°C 30 min	endo-1,4-xylanase (GH11) + α-L-arabinofuranosidase (GH51) + feruloy-esterase (CE1) + acetylxylan-esterase (CE6) pH 4.8, 50°C	588	Álvarez et al. (2021)
Sugarcane bagasse	Ionic liquid 1-ethyl-3-methylimidazolium acetate 100°C 30 min	Endo-1,4-xylanase (*NS50030*) + α-L-arabinofuranosidase (GH51) pH 5, 50°C	530	Ávila et al. (2020)
Sugarcane bagasse	Glacial acetic acid 8.74 M / hydrogen peroxide 2.6 M 60°C during 7 h	Endo-1,4-xylanase recombinant (GH11) + feruloyl esterase (CE1) pH 5, 50°C	356	Goldbeck et al. (2016)

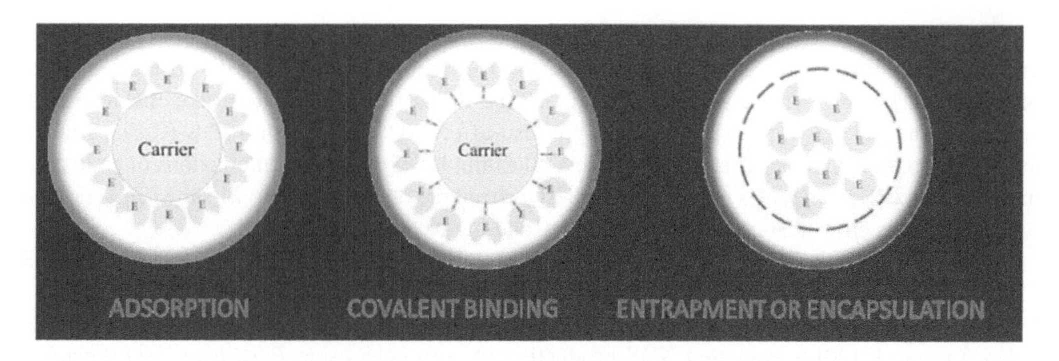

FIGURE 7.6 Different immobilization methods. (E = Enzyme.) (Adapted from Lyu et al., 2021.)

A yield of 11.1 gXOS/100 g barley straw were obtained from steam-exploded barley straw obtained at 180°C for 30 min and treated with twoendo-β-(1,4)-D-xylanases enzymes (GH10) and (GH11) (Álvarez et al., 2018). With the supplementation of accessory enzymes (α-L-arabinofuranosidase, acetyl-xylan esterase and feruloyl esterase),the yield increased up to 13.0 g XOS/100 g barley straw(Álvarez et al., 2021).In the same line, 8.9 g of XOS/100 g of wheat straw were generated after subjecting the feedstock to steam explosion at 200°C for 4 min and hydrolysing the residue with xylanolytic enzymes (Álvarez et al., 2017). The enzymatic hydrolysis can be carried out by adding enzymes together with the substrate, producing enzymes *in-situ* by some microbial system or using the immobilized enzymes. When enzymes are free in submerged cultivation, hydrolysis often exhibit difficulties. Free-form enzyme systems are more susceptible to salts, inhibitory substances, surfactants, etc. Some of the drawbacks identified in the case of free-form enzymes include substrate/product inhibition, enzymes instability or inactivity and difficulties for product recovery (Krajewska, 2004). Because of that, efforts have been directed towards enzyme immobilization. Xylanases immobilization promotes greater hydrolytic efficiency by supporting the enzyme into materials that guarantee proper flow mechanics (*i.e.* adequate pore size) and sufficient active sites for adhesion (Ariaeenejad et al., 2020). When selecting the most suitable immobilization strategy, immobilized enzymes should behave similarly to free-enzymes in terms of activity and selectivity. Physical and chemical interactions (physical adsorption, entrapment and covalent linkage) support this enzyme-material interaction. The scheme of those 3 types of immobilization methods can be seen in Figure 7.6.

Adsorption is based on van der Waals interactions, hydrophobic and electrostatic interactions. In this manner, adsorption does not entail a real chemical bonding and does not need functionalized materials for bindings. Enzymes entrapment does not need chemical bonds since this method includes the physical trapping of enzymes within a matrix, thereby keeping them confined and protected from harsh environments. In contrast, covalent bonds are established between the supporting material and the enzyme in the case of the covalent binding method. In some cases, covalent binding results in the need of chemical modifications for the functional groups of enzymes to efficiently bond. Each of those immobilization methods have advantages and disadvantages (Table 7.2, adapted from Lyu et al., 2021) and therefore, proper selection should be made according to the set-up when conducting this immobilization.

Apparently, the covalent linkage is preferred because the stiffening of the protein structure enhances the thermal/operational stability of the enzyme (Milessi et al., 2016). Immobilization is regarded as a strategy with two-fold benefits: on the one hand, enzymes might be recovered and reused and, on the other hand, enzyme-free products are attained and therefore, purification steps are facilitated. According to Santibañez et al. (2021), xylanases immobilization efficiency reached 90% while keeping 70% of their activity yields. In the case of *in-situ* production of enzymes by some microbial systems, a xylanase-producing microorganism is cultivated and secretes the enzymes in

TABLE 7.2
Advantages and Disadvantages of Most Commonly Used Methods for Enzyme Immobilization

	Advantages	Disadvantages
Adsorption	Simple and cheap	Low stability
	No reagents needed	Weak binding
	High catalytic activity	
Covalent binding	Strong binding	Reduced enzyme mobility
	Prevent enzyme leaching	Conformational restriction
	High thermostability	
Entrapment	Protected enzyme activity	Mass transfer limitations
	Simple downstream processing	Low enzyme loading

the culture broth in the presence of a carbon source. This is the case, for instance, for the investigation conducted by Amorim et al. (2019) who conducted direct microbial fermentation of brewer's spent grain using *Trichoderma* sp. for XOS production purposes. An advanced alternative is the use of recombinant enzymes produced in non-xylanolytic hosts. Some of these attempts include XOS production by using *Bacillus subtilis* and a clone containing xylanase gene from *T. reesei* (the clone increased the XOS production by 16%, Amorim et al., 2018) and the use of *Pichia pastoris* to express the xylanase GH11 isolated from *Aspergillus niger* (Aiewviriyasakul et al., 2021). Indeed, since the 2010s, *P. pastoris* has successfully replaced *S. cerevisiae* when targeting the production of eukaryotic recombinant proteins. This is due to the fact that *Pichia* has been shown to grow at high cell densities while producing active and functional enzymes (Wan Azelee et al., 2016).

It should also be highlighted that some protein engineering approaches have been accomplished to improve endo-xylanases. These approaches aim to increase the catalytic activity, improve their thermostability, and increase the XOS/xylose ratio (Santibañez et al., 2021). For instance, Wu et al., (2018) improved GH11 xylanases activity and stability by changing its conformation and Bu et al. (2018) improved its thermostability by mutagenesis via computational library design. In general, many microorganisms produce a variety of xylanolytic enzymes with different properties. However, not all microorganisms produce all the enzymatic activities required for total xylan depolymerization. For this reason, the formulation of suitable and economically cost-effective enzyme cocktails is one of the main challenges in the hydrolysis of lignocellulosic materials. To overcome this drawback, considerable research efforts are now being made to develop new multifunctional enzymes that can hydrolyse a wide variety of bonds in polysaccharides. Alternatively, adaptive evolution and genetic/metabolic engineering are being also widely employed to improve enzyme properties and stabilization (Binod et al., 2018).

7.6 DOWNSTREAM PROCESS

In the enzymatic hydrolysis medium, monosaccharides and other non-sugar compounds are generated together with XOS. For this reason, the hydrolysate must be refined to reach high XOS recoveries and separation efficiencies.When XOS are used in the food industry, 75–95% purity levels are required and, thereby, an efficient purification step is mandatory (Gullón et al., 2009). Extracting XOS in a water-miscible organic solvent is often unfeasible. The subsequent water removal for product concentration and drying involves a high energy demand. For these reasons, several XOS purification methods have been tested in the last decade, namely membrane separation, adsorption affinity, and chromatography. Figure 7.7 summarizes the main advantages and disadvantages of the XOS purification techniques described in the literature (Kruschitz and Nidetzky, 2020).

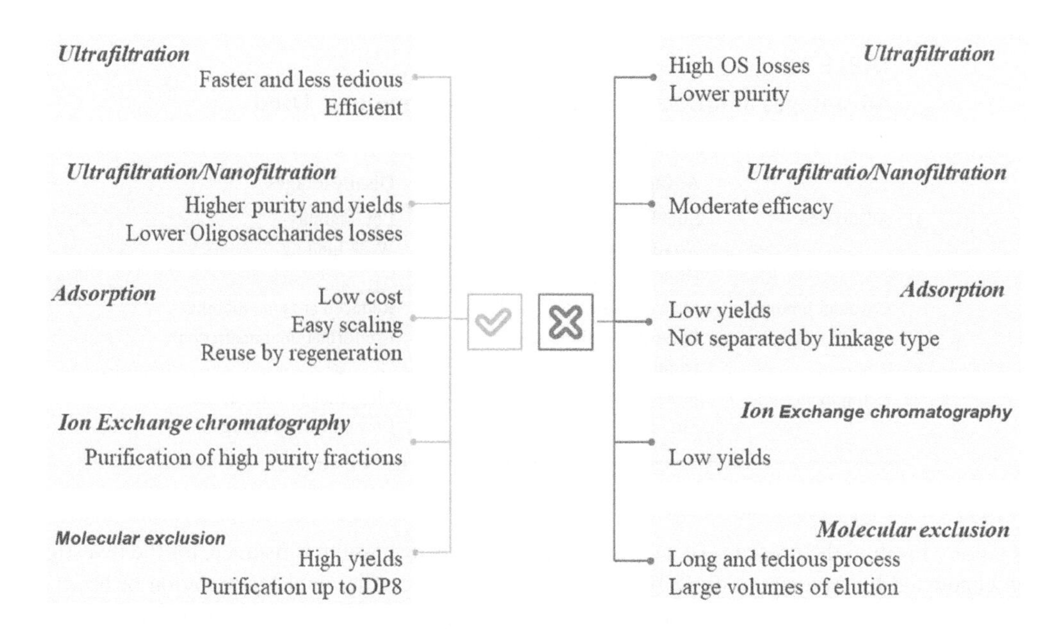

FIGURE 7.7 Advantages and disadvantages of fractionation and purification techniques used with XOS.

7.6.1 Membrane Separation

For XOS separation, there is a great variety of membranes based on different materials (cellulose, ceramics, organic polymers, etc.) and pore size. Polymeric composites with a support layer of polysulfone have been preferably used for XOS purification (Kruschitz and Nidetzky, 2020). Ultrafiltration is a particular type of filtration that requires porous membranes. This process requires a driving force, which increases with decreasing membrane pore size. Among other factors, ultrafiltration efficiency is mainly affected by substitutions, the degree of polymerization, branching, and the viscosity of the solution (Naidu et al., 2018). For instance, when using a combination of ultrafiltration (50 kDa) and nanofiltration (1 kDa), a 74.5% XOS recovery was obtained from almond shells (Singh et al., 2019). The main advantages of this technique are the low energy requirement and the easy manipulation and scale-up. In contrast, this method is inefficient for eliminating small molecules such as monosaccharides. Combining nanofiltration with ultrafiltration can provide an efficient approach for removing unwanted solutes and obtaining highly pure saccharides. Usually, the ultrafiltration process is used to separate polysaccharides and protein together with other impurities. Then, the permeate from ultrafiltration is handled with nanofiltration. The resulting XOS are concentrated in the retentate and the smaller monosaccharides and disaccharides are removed. As a disadvantage, membranes are often hydrophobic and therefore very susceptible to fouling. Membrane fouling refers to deposition or adsorption of biological, organic, inorganic, and colloidal matter onto the surface of the membrane and into the membrane pores. These depositions ultimately cause selectivity loss and permeate flux deterioration, consequently resulting in higher production costs (Li et al., 2021b). Because of this, membrane fouling remains a major issue in achieving efficient saccharide separation with nanofiltration systems. Foulants deposited on the surface of the nanofiltration membrane during saccharide purification are classified into saccharide and non-saccharide components. The former consists mainly of polysaccharides, crystallization of saccharides, etc. The latter includes high molecular weight compounds such as pigments, proteins, and colloidal complexes, and low molecular weight compounds such as phenolic and acetic acid. Hydrophilic modification effectively alleviates membrane fouling through physical and chemical methods (Li et al., 2021b). This method has been used to concentrate, purify and fractionate XOS from hemicellulose from wheat bran (Arkell et al., 2013), alkali-pretreated empty fruit bunch hydrolysate (Wijaya et al., 2020), and alkali extracted from corn straw (Lian et al., 2020).

7.6.2 Adsorption Affinity

Adsorption is another simple separation technique widely used for XOS separation. It is based on the difference in affinity of the compounds to physically bind to the adsorbent. There are multiple adsorbents on the market capable of trapping heavy metals, organic acids, and even dyes. Activated carbon has been used to purify XOS from corn cobs and barley hulls (Amorim et al., 2018), and is even used on an industrial scale (Sato et al., 2010).

7.6.3 Chromatographic Separation

Compared to membrane separation, chromatographic separation efficiently removes undesirable small products such as furfural, acetic acid, and monosaccharides (Moniz et al., 2014). Both ion exchange and gel permeation chromatography can be applied for XOS purification. Ion-exchange chromatography uses ion-exchange resins to remove organic and inorganic ions present in the hydrolysis media. Ion exchange resins are composed of a polymeric matrix with positively or negatively charged functional groups that act as anion and cation exchangers depending on the affinity for the counter ions. Weak cation exchangers can be compounds containing sulfite functional groups and strong exchangers with sulfonic acid. On the other hand, weak cation exchange resins have functional groups such as carboxylic acids. In contrast, weak cation exchange resins have groups functional such as carboxylic acids and strong exchangers have sales of quaternary ammonium (Soto et al., 2011).

In gel permeation chromatography or molecular exclusion chromatography, the separation takes place by passing the mixture of compounds through a stationary phase formed by gel with spherical particles with pores of a specific size, which arranges the solutes in decreasing order of molecular size.The most commonly used gels are Bio-Gel or Sephadex types (Palaniappan et al., 2021). In the last years, a simulated moving bed has been patented for the purification of XOS. However, the application of this method has not been reported. This technology reduces solvent consumption and improves separation performance in systems with low selectivity and resolution (Kruschitz and Nidetzky, 2020; Santibañez et al., 2021).

7.7 CONCLUSIONS

XOS are emerging prebiotics that can be sustainably produced from hemicellulose after subjecting this polymer to hydrolysis. Although chemical hydrolysis of hemicellulose has been proposed, enzymatic hydrolysis allows milder reaction conditions producing fewer degradation products that can be potentially toxic and compromise XOS application. The pretreatment step is crucial for the efficient decomposition of xylans. Among the different available pretreatments, hydrothermal treatment seems to be the most promising at an industrial scale. However, conditions minimizing xylose release and favouring XOS production should be sought for effective XOS production. Both the need for LB pretreatment and the relative low efficiency of enzymatic hydrolysis currently achieved may limit the production of XOS at industrially relevant conditions. Thus, for the successful production of XOS from hemicellulose, the formulation of suitable and economically cost-effective enzyme cocktails is of paramount importance. Because the composition and structure of xylan determines the type and yield of the produced XOS, the optimum enzymatic hydrolysis conditions for each feedstock should be determined. The discovery of new enzymes will definitively boost the progress of the biotransformation of xylan into these high value-added products. Lastly, a purification step is necessary in XOS production. This latter step represents a big challenge for the overall efficiency of the process. Indeed, the application of the produced compounds (pharmaceutical, food, etc.) and the economic viability of the process should be considered to make an adequate selection of the purification and separation step.

ACKNOWLEDGEMENT

This research has been supported by the project BIOMIO (PID2020-119403RB-C21 and PID2020-119403RB-C22) funded by MCIN/AEI/10.13039/501100011033 and "European Union NextGenerationEU/PRTR". The grants RYC2019-027773-I funded by MCIN/AEI/10.13039/501100011033 and by "ESFInvesting in your future" is also acknowledged.

REFERENCES

Agbor, V.B., Cicek, N., Sparling, R., Berlin, A., Levin, D.B. 2011. Biomass pretreatment: Fundamentals toward application. *Biotechnology Advances* 29:675–685.

Aiewviriyasakul, K., Bunterngsook, B., Lekakarn, H., Sritusnee, W., Kanokratana, P., Champreda, V. 2021. Biochemical characterization of xylanase GH11 isolated from *Aspergillus niger* BCC14405 (XylB) and its application in xylooligosaccharides production. *Biotechology Letters* 43(12):2299–2310.

Álvarez, C., González, A., Ballesteros, I., Negro, M.J. 2021. Production of xylooligosaccharides, bioethanol, and lignin from structural components of barley straw pretreated with a steam explosion. *Bioresource Technology* 342:125953. https://doi.org/10.1016/j.biortech.2021.125953

Álvarez, C., González, A., Negro, M.J., Ballesteros, I., Oliva, J.M., Sáez, F. 2017. Optimized use of hemicellulose within a biorefinery for processing high value-added xylooligosaccharides. *Industrial Crops and Products* 99:41–48. https://doi.org/10.1016/j.indcrop.2017.01.034

Álvarez, C., Sáez, F., González, A., Ballesteros, I., Oliva, J.M., Negro, M.J. 2018. Production of xylooligosaccharides and cellulosic ethanol from steam-exploded barley straw. *Holzforschung* 73:35–44. https://doi.org/10.1515/hf-2018-0101

Amorim, C., Silvério, S.C., Rodrigues, L.R. 2019. One-step process for producing prebiotic arabino-xylooligosaccharides from brewer's spent grain employing *Trichoderma* species. *Food Chemistry* 270:86–94. https://doi.org/10.1016/j.foodchem.2018.07.080

Amorim, C., Silvério, S.C., Silva, S.P., Coelho, E., Coimbra, M.A., Prather, K.L.J., Rodrigues, L.R. 2018. Single-step production of arabino-xylooligosaccharides by recombinant *Bacillus subtilis* 3610 cultivated in brewers' spent grain. *Carbohydrates Polymers* 199:546–554. https://doi.org/10.1016/j.carbpol.2018.07.017

Ariaeenejad, S., Jokar, F., Hadian, P., Mamani, L., Gharaghani, S., Fereidoonnezhad, M., Salekdeh, G.H. 2020. An efficient nano-biocatalyst for lignocellulosic biomass hydrolysis: xylanase immobilization on organically modified biogenic mesoporous silica nanoparticles. International. *Journal of Biological Macromolecules* 164:3462–3473. https://doi.org/10.1016/j.ijbiomac.2020.08.211

Arkell, A., Krawczyk, H., Jönsson, A.S. 2013. Influence of heat pretreatment on ultrafiltration of a solution containing hemicelluloses extracted from wheat bran. *Separation and Purification Technology* 119:46–50. https://doi.org/10.1016/j.seppur.2013.09.001

Ávila, P.F., Franco Cairo, J.P.L., Damasio, A., Forte, M.B.S., Goldbeck, R. 2020. Xylooligosaccharides production from a sugarcane biomass mixture: Effects of commercial enzyme combinations on bagasse/straw hydrolysis pretreated using different strategies. *Food Research International* 128:108702. https://doi.org/10.1016/j.foodres.2019.108702

Bhardwaj, N., Kumar, B., Verma, P. 2019. A detailed overview of xylanases: an emerging biomolecule for current and future prospective. *Bioresources and Bioprocessing* 6:40. https://doi.org/10.1186/s40643-019-0276-2

Biely, P., Singh, S., Puchart, V., 2016. Towards enzymatic breakdown of complex plant xylan structures: State of the art. *Biotechnology Advances* 34:1260–1274. https://doi.org/10.1016/j.biotechadv.2016.09.001

Binod, P., Gnansounou, E., Sindhu, R., Pandey, A. 2018. Enzymes for second generation biofuels: Recent developments and future perspectives. *Bioresource Technology Reports*:317–325. https://doi.org/10.1016/j.biteb.2018.06.005

Borsenberger, V., Dornez, E., Desrousseaux, M.-L., Massou, S., Tenkanen, M., Courtin, C.M., Dumon, C., O'Donohue, M.J., Fauré, R. 2014. A ^{1}H NMR study of the specificity of α-L-arabinofuranosidases on natural and unnatural substrates. *Biochimica Biophysica Acta* 180:3106–3114. https://doi.org/10.1016/j.bbagen.2014.07.001

Bouiche, C., Boucherba, N., Benallaoua, S., Martinez, J., Diaz, P., Pastor, F.J., Valenzuela, S.V. 2020. Differential antioxidant activity of glucuronoxylooligosaccharides (UXOS) and arabinoxylooligosaccharides (AXOS) produced by two novel xylanases. *International Journal of Biological Macromolecules* 155:1075–1083. https://doi.org/10.1016/j.ijbiomac.2019.11.073

Brodeur, G., Yau, E., Badal, K., Collier, J., Ramachandran, K.B., Ramakrishnan, S. 2011. Chemical and physicochemical pretreatment of lignocellulosic biomass: A review. *Enzyme Research* 2011:787532. https://doi.org/10.4061/2011/787532

Bu, Y., Cui, Y., Peng, Y., Hu, M., Tian, Y., Tao, Y., Wu, B. 2018. Engineering improved thermostability of the GH11 xylanase from *Neocallimastixpatriciarum* via computational library design. *Applied Microbiology and Biotechnology* 2(8):3675–3685. https://doi:10.1007/s00253-018-8872-1

Capetti, C.C. de M., Vacilotto, M.M., Dabul, A.N.G., Sepulchro, A.G.V., Pellegrini, V.O.A., Polikarpov, I. 2021. Recent advances in the enzymatic production and applications of xylooligosaccharides. *World Journal of Microbiology and Biotechnology* 37:1–12. https://doi.org/10.1007/s11274-021-03139-7

Chemin, M., Wirotius, A.L., Ham-Pichavant, F., Chollet, G., Da Silva Perez, D., Petit- Conil, M., Cramail, H., Grelier, S. 2015. Well-defined oligosaccharides by mild acidic hydrolysis of hemicelluloses. *European Polymer Journal* 66:190–197. https://doi.org/10.1016/j.eurpolymj.2015.02.008

Childs, C.E., Roytio, H., Alhoniemi, E., Fekete, A.A., Forssten, S.D., Hudjec, N., Lim, Y.N., Steger, C.J., Yaqoob, P., Tuohy, K.M., Rastall, R.A., Ouwehand, A.C., Gibson, G.R. 2014. Xylo-oligosaccharides alone or in synbiotic combination with *Bifidobacterium animalis* subsp. lactis induce bifidogenesis and modulate markers of immune function in healthy adults: A double-blind, placebo-controlled, randomised, factorial cross-over study. *British Journal of Nutrition* 111:1945–1956. https://doi:10.1017/S0007114513004261

Corim Marim, A.V., Gabardo, S. 2021. Xylooligosaccharides: prebiotic potential from agro-industrial residue, production strategies and prospects. *Biocatalysis and Agricultural Biotechnology* 37:02190. https://doi.org/10.1016/j.bcab.2021.102190

Dahmen, N., Lewandowski, I., Zibek, S., Weidtmann, A. 2019. Integrated lignocellulosic value chains in a growing bioeconomy: Status quo and perspectives. *GCB Bioenergy* 11:107–117. https://doi.org/10.1111/gcbb.12586

del Rio, P.G., Gullón, B., Wu, J., Saddler, J., Garrote, G., Romaní, A. 2022. Current breakthroughs in the hardwood biorefineries: Hydrothermal processing for the co-production of xylooligosaccharides and bioethanol. *Bioresource Technology* 343:126100. 10.1016/j.biortech.2021.126100

Finegold, S.M., Li, Z., Summanen, P.H., Downes, J., Thames, G., Corbett, K., Dowd, S., Krak, M., Heber, D. 2014. Xylooligosaccharide increases bifidobacteria but not lactobacilli in human gut microbiota. *Food & Function Journal* 5:436–445. https://doi.org/10.1039/c3fo60348b

Gibson, G.R., Hutkins, R., Sanders, M.E., Prescott, S.L., Reimer, R.A., Salminen, S.J., Scott, K., Stanton, C., Swanson, K.S., Cani, P.D., Verbeke, K., Reid, G. 2017. Expert consensus document: The International Scientific Association for Probiotics and Prebiotics (ISAPP) consensus statement on the definition and scope of prebiotics. *Nature Reviews Gastroenterology & Hepatology* 14:491–502. https://doi.org/10.1038/nrgastro.2017.75

Gibson, G.R., Roberfroid, M.B. 1995. Dietary modulation of the human colonic microbiota: Introducing the concept of prebiotics. *Journal of Nutrition* 125:1401–1410. https://doi.org/10.1079/NRR200479

Goldbeck, R., Gonçalves, T.A., Damásio, A.R.L., Brenelli, L.B., Wolf, L.D., Paixão, D.A.A., Rocha, G.J.M., Squina, F.M. 2016. Effect of hemicellulolytic enzymes to improve sugarcane bagasse saccharification and xylooligosaccharides production. *Journal of Molecular Catalysis B: Enzymatic* 131:36–46. https://doi.org/10.1016/j.molcatb.2016.05.013

Gullón, B., Eibes, G., Dávila, I., Moreira, M.T., Labidi, J., Gullón, P. 2018. Hydrothermal treatment of chestnut shells (*Castanea sativa*) to produce oligosaccharides and antioxidant compounds. *Carbohydrates Polymers* 192:75–83. https://doi.org/10.1016/j.carbpol.2018.03.051

Gullón, P., Gullón, B., Moure, A., Alonso, J.L., Domínguez, H., Parajó, J.C. 2009. Manufacture of prebiotics from biomass sources. In: *Prebiotics and Probiotics Science and Technology*, ed. D. Charalampopoulos and R.A. Rastall. Springer New York, New York, NY, 535–589. https://doi.org/10.1007/978-0-387-79058-9_14

Han, J., Cao, R., Zhou, X., Xu, Y. 2020. An integrated biorefinery process for adding values to corncob in co-production of xylooligosaccharides and glucose starting from pretreatment with gluconic acid. *Bioresource Technology* 307:123200. https://doi.org/10.1016/j.biortech.2020.123200

Hettiarachchi, S. A., Kwon, Y.-K., Lee, Y., Jo, E., Eom, T.-Y., Kanf, Y.H., Kang, D.H., De Zoysa, M., Marasinghe, S.D., Oh, C. 2019. Characterization of an acetyl xylan esterase from the marine bacterium *Ochrovirgapacifica* and its synergism with xylanase on beechwood xylan. *Microbial Cell Factories* 18(1):122. https://doi.org/10.1186/s12934-019-1169-y

Hu, F., Ragauskas, A. 2012. Pretreatment and lignocellulosic chemistry. *Bioenergy Research* 5:1043–1066. https://doi.org/10.1007/s12155-012-9208-0

IUPAC, 1982. Abbreviated terminology of oligosaccharide chains. *Pure Applied Chemistry* 54:1517–1522. https://doi.org/10.1016/j.cvex.2010.01.004

Khat-Udomkiri, N., Sivamaruthi, B.S., Sirilun, S., Lailerd, N., Peerajan, S., Chaiyasut, C. 2018. Optimization of alkaline pretreatment and enzymatic hydrolysis for the extraction of xylooligosaccharide from rice husk. *AMB Express* 8 (1):1–10. https://doi.org/10.1186/s13568-018-0645-9.

Khat-Udomkiri, N., Toejing, P., Sirilun, S., Chaiyasut, C., Lailerd, N. 2020. Antihyperglycemic effect of rice husk derived xylooligosaccharides in high-fat diet and low-dose streptozotocin-induced type 2 diabetic rat model. *Food Science Nutrition* 8:428–444. https://doi.org/10.1002/fsn3.1327

Kim, S.J., Chandrasekar, B., Rea, A.C., Danhof, L., Zemelis-Durfee, S., Thrower, N., Shepard, Z.S., Pauly, M., Brandizzi, F., Keegstra, K. 2020. The synthesis of xyloglucan, an abundant plant cell wall polysaccharide, requires CSLC function. *Proceedings of the National Academy of Sciences of the United States of America* 177:20316–20324. https://doi.org/10.1073/pnas.2007245117

Krajewska, B. 2004. Application of chitin-and chitosan-based materials for enzyme immobilizations: A review. *Enzyme and Microbiology Technology* 35 (2–3):126–139. https://doi.org/10.1016/j.enzmictec.2003.12.013

Kruschitz, A., Nidetzky, B. 2020. Downstream processing technologies in the biocatalytic production of oligosaccharides. *Biotechnology Advances* 43:107568. https://doi.org/10.1016/j.biotechadv.2020.107568

Le, B., Yang, S.H. 2019. Production of prebiotic xylooligosaccharide from aqueous ammonia-pretreated rice straw by β-xylosidase of *Weissella cibaria*. *Journal of Applied Microbiology* 126:1861–1868. https://doi.org/10.1111/jam.14255

Li, X., Dilokpimol, A., Kabel, M.A., de Vries, R.P. 2022. Fungal xylanolytic enzymes: Diversity and applications. *Bioresource Technology* 344:126290. https://doi.org/10.1016/j.biortech.2021.126290

Li, X., Tan, S., Luo, J., Pinelo, M. 2021b. Nano filtration for separation and purification of saccharides from biomass. *Frontiers of Chemical Science and Engineering* 15:837–853. https://doi.org/10.1007/s11705-020-2020-z

Li, Z., Li, Z., Zhu, L., Dai, N., Sun, G., Peng, L., Wang, X., Yang, Y. 2021a. Effects of xylo-oligosaccharide on the gut microbiota of patients with ulcerative colitis in clinical remission. *Frontiers in Nutrition* 8:1–11. https://doi.org/10.3389/fnut.2021.778542

Lian, Z., Wang, Y., Luo, J., Lai, C., Yong, Q., Yu, S. 2020. An integrated process to produce prebiotic xylo-oligosaccharides by autohydrolysis, nanofiltration and endo-xylanase from alkali-extracted xylan. *Bioresource Technology* 314:123685. https://doi.org/10.1016/j.biortech.2020.123685

Lim, S.M., Kim, E., Shin, J.H., Seok, P.R., Jung, S., Yoo, S.H., Kim, Y. 2018. Xylobiose prevents high-fat diet induced mice obesity by suppressing mesenteric fat deposition and metabolic dysregulation. *Molecules* 23(3):705. https://doi.org/10.3390/molecules23030705

Linares-Pasten, J.A., Aronsson, A., Karlsson, E.N. 2018. Structural considerations on the use of endo-xylanases for the production of prebiotic xylooligosaccharides from biomass. *Current Protein & Peptide Science* 19:48–67. https://doi.org/10.2174/1389203717666160923155209

Liu, X., Liu, Y., Jiang, Z., Liu, H., Yang, S., Yan, Q. 2018. Biochemical characterization of a novel xylanase from Paenibacillusbarengoltzii and its application in xylooligosaccharides production from corncobs. *Food Chemistry* 264:310–318. https://doi.org/10.1016/j.foodchem.2018.05.023

Lombard, V., GolacondaRamulu, H., Drula, E., Coutinho, P.M., Henrissat, B. 2014. The carbohydrate-active enzymes database (CAZy) in 2013. *Nucleic Acids Research* 42:490–495. https://doi.org/10.1093/nar/gkt1178

Long, J., Yang, J., Henning, S.M., Woo, S.L., Hsu, M., Chan, B., Heber, D., Li, Z. 2019. Xylooligosaccharide supplementation decreases visceral fat accumulation and modulates cecum microbiome in mice. *Journal of Functional Foods* 52:138–146. https://doi.org/10.1016/j.jff.2018.10.035

Lyu, X., Gonzalez, R., Horton, A., Li, T. 2021. Immobilization of enzymes by polymeric materials. *Catalysts* 11:1211. https://doi.org/10.3390/catal11101211

Malgas, S., Mafa, M.S., Mkabayi, L., Pletschke, B.I. 2019. A mini review of xylanolytic enzymes with regards to their synergistic interactions during hetero-xylan degradation. *World Journal of Microbiology and Biotechnology* 35:1–13. https://doi.org/10.1007/s11274-019-2765-z

Manning, S.T., Gibson, G.R. 2004. Prebiotics. *Best Practice and Research in Clinical Gastroenterology* 18:287–298. https://doi.org/10.1053/ybega.2004.445

Mikkonen, K.S., Tenkanen, M. 2012. Sustainable food-packaging materials based on future biorefinery products: Xylans and mannans. *Trends in Food Science & Technology* 28:90–102.

Milessi, T. S. S., Kopp, W., Rojas, M. J., Manrich, A., Baptista-Neto, A., Tardioli, P.W., Giordano, R.C., Fernandez-Lafuente, R., Guisa, J.M., Giordano, R.L.C. 2016. Immobilization and stabilization of an endoxylanase from *Bacillus subtilis* (XynA) for xylooligosaccharides (XOs) production. *Catalysis Today* 259(Part 1):130–139. https://doi.org/10.1016/j.cattod.2015.05.032

Moniz, P., Pereira, H., Duarte, L.C., Carvalheiro, F. 2014. Hydrothermal production and gel filtration purification of xylo-oligosaccharides from rice straw. *Industrial Crops and Products* 62:460–465. https://doi.org/10.1016/j.indcrop.2014.09.020

Montipó, S., Ballesteros, I., Fontana, R.C., Liu, S., Martins, A.F., Ballesteros, M., Camassola, M. 2018. Integrated production of second generation ethanol and lactic acid from steam-exploded elephant grass. *Bioresource Technology* 249:1017–1024. 10.1016/j.biortech.2017.11.001

Moreno, A.D., Tomás-Pejó, E., Ballesteros, M., Negro, M.J. 2019. Pretreatment technologies for lignocellulosic biomass deconstruction within a biorefinery perspective. In: Biofuels: Alternative Feedstocks and Conversion Processes for the Production of Liquid and Gaseous Biofuels, ed. A. Pandey, C. Larroche, C.G., Dussap, E. Gnansounou, S.K. Khanal, and S. Ricke S. 379–399. Amsterdam: Academic Press. https://doi.org/10.1016/c2018-0-00957-3

Motta, F.L., Andrade, C.C.P., Santana, M.H.A. 2013. A review of xylanase production by the fermentation of xylan: Classification, characterization and applications. In: *Sustainable Degradation of Lignocellulosic Biomass Techniques, Applications and Commercialization*. Intech. https://doi.org/10.5772/53544

Naidu, D.S., Hlangothi, S.P., John, M.J. 2018. Bio-based products from xylan: A review. *Carbohydrates Polymers* 179:28–41. https://doi.org/10.1016/j.carbpol.2017.09.064

Neumüller, K.G., De Souza, A.C., Van Rijn, J.H.J., Streekstra, H., Gruppen, H., Schols, H.A. 2015. Positional preferences of acetyl esterases from different CE families towards acetylated 4-O-methyl glucuronic acid-substituted xylo-oligosaccharides. *Biotechnology for Biofuels* 8:1–11. https://doi.org/10.1186/s13068-014-0187-6

Nordberg Karlsson, E., Schmitz, E., Linares-Pastén, J.A., Adlercreutz, P. 2018. Endo-xylanases as tools for production of substituted xylooligosaccharides with prebiotic properties. *Applied Microbiology and Biotechnology* 102:9081–9088. https://doi.org/10.1007/s00253-018-9343-4

Palaniappan, A., Antony, U., Emmambux, M.N., 2021. Current status of xylooligosaccharides: Production, characterization, health benefits and food application. *Trends in Food Science & Technology* 111:506–519. https://doi.org/10.1016/j.tifs.2021.02.047

Pielhop, T., Amgarten, J., Studerm M.H., Von Rohr, P.R. 2017. Pilot-scale steam explosion pretreatment with 2-naphthol to overcome high softwood recalcitrance. *Biotechnology for Biofuels* 10:130. https://doi.org/10.1186/s13068-017-0816-y

Pinales-Márquez, C.D., Rodríguez-Jasso, R.M., Araújo, R.G., Loredo-Treviño, A., Nabarlatz, D., Gullón, B., Ruiz, H.A. 2021. Circular bioeconomy and integrated biorefinery in the production of xylooligosaccharides from lignocellulosic biomass: A review. *Industrial Crops and Products* 162: 113274. https://doi.org/10.1016/j.indcrop.2021.113274

Poletto, P., Pereira, G.N., Monteiro, C.R.M., Pereira, M.A.F., Bordignon, S.E., de Oliveira, D., 2020. Xylooligosaccharides: Transforming the lignocellulosic biomasses into valuable 5-carbon sugar prebiotics. *Process Biochemistry* 91:352–363. https://doi.org/10.1016/j.procbio.2020.01.005

Puchart, V., Šuchová, K., Biely, P., Jana, U.K., Kango, N., Pletschke, B. 2021. Xylanases of glycoside hydrolase family 30 – An overview. *Frontiers in Nutrition* 47:1–13. https://doi.org/10.3389/fnut.2021.670817

Rico, X., Gullón, B., Alonso, J.L., Parajó, J.C., Yáñez, R. 2018. Valorization of peanut shells: Manufacture of bioactive oligosaccharides. *Carbohydrates Polymers* 183, 21–28. https://doi.org/10.1016/j.carbpol.2017.11.009

Santibañez, L., Herníquez, C., Corro-Tejeda, R., Bernal, S., Armijo, B., Salazar, O., 2021. Xylooligosaccharides from lignocellulosic biomass: A comprehensive review. *Carbohydrates Polymers* 251:117118. https://doi.org/10.1016/j.carbpol.2020.117118.

Sato, N., Shinji, K., Mizuno, M., Nozaki, K., Suzuki, M., Makishima, S., Shiroishi, M., Onoda, T., Takahashi, F., Kanda, T., Amano, Y. 2010. Improvement in the productivity of xylooligosaccharides from waste medium after mushroom cultivation by hydrothermal treatment with suitable pretreatment. *Bioresource Technology* 101:6006–6011. https://doi.org/10.1016/j.biortech.2010.03.032

Scheller, H.V., Ulvskov, P. 2010. Hemicelluloses. *Annual Review of Plant Biology* 61:263–289.

Singh, R.D., Nadar, C.G., Muir, J., Arora, A. 2019. Green and clean process to obtain low degree of polymerisation xylooligosaccharides from almond shell. *Journal of Cleaner Production* 241:118237. https://doi.org/10.1016/j.jclepro.2019.118237

Singh S., Singh, G., Arya, S.K. 2018. Mannans: An overview of properties and application in food products. *Journal of Biological Macromolecules* 119:79–95. https://doi.org/10.1016/j.ijbiomac.2018.07.130

Soto, M.L., Moure, A., Domínguez, H., Parajó, J.C. 2011. Recovery, concentration and purification of phenolic compounds by adsorption: A review. *Journal of Food Engineering* 105:1–27. https://doi.org/10.1016/J.JFOODENG.2011.02.010

St John, F.J., González, J.M., Pozharski, E. 2010. Consolidation of glycosyl hydrolase family 30: A dual domain 4/7 hydrolase family consisting of two structurally distinct groups. *FEBS Letters* 584:4435–4441. https://doi.org/10.1016/j.febslet.2010.09.051

St John, F.J., Rice, J.D., Preston, J.F. 2006. Characterization of XynC from *Bacillus subtilis* subsp. subtilis strain 168 and analysis of its role in depolymerization of glucuronoxylan. *Journal of Bacteriology* 188:8617–8626. https://doi.org/10.1128/JB.01283-06

Šuchová, K., Puchart, V., Spodsberg, N., Mørkeberg Krogh, K.B.R, Biely, P. 2020. A novel GH30 xylobiohydrolase from *Acremonium alcalophilum* releasing xylobiose from the non-reducing end. *Enzyme and Microbial Technology* 134:109484. https://doi.org/10.1016/j.enzmictec.2019.109484

Turck, D., Bresson, J.L., Burlingame, B., Dean, T., Fairweather-Tait, S., Heinonen, M., Hirsch-Ernst, K.I., Mangelsdorf, I., McArdle, H.J., Naska, A., Neuhäuser-Berthold, M., Nowicka, G., Pentieva, K., Sanz, Y., Siani, A., Sjödin, A., Stern, M., Tomé, D., Vinceti, M., Willatts, P., Engel, K.H., Marchelli, R., Pöting, A., Poulsen, M., Schlatter, J.R., Turla, E., van Loveren, H. 2018. Safety of xylo-oligosaccharides (XOS) as a novel food pursuant to Regulation (EU) 2015/2283. *EFSA Journal* 16(7):e05361. https://doi.org/10.2903/j.efsa.2018.5361

Vuong, T.V., Master, E. 2022. Enzymatic upgrading of heteroxylans for added-value chemicals and polymers. *Current Opinion in Biotechnology* 73:51–60. https://doi.org/10.1016/j.copbio.2021.07.001

Wan Azelee, N.I., Jahim, J.M., Ismail, A.F., Fuzi, S.F.Z.M., Rahman, R.A., Md Illias, R., 2016. High xylooligosaccharides (XOS) production from pretreated kenaf stem by enzyme mixture hydrolysis. *Industrial Crops and Products* 81:11–19. https://doi.org/10.1016/j.indcrop.2015.11.038

Wan, C., Zhou, Y., Li, Y., 2011. Liquid hot water and alkaline pretreatment of soybean straw for improving cellulose digestibility. *Bioresource Technology* 102:6254–6259. https://doi.org/10.1016/j.biortech.2011.02.075

Wang, W., Yan, R., Nocek, B.P., Vuong, T.V., Di Leo, R., Xu, X., Cui, H., Gatenholm, P., Toriz, G., Tenkanen, M., Savchenko, A., Master, E.R., 2016. Biochemical and structural characterization of a five-domain GH115 α-glucuronidase from the marine *bacterium Saccharophagus degradans* 2-40T. *Journal of Biological Chemistry* 291 (1):14120–33. https://doi.org/10.1074/jbc.M115.702944

Watson, J. 2019. Prebiotic ingredients market to reach USD 8.34 billion by 2026. Reports and data. https://www.globenewswire.com/news-release/2019/10/16/1930796/0/en/Prebiotic-Ingredients-Market-To-Reach-USD-8-34-Billion-By-2026-Reports-And-Data.html.

Wijaya, H., Sasaki, K., Kahar, P., Rahmani, N., Hermiati, E., Yopi, Y., Ogino, C., Prasetya, B., Kondo, A., 2020. High enzymatic recovery and purification of xylooligosaccharides from empty fruit bunch via nanofiltration. *Processes* 8(6):619. https://doi.org/10.3390/PR8050619

Wu, X., Tian, Z., Jiang, X., Zhang, Q., Wang, L. 2018. Enhancement in catalytic activity of *Aspergillus niger* XynB by selective site-directed mutagenesis of active site amino acids. *Applied Microbiology and Biotechnology* 102:249–26. https://doi:10.1007/s00253-017-8607-8

Yang, J., Summanen, P.H., Henning, S.M., Hsu, M., Lam, H., Huang, J., Tseng, C.H., Dowd, S.E., Finegold, S.M., Heber, D., Li, Z. 2015. Xylooligosaccharide supplementation alters gut bacteria in both healthy and prediabetic adults: A pilot study. *Frontiers in Physiology* 6:1–11. https://doi.org/10.3389/fphys.2015.00216

8 Microbial Enzymes for Production of Fructooligosaccharides

Kim Kley Valladares-Diestra[1], Luciana Porto de Souza Vandenberghe[1], Dão Pedro de Carvalho Neto[2], Luis Daniel Goyzueta-Mamani[3], and Carlos Ricardo Soccol[1]

[1] Department do Bioprocess Engineering and Biotechnology, Federal University of Paraná, Curitiba, Paraná, Brazil

[2] Federal Institute of Education, Science and Technology of Paraná (IFPR), Londrina, Paraná, Brazil

[3] Vicerrectorado de Investigación, Universidad Católica de Santa María, Urb. San José s/n—Umacollo, Arequipa, Peru

CONTENTS

8.1 INTRODUCTION

Fructooligosaccharides (FOs) are small non-digestible carbohydrates, which are classified as prebiotics due to their selective capacity to stimulate the activity and growth of beneficial bacteria that colonize the digestive tract and make up the intestinal microbiota, mainly from the genus *Bifidobacterium* and *Lactobacillus* (van Wyk et al. 2013; Ganaie et al. 2014). FOs are classified "generally recognized as safe" (GRAS) by the US Food and Drug Administration (Picazo et al. 2019). Due to these characteristics, which highlight their great nutritional value, FOs are of potential interest in the food and pharmaceutical industries.

The chemical composition and architectural conformation of FOs follows a characteristic pattern, having in its main structure a chain of fructose units linked by β-(1-2)-fructofuranosyl linkages with a terminal glucose molecule through glycosidic bridges β-(2-1). The classification of FOs is mainly based on the number of fructose molecules present in the main chain. For example, inulin is one of the main types of FOs, with a degree of polymerization of up to 60 fructose units (Morris and Morris 2012). On the other hand, FOs with a lower degree of polymerization such as 1-kestose, nystose, and 1-fructofuranosyl-nystose present 2, 3 and 4 fructose units in their main chain, respectively, and are

DOI: 10.1201/9781003311164-11

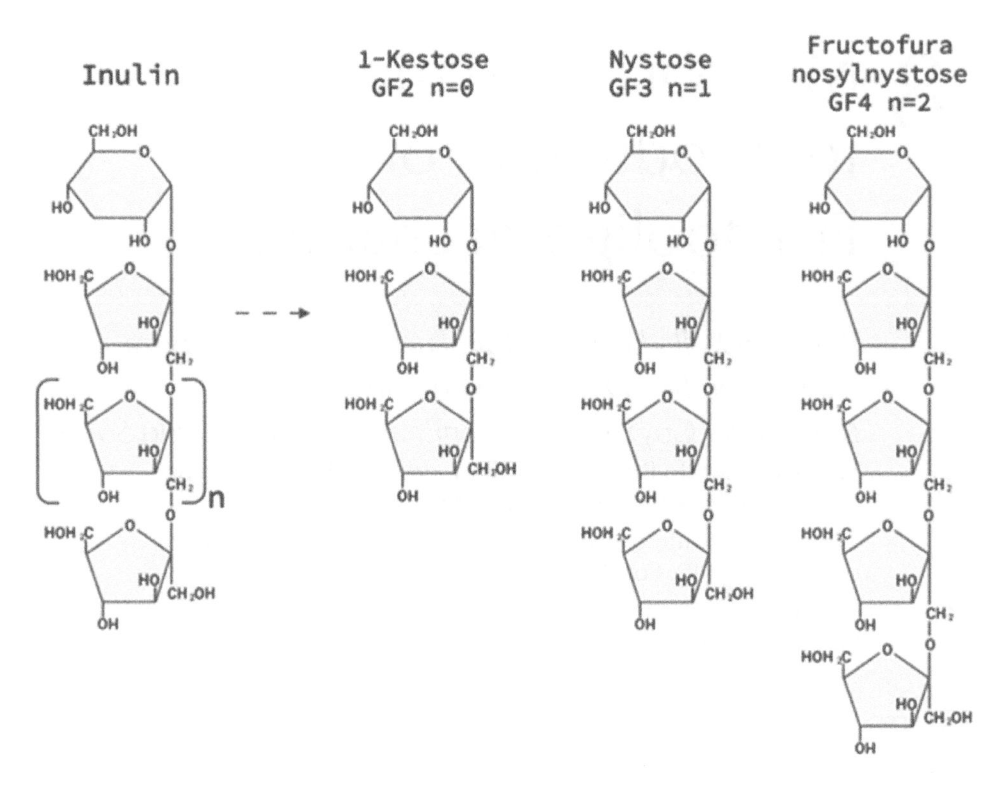

FIGURE 8.1 Chemical structures of the most important FOs found in nature according to their polymerization degree.

known as small FOs (Figure 8.1). The degree of polymerization of FOs depends mainly on the source from which it was extracted and the production process (Roberfroid 2008).

FOs have commonly been extracted from plant sources such as bananas, onions, chicory root, garlic, asparagus, barley, wheat, tomatoes, leeks, *Stevia rebaudiana*, *Jerusalem artichokes*, and yacon (Maiorano et al. 2008) in the form of inulin. However, the production of small FOs has also been reported through the action of microbial-derived enzymes, which perform the transfructosylation of sucrose (Ganaie et al. 2014; Picazo et al. 2019). The use of bacterial and fungal strains allows the enzymatic production under controlled conditions, at laboratory or industrial scale. The biochemical synthesis of FOs occurs through two different pathways: (1) the FOs production through enzymatic hydrolysis of inulin, where the main applied enzymes are endo-inulinases and exo-inulinases; and (2) the production of FOs from the transfructosylation of sucrose, where fructosyltransferases or fructofuranosidases are the main enzymes applied in this process. In both production pathways, enzymes are the main catalyst tools for the biosynthesis of FOs. The enzymes involved in the production of FOs are obtained from plants, fungi, or bacteria, microbial enzymes are the most studied and applied due to their ease of obtaining and handling on a laboratory and industrial scale. In this sense, the development of enzymes with greater stability and enzymatic activity are the objectives of real attention in the biotechnological field (Ganaie et al. 2014).

8.2 ENZYMES FOR FOs PRODUCTION

The production of food products for the benefit of human health with the addition of prebiotics is becoming a popular trend, due to their contributions in the treatment and prevention of diseases. For this reason, the food industry shows great interest in the production of FOs on a large scale.

Enzyme-producing microorganisms have been used ancestrally in the empirical production of fermented beverages and food, such as cheese, bread, wine, and beer (Ganaie et al. 2014). The synthetic synthesis of FOs was first carried out in 1980 in Japan by Meija Seika Kaisha under the trade name of Meiligo. Normally, the synthesis of FOs is a two-stage process, in which the first stage is the production of enzymes through a biotechnological process. In the second stage, the produced enzymes are used in the biotransformation and production of FOs under controlled conditions for a better performance and yield (Sangeetha et al. 2004; Flores-Maltos et al. 2016). The most widely used commercial enzymes in the production of FOs can be produced intracellularly or extracellularly by different microorganisms from sucrose or inulin.

The most used enzymes in the production of FOs are fructosyltransferase (FTase; 2.4.1.9), β-fructofuranosidase (FFase; 3.2.1.26) and endo-inulinase (3.2.1.7). FTases act on sucrose by cleaving the β-1,2 bond between glucose and fructose to transfer a fructosyl group to an acceptor molecule that can be sucrose or small FOs in formation. This enzymatic reaction releases glucose as a co-product of the reaction, which is known as transfructosylating activity. On the other hand, the FFases show a hybrid behaviour depending on the concentration of sucrose. At low concentrations of sucrose, these enzymes hydrolyse the β-1,2 bonds, releasing glucose and fructose. However, at high concentrations of sucrose, they have a transfructosylating activity, presenting the same behaviour as a FTases, producing FOs and releasing glucose (Ganaie et al. 2014). Finally, endo-inulinases act randomly and hydrolyse internal β-1,2 linkages of inulin, releasing FOs with different degrees of polymerization (Flores-Gallegos et al. 2015).

According to the substrate used to obtain FOs, the production of this prebiotic can be divided into two enzymatic pathways: (1) inulin hydrolysis and (2) sucrose transfructosylation. Due to these production ways, the enzymes applied will also be different.

8.2.1 Enzyme for Inulin Hydrolysis

Inulin hydrolysis is mediated by enzymes inulinases (also called inulases), that are classified as endo- and exo-inulinases that act by cleaving β-2,1 bonds into inulin. Endo-inulinases (2,1-β-D-fructan fructanohydrolase; EC 3.2.1.7) randomly hydrolyse the internal bonds of inulin and can produce FOs, and oligo-fructose (oligosaccharides formed only by fructose units). On the other hand, exo-inulinases (β-D-fructan fructohydrolase; EC 3.2.1.80) cleave the β-2.1 bonds sequentially from the non-reducing end of inulin and separate the terminal fructosyl units, releasing fructose (Figure 8.2) (Flores-Gallegos et al. 2015).

When endo-inulinases and exo-inulinases are applied together, the FOs obtained has a decrease in prebiotic activity. This is mainly due to the action of exo-inulinases that hydrolyse the non-reducing end of inulin, releasing sugars such as glucose, fructose, and sucrose. So, in this case it is necessary to remove these non-prebiotic sugars, with nanofiltration technologies or ion exchange chromatography, which renders the final process more expensive. An alternative to solve this problem is the genetic modification for the production of a recombinant endo-inulinase that has a greater affinity to the substrate with higher production of FOs (Picazo et al. 2019).

The comparison of amino acid sequences showed that inulinases (exo- and endo-inulinases) belong to the group of glycoside hydrolase (GH) enzymes. This group is the most diverse, consisting of microbial enzymes that degrade biomass. Inulinases are members of the GH32 family, an enzymatic family that not only shares the similarity in the amino acid sequence, but also has a specific activity towards polysaccharides and oligosaccharides transformed by fructose (van Wyk et al. 2013). According to Flores-Gallegos et al. (2015) the most conserved protein motifs in enzymes of the GH32 family are WMN**D**PNG (block A), YHLFYQ (block B), WGHATS (block C), FTGS (block D), R**D**P (block E), and **E**CP (block G). These enzymes present within the catalytic domain three well-conserved amino acid residues: the nucleophilic aspartic acid residue (A block), the glutamic acid residue that acts as acid-base catalyst (G block), and finally another aspartic acid residue (E block) that acts as a stabilizer of the transition state. In the case of exo-inulinases, these motifs

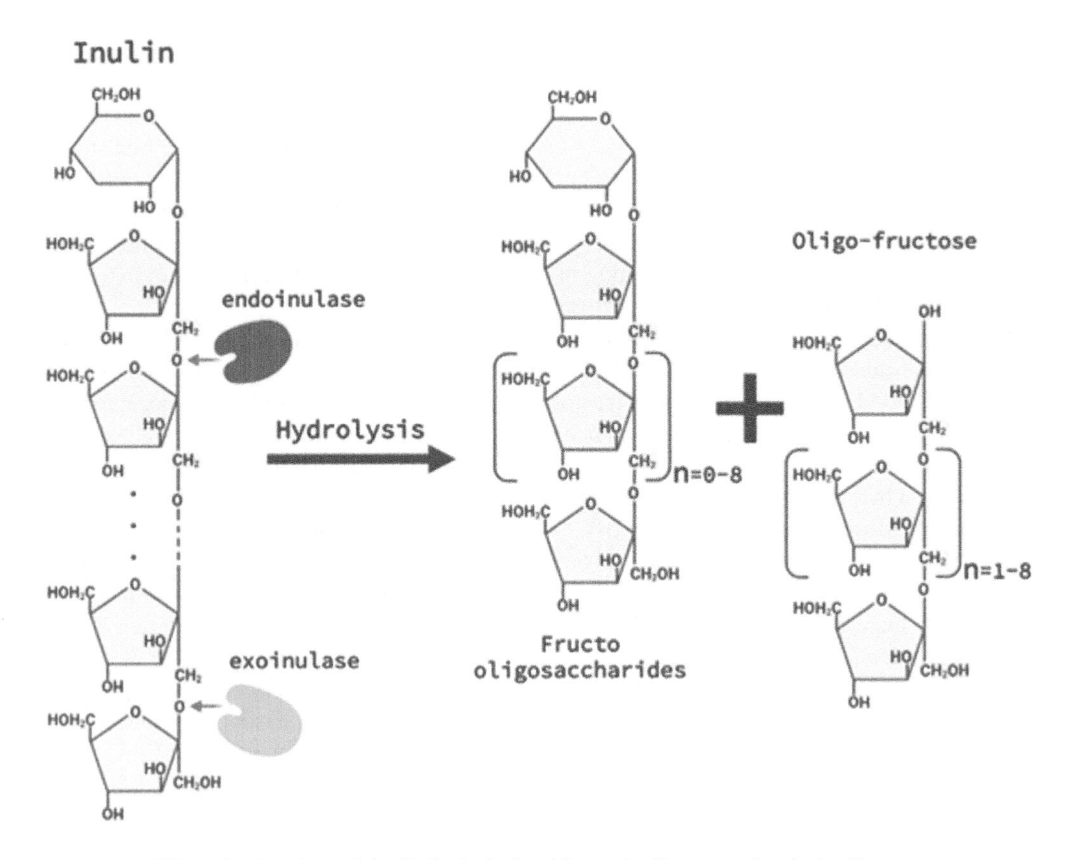

FIGURE 8.2 FO production through inulin hydrolysis with exo-inulinases and endo-inulinases.

and residues within the catalytic domain are fairly well conserved. In endo-inulinases a higher affinity for the glutamic acid residue is observed compared to aspartic acid (as a nucleophilic residue) within the WMNEPNG motif (block A). Although exo-inulinases and endo-inulinases are considered inulinases and synergistically hydrolyse inulin, the genes encoding their expression show that these two types of enzymes have evolved independently.

The production of inulinases has been reported in different organisms from plants to bacteria. However, filamentous fungi and yeasts produce these enzymes known for their ability to degrade inulin present in plant cell walls (Flores-Gallegos et al. 2015). Inulinases obtained from filamentous fungi present higher inulinases activity and high thermal stability (up to 60°C), which facilitates their use at high temperatures and prevents contamination by other microorganisms. The genera *Aspergillus* and *Penicillium,* and yeasts of the genera *Kluyveromyces* are efficient producers of inulinases. Bacteria of the genera *Bacillus*, *Pseudomonas* and *Streptomyces* have been reported as high yielding inulinases strains (Ganaie et al. 2014).

8.2.2 Enzyme for Sucrose Transfructosylation

The transfructosylation of sucrose for the formation of small FOs is mediated by fructosyltransferase (FTase; 2.4.1.9) and β-fructofuranosidase (FFase; 3.2.1.26). The biotransformation process occurs by the cleavage of the β-1,2 bond of sucrose and the transfer of a fructosyl group to an acceptor that could be another sucrose or a small FOs; releasing a glucose molecule as by-product. Depending on the type of bond to which the fructosyl group is transferred to sucrose, different types of neo-FOs can be forming such as neokestose or 6-kestose (Figure 8.3) (Ganaie et al. 2014).

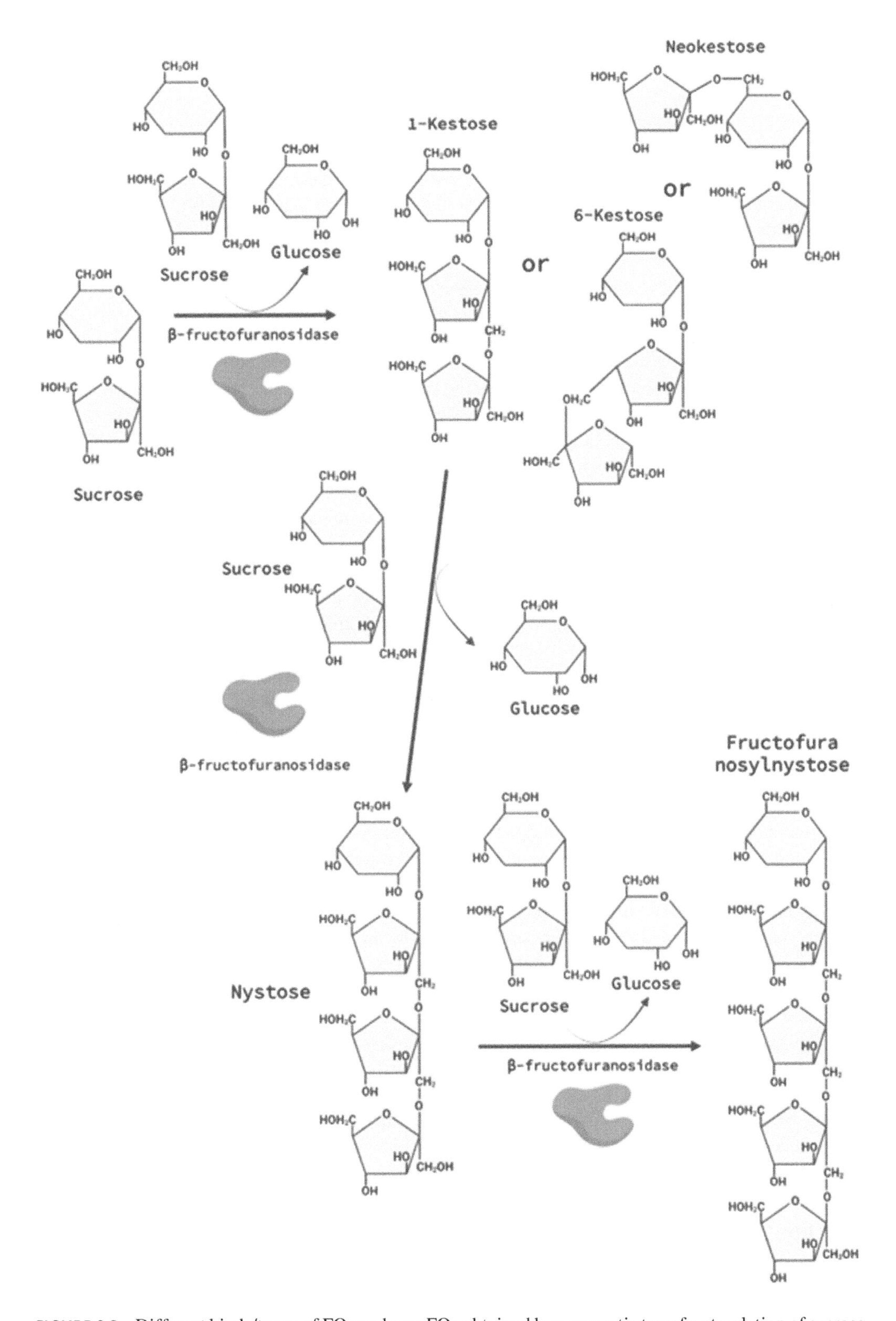

FIGURE 8.3 Different kinds/types of FOs and neo-FOs obtained by enzymatic transfructosylation of sucrose.

Enzymes that exhibit transfructosylation activity produce 1-kestose, nystose, and 1-fructofuranosyl-nystose. Some specific FFase enzymes from microorganisms such as the yeast *Xanthophyllomyces dendrorhous* (ATCC-MYA-131) produce neokestose (Linde et al. 2012). On the other hand, β-fructofuranosidase enzymes from *Schwanniomyces occidentalis* produce 6-kestose and the modification by molecular techniques (such as direct evolution) can increase fructose transferase activity up to 5 times for increased production of 6-kestose (de Abreu et al. 2013). The structural characteristics of the FTase or FFase will determine the synthesis of β-(2→6) or β-(2→1) bonds, the degree of polymerization of FOs and the types of molecules, which are present in their branches (Hijum et al. 2006).

The mechanism of the transfructosylation enzymatic reaction for FOs production from sucrose was proposed by Chambert et al. (1974) who suggested that the behaviour of enzyme kinetics follows a Ping-Pong BiBi type model. The enzyme and sucrose form an intermediate enzyme-fructose complex (Ping) releasing a glucose molecule (Pong). The intermediate complex can then interact with another sucrose molecule (Ping) transferring the fructose molecule for the formation of FOs, and releasing its second product (Pong) (Flores-Maltos et al. 2016).

The transfructosylation biochemical mechanisms can be separated into two stages. The first stage is defined by the interaction between sucrose and the active centre of the enzyme. This interaction generates the formation of a covalent bond between the fructose residue (belonging to sucrose) with the aspartic residue 86 (belonging to the enzyme), releasing a glucose molecule. In the second stage, a new sucrose molecule rearranges within the active site, in the place of the released glucose. The fructose is then transferred to the sucrose molecule for the formation of 1-kestose. Between these two stages the aspartic residue 247 of the enzyme plays an important role serving as a stabilizer between the transition from one stage to another biochemical mechanism within the enzyme (Kim et al. 1996). To obtain nystose, 1-kestose is coupled in the second stage instead of sucrose and the same procedure is applied to obtain 1-fructofuranosyl-nystose, using nystose as acceptor in the second stage of the biochemical mechanism of transfructosylation. For this reason and due to the ease of the FTases enzyme to couple to different substrates such as FOs themselves, several simultaneous reactions occur in parallel and in series (Vega and Zuniga-Hansen 2014).

FFases and FTases belong to the group of glycoside hydrolases which belong to the GH32 and GH68 families. In both cases these enzymes´ families have in common a five-bladed β-helix catalytic domain type with a negatively charged deep central pocket and three identical catalytic residues (Tonozuka et al. 2012). The three catalytic residues are the most conserved: aspartic acid the residue that leads the nucleophilic attack, glutamic acid the residue that acts as the general catalytic acid/base, and, finally, the other aspartic acid that fulfills the function of stabilizer of the state of transition. These three residues are normally located within the conserved regions of the WMNDPNG, EC, and RDP motifs, respectively (Pons et al. 2004). According to Trollope et al. (2015), the alignment and structural analysis of a fungal fructosyltransferase belonging to the GH32 family indicate that enzymes containing a WMNDPNG domain would produce a lower amount of FOs, while enzymes with the GQIGDPC domain will have a greater performance for a high production of FOs.

Like many other enzymes, FTases and FFases can be produced by a wide variety of organisms ranging from plants to microorganisms. The FTases and FFases belonging to the GH32 family are produced mainly by plants and fungi, while those belonging to the GH68 family are produced mainly by bacteria (Pons et al. 2004).

8.3 BIOTECHNOLOGICAL PRODUCTION OF ENZYMES

The production of FOs can be achieved using sucrose and inulin as the main substrate source through conventional chemical processes or biotechnological synthesis. Although the easy scale-up of chemical processes can be attractive, several factors limit its implementation, such as: (1) the use of chemical catalysts with low specificity, hindering the application of the FOs to the pharmaceutical and food industries (most profitable markets) and generating pollutant wastewater or by-products; (2) high chemical sensitivity of sugar substrate, impacting on the final purity of the FOs; (3) and high energy cost, due to operational conditions at high temperatures (75–80°C) and acidic pH (Tomotani and Michelle 2007; Tomotani and Vitolo 2007; Flores-Maltos et al. 2016). In this sense, the biotechnological approach is considered a viable avenue for a less costly, selective, and more sustainable FOs production.

Industrial production of microbial-derived β-D-fructofuranosidases, fructosyltransferase, and inulinases can be performed in two configurations: discontinuous (batch) or continuous. On the former process, an initial cultivation step is required for the target enzyme production. Fermentation parameters, such as aeration, agitation, temperature, pH, and time must be optimized for each microorganism, but these conditions are commonly reported in literature (Flores-Maltos et al. 2016). The produced enzyme must then be recovered from the fermentation media through sterile filtration (0.22–0.11 μm pore size), centrifugation, or a combination of both techniques (Vaňková et al. 2008; Sánchez-Martínez et al. 2020). As previously described, these enzymes can also be retained as an intracellular protein in some bacterial and fungal species, thus requiring an additional intermediary downstream process for its release (Gomes et al. 2020). Due to its low operational costs, easy-handling technology, higher scalability, selectivity, and stability of the enzymes after the release, ultrasonication is widely indicated (Özbek and Ülgen 2000; Gomes et al. 2020). After the recovery, the purified enzyme is added to a sucrose-rich medium and incubated at 40–55° C for up to 25 h, followed by an inactivation step using high temperatures. The continuous process, on the other hand, is mainly characterized by the immobilization of the biocatalyst particle (*i.e.*, whole cell) onto inert materials (*e.g.*, calcium alginate beads, porous glass, and silica) or insoluble substrates (*i.e.*, agro-industrial by-products), which minimizes pH and temperature fluctuations, allows the reuse of the biocatalyst particle, and improves significantly the volumetric production of the bioreactor (Sánchez-Martínez et al. 2020).

Despite the numerous and expensive downstream processes associated to enzymatic preparation, this market is experiencing a tangible growth due to advances provided by a large volume of applied researches and patents recently developed. Amongst the main factors that influence the industrial enzyme preparation for FOs production, it is possible to highlight the microbial source, operational conditions, and fermentation strategy. Table 8.1 summarizes recently published results for the production of FOs-catalytic enzymes under different fermentation processes and conditions.

Bacterial and fungal species are the main source of FTases, FFases, and inulinases for FO production. Although enzyme extraction from plant materials is a possibility, it is considered unfeasible due to low yields and variations of plant mass production as a consequence of edaphoclimatic conditions (Singh et al. 2016; Kumar and Dubey 2019). In this sense, screening of microbial sources of these enzymes has been extensively reported in the literature (Fernandez et al. 2007; Arrizon et al. 2012; Paixão et al. 2013; Nascimento et al. 2016). *Aspergillus niger, Aspergillus oryzae, Aspergillus japonicus, Fusarium oxysporum, Penicillium rugulosum, Arthrobacter* sp., *Bacillus subitilis, Bifidobacterium breve,* and *Lactiplantibacillus plantarum* (former *Lactobacillus plantarum*) are considered promising producers of FOs. Yeasts and filamentous fungi are the most preferable microorganisms due to high sucrose-fermenting ability, production of high levels of extracellular enzymes, high osmose tolerance, and versatility to different fermentation processes (*e.g.*, submerged and solid state) (Venkateshwar et al. 2010).

Although these are important features, industrial FOs production imposes harsh conditions (*e.g.*, high temperatures, alkali or acidic pH, shear force), which require a constant search for well-adapted strains. Sandoval-González et al. (2018) evaluated the potential of yeasts isolated from aguamiel, a Mexican indigenous alcoholic beverage from *Agave* sp., for inulinase synthesis. After initial trials, the selected *Saccharomyces paradoxus* strain was used for an experimental design for optimization on a stirred-tank reactor (STR). The study revealed that optimal conditions were achieved at 40°C and alkaline pH (7.7), which are consistent with the temperatures of semi-arid desert areas of Mexico where *Agave* sp. plants are harvested and the average pH of aguamiel

TABLE 8.1
Overview of Microbial-derived Enzymes Production for Fructooligosaccharides

Fermentation	Enzyme	Microorganism	Carbon Source	Operational Conditions	Bioreactor Type	Enzyme Production/ Activity	Reference
Submerged	FTase	*Pichia pastoris*	Glycerol (50%)	28°C (20 h) pH = 5.5 1-2 vvm 500–900 rpm	STR	102.1 U/mL	(Hernández et al. 2018)
		Yarrowia lipolytica	Sucrose (2–20%)	30°C (48 h) pH = 6	–	0.85 U/mg of DCW	(Zhang et al. 2016)
		Aspergillus oryzae	Sucrose (15%)	30° C (76 h) 200 rpm	–	19.76 U/mL	(Cunha et al. 2019)
		Ceratotherium simum	Sucrose (30%)	28°C (7 days) 200 rpm	–	529.5 U/mL	(Ojwach et al. 2020)
	Inulinase	*Aspergillus tritici*	Raw inulin from *Asparagus racemosus* root tubers (2.5%)	37°C (8 days) pH = 5.5	–	25.39 IU/mL	(Singh et al. 2020)
		Aspergillus tritici	Raw inulin from *Asparagus racemosus* root tubers (3%)	37°C (8 days) pH = 5.5 1 vvm 166 rpm	STR	40.27 IU/mL	(Singh et al. 2021)
		Saccharomyces cerevisiae	Sucrose (45%)	30°C (45 h) pH = 5.5 1 vvm 900 rpm	Jar fermenter	220 U/mL (from 39 g DCW)	(Ko et al. 2019)
		Rhodotorula mucilaginosa	Glucose (1.25%)	28°C (48 h) pH = 5.0 100 rpm	–	0.62 U (activity)	(Ribeiro et al. 2021)

(Continued)

TABLE 8.1 ***(Continued)***
Overview of Microbial-derived Enzymes Production for Fructooligosaccharides

Fermentation	Enzyme	Microorganism	Carbon Source	Operational Conditions	Bioreactor Type	Enzyme Production/ Activity	Reference
	FFase	*Aspergillus thermomutans*	Sucrose (10%)	30°C (72 h) pH = 6.0 100 rpm	–	6.5 U/mg of protein	(Tódero et al. 2019)
		Aureobasidium melanogenum	Sucrose (18%)	29°C (24 and 72 h) 180 rpm	STR	2100 U/mL	(Zhang et al. 2019)
	Levansucrase	*Bacillus subtilis natto*	Sucrose (40%)	35°C (48 h) 0.2 vvm	STR	NS	(Magri et al. 2020)
Solid state	FFase	*Aspergillus tamarii*	Soy bran (3g substrate/15% sucrose)	30°C (96–120 h) 60% moisture	–	229.43 U/mL (hydrolytic) 66.93 U/mL (transfructosylating)	(Oliveira et al. 2020)
	FTase	*Aspergillus flavus*	Sugarcane bagasse enriched with yeast extract	28°C (96 h) pH = 4.8	–	423.18 U per g of dry substrate	(Ganaie et al. 2017)
		Aspergillus oryzae	Agave sap (35)	32° C (48 h)	Column reactor	1347 U/L	(Muñiz-Márquez et al. 2016)

NS: not shown.

fermentation, respectively (Sandoval-González et al. 2018). Although the selection of new strains is a viable avenue for stress tolerance imposed by processing conditions, a common issue associated with wild strains still lingers: low yield. This fact can be associated with high concentrations of glucose in the fermentation medium, which represses the synthesis of some enzymes/proteins (Gancedo 1998). In this sense, development of genetically engineered microorganisms through recombination or *knockout* is an efficient tool for enzymatic synthesis improvement. Zhang et al. (2019) created a mutant of the β-fructofuranosidase-producing strain *Aureobasidium melanogenum* 33 by silencing the *CREA* gene, which is responsible for the synthesis of the transcriptional-binding repressor CreA. The results revealed that the disruptant mutant presented a 3.5-fold increase in production of β-fructofuranosidase, being able to sustain a high productivity despite the increase of glucose concentration in the fermentation medium. Ko et al. (2019) used the model *Saccharomyces cerevisiae* for the expression of the inulosucrase from *Limosilactobacillus reuteri* (former *Lactobacillus reuteri*), due to high protein secretion capability and degradation of remaining mono- and di-saccharides that may affect the prebiotic effect of FOs. The increased secretion of the enzyme was achieved by truncating both amino- and carboxy-terminal sections of the gene (LrInu) and introduction of a 98aa signal peptide from N-terminus of the mitochondrial inner membrane protein. After optimization assays, authors achieved a production of 220 U/mL from 39g of dried yeast.

Besides the isolation of the enzyme-producing microorganism, the selection of appropriate carbon/nitrogen sources, micronutrients, and operational conditions is also of the upmost importance for avoiding catabolite repression or low productivity (Kumar and Dubey 2019). Sucrose is the most suitable substrate for FTase and levansucrase due to positive linear correlations between carbon source concentration and enzymatic activity (Dhake and Patil 2007), despite showing inhibitory effects at concentrations ≥300 g/L (Belghith et al. 2012). Agro-industrial residues have also been investigated as potential carbon sources due to low cost, great availability, suitable supports for solid state fermentation (SSF), and sustainability aspects. De la Rosa et al. (2020) proposed the optimized FO production by *A. oryzae* in SSF using sugar cane bagasse, coffee husk, and peels of pineapple, prickle pear, and banana enriched with maguey sap. The results showed that cane bagasse was a suitable support for enabling continuous enzymatic/FOs production, as it enables superior aeration and heat/mass transfer due to absorption capacity. Mineral salts can also play important roles during enzymatic synthesis, as they influence the modulation of environmental stress, acting as co-factors or inhibiting microbial growth due to toxicity. A study conducted by Venkateshwar et al. (2010) evaluated the effects of 11 mineral nutrients over the *Saccharomyces* sp. GVT263 growth and FFase enzyme production. It was observed that monopotassium phosphate (KH_2PO_4) displayed inhibition for yeast growth, although having displayed the higher stimulation over FFase production. This discrepancy could be associated with the following factors: (1) disruption of the low sodium/potassium ratios when KH_2PO_4 was used concomitantly with NaCl and Na_2MoO_4 during the experimental design, which is detrimental for yeast growth due to ionic and hyperosmotic stresses (Yenush 2016); (2) potassium is an essential co-factor to enzymes involved in protein biosynthesis and carbohydrate metabolism, thus increasing the FFase production (Venkateshwar et al. 2010). In addition, $MnSO_4$, $ZnSO_4$, and $CoCl_2$ showed positive effects on the FFase, while $CaCl_2$ and Na_2MoO_4 inhibited the enzyme production.

Finally, pH, agitation, temperature, and aeration must be optimized during submerged fermentation (SF), while moisture content and solid support are crucial aspects for SSF. The majority of the studies conducted on the production of enzymes directed for FOs are conducted under SF conditions and, despite how the conditions may vary according to the needs of the source microorganism, general conditions are well-established: highly concentrated sucrose medium (150–450 g/L), temperature range 28–37°C during 20 h – 8 days, according to the data displayed in Table 8.1. This strategy is attractive because it facilitates the product purification and is easily scalable (Kumar and Dubey 2019). However, the constant accumulation of by-products (mainly glucose) reduces enzyme synthesis and FOs production efficiency. In order to avoid such losses, membrane reactor models are being proposed as a solution, where a continuous removal of glucose occurs along with a simultaneous influx of sucrose-rich medium, allowing a selective retention of FOs (Nishizawa et al. 2001; Burghardt et al. 2019). SSF, on the other hand, works on a strict control of moisture content, which significantly reduces the contamination risks, generation of residual wastewater, and the catabolic repression (Hölker et al. 2004). Additionally, SSF allows a significant cost reduction, since the extraction and immobilization phases can be considered unnecessary. Ganaie et al. (2017) proposed the use of sugarcane bagasse as a substrate/support for the *Aspergillus flavus* NFCCI 2364 spore suspension for the FTase. After optimization design, the authors observed that a sugarcane bagasse particle of 2.5 mm was able to provide an average of 423.18 U per gram of dry substrate at 96 h of SSF. In the same way, Oliveira et al. (2020) showed that FFase can be produced on SSF using 3 g of soy bean as support/substrate at a 60% moisture content and was able to achieve higher hydrolytic and transfructosylating activity (229.43 and 66.93 U/mL) for FFase after 120 and 96 h, respectively. These results demonstrate the feasibility of this fermentation strategy for industrial-scale application; however, further pilot optimization studies are needed.

8.4 ENZYME APPLICATION FOR FOs PRODUCTION

The application of enzymes in the production of FOs follows different approaches and is quite diverse. The traditional process is the application of purified enzymes together with the substrate in a discontinuous (batch) process. After the enzymatic reaction for the biotransformation of the substrate into FOs, the system is inactivated by the action of temperature, denaturing the present enzymes. Thus, after the incubation process and enzymatic action, the FOs are recovered by nanofiltration or chromatography processes with the use of high-affinity resins. However, in recent years different methodologies have been developed for better performance in the production of FOs. Processes such as the immobilization of enzymes (Ribeiro et al. 2021) or the use of membrane bioreactors (Fan et al. 2020) are applied in the search for higher yields and lower costs in the production of FOs on an industrial scale (Sánchez-Martínez et al. 2020).

Conventionally, at the industrial level, the enzymatic production of FOs at food-grade follows two process steps: the enzyme production and the enzymatic synthesis of FOs from sucrose. Figure 8.4 shows the different processes for FOs production using microbial FTases and sucrose. As can be seen, the production of these prebiotics can be carried out with free or immobilized enzymes, as well as with the use of whole microbial cells. Normally extracellular enzymes are applied directly in the reaction with the substrate. Whereas with microbial cells, with intracellular enzymes, the substrate is biotransformed to FOs within the microbial cells. On the other hand, the immobilization strategy (enzymes or microbial cells) presents a high potential because the biocatalysts can be reused for several cycles in FOs production. The use of microbial cells eliminates the need for cell rupture for the release of enzymes, which can contribute to the reduction of final process costs (Sánchez-Martínez et al. 2020).

The most used enzymes in the industrial production of FOs are the FTases, through the transfructosylation of sucrose reaching an average yield of 50–60%. This process also generates the co-production of glucose and fructose between 30% and 40%, and 10% and 20%, respectively (Nobre et al. 2016). The presence of monosaccharides such as glucose or fructose inhibits FTases activity, decreasing the final yield of FOs production. Due to this, different alternatives have been proposed for the elimination of these monosaccharides (mainly glucose) such as the use of enzymes for the oxidation of glucose or the application of organisms capable of fermenting glucose and biotransforming it into other biomolecules. Sheu et al. (2001) used an enzymatic complex of β-fructofuranosidase and commercial glucose oxidase to improve the production yield of FOs. The strategy involved the production of FOs through sucrose transfructosylation, while the glucose oxidase enzyme oxidized the co-produced glucose, converting it into gluconic acid. In addition, $CaCO_3$ was used to generate the precipitation of gluconic acid. The application of this strategy prevented the inhibition of the enzyme β-fructofuranosidase, which meant an increase in the production of FOs from 58% to 90%.

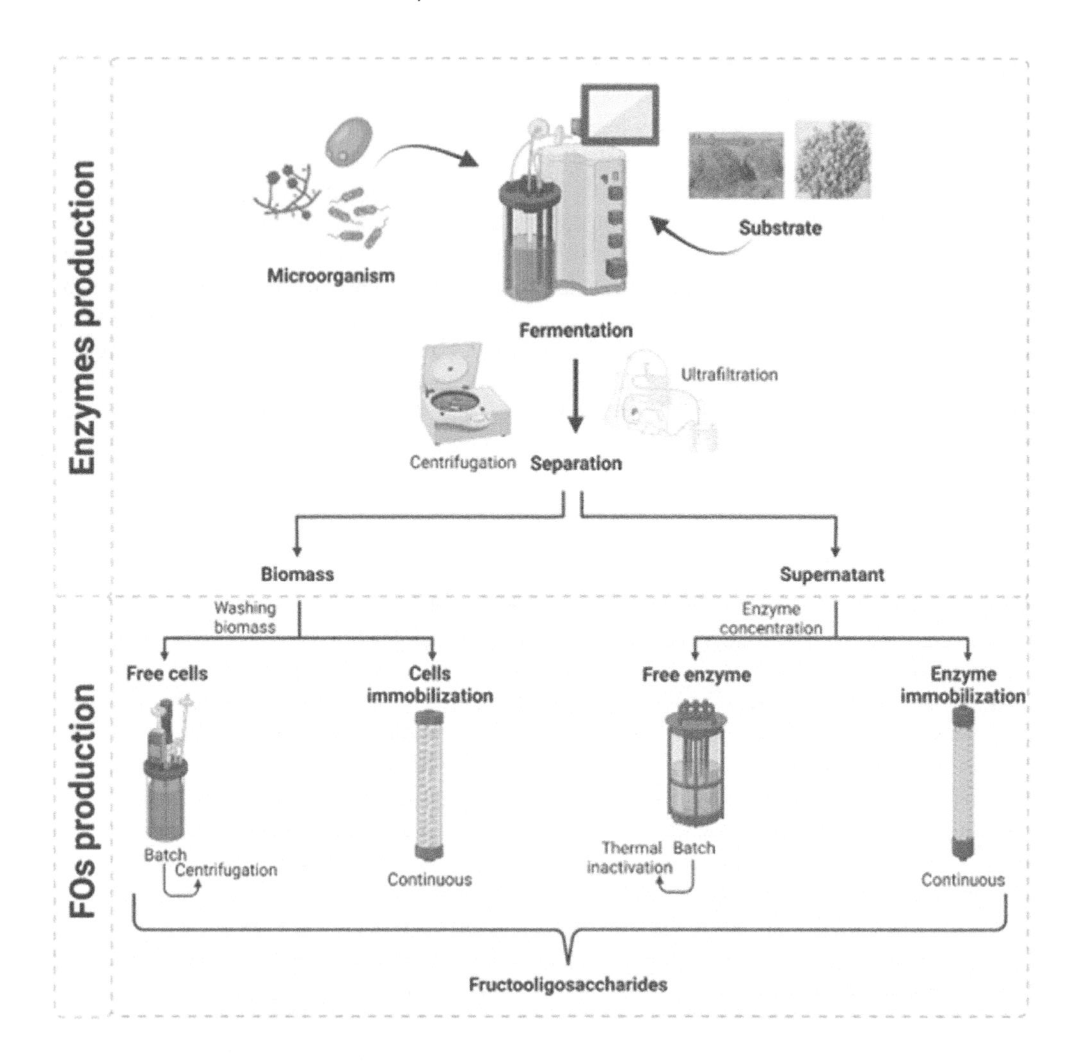

FIGURE 8.4 Different strategies of industrial FOs production applying free or immobilized enzymes/cells and under batch or continuous process. (Adapted from Sánchez-Martínez et al. 2020.)

Another enzyme that can be used in this sense is the glucose isomerase, which generates structural changes in the co-produced glucose, thus avoiding its inhibitory capacity (Ganaie et al. 2014).

On the other hand, the use of microorganisms capable of consuming glucose is a strategy that is being studied and generates high expectations for its large-scale application. Castro et al. (2019) used immobilized *A. pullulans* cells for the production of FOs, and after 10 h of fermentation encapsulated cells of *S. cerevisiae* were added. This strategy allowed the production of 119 g/L of FOs, which represented an increase in the initial productivity from 4.9 to 5.9 g/L h. The addition of *S. cerevisiae* encapsulated cells not only increased the productivity of FOs, but also increased the degree of purity of more than 67% (w/w). Finally, a concentration of 58 g/L of ethanol was reached through glucose fermentation by *S. cerevisiae*. Other microorganisms such as *Pachysolen* sp are capable of fermenting glucose and fructose to produce ethanol. In addition to being a highly appreciated molecule by the biotechnological industry, ethanol facilitates the purification process of co-produced FOs. Due to its low boiling point, ethanol evaporates giving rise to a purified FOs syrup (Sánchez-Martínez et al. 2020).

Enzyme packaging, immobilization, or encapsulation processes for the production of FOs are widely studied strategies. The most common goals for conducting these types of studies are enzyme recyclability and use in continuous processes. Singh and Singh (2021) used halloysite nanoclay to immobilize a fungal origin endo-inulinases, obtaining a maximum yield of 57.2% enzymatic activity. The immobilization of endo-inulinases allowed a 13-fold increase in the half-life of the enzyme. However, a decrease in its V_{max} and increase in its K_m was also observed after immobilization. Under optimal conditions, a yield of 99.6% of FOs could be obtained from 70 g/L of inulin, reaching 74.1% of FOs with 2–4 of polymerization degree and 25.4% FOs with 5–9 of polymerization degree. In addition, the enzymatic activity of the immobilized endo-inulinases was maintained at 50% until the third cycle of reuse. Other endo-inulinases immobilization matrices for the production of cheaper FOs, such as chicken eggshell, have also been evaluated, showing great potential for their use on a large scale (Ribeiro et al. 2021).

The yields of immobilized enzymes compared to the free enzyme are lower because, in the immobilization process, the enzymes are damaged and/or denatured, thus losing part of their enzymatic activity. Another cause of the low yield of immobilized enzymes is the limitation of mass transfer, which hinders enzymatic action against the substrate (Sánchez-Martínez et al. 2020). Bedzo et al. (2019) carried out a technoeconomic study for the production of FOs with the enzyme β-fructofuranosidase from sucrose. Three strategies were evaluated: free enzymes, enzymes immobilized with calcium alginate, and those with amberlite IRA 900. For an annual production of 2000 tons of FOs, the results showed that a minimum sales price of US$ 1.82 was obtained with the use of free enzyme. While for the enzyme immobilized in calcium alginate (6 cycles of reuse) and amberlite IRA 900 (12 cycles of reuse) they were US$ 1.87 and US$ 2.06, respectively, showing that at industrial scale the enzymatic immobilization stage turns out to be costly and greatly influenced by the final price of the product.

Finally, a more recent strategy is the use of genetically modified enzymes and their recycling within their own host cell for their next reuse. Wang et al. (2021) expressed a β-fructofuranosidase in the filamentous fungus *A. niger* ATCC 20611, by means of directed evolution techniques. The thermal stability of the enzyme was increased, which allowed it to maintain 91% of its initial enzymatic activity after 30 h at 50°C. In addition, it maintained a 79.3% of its initial enzyme activity and an efficiency of 57% of FOs production after 6 reuse cycles. The recyclability process of the enzyme occurred through the recovery of mycelium by means of filtration, which was subjected to another cycle of FOs production. A total of 6 re-cycles of mycelium were reached, showing great efficiency and potential for its use at industrial scale.

As can be seen, there are different strategies for the application of microbial enzymes in the production of FOs; these strategies are gradually being optimized and adapted for application at industrial scale. The aim is to reduce costs in the production process, increase productivity, develop less-polluting processes, and avoid the overproduction of by-products.

8.5 GLOBAL MARKET AND INDUSTRY

The FOs production industry has grown rapidly in the last two decades due to its application as an additive in functional foods. The market of FOs is segmented according to their applications, such as dietary supplements, in animal feed, pharmaceuticals, and food and beverages (Khanvilkar and Arya 2015). Among these segments, the demand for infant formula supplementation and prebiotic or fibre supplement represented over of 40% of global market volume in 2015, with a demand was over 31 kilotons (Grand View Research 2016). In addition, an expansion of 18.9% in the food and beverage segment of the global market is expected by 2028.

In fact, the global food industry of FOs had an estimated value of up to US$2.37 billion in 2020, and is projected to reach US$5.22 billion by 2028 with a compound annual growth rate (CAGR) prevision of 10.39% from 2021 to 2028 (Reports and Data 2020). The earnings in the USA were up to US$198.3 million in 2014. The CAGR prevision for European and Asian markets by 2027 is 10.5% and 10.9%, respectively (Trujillo et al. 2014). According to the demands, the FOs global market has been divided into North America, Europe, Asia, and the rest of the world. The top three consumer countries are the USA, China, and Japan (Kumar et al. 2018). According to the Verified Market Research report about FOs, the top nine key players in the market are Prebiotin, Beneo-Orafti SA, Cosucra-Groupe Warcoing SA, Quantum Hi-Tech, Shadong Bailong Chuangyua, CJ CheilJedang, GTC Nutrition, Jarrow Formulas, and Victory Biology Engineering Co. Ltd (Verified Market Research 2020).

In industry, the production of FOs occurs through the degradation of inulin or polyfructose, or by enzymatic transfructosylation (Picazo et al. 2019). The novel processes for FOs production using microbial enzymes, such as recombinant enzymes, engineered enzymes, and new substrates, have reduced the production cost. The future prospection is based on the use of bioprocess as a key factor, screening new potential transgenic microorganisms with high conversion yields (transfructosylating activity) and fewer FOs by-products, such as glucose and sucrose (Ganaie et al. 2014).

Nowadays, customers tend to care about health and chronic disease prevention, such as diabetes or cancer, demanding more FOs supplements or nutraceuticals; thus, further research, innovation, and investment will be required to fulfil the demand.

8.5.1 Industry and Patent Landscape

According to the Derwent patents database Clarivate Analytics (2018), inventions on FOs started in 2001 and currently, a total of 2100 patents are registered. Among these patents, 114 documents are specific to the use of microbial enzymes for FO production. Most companies have registered FOs-based formulations since 2001 (Figure 8.5a). Nestle and Novartis innovated dietary fibre formulations to solve the problems of gut microbiota, vascular, and gastric disorders (Haschke et al. 2001; Troup and Falk 2007). Other companies, such as IAMS Co (Reinhart 1998), found potential in FO-enriched nutrition for pets. The principal and most referenced companies are shown in Figure 8.5(a). The patent landscape (Figure 8.5b) based on FOs production by microbial enzymes was launched in 2007 by the Council Sci Industry (Mysore and Siddalingaiya 2006); they patented the FOs production involving extracellular fructosyl transferase obtained from *Aspergillus oryzae* by using residues.

In 2010, Zhen-AO Bio-Tech patented an enzyme preparation to improve the abnormal condition of blood cells such as erythrocyte and leucocytes. In the following years, companies such as Daxinganling Lingoberry Organic Foods (Wan et al. 2012) and the University of Nanjing in China (Jiang et al. 2014) patented processes for improving FO production yields over 98%. According to the number of citations, these patents were used as bases and references for the innovation of many others. To date, patents about FO products based on the enzymatic process were granted to Sinder (Shao 2018) and Samyang (Lu 2019) – companies where ready-made products are prepared with enzyme and probiotic formulations for nutrition and medicine. Thus, the use of fungal hyphae from the genus *Aspergillus* to enhance FO production is still being developed and patented by enterprises such as Quantum Hi Tech Co LTD (Zeng et al. 2020), Shandong Tainxing Biotechnology (Gan et al. 2015), and Revelations Biotech LTD (Beeram et al. 2021).

8.6 CONCLUSIONS

Microbial enzymes are powerful tools for application in the industrial development of FOs production. Fructosyltransferase, β-fructofuranosidase, and endo-inulinase are the main types of enzymes used in the biotechnological production of FOs. Filamentous fungi are the preferred producers of these enzymes in SF and SSF. SSF can take advantage of the use of agro-industrial residues, reducing costs in the production of enzymes. The development of new technology strategies, such as the use of microbial cells and immobilized enzymes, improve the production of FOs allowing the reuse of biocatalysts in different cycles, which is an economic strategy that could be employed on a large scale. These approaches contribute to the bioeconomy and the definition of sustainable processes in the biotechnology production of FOs at an industrial scale.

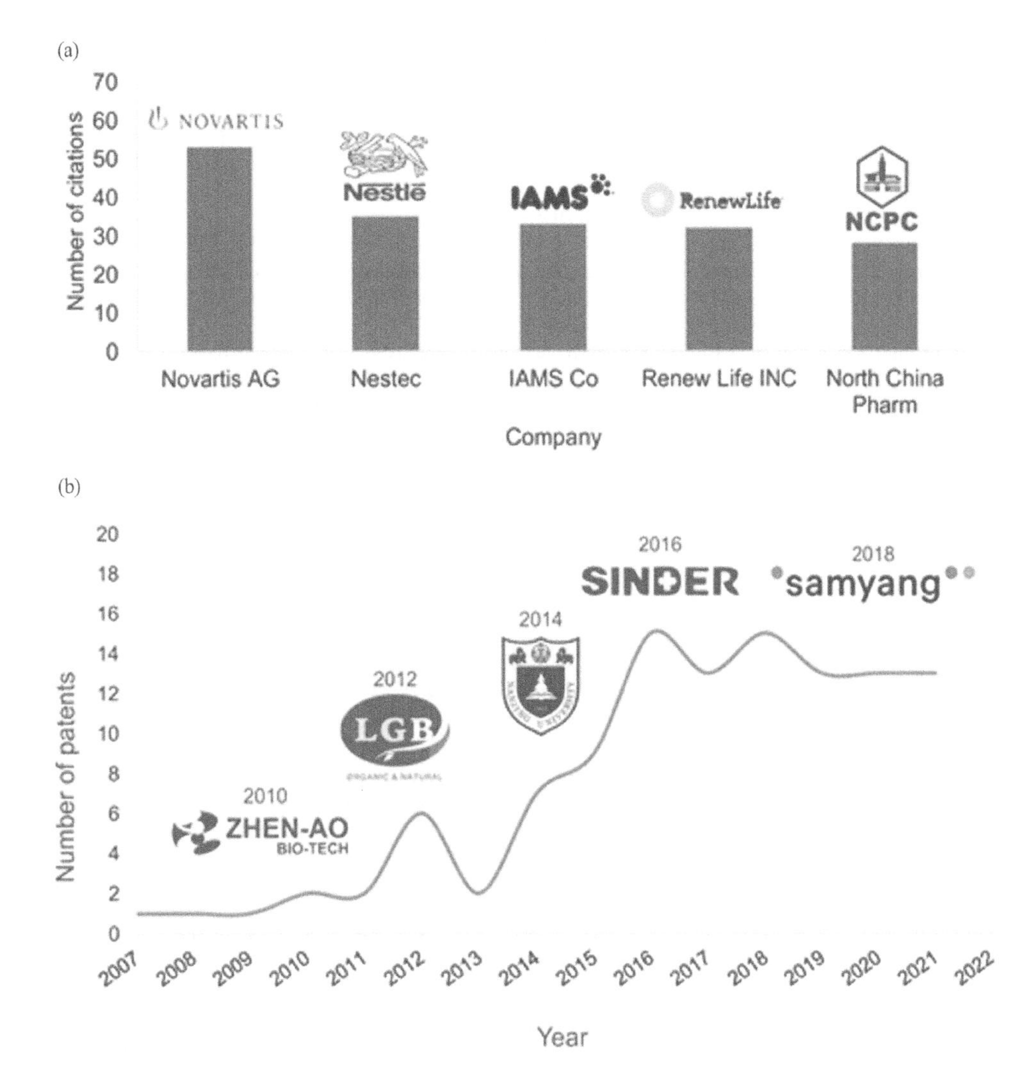

FIGURE 8.5 (a) Most cited patents about FOs-based products by key players worldwide; (b) Timeline of worldwide deposited patents on enzymatic production of FOs. (From 2007 until 2022.)

REFERENCES

Arrizon, Javier, Sandrine Morel, Anne Gschaedler, and Pierre Monsan. 2012. Fructanase and Fructosyltransferase Activity of Non-*Saccharomyces* Yeasts Isolated from Fermenting Musts of Mezcal. *Bioresource Technology* 110: 560–65. doi:10.1016/j.biortech.2012.01.112.

Bedzo, Oscar K. K., Mohsen Mandegari, and Johann F. Görgens. 2019. Comparison of Immobilized and Free Enzyme Systems in Industrial Production of Short-Chain Fructooligosaccharides from Sucrose Using a Techno-Economic Approach. *Biofuels, Bioproducts and Biorefining* 13 (5): 1274–88. doi:10.1002/bbb.2025.

Beeram, R. C., D. Sinhaa, B. B. Musuku, C. Are, and D. Kumar. 2021. New Modified Polypeptide i.e Beta-Fructofuranosidase of Aspergillus niger Comprising Specific Amino Acid Sequence, Useful for Producing Fructooligosaccharides. WO2021106016-A1, issued 2021.

Belghith, Karima Srih, Imen Dahech, Hafedh Belghith, and Hafedh Mejdoub. 2012. Microbial Production of Levansucrase for Synthesis of Fructooligosaccharides and Levan. *International Journal of Biological Macromolecules* 50 (2): 451–58. doi:10.1016/j.ijbiomac.2011.12.033.

Burghardt, Jan Philipp, Luca Antonio Coletta, Ramona van der Bolt, Mehrdad Ebrahimi, Doreen Gerlach, and Peter Czermak. 2019. Development and Characterization of an Enzyme Membrane Reactor for Fructo-Oligosaccharide Production. *Membranes* 9 (11): 148. doi:10.3390/membranes9110148.

Castro, Cristiana C, Clarisse Nobre, Guy De Weireld, and Anne-Lise Hantson. 2019. Microbial Co-Culturing Strategies for Fructo-Oligosaccharide Production. *New Biotechnology* 51 (July): 1–7. doi:10.1016/j.nbt.2019.01.009.

Chambert, Regis, Genevieve Treboul, and Raymond Dedonder. 1974. Kinetic Studies of Levansucrase of Bacillus Subtilis. *European Journal of Biochemistry* 41 (2): 285–300. doi:10.1111/j.1432-1033.1974.tb03269.x.

Clarivate Analytics. 2018. Derwent World Patent Index.

Cunha, Josivan S., Cristiane A. Ottoni, Sergio A. V. Morales, Elda S. Silva, Alfredo E. Maiorano, and Rafael F. Perna. 2019. Synthesis and Characterization of Fructosyltransferase from *Aspergillus Oryzae* IPT-301 for High Fructooligosaccharides Production. *Brazilian Journal of Chemical Engineering* 36 (2): 657–68. doi:10.1590/0104-6632.20190362s20180572.

de Abreu, Miguel, Miguel Alvaro-Benito, Julia Sanz-Aparicio, Francisco J. Plou, Maria Fernandez-Lobato, and Miguel Alcalde. 2013. Synthesis of 6-Kestose Using an Efficient β-Fructofuranosidase Engineered by Directed Evolution. *Advanced Synthesis & Catalysis* 355 (9): 1698–1702. doi:10.1002/adsc.201200769.

de la Rosa, Orlando, Diana B. Múñiz-Marquez, Juan C. Contreras-Esquivel, Jorge E. Wong-Paz, Raúl Rodríguez-Herrera, and Cristóbal N. Aguilar. 2020. Improving the Fructooligosaccharides Production by Solid-State Fermentation. *Biocatalysis and Agricultural Biotechnology* 27: 101704. doi:10.1016/j.bcab.2020.101704.

Dhake, A. B., and M. B. Patil. 2007. Effect of Substrate Feeding on Production of Fructosyltransferase by Penicillium Purpurogenum. *Brazilian Journal of Microbiology* 38 (2): 194–99. doi:10.1590/S1517-83822007000200002.

Fan, Rong, Jan P. Burghardt, Florian Prell, Holger Zorn, and Peter Czermak. 2020. Production and Purification of Fructo-Oligosaccharides Using an Enzyme Membrane Bioreactor and Subsequent Fermentation with Probiotic Bacillus Coagulans. *Separation and Purification Technology* 251 (February): 117291. doi:10.1016/j.seppur.2020.117291.

Fernandez, Rubén Cuervo, Cristiane Angélica Ottoni, Elda Sabino Da Silva, Rosa Mitiko Saito Matsubara, José Márcio Carter, Luis Roberto Magossi, Maria Alice Alves Wada, Maria Filomena De Andrade Rodrigues, Beatriz Guilarte Maresma, and Alfredo Eduardo Maiorano. 2007. Screening of β-Fructofuranosidase-Producing Microorganisms and Effect of PH and Temperature on Enzymatic Rate. *Applied Microbiology and Biotechnology* 75 (1): 87–93. doi:10.1007/s00253-006-0803-x.

Flores-Gallegos, Adriana C., Jesús A. Morlett-Chávez, Cristóbal N. Aguilar, Marta Riutort, and Raúl Rodríguez-Herrera. 2015. Gene Encoding Inulinase Isolated from Penicillium Citrinum ESS and Its Molecular Phylogeny. *Applied Biochemistry and Biotechnology* 175 (3): 1358–70. doi:10.1007/s12010-014-1280-9.

Flores-Maltos, Dulce A., Solange I. Mussatto, Juan C. Contreras-Esquivel, Raúl Rodríguez-Herrera, José A. Teixeira, and Cristóbal N. Aguilar. 2016. Biotechnological Production and Application of Fructooligosaccharides. *Critical Reviews in Biotechnology* 36 (2): 259–67. doi:10.3109/07388551.2014.953443.

Gan, Z., G. Dou, Li F., X. Shao, and T. Yang. 2015. Preparing Fructooligosaccharide Feed Additive Involves Preparing Sucrose Solution, Inoculating Aspergillus niger, Fermenting Culture, Adding Yeast, Continuing Fermentation and Separating Fructooligosaccharide Feed Additive Solution. CN104116000-A, issued 2015.

Ganaie, Mohd Anis, Agbaje Lateef, and Uma Shanker Gupta. 2014. Enzymatic Trends of Fructooligosaccharides Production by Microorganisms. *Applied Biochemistry and Biotechnology* 172 (4): 2143–59. doi:10.1007/s12010-013-0661-9.

Ganaie, Mohd Anis, Hemant Soni, Gowhar Ahmad Naikoo, Layana Taynara Santos Oliveira, Hemant Kumar Rawat, Praveen Kumar Mehta, and Narendra Narain. 2017. Screening of Low Cost Agricultural Wastes to Maximize the Fructosyltransferase Production and Its Applicability in Generation of Fructooligosaccharides by Solid State Fermentation. *International Biodeterioration and Biodegradation* 118: 19–26. doi:10.1016/j.ibiod.2017.01.006.

Gancedo, Juana M. 1998. Yeast Carbon Catabolite Repression. *Microbiology and Molecular Biology Reviews* 62 (2): 334–61. doi:10.1128/mmbr.62.2.334-361.1998.

Gomes, Tatiane Aparecida, Cristina Maria Zanette, and Michele Rigon Spier. 2020. An Overview of Cell Disruption Methods for Intracellular Biomolecules Recovery. *Preparative Biochemistry and Biotechnology* 50 (7): 635–54. doi:10.1080/10826068.2020.1728696.

Grand View Research. 2016. Fructooligosaccharides (FOs) Market Report Fructooligosaccharides (FOs) Market Analysis By Source (Inulin, Sucrose) By Application (Food & Beverages, Infant Formula, Dietary Supplements, Animal Feed, Pharmaceuticals) and Segment Forecasts to 2024.

Haschke, F., A. Carrie, Z. Kratky, H. Link-Amster, and F. Rochat. 2001. Enhancing Immune Responses by Administration of a Composition, e.g., A Food, Which Includes a Prebiotic, Especially a Combination of Fructooligosaccharide and Ilunin. WO200164225-A1, issued 2001.

Hernández, Lázaro, Carmen Menéndez, Enrique R. Pérez, Duniesky Martínez, Dubiel Alfonso, Luis E. Trujillo, Ricardo Ramírez, et al. 2018. Fructooligosaccharides Production by *Schedonorus Arundinaceus* Sucrose:Sucrose 1-Fructosyltransferase Constitutively Expressed to High Levels in *Pichia Pastoris*. *Journal of Biotechnology* 266: 59–71. doi:10.1016/j.jbiotec.2017.12.008.

Hijum, Sacha A. F. T. van, Slavko Kralj, Lukasz K. Ozimek, Lubbert Dijkhuizen, and Ineke G. H. van Geel-Schutten. 2006. Structure-Function Relationships of Glucansucrase and Fructansucrase Enzymes from Lactic Acid Bacteria. *Microbiology and Molecular Biology Reviews* 70 (1): 157–76. doi:10.1128/MMBR.70.1.157-176.2006.

Hölker, U., M. Höfer, and J. Lenz. 2004. Biotechnological Advantages of Laboratory-Scale Solid-State Fermentation with Fungi. *Applied Microbiology and Biotechnology* 64 (2): 175–86. doi:10.1007/s00253-003-1504-3.

Jiang, M., M. Zhang, C. Wei, H. Lv, J. Ma, H. Wu, and G. Li. 2014. Purifying Fructooligosaccharide, By Mixing Inulin and Inulinase, Adding Mixture, Phosphate Buffered Saline and Glycosyltransferase to Reactor, Enzymolyzing, and Passing Enzymolysate through Nanofiltration Membrane. CN103980382-A, issued 2014.

Khanvilkar, Shubhangi S., and Shalini S. Arya. 2015. Fructooligosaccharides: Applications and Health Benefits. A Review. *Agro Food Industry Hi Tech* 26 (6): 8–12.

Kim, Min-Hong, Man-Jin In, Hyung Joon Cha, and Young Je Yoo. 1996. An Empirical Rate Equation for the Fructooligosaccharide-Producing Reaction Catalyzed by β-Fructofuranosidase. *Journal of Fermentation and Bioengineering* 82 (5): 458–63. doi:10.1016/S0922-338X(97)86983-8.

Ko, Hyunjun, Jung Hoon Bae, Bong Hyun Sung, Mi Jin Kim, Hyun Ju Park, and Jung Hoon Sohn. 2019. Microbial Production of Medium Chain Fructooligosaccharides by Recombinant Yeast Secreting Bacterial Inulosucrase. *Enzyme and Microbial Technology* 130 (May): 109364. doi:10.1016/j.enzmictec.2019.109364.

Kumar, Chityal Ganesh, Sarada Sripada, and Yedla Poornachandra. 2018. Status and Future Prospects of Fructooligosaccharides as Nutraceuticals. In *Role of Materials Science in Food Bioengineering*, 451–503. Elsevier. doi:10.1016/B978-0-12-811448-3.00014-0

Kumar, Punit, and Kashyap Kumar Dubey. 2019. Current Perspectives and Future Strategies for Fructooligosaccharides Production through Membrane Bioreactor. In *Applied Microbiology and Bioengineering*, edited by Pratyoosh Shukla, 185–202. Boca Raton: Academic Press. doi:10.1016/b978-0-12-815407-6.00010-1.

Linde, Dolores, Barbara Rodríguez-Colinas, Marta Estévez, Ana Poveda, Francisco J. Plou, and María Fernández Lobato. 2012. Analysis of Neofructooligosaccharides Production Mediated by the Extracellular β-Fructofuranosidase from Xanthophyllomyces Dendrorhous. *Bioresource Technology* 109 (April): 123–30. doi:10.1016/j.biortech.2012.01.023.

Lu, R. 2019. Enzyme Jelly Useful for Promoting Excretion of Intestinal Toxin and Reducing Fat, Comprises Cubilose, Fruit Juice, Fructooligosaccharide, Isomisomaltose, Honey, Sodium Citrate, Potassium Sorbate, Konjaku and Carrageenin. CN107950955-A, issued 2019.

Magri, A., M. R. Oliveira, C. Baldo, C. A. Tischer, D. Sartori, M. S. Mantovani, and M. A. P. C. Celligoi. 2020. Production of Fructooligosaccharides by *Bacillus Subtilis Natto* CCT7712 and Their Antiproliferative Potential. *Journal of Applied Microbiology* 128 (5): 1414–26. doi:10.1111/jam.14569.

Maiorano, Alfredo Eduardo, Rosane Moniz Piccoli, Elda Sabino da Silva, and Maria Filomena de Andrade Rodrigues. 2008. Microbial Production of Fructosyltransferases for Synthesis of Pre-Biotics. *Biotechnology Letters* 30 (11): 1867–77. doi:10.1007/s10529-008-9793-3.

Morris, Cécile, and Gordon A. Morris. 2012. The Effect of Inulin and Fructo-Oligosaccharide Supplementation on the Textural, Rheological and Sensory Properties of Bread and Their Role in Weight Management: A Review. *Food Chemistry* 133 (2): 237–48. doi:10.1016/j.foodchem.2012.01.027.

Muñiz-Márquez, Diana B., Juan C. Contreras, Raúl Rodríguez, Solange I. Mussatto, José A. Teixeira, and Cristóbal N. Aguilar. 2016. Enhancement of Fructosyltransferase and Fructooligosaccharides Production by *A. Oryzae* DIA-MF in Solid-State Fermentation Using Aguamiel as Culture Medium. *Bioresource Technology* 213: 276–82. doi:10.1016/j.biortech.2016.03.022.

Mysore, N. R., and G. P Siddalingaiya. 2006. Preparing Fructooligosaccharide (FOS) Edible Films, to Deliver e.g. FOS, Comprises Preparing FOS Using Aspergillus oryzae Mixing FOS, Carboxymethyl Cellulose and Glycerol in Water Adding Gethayaegnts, Boiling and Dipping Food in Solution. WO2006103698-A1, issued 2006.

Nascimento, A. K. C., C. Nobre, M. T. H. Cavalcanti, J. A. Teixeira, and A. L. F. Porto. 2016. Screening of Fungi from the Genus *Penicillium* for Production of β- Fructofuranosidase and Enzymatic Synthesis of Fructooligosaccharides. *Journal of Molecular Catalysis B: Enzymatic* 134: 70–78. doi:10.1016/j.molcatb.2016.09.005.

Nishizawa, Koji, Mitsutoshi Nakajima, and Hiroshi Nabetani. 2001. Kinetic Study on Transfructosylation by β-Fructofuranosidase from *Aspergillus Niger* ATCC 20611 and Availability of a Membrane Reactor for Fructooligosaccharide Production. *Food Science and Technology Research* 7 (1): 39–44. doi:10.3136/fstr.7.39.

Nobre, C., C. C. Castro, A.-L. Hantson, J. A. Teixeira, G. De Weireld, and L. R. Rodrigues. 2016. Strategies for the Production of High-Content Fructo-Oligosaccharides through the Removal of Small Saccharides by Co-Culture or Successive Fermentation with Yeast. *Carbohydrate Polymers* 136 (January): 274–81. doi:10.1016/j.carbpol.2015.08.088.

Ojwach, Jeff, Ajit Kumar, Taurai Mutanda, and Samson Mukaratirwa. 2020. Fructosyltransferase and Inulinase Production by Indigenous Coprophilous Fungi for the Biocatalytic Conversion of Sucrose and Inulin into Oligosaccharides. *Biocatalysis and Agricultural Biotechnology* 30 (October): 101867. doi:10.1016/j.bcab.2020.101867.

Oliveira, Rodrigo Lira de, Marcos Fellipe da Silva, Attilio Converti, and Tatiana Souza Porto. 2020. Production of β-Fructofuranosidase with Transfructosylating Activity by *Aspergillus Tamarii* URM4634 Solid-State Fermentation on Agroindustrial by-Products. *International Journal of Biological Macromolecules* 144: 343–50. doi:10.1016/j.ijbiomac.2019.12.084.

Özbek, Belma, and Kutlu Ö. Ülgen. 2000. The Stability of Enzymes after Sonication. *Process Biochemistry* 35 (9): 1037–43. doi:10.1016/S0032-9592(00)00141-2.

Paixão, Susana M., Pedro D. Teixeira, Tiago P. Silva, Alexandra V. Teixeira, and Luís Alves. 2013. Screening of Novel Yeast Inulinases and Further Application to Bioprocesses. *New Biotechnology* 30 (6): 598–606. doi:10.1016/j.nbt.2013.02.002.

Picazo, Brian, Adriana C. Flores-Gallegos, Diana B. Muñiz-Márquez, Abril Flores-Maltos, Mariela R. Michel-Michel, Orlando de la Rosa, Rosa Maria Rodríguez-Jasso, Raúl Rodríguez-Herrera, and Cristóbal Noé Aguilar-González. 2019. Enzymes for Fructooligosaccharides Production: Achievements and Opportunities. In *Enzymes in Food Biotechnology*, 303–20. Elsevier. doi:10.1016/B978-0-12-813280-7.00018-9.

Pons, Tirso, Daniil G. Naumoff, Carlos Martínez-Fleites, and Lázaro Hernández. 2004. Three Acidic Residues Are at the Active Site of a β-Propeller Architecture in Glycoside Hydrolase Families 32, 43, 62, and 68. *Proteins: Structure, Function, and Bioinformatics* 54 (3): 424–32. doi:https://doi.org/10.1002/prot.10604.

Reinhart, G. A. 1998. Treating Small Intestine Bacterial Overgrowth in Animals, Particularly Dogs. US5776524-A, issued 1998.

Reports and Data. 2020. Fructooligossacharides (FOS) Market.

Ribeiro, Geise Camila Araujo, Pedro Fernandes, Dayse Alessandra Almeida Silva, Hugo Neves Brandão, and Sandra Aparecida de Assis. 2021. Inulinase from *Rhodotorula Mucilaginosa*: Immobilization and Application in the Production of Fructooligosaccharides. *Food Science and Biotechnology* 30 (7): 959–69. doi:10.1007/s10068-021-00931-x.

Roberfroid, Marcel. 2008. Prebiotics. In *Handbook of Prebiotics*, 39–68. CRC Press. doi:10.1201/9780849381829.ch3.

Sánchez-Martínez, María José, Sonia Soto-Jover, Vera Antolinos, Ginés Benito Martínez-Hernández, and Antonio López-Gómez. 2020. Manufacturing of Short-Chain Fructooligosaccharides: From Laboratory to Industrial Scale. *Food Engineering Reviews* 12 (2): 149–72. doi:10.1007/s12393-020-09209-0.

Sandoval-González, R. S., H. Jiménez-Islas, and J. L. Navarrete-Bolaños. 2018. Design of a Fermentation Process for Agave Fructooligosaccharides Production Using Endo-Inulinases Produced *In Situ* by *Saccharomyces Paradoxus*. *Carbohydrate Polymers* 198: 94–100. doi:10.1016/j.carbpol.2018.06.075.

Sangeetha, P. T., M. N. Ramesh, and S. G. Prapulla. 2004. Production of Fructo-Oligosaccharides by Fructosyl Transferase from Aspergillus Oryzae CFR 202 and Aureobasidium Pullulans CFR 77. *Process Biochemistry* 39 (6): 755–60. doi:10.1016/S0032-9592(03)00186-9.

Shao, S. 2018. Complex Probiotic Fungus Functional Food Used in Healthcare Area, Includes Matsutake Enzymolysis Powder, Radix Ophiopogonis Extract, Lentinan, Fructooligosaccharide, Wheat Fiber, Soybean Fiber, Tea Polyphenols and Mal. CN105852099-A, issued 2018.

Sheu, Dey Chyi, Po Jang Lio, Shih Tse Chen, Chi Tsai Lin, and Kow Jen Duan. 2001. Production of Fructooligosaccharides in High Yield Using a Mixed Enzyme System of β-Fructofuranosidase and Glucose Oxidase. *Biotechnology Letters* 23 (18): 1499–1503. doi:10.1023/A:1011689531625.

Singh, R. S., Rupinder Pal Singh, and John F. Kennedy. 2016. Recent Insights in Enzymatic Synthesis of Fructooligosaccharides from Inulin. *International Journal of Biological Macromolecules* 85: 565–72. doi:10.1016/j.ijbiomac.2016.01.026.

Singh, R. S., and Taranjeet Singh. 2021. Fructooligosaccharides Production from Inulin by Immobilized Endoinulinase on 3-Aminopropyltriethoxysilane Functionalized Halloysite Nanoclay. *Catalysis Letters*, September. doi:10.1007/s10562-021-03803-5.

Singh, R. S., Taranjeet Singh, and Ashok Pandey. 2020. Fungal Endoinulinase Production from Raw *Asparagus* Inulin for the Production of Fructooligosaccharides. *Bioresource Technology Reports* 10 (January): 100417. doi:10.1016/j.biteb.2020.100417.

Singh, R. S., Taranjeet Singh, and Ashok Pandey. 2021. Production of Fungal Endoinulinase in a Stirred Tank Reactor and Fructooligosaccharides Preparation by Crude Endoinulinase. *Bioresource Technology Reports* 100743. doi:10.1016/j.biteb.2021.100743.

Tódero, L. M., Carem G. V. R., and Luis H. S. G. 2019. Production of Short-Chain Fructooligosaccharides (ScFOS) Using Extracellular β-D-Fructofuranosidase Produced by *Aspergillus Thermomutatus*. *Journal of Food Biochemistry* 43 (8): e12937. doi:10.1111/jfbc.12937.

Tomotani, E. J., and Michele, V. 2007. Production of High-Fructose Syrup Using Immobilized Invertase in a Membrane Reactor. *Journal of Food Engineering* 80 (2): 662–67. doi:10.1016/j.jfoodeng.2006.07.002.

Tonozuka, T., Akiko, T., Gaku, Y., Takatsugu, M., et al. 2012. Crystal Structure of a Lactosucrose-Producing Enzyme, Arthrobacter Sp. K-1 β-Fructofuranosidase. *Enzyme and Microbial Technology* 51 (6–7): 359–65. doi:10.1016/j.enzmictec.2012.08.004.

Trollope, Kim M., Niël van Wyk, Momo A. Kotjomela, and Heinrich Volschenk. 2015. Sequence and Structure-Based Prediction of Fructosyltransferase Activity for Functional Subclassification of Fungal GH32 Enzymes. *FEBS Journal* 282 (24): 4782–96. doi:10.1111/febs.13536.

Troup, J. P., and A. L. Falk. 2007. Dietary Fiber Formulation for Treating, e.g. Irritable Bowel Syndrome, Hypertension, Dyslipidemia, Obesity, Heart Disease, or Stroke, Includes Soluble Fibers Including Partially-Hydrolized Guar Gum and Fructooligosaccharide. WO2007050656-A2, issued 2007.

Trujillo, L. E., Marcillo, V. E., Avalos, R., Ponce, L. K., and Ramos, T. 2014. From the Laboratory to the Industry: Enzymatic Production and Applications of Shortchain Fructooligosaccharides (Fos). Recent Advances and Current Perspectives. Bionatura.

Van Wyk, N, Kim, M. T., Emma, T. S., Brenda, D. W., and Heinrich, V. 2013. Identification of the Gene for β-Fructofuranosidase from Ceratocystis Moniliformis CMW 10134 and Characterization of the Enzyme Expressed in *Saccharomyces cerevisiae*. *BMC Biotechnology* 13 (1): 100. doi:10.1186/1472-6750-13-100.

Vaňková, K., Zdenka, O., Monika, A., and Milan, P. 2008. Design and Economics of Industrial Production of Fructooligosaccharides. *Chemical Papers* 62 (4): 375–81. doi:10.2478/s11696-008-0034-y.

Vega, R., and Zuniga-Hansen, M. E. 2014. A New Mechanism and Kinetic Model for the Enzymatic Synthesis of Short-Chain Fructooligosaccharides from Sucrose. *Biochemical Engineering Journal* 82 (January): 158–65. doi:10.1016/j.bej.2013.11.012.

Venkateshwar, M., K. Chaitanya, Md Altaf, E. J. Mahammad, Hameeda Bee, and Gopal Reddy. 2010. Influence of Micronutrients on Yeast Growth and β-d-Fructofuranosidase Production. *Indian Journal of Microbiology* 50 (3): 325–31. doi:10.1007/s12088-010-0005-1.

Verified Market Research. 2020. Global Fructooligosaccharide (FOS) Market Size By Source, By Application, By Geographic Scope and Forecast.

Wan, L., D. Yao, and Y. Zhang. 2012. Purifying Chicory Fructooligosaccharide, Useful in the Field of Natural Organic Chemistry, Comprises e.g. Crushing Chicory, Mixing, Performing Enzymolysis by Adding Aspergillus niger and Endo-Inulinase Powder, Filtering, and Extracting. CN102732585-A, issued 2012.

Wang, J., Jing, Z., Lushan, W., et al. 2021. Continuous Production of Fructooligosaccharides by Recycling of the Thermal-Stable β-Fructofuranosidase Produced by Aspergillus Niger. *Biotechnology Letters* 43 (6): 1175–82. doi:10.1007/s10529-021-03099-w.

Yenush, L. 2016. Potassium and Sodium Transport in Yeast. In *Yeast Membrane Transport*, edited by José Ramos, Hana Sychrová, and Maik Kschischo, 892:187–228. London: Springer. doi:10.1007/978-3-319-25304-6_8.

Zeng, X., B. Deng, Y. Wei, X. Yang, and Z. Xie. 2020. New Aspergillus japonicus Strain QHT-40-U475 Used in the Preparation of Brown Sugar Fructooligosaccharides Is Deposited at China Centre for Type Culture Collection. CN111088170-A, issued 2020.

Zhang, L., Jin, An, Lijuan, Li, Hengwei, Wang, Dawen, Liu, Ning, Li, Hairong, C., and Zixin, D. 2016. Highly Efficient Fructooligosaccharides Production by an Erythritol-Producing Yeast *Yarrowia Lipolytica* Displaying Fructosyltransferase." *Journal of Agricultural and Food Chemistry* 64 (19): 3828–37. doi:10.1021/acs.jafc.6b00115.

Zhang, S., Hong, J., Sijia, Xue, Na, Ge, Yi, Sun, Zhe, Chi, Guanglei, Liu, and Zhenming, Chi. 2019. Efficient Conversion of Cane Molasses into Fructooligosaccharides by a Glucose Derepression Mutant of *Aureobasidium Melanogenum* with High β-Fructofuranosidase Activity. *Journal of Agricultural and Food Chemistry* 67 (49): 13665–72. doi:10.1021/acs.jafc.9b05826.

9 Enzymes for Lactose Hydrolysis and Transformation

Ariane Fátima Murawski de Mello[1], Luciana Porto de Souza Vandenberghe[1], Clara Matte Borges Machado[1], Agnes de Paula Scheer[2], Aline B. Argenta[2], Gilberto Vinicius de Melo Pereira[1], Alexander da Silva Vale[1], and Carlos Ricardo Soccol[1]

[1] Department do Bioprocess Engineering and Biotechnology, Federal University of Paraná, Curitiba, Paraná, Brazil

[2] Department of Chemical Engineering, Federal University of Paraná, Centro Politécnico, Curitiba, Paraná, Brazil

CONTENTS

9.1 INTRODUCTION

β-galactosidases (β-D-galactohydrolase, EC 3.2.1.23) or lactases, are enzymes that hydrolyse lactose to glucose and galactose (DeCastro et al., 2018) and are the most studied and marketed enzymes (Vera et al., 2020). They can be used in the industrial production of dairy products (Ansari and Husain, 2012). Milk-derived products are not consumed by lactase deficient people and the enzymatic hydrolysis of lactose can be a solution to this problem (Mahdian et al., 2016). β-galactosidases can also catalyse transgalactosylation reaction when lactose is present (Wang et al., 2014), which gives these enzymes the ability to synthesize prebiotics, such as galactooligosaccharides (GOS) (Aulitto et al., 2021; Kumari et al., 2021).

β-galactosidases are classified into four groups of the glycosyl hydrolases family (GH 1, GH 2, GH 35, and GH 42) (Soto et al., 2017). These important commercial enzymes can be obtained from different sources such as plants, animals, and microorganisms including fungi, bacteria, and yeasts (Albuquerque et al., 2021). However, β-galactosidases from *Kluyveromyces lactis* are among the most reported (Pereira-Rodríguez et al., 2010, Katrolia et al., 2011), with diverse applications in the environmental, food, and biotechnological industries (Albuquerque et al., 2021). In the food industry, this enzyme is essential in the processing of lactose. Aside from the employment of β-galactosidases in dairy to obtain lactose-free products with high efficiencies over 99.5%, they have also been applied mainly in the conversion of lactose into other industrial products, in the

DOI: 10.1201/9781003311164-12

treatment of commercially discarded whey (Karasova et al., 2002), and the production of GOS (González-Cataño et al., 2017). This chapter describes the main β-galactosidases production processes, their applications, immobilization techniques, recombinant β-galactosidases production, and patents and innovation involving these consolidated enzymes.

9.2 MICROBIAL PRODUCTION OF β-GALACTOSIDASES

β-galactosidases enzymes can be produced by animal, plant, or microbial cells (Panesar et al., 2006); however, microbial sources show higher productivity (Ansari and Satar, 2012) and easier manipulation (DeCastro et al., 2018). *Aspergillus* spp. And *Kluyveromyces* spp. Are the most used microorganisms to obtain β-galactosidases, since they present high yields and are "generally recognized as safe" (GRAS) for human consumption (Oliveira et al., 2011). The fungus *Aspergillus* spp. produces and excretes β-galactosidase to the extracellular medium (Oliveira et al., 2011), facilitating downstream operations. Fungi-derived β-galactosidase has an optimal pH between 2.5 and 5.4 and an optimal temperature up to 50°C, allowing its use in the processing of acid whey and its ultrafiltration permeate (Panesar et al., 2006). The yeast *Kluyveromyces* spp. produces β-galactosidase intracellularly (Oliveira et al., 2011) and has an optimal pH in the neutral range making it suitable to hydrolyse lactose in milk (Panesar et al., 2006).

Mesophilic, psychrophilic, and thermophilic microorganisms produce β-galactosidases with environmental adaptations that can be useful in the biotechnology industry (Ansari and Satar, 2012). β-galactosidases can be produced from different microbial sources and media under a range of different conditions (Table 9.1). β-galactosidases produced by thermophilic microorganisms have higher initial enzyme production, promote high substrate solubilization, with lower microbial contamination due to the higher operation temperature conditions (DeCastro et al., 2018). They can also be employed at different temperatures due to their thermostability (Jensen et al., 2017). This characteristic is probably because of the compact structure, tight helix, increased interaction of thermophilic β-galactosidases such as salt bridges and hydrogen bonds among the amino acid residues (Kumar et al., 2014). Shuanggui et al. (2018) (CN109055336) claimed a method to produce high-temperature-resistant β-galactosidases. The inventors used a magnetic field mutagenesis to induce the enzyme production that occurred after 2 h at 85°C.

Cold-adapted enzymes do not require a heating step, which reduces the production costs and avoids mesophilic contamination (Rutkiewicz et al., 2019). They also have high catalytic efficiency possibly because of the enhanced flexibility (of the whole protein or just the active site), due to their low sensitivity to cold temperatures (Kumar et al., 2014). This higher flexibility can increase the enzyme promiscuity – the ability of the enzyme to catalyse different reactions (Mangiagalli and Lotti, 2021). Qingfang et al. (2012) (CN102653747) claimed a method for cold-adapted enzyme production through the use of a synthetic media supplemented with lactose during 24–48 h at 10–16°C, reaching an enzyme yield of 120 U/mL. Naiyu et al. (2012) (CN102653746) also claimed a process for low-temperature β-galactosidase production that also utilizes a synthetic media supplemented with lactose at 10–16°C; however, the fermentation was carried out during 72–144 h, reaching an activity of 115 U/mL.

Increased demand for β-galactosidase production requires higher production yields and lower costs (Manera et al., 2008). In this sense, Daavu and Potireddy (2020) (IN201941010345) claimed a method to produce β-galactosidases by a marine isolate *Bacillus aryabhattai*. The fermentation was carried out with a synthetic medium at 30–35°C and pH 7.0-7.3 for 24 h and 110–150 rpm, achieving an enzymatic yield of 8 U/L.

With the current technological advances such as recombinant DNA technology (Oliveira et al., 2011), protein engineering (Ansari and Satar, 2012), optimization of media and cultivation conditions (Raol et al., 2014), high yields and productivities of β-galactosidase can be obtained, such as in the research reported by Al-Jazairi et al. (2015) (4,997 U/mL.min). In this case a synthetic media was used, which can increase production costs, but facilitate enzyme recovery and purification.

TABLE 9.1
Microbial β-Galactosidases Produced under Different Conditions

Microbial Source	Microorganism	Activity	T (°C)	pH	T (h)	Media	Activity Detection	Reference
Fungus	*Teratosphaeriaacidotherma*	54.6 U	30	5	48	Lactose medium (2% of lactose)	NPGA Glucose and oxidase method TLC analysis	(Yamada et al., 2016)
Fungus	*Aspergillus niger*	24.64 U/mL	28	7	168	2% of soybean residue	ONPG	(Martarello et al., 2019)
Fungus	*Aspergillus lacticoffeatus*	460 U/mL	28	6.5	216	Lactose medium (2% of lactose)	ONPG	(Cardoso et al., 2017)
Fungus	*Aspergillus niger*	8,809 U/mg	30	–	96	Wheat straw	ONPG	(Kazemi et al., 2016)
Fungus	*Aspergillus tubingensis*	7,890 U/gds	28	–	192	Different agro-residues	ONPG	(Raol et al., 2015)
Fungus	*Aspergillus alliaceus*	0.0486 IU/mL	30	4.5	144	Synthetic media supplemented with lactose (1 g/L)	Glucose oxidase	(Sen et al., 2012)
Fungus	*Thermomyceslanuginosus*	3.94 U/mg	45	5.5	168	Synthetic media supplemented with lactose (20 g/L)	ONPG	(El-Gindy et al., 2009)
Bacteria	*Enterobacter ludwigii*	34.37 U/mL	20	7.3	48	LB broth supplemented with lactose (2%)	ONPG	(Alikunju et al., 2018)
Bacteria	*Bacillus* sp.	50 U/mL	37	7	48	Lactose medium supplemented with MgCl2 (1% of lactose)	X gal ONPG	(Kamran et al., 2016)
Bacteria	*Picrophilustorridus*	94 IU/mL	70	5–5.5	72	Synthetic media	ONPG	(Murphy and Walsh, 2019)
Bacteria	*Streptococcus thermophilus*	78.85 U/mg	40	6.8	5	Synthetic media supplemented with lactose (1%)	ONPG	(Sangwan et al., 2015)
Bacteria	*Streptococcus thermophilus*	7.76 U/mL	40	7.2	24	Acid whey	ONPG	(Princely et al., 2013)
Yeast	*Kluyveromycesmarxianus*	14.19 U/mL	30	–	25	Porungo cheese whey	ONPG	(Marim et al., 2021)
Yeast	*Kluyveromycesmarxianus*	589 U/mg	30	5	18	Yeast extract, peptone and glucose supplemented with lactose at 2%	ONPG	(Boudjema et al., 2016)
Yeast	*Kluyveromycesmarxianus*	40.7 U/mL	30	5	96	Lactose (20 g/L) and residual glycerol (60 g/L)	ONPG	(Machado et al., 2015)
Yeast	*Kluyveromycesmarxianus*	8.82 U/mL	37	–	18	Synthetic media supplemented with lactose (20 g/L)	ONPG	(Am-Aiam and Khanongnuch, 2015)
Yeast	*Kluyveromycesmarxianus*	4,997 U/mL.min	20	3	64	Synthetic media supplemented with lactose (10%)	ONPG	(Al-Jazairi et al., 2015)

Furthermore, agro-industrial wastes can be used to reduce costs and environment impact (Raol et al., 2015; Marim et al., 2021). For example, Kazemi et al. (2016) reached an enzymatic activity of 8,809 U/mg using wheat straw as substrate that is obtained after harvesting and removing the edible products (Suopajärvi et al., 2020). Another example was reported by Raol et al. (2015), which obtained 7,890 U/gds of β-galactosidase yield, using wheat bran and deproteinized acid cheese whey at a 1:4 ratio. Wheat bran is an agricultural residue rich in carbohydrates well suited for the growth of microorganisms (Yunus et al., 2015) and acid whey is a by-product of cheese production obtained through acid protein coagulation (Tsanasidou et al., 2021). In both studies the media was not supplemented with lactose, which contributes to the reduction of costs of β-galactosidase production (Raol et al., 2014). Furthermore, Wei et al. (2020) (CN111378636) claimed a method for β-galactosidase production using whey powder as substrate at 16–37°C, 200–500 rpm and pH 8.5–10, achieving an enzyme yield of 2.6–5.2 U/mg.

For β-galactosidase enzymatic activity determination, the *O*-nitrophenyl-β-D-galactopyranoside (ONPG) method is the most employed (Table 9.1). This method is based on the enzymatic reaction, which converts ONPG to *O*-nitrophenol (ONP) (Kazemi et al., 2016). The product formation is detected spectrophotometrically at 420 nm using an ONP calibration curve prepared in the same conditions as the enzymatic reaction (Raol et al., 2015). The reaction is stopped by the addition of Na_2CO_3 (Marim et al., 2021) or NaOH (Raol et al., 2015). One unit of enzyme activity is defined as the amount of enzyme that liberates 1 μmole of ONP in one minute under assay conditions (Am-Aiam and Khanongnuch, 2015).

9.3 IMMOBILIZATION OF β-GALACTOSIDASES

The hydrolysis process can be carried out with the enzyme in free or immobilized form, but in the free mode, the enzyme has reduced stability for prolonged use. Enzyme immobilization consists of physical confinement with retention of its catalytic activity, which can be used repeatedly and continuously. The possibility of reusing the enzyme offered by the immobilization system makes this technique very advantageous, as it allows for a reduction in operating costs, in addition to providing greater stability to the enzyme. Furthermore, to operational stability, immobilization provides storage stability, both of which are important process parameters. In the study of Vasileva et al. (2016), for example, storage of the enzyme immobilized β-gal in membrane and in free form, under conditions 4°C, pH 6.8 for 30 days, resulted in a remaining enzyme activity of 75% and 38%, respectively.

Furthermore, other advantages exist such as these given by Grosová et al. (2008), Szymańska et al. (2007), and Verma et al. (2012): possibility of operating in continuous processes and multienzyme reaction systems, promotion of reuse without a significant decrease in activity, use of a smaller reactor volume, as the immobilized enzyme can be used in high concentration in a smaller volume; no contamination of product by the enzyme (especially useful in the food industry); greater ease of process control; ease of separation of the final product, and ease of stopping the reaction.

9.3.1 Techniques and Matrices for Immobilization

The techniques used for enzyme immobilization are classified into physical and chemical. Physical retention can be through the entrapment of the enzyme in a polymeric matrix, capsule, or membrane; or by adsorption of the enzyme to the surface of a support, which is based on the physical interactions between the enzyme and adsorbent, such as hydrogen bonding, hydrophobic interactions, and van der Waals force. In the chemical technique, immobilization can occur by covalent bonding, which is the most used method for immobilization and is characterized by enzymes covalently linked to the support through functional groups on the enzymes that are not essential for catalytic activity, or by cross linking that uses bi- or multifunctional compounds, which serve as the reagent for intermolecular cross linking of the biocatalyst. The cross linking method is often used

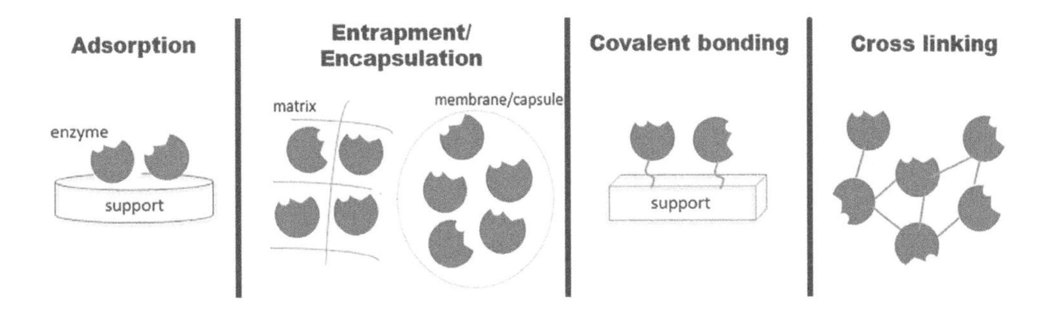

FIGURE 9.1 Enzyme immobilization methods. (Based on Nguyen and Kim, 2017.)

in combination with adsorption and entrapment to β-galactosidase immobilization (Grosová et al., 2008; Nguyen and Kim, 2017; Liu et al., 2018). Figure 9.1 illustrates the main methods of enzyme immobilization.

The choice of the most appropriate immobilization technique will depend on the enzyme used, the process to be carried out, and the properties of the adopted support. Furthermore, in some cases, different methods can be combined. For example, an enzyme can be pre-immobilized on spheres by adsorption or covalent bonding and, subsequently, trapped in a polymeric matrix (Guisan, 2013). Each of the immobilization methods has advantages and disadvantages that influence the decision-making technique to be used, as shown in Table 9.2.

The supports used for immobilization are classified, according to their chemical composition, into inorganic and organic. Among the inorganic materials, the main ones are bentonite, silica, glass, and metallic oxides. Organic supports can be natural (cellulose, agar, chitosan, alginate, and collagen) and synthetic (polystyrene, polyacrylamide, and other polymers). Inorganic matrices have advantages such as high stability against physical, chemical, and microbial degradation. However, most industrial applications are carried out with organic matrices, due to its hydrophilic character, which is one of the most important factors for enzymatic activity (Guisan, 2013). The choice of immobilization matrix must consider some characteristics such as particle diameter, physical resistance to compression, inertness for enzymes, hydrophilic character, microbial resistance, and availability at low cost (Trevan, 1980; Buchholz and Klein, 1987; Khan and Alzohairy, 2010). Some materials such as chitosan, cellulose-gelatin, membranes, silica nanoparticles, sodium, and calcium alginate spheres stand out among the researches and have been commonly used to immobilize enzymes of interest in the food area (Ansari and Husain, 2011; Verma et al., 2012; Das et al., 2015).

9.3.2 Applications of the Immobilized β-Galactosidase

The β-galactosidase is one of the enzymes that has been immobilized on supports such as those previously mentioned, especially for the hydrolysis of lactose present in milk and cheese whey or for the synthesis of GOS, which have aroused interest due to their prebiotic characteristics. In several studies using β-galactosidase, the systems that use a free enzyme have been compared with those that use an immobilized enzyme. In addition, different materials for the enzyme immobilization have been studied to make the process of hydrolysis or synthesis of GOS more economical. The main and most recent studies using immobilized β-galactosidase are compiled in Table 9.3.

Ansari and Husain (2010, 2011, 2012) studied the hydrolysis of lactose present in milk and whey using β-galactosidase from *Aspergillus oryzae* immobilized on a support and compared this approach to the process employing the soluble enzyme. The authors observed that immobilized β-galactosidase exhibited significantly higher stability than the free enzyme, against various types of denaturants and during storage. Furthermore, the immobilized enzyme could be used to hydrolyse lactose from milk and whey, in both batch and continuous processes.

TABLE 9.2
Advantages and Disadvantages of Enzyme Immobilization

Methods	Advantages	Disadvantages
Adsorption	Simple, low cost, small enzyme changes, no additional agents needed	Enzyme-adsorbent interaction is weak (desorption, leaching occurs), susceptible to pH and temperature
Entrapment	Low cost, high catalytic activity, good stability as it minimizes leaching and enzymatic denaturation	Mass transfer limitation, support may break
Covalent bonding	Strong bonds between enzymes and matrix, little leakage of enzyme from support	More complex process, high costs and risk of protein denaturation, decreased enzyme activity
Cross linking	Simple method, with strong chemical bond between enzyme and matrix, so enzymatic leakage is minimal	Limitation of mass transfer, decreased enzyme activity and conformational change of the enzyme

Sources: Guisan (2013); Liu et al. (2018); and Nguyen and Kim (2017).

The immobilization of the β-galactosidase enzyme from *Escherichia coli* also was investigated using a membrane as the support. The authors evaluated the hydrolysis of lactose present in whey using the immobilized β-galactosidase enzyme in a reactor with a spiral membrane and compared this procedure to the process performed with the free enzyme. The authors reported that the hydrolysis of lactose by the immobilized enzyme was 1.6 times more efficient than the hydrolysis using the free enzyme. In addition, the immobilized enzyme showed a high stability during the process, even after several cycles (Vasileva et al., 2016).

TABLE 9.3
Different β-Galactosidase Immobilization Techniques

Enzyme Origin	Immobilization Matrix	References
β-Galactosidase from *Aspergillus oryzae*	Calcium alginate spheres, Concanavalin A, zinc oxide nanoparticles	(Haider and Husain, 2009; Ansari and Husain, 2010, 2011, 2012; Husain et al., 2011)
β-galactosidase from *Kluyveromyces marxianus*	Sodium alginate gel	(Singh and Singh, 2012)
β-galactosidase from *Kluyveromyces fragilis*	Polyethersulfone membrane	(Regenhardt et al., 2013)
β-Galactosidase from *Kluyveromyces lactis*	Glutaraldehyde-activated chitosan macroparticles	(Klein et al., 2013)
β-Galactosidase from *Bacillus circulans*	Calcium alginate spheres	(Sen et al., 2014)
β-galactosidase from *Kluyveromyces marxianus*	Chitosan macroparticles	(Kokkiligadda et al., 2016)
β-Galactosidase from *Escherichia coli*	Polypropylene membrane	(Vasileva et al., 2016)
β-galactosidase from *Kluyveromyces lactis*	Calcium alginate spheres and gelatin	(Mörschbächer et al., 2016)
β-galactosidase from dairy yeast	Silicon dioxide nanoparticles	(Beniwal et al., 2018)
β-Galactosidase from *Aspergillus oryzae*	Glyoxyl agarose	(Suárez et al., 2018)
β-galactosidase from *Kluyveromyces lactis* and *Aspergillus oryzae*	Magnetic particles and amino methacrylate beads	(Todea et al., 2021)
β-galactosidase from *Kluyveromyces lactis*	Calcium alginate spheres	(Argenta et al., 2021)
β-Galactosidase from *Aspergillus oryzae*	Cotton cloth pretreated with polyethyleneimine (PEI)	(Wang et al., 2021)

Kokkiligadda et al. (2016) tested *Kluyveromyces marxianus* for the production of β-galactosidase using whey. The authors then immobilized the β-galactosidase on chitosan macroparticles to obtain maximum hydrolysis (89%) of the concentrated whey. Lastly, the β-galactosidase immobilized on chitosan and *Saccharomyces cerevisiae* immobilized on calcium alginate were used for ethanol production from whey, which was reportedly 28.9 $g.L^{-1}$. A similar study was conducted by Beniwal et al. (2018), where β-galactosidase from dairy yeast was immobilized on silicon dioxide nanoparticles for the hydrolysis of whey, with the subsequent bioconversion of the hydrolyzed whey into ethanol, using *Saccharomyces cerevisiae* and *Kluyveromyces marxianus* for the fermentation.

In the production of GOS, the enzyme β-galactosidase from *Aspergillus oryzae* was also evaluated in its free form and cotton cloth-immobilized, during research by Wang et al. (2021). The authors verified in an economic evaluation that there was almost no difference in the manufacturing cost of GOS between the free and immobilized enzyme systems, as well as for the maximum yields, which were 32.62% and 32.48%, respectively. However, the immobilized enzyme showed desirable reuse, with maintenance of the initial enzymatic activity above 63% after five consecutive enzymatic reactions.

The results of these studies demonstrate the use of the immobilized β-galactosidase is a strategy that provides biotechnological processes with high stability and presents new possibilities, such as the use of different matrices, for the immobilization enzyme from different sources. Furthermore, it shows the application of immobilized β-galactosidase for lactose hydrolysis, whey processing, GOS synthesis, and participation in ethanol production.

9.4 RECOMBINANT β-GALACTOSIDASES

The ability to recover microorganisms into pure liquid cultures allowed the production of β-galactosidases and other enzymes in large scales. However, enzyme production by wild microorganisms is generally very limited since the main metabolic functions of microbial cells in natural habitats are related to reproduction and growth. Thus, the adaptation of microorganisms to their optimal nutrient sources, temperature, humidity, and so on, became an important strategy to turn enzyme production industrially viable. Another ally in increasing microbial enzyme production is the use of genetic engineering techniques (Oliveira et al., 2007; Borrelli and Trono, 2015). The recombinant DNA technology has been widely used to generate microbial strains producing high amounts of β-galactosidase (Oliveira et al., 2011). This technique consists of transferring genes from one organism to another, i.e., the recombination of DNA from different sources. Generally, recombination involves the following steps: (1) gene identification and isolation; (2) expression vector construction; (3) transformation of the host strain; (4) screening of the best recombinant strain; (5) production optimization of the selected strain; and (6) enzyme purification. It is important to emphasize that each of these steps present specific characteristics related mainly to the host cells, as described by Olempska-Beer et al. (2006).

For a long time, industrial β-galactosidases have been obtained mainly from *Aspergillus* spp. and *Kluyveromyces* spp. because these microorganisms showed acceptable productivity and yield. In addition, products from these genera are classified as safe (GRAS status) for human consumption (Panesar et al., 2006; Kosseva et al., 2009; Husain, 2010). Another advantage of the β-galactosidase produced by *Aspergillus niger* is that this fungus produces enzymes extracellularly, which facilitates and reduces downstream costs. On the other hand, the β-galactosidase produced by *Kluyveromyces lactis* is intracellular, which limits industrial production due to the costs associated with the purification process. However, the yeast *K. lactis* usually synthesizes more enzyme units compared to *A. niger* (Oliveira et al., 2011). Therefore, the first recombinant β-galactosidases were produced from the gene sequences obtained from these two genera, and the yeast *Saccharomyces cerevisiae* was used as the host cell, as shown in Table 9.4.

Kumar et al. (1992) constructed a vector (plasmid pVK1.1) containing the *lacA* gene (coding for *A. niger* β-galactosidase) under the control of the *ADH1* promoter. This vector was used in the

TABLE 9.4
β-Galactosidase from Different Microbial Origins

Microbial Source of β-Galactosidase	Vector/Plasmid	Expression Host	Reference
Aspergillus niger	pVK1.1	*S. cerevisiae* GRF167	(Ramakrishnan and Hartley, 1993)
Aspergillus niger	pLD1	*S. cerevisiae W204-FLO1L*	(Domingues et al., 2000a
Aspergillus niger	pVK1.1	*S. cerevisiae* NCYC869-A3	(Domingues et al., 2002)
Aspergillus niger	pVK1.1	*S. cerevisiae* NCYC869-A3	(Domingues et al., 2004)
Aspergillus niger	pVK1.1	*S. cerevisiae NCYC869-A3*	(Domingues et al., 2005)
Aspergillus niger	pCO1	*S. cerevisiae* NCYC 869	(Oliveira et al., 2007)
Kluyveromyces lactis	YEplac118/LAC4	*S. cerevisiae* LD1	(Becerra et al., 1997)
Kluyveromyces lactis	YEplac118/LAC4	*S. cerevisiae* LHDP1	(Becerra et al., 2001a)
Kluyveromyces lactis	pSPGK1/LAC4	*K. lactis* MW 190-9B	(Becerra, et al., 2001b)
Kluyveromyces lactis	YEpFLAG1/LAC4	S. cerevisiae BJ3505	(Beccerra et al., 2002)
Kluyveromyces lactis	YEplac181-*LAC4*	*S. cerevisiae* LD1	(Becerra et al., 2004)
Kluyveromyces lactis	pSPGK1-LAC4	*K. lactis* MW 190-9B	(Rodríguez et al., 2006)
Pyrococcuswoesei	pET15bb-Ga pET30b-Gal pUET1b-Gal	*E. coli* TOP10F9	(Daabrowski et al., 2000)
Rhizobium meliloti	pKK223–3	*E. coli* JM109	(Leahy et al., 2001)
Lactobacillus reuteri	pHA1031	*E. coli* BL21	(Nguyen et al., 2007)
Alicyclobacillus acidocaldarius	pPIC9	*P. pastoris* GS115	(Yuan et al., 2008)
Lactobacillus reuteri L103 *Lactobacillus acidophilus* R22 *Lactobacillus plantarum* WCFS1 *Lactobacillus sakei* Lb790	pSIP403 and pSIP409	*L. plantarum* WCFS1 and *L. sakei* Lb790	(Halbmayr et al., 2008)
Paenibacillusthiaminolyticus	pET16b-βgal-2M	*E. coli BL21*	(Benešová et al., 2009)

construction of several recombinant *S. cerevisiae* strains (Kumar et al., 1992; Ramakrishnan and Hartley, 1993; Domingues et al., 2000a, 2002; Oliveira et al., 2007). Interestingly, Domingues et al. (2002) used a yeast strain with flocculant properties and the pVK1.1 vector containing the *lacA* gene to construct a new phenotype strain. This approach allowed a new method of extracellular β-galactosidase production in a continuous fermentation system with high cell density and productivity. Flocculating yeasts are widely used for ethanol production and their growth occurs in clusters, forming flakes or agglomerates of visible sizes. It facilitates cells separation and, consequently, their reuse by decanting process. In addition, the use of flocculating cells makes the enzyme purification process simpler and less aggressive, since the concentration of cells to be separated from the culture medium by centrifugation is reduced (Domingues et al., 2000b).

Domingues et al. (2005) evaluated the production of β-galactosidase by the recombinant flocculent *S. cerevisiae* strain NCYC869-A3/pVK1.1 in a continuous fermentation process using lactose as a carbon source. The best result for enzyme production (6.2×10^5 U/1 h) was obtained when the fermentation was carried out at 0.24 h^{-1} dilution rate and a 50 g/L feed lactose concentration. Furthermore, the results also demonstrated that higher dilution rates favour an increase in biomass and enzyme activity. However, this increase in extracellular β-galactosidase activity ended up generating a higher rate of hydrolysis and accumulation of lactose into glucose and galactose in the culture medium. High concentrations of simple sugars can repress the production of β-galactosidase and induce the growth of cells that lost the vector responsible for the production of the recombinant enzyme (Domingues et al., 2005).

An alternative to improve the genetic stability of recombinant strains is to integrate multiple copies of the gene encoding β-galactosidase into the host cell genome (Oliveira et al., 2011). Studies suggest that some repeated regions in the yeast genome can be used for vector insertion by a process of homologous recombination (Sakai et al., 1990; Lopes et al., 1991). These regions include ribosomal DNA and δ sequences. For instance, more than 100 δ-sequences flanking the Tyl retrotransposons have been identified in the *S. cerevisiae* genome (Wyrick et al., 2001). However, some factors, such as the size and stability of the vector, can influence the number of copies to be inserted into the genome. For example, Cho et al. (1999) were able to integrate about 44 copies of the gene of interest into the *S. cerevisiae* genome, while Sakai et al. (1990) inserted only 3 copies. This low integration rate was associated with possible instability of the vector (Sakai et al., 1990).

Oliveira et al. (2007) constructed an integrative vector (pCO1) for the δ-region of *S. cerevisiae* NCYC869-wt containing β-galactosidase gene from *A. niger*. The results showed that, after transformation, four clones with about seven copies of the *lacA* gene were obtained. After screening, the authors selected a recombinant strain to evaluate the productivity in a continuous fermenter. As a control, the authors used *S. cerevisiae* NCYC869-A3/pVK1.1 built by (Domingues et al., 2005). Under all conditions evaluated, *S. cerevisiae* NCYC869-wt/pCO1 showed higher β-galactosidase activity than the control, suggesting that the new recombinant yeast exhibits greater genetic stability.

β-galactosidase from *K. lactis* (encoded by the *Lac4* gene) is another important enzyme used in the food industry. Because it is an intracellular protein, some strategies have been used to generate recombinant cells capable of secreting the enzyme into the medium (Oliveira et al., 2011; Movahedpour et al., 2021). An interesting alternative is to add in the gene a sequence that encodes the signal peptide for secretion of the recombinant enzyme. However, some factors, such as enzyme size, three-dimensional structure, isoelectric point, culture medium composition, and cell wall structure, are determinants for this process to occur efficiently (Nobel and Barnett, 1991; Rossini et al., 1993; Rodríguez et al., 2006).

For instance, Becerra et al. (2001a) constructed a vector (YEplac181-*Lac4*) containing the secretion signal corresponding to the pre-sequence (16 amino acids) of the *K. lactis* killer toxin (α-subunit) and the *Lac4* gene. *S cerevisiae* CGY1585 (sscl^{-1}), which exhibits a super-secreting phenotype, was used as the host cell. Analysis of the recombinant strains showed that, although they were able to grow in a culture medium containing lactose as the sole carbon source, the secretion rate of the recombinant enzyme (1.7 EU mL^{-1}) was relatively low. However, it is important to note that β-galactosidase from *K. lactis* was the first and largest intracellular protein expressed by the strain CGY1585 (sscl-1). Later, Becerra et al. (2004) also used the YEplac181-LAC4 vector and two mutant *S. cerevisiae* strains (LD1 and LHDP1) as the host cell. Due to mutations in genes responsible for cell wall formation, these strains exhibited an osmotic-corrective thermosensitive autolytic phenotype. This feature allows cell lysing to be carried out through a thermal process at temperatures higher than 37°C. Thus, intracellular enzymes can be easily recovered from the culture medium. The results showed that strain LHDP1 had the highest growth rate and production of β-galactosidase, with the release of the enzyme into the extracellular medium reaching about 51% (Becerra et al., 2004).

Bacteria, including *Lactobacillus reuteri, Rhizobium meliloti, Pyrococcus woesei, Lactobacillus sakei, Paenibacillus thiaminolyticus, Thermotogamarima, Caldicellulosiruptor saccharolyticus,* and *Thermotoga maritima*, have also been widely studied as vectors of recombinant β-galactosidases (Table 9.4). Interestingly, β-galactosidases from extremophilic bacteria, such as *T. maritima, C. saccharolyticus*, and *T. maritima* (Li et al., 2009; Schmidt and Stougaard, 2010; Park and Oh, 2010), have shown great biotechnological potential for being thermostable, having low product inhibition, and exhibiting longer half-lives when compared to enzymes obtained from mesophilic microorganisms (Oliveira et al., 2011).

Pichia pastoris is an important host of bacterial enzymes because the products from this yeast have GRAS status. This yeast also produces a low amount of native extracellular proteins, which facilitates and reduces the cost of the purification process (Cregg et al., 2000). However, it is important to note that the expression of bacterial genes in *P. pastoris* may be low due to codon bias since the cDNA encoding the enzyme of interest contains codons corresponding to tRNAs that are not abundant in the host cell (Gustafsson et al., 2004). Optimizing the codons of the gene of interest is one strategy to improve the expression of the recombinant enzyme. Sequence analysis of the gene (*bgal*) encoding β-galactosidases from *Planococcus* sp-L4 performed by Mahdian et al. (2016) showed that several codons were not recommended for expression in *P. pastoris*. For instance, about 81% of the codons corresponding to Proline in the bgal sequence are CCG, while the tRNA of this codon in *P. pastoris* is approximately 10%. Generally, to obtain higher expression rates, the tRNA corresponding to a given codon should be higher than 30%. Thus, the authors use bioinformatics tools to substitute the CCG codon for CCT and CCA which represent about 42% and 31%, respectively, of the tRNA present in *P. pastoris* (Mahdian et al., 2016).

9.5 APPLICATIONS OF β-GALACTOSIDASES

Due to β-galactosidase's hydrolytic and transgalactosylation activities (Abbasi and Saeedabadian, 2015; Thakur et al., 2022), its hydrolytic potential is largely exploited by the dairy industry to produce compounds with low or no lactose content. Other studies have also demonstrated its potential for producing GOS through transgalactosylation activity (Montalto et al., 2005; Moloughney, 2014).

According to Vera et al. (2020) β-galactosidase applications can be classified as conventional and non-conventional. Conventional applications are related to its hydrolytic activity, which have been used for a long time by the industry. Non-conventional applications involve new applications based on their transgalactosylation activity. To clarify this classification, it is important to understand the catalytic mechanism of these enzymes. β-galactosidases have a catalytic mechanism with retention of the β-anomeric configuration of the substrate (CAZy, 2019), which is relevant in the synthesis of glycosides and oligosaccharides by transgalactosylation. In fact, the β-glycosidic bonds in prebiotic compounds, such as GOS, fGOS, and lactulose, synthetized by β-galactosidases helps to prevent their hydrolysis by the animal digestive enzymes in the gastrointestinal tract (Carlson et al., 2018; Vera et al., 2020). β-galactosidases catalyse both transgalactosylation and hydrolysis reactions at different proportions that depend on reaction conditions, which are strongly influenced by lactose concentration. Usually, at lactose concentrations lower than 300 mM and high water activities (Aw), hydrolytic reactions prevail for most of the β-galactosidases. On the other hand, transgalactosylation dominates at concentrations higher than 980 mM and lower Aw. In this case, significant fractions of lactose act as nucleophile acceptors producing galactosyllactose.

The main conventional applications of β-galactosidases are linked to lactose hydrolysis for the production of low-lactose and lactose-free milk and dairy products, and lactase tablets for lactose-intolerant people. The production of whey syrups, to ameliorate the sensory properties and bioavailability of whey-derived products, and in dairy waste management (Mlichová and Rosenberg, 2006) are also reported.

Non-conventional applications of β-galactosidases involve the catalysis of transgalactosylation reactions, which occurs at high substrate concentrations (Vera et al., 2012), use of organic (co-) solvents (Vera et al., 2017), use of ionic liquids (Lang et al., 2006), or with low Aw media (Manera et al., 2012). One important point is that the use of non-aqueous media may inhibit the enzyme activity and reduce stability and lactose solubility. So, the solvent must be separated from the product and recovered.

Figure 9.2 presents important applications of β-galactosidases. In fact, the lack of this enzyme in the body causes problems in lactose digestion for a vast number of people who suffer from lactose intolerance, which is almost 65% of the world population (Ingram et al., 2009). Some symptoms of lactose intolerance include diarrhea, nausea, muscular spasm, swelling, borborygmi, and chronic flatulence (Leonardi et al., 2012; Misselwitz et al., 2013). That is why many industries are searching for the development of reliable and robust β-galactosidases biosensors with high response rate, increased detection limit and higher lifetime. In recent years, different strategies have been developed to immobilize β-galactosidases for their utilization in biosensors (Sharma and Leblanc,

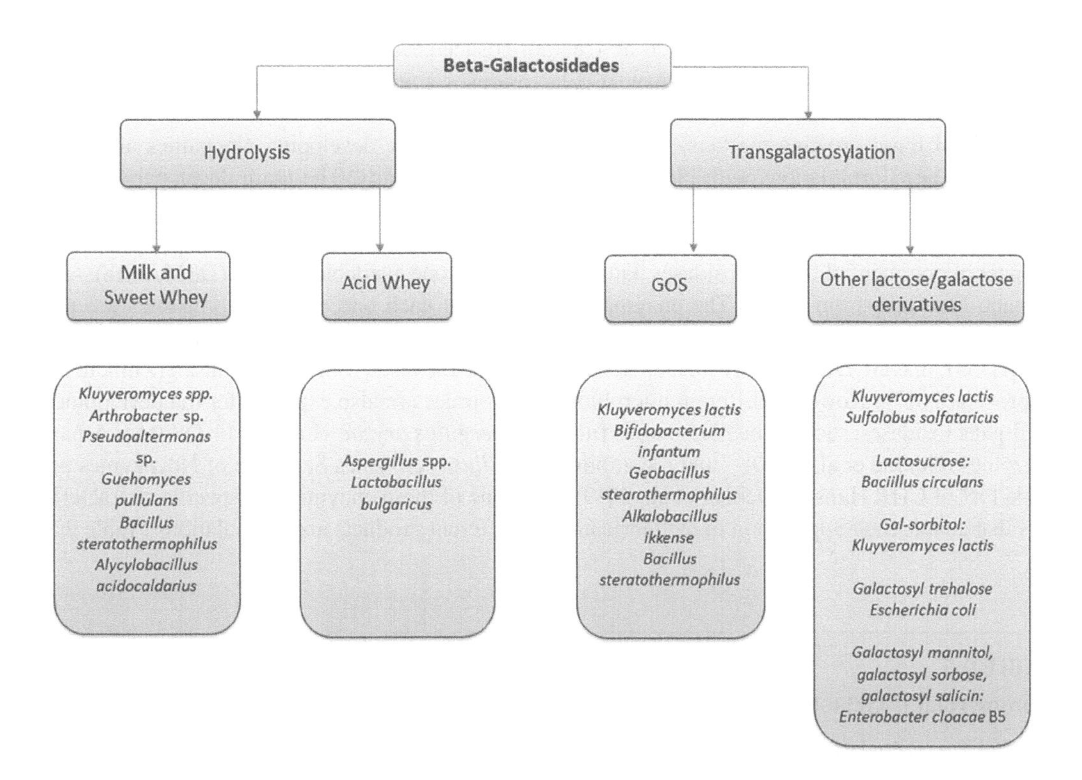

FIGURE 9.2 Applications of β-galactosidases.

2017). In the design of biosensors it is crucial to immobilize the enzyme on the transducer surface. Different benefits of enzyme immobilization can be observed such as the enhancement of enzyme properties, ease of separation, reduction of product cost, and significant increase of thermal and pH stability (Bučko et al., 2012). β-galactosidases have been largely explored in biosensors for lactose due to their high adaptability. The main enzyme immobilization techniques have already been discussed in this chapter.

9.6 COMMERCIAL PRODUCTS

As discussed throughout the chapter, β-galactosidases can be used in several industries and can be applied in the manufacture of a great diversity of products. This variety of possible applications generates the need for the development of commercial products and formulations with high purity and specificity, with characteristics that can meet the conditions of each process. Therefore, the development of new enzymes is boosted by the growing markets in which β -galactosidases can be applied, such as the markets of lactose-free products and prebiotics (GOS), but also the area of residues' processing, such as whey.

The market for lactose-free products is in constant growth, it was valued at USD 12.1 billion in 2020 and it is projected to reach USD 18.4 billion in 2025, representing a compound annual growth rate (CAGR) of 8.7% (Markets and Markets, 2020). The main drivers of growth are the rising awareness of lactose intolerance around the world and the raising concern about healthier food products, with no added sugar. Therefore, the high market demand for these goods opens an opportunity for the development of new enzymes that can reduce operation time and costs. As an example, the GOS market was valued at USD 570 million in 2021 and may reach USD 850 million by 2026, with a CAGR of 6% (Market Data Forecast, 2021). Even if the market drivers are related to the growing need and concern for healthier and nutritious food, it can be held back by competition from other prebiotics such as fructooligosaccharides, xylooligosaccharides, and isomalt. Therefore, the development of new β-galactosidases could make GOS production competitive and more advantageous when compared to other alternatives. Thereby, in order to meet the demands of these growing markets, several specialized enzyme-developing companies are acting to provide new formulations with high levels of purity and quality. The main developers of commercial β-galactosidases are Novozymes and CHR Hansen from Denmark, Danisco Dupont and Enzyme Development Company (EDC) from the USA, DSM from the Netherlands – which was the pioneer of commercial β-galactosidases, launching the first one available in 1964 (DSM, 2018) – and Amano Enzymes from Japan. The enzymes developed by each one of these companies are produced mainly by the yeasts of *Kluyveromyces*genre, such as Lactozym® from Novozymes, Maxilact® from DSM, and GODO-YLN2 from Danisco Dupont, HA-Lactase® from CHR Hansen (Zolnere and Ciprovica, 2017)). However, different microbiological sources are also explored for the development of β-galactosidases, such as the filamentous fungus *Aspergillus oryzae* –Lactase 14-DS from Amano Enzymes (Genina et al., 2010) – and recombinant *Bacillus* – enzymes Saphera® of Novozymes and Nola Fit® of CHR Hansen (Dekker et al., 2019). Each one of these enzymes has specific characteristics that enable their application in the fabrication of different products and formulations (Table 9.5).

TABLE 9.5
Commercial β-Galactosidases and Their Characteristics

Product	Producing Microorganism	β-Galactosidase Activity	Characteristics	Company
Saphera® 2600L	*Bacillus licheniformis*	2600 LAU-B/g	Can enhance sweetness in food products	Novozymes [www.novozymes.com]
Lactozym® Pure 6500L	*Kluyveromyces lactis*	6500 LAU/g	Guarantees clean taste and can be used in organic products	Novozymes [www.novozymes.com]
Lactozym® Pure Conc. G	*Kluyveromyces lactis*	>100000 LAU/g	Is suitable for use in infant formula. Comes in granulate form.	Novozymes [www.novozymes.com]
Maxilact® LGI 5000	*Kluyveromyces lactis*	5000 U/g	Can reduce hydrolysis time in 33%, ensuring higher process productivity	DSM [www.dsm.com]
GODO-YLN2	*Kluyveromyces lactis*	5000 NLU/g	Is highly purified and acts better on a neutral pH range	Danisco Dupont [www.food.dupont.com]
NOLA® Fit 5500	*Bacillus licheniformis*	5500 BLU/g	Can act on a wide range of both pH and temperatures	CHR Hansen [www.chr-hansen.com]
HA-Lactase® 5200	*Kluyveromyces lactis*	5200 NLU/g	Can act on several dairy products: milk, yoghurt, cheese and others	CHR Hansen [www.chr-hansen.com]
Enzeco® fungal lactase	*Aspergillus oryzae*	–	Stable at acidic media and higher temperatures	Enzyme Development Company [www.enzymedevelopment.com]
Enzeco® lactase NL	*Kluyveromyces* sp.	–	Can be applied in food products and GOS production	Enzyme Development Company [www.enzymedevelopment.com]
Lactase 14-DS	*Aspergillus oryzae*	14 U/mg	It is majorly used in food processing applications. Comes in powder form.	AmanoEnzyme [www.amano-enzyme.com]

9.7 PATENTS AND INNOVATION

In the bioprocess engineering and biotechnology areas, new solutions with high technology and innovation are constantly being developed in order to meet the needs of the new bio-based economies and society (Kardung et al., 2021). The development of new enzyme formulations is, therefore, surrounded by innovation, with new production and purification processes and recombinant microorganism strains or polypeptides that can enhance an enzyme with higher activity or new properties. One of the ways of measuring the innovation in biotechnological processes development is through the search and analysis of patent data since companies, universities, and inventors are prone to protect their technologies with this system. Therefore, in order to reveal the innovation and patent scenery involving new β-galactosidases production processes and applications, a patent search and analysis was conducted.

The selected terms for searching in the patent databases were β-galactosidase and its variations, lactases, production and immobilization. The International Patent Codes (IPCs) selected for the search were C12N-009 (refers to enzymes in general), C12P (refers to fermentation and enzyme-using processes), and C13 (refers to the sugar industry). Terms involving systems, apparatus, and devices were removed from the search, in order to reduce the utility model residues that could appear. Terms involving immunology and diagnosis were also removed since the focus of this chapter is the microbial production of β-galactosidases and their direct application in industries. The search was made in the Derwent Innovation Index (DII) database using Boolean language for the combination of terms and (*) for truncation of terms when necessary. The search was conducted relating to the last 10 years (since 2011) and 574 results were obtained in January 2022. After manual revision and sorting of the documents, a total of 301 documents were analysed regarding their region and country of origin, main applicants, IPCs and author-elaborated classifications, and year of application, as follows.

The countries that have the highest number of documents applied are directly related to the location of the companies that supply commercial β-galactosidases (Figure 9.3). Denmark holds the headquarters of Novozymes (major applicant of patent documents in this analysis, with 24 documents) and CHR Hansen (with 7 documents applied). The USA, on the other hand, has Dupont in

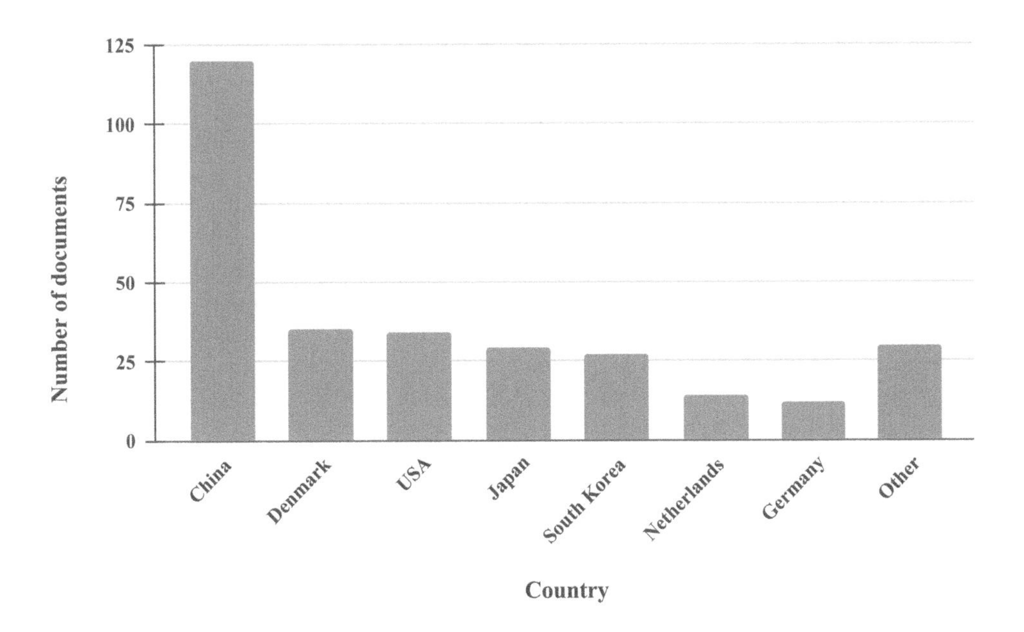

FIGURE 9.3 Number of applied patent documents about production and application of β-galactosidases by country.

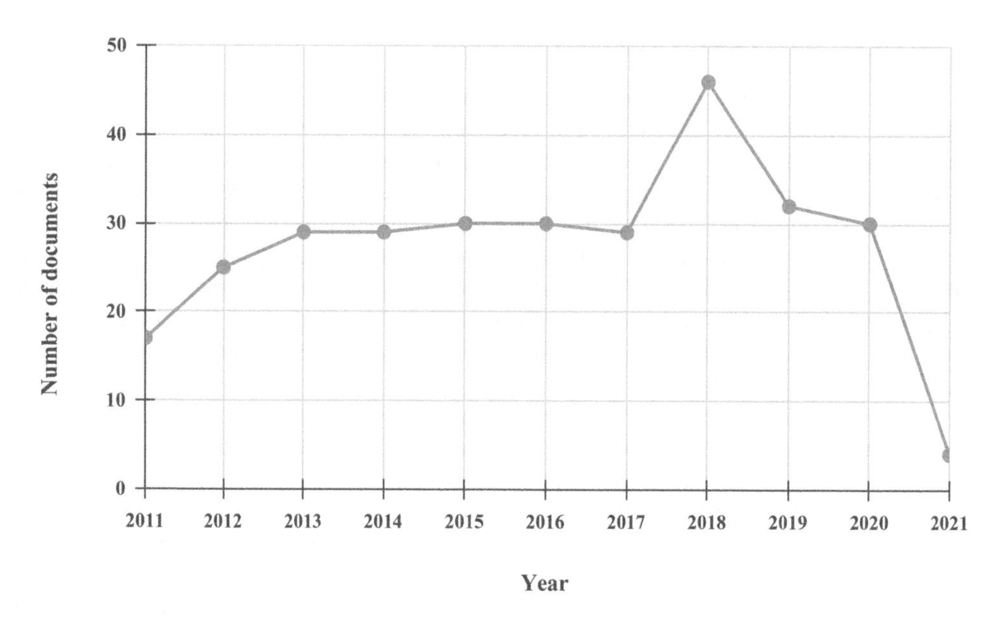

FIGURE 9.4 Evolution on the number of applied patent documents about production and application of β-galactosidases since 2011.

its territory, a company that has 10 applied documents. Lastly, the Dutch enterprise DSM holds 11 documents, contributing to the Netherlands among the 10 largest applicants. However, China – the country that has more applied patent documents – does not hold any enzyme-specialized enterprise in its territory, but is a country that has great governmental incentives for research and development of new products and technologies, surpassing the United States in 2019, becoming the largest international patent filler (WIPO, 2020). Besides, the Chinese University of Jiangnan appears together with CHR Hansen in fifth place among the major applicants with 7 documents.

Regarding world regions, Asia has the majority of applied documents (61%) on β-galactosidases, followed by Europe with 25%, and North America with 11%. Besides large governmental funding of R&D, the Asiatic region is the one that has a large prevalence of lactose intolerant people (about 64% of its population, with a wide range among countries). Therefore, new food products that don't contain this sugar in their formulas are being constantly developed for meeting the needs of the population. Furthermore, the Asiatic region is known for the production and development of several fermented products such as fermented milk (Yakult also appears as one of the major patent applicants in the analysis), *kimchi* (chard fermented with lactic acid bacteria) (Noh et al., 2016), *miso* (soybean and rice paste fermented with *Aspergillus oryzae*) (Nout, 2015) among others.

Over the last years, β-galactosidase-related patents have been following a constant line with few variations (Figure 9.4). This may reveal that the technological development of β-galactosidase is in its mature phase, with consolidated and punctual progress (Zartha et al., 2016). Only in the beginning of the studied period, and in 2018, is it possible to observe a positive variation. This could have happened due to the release to the market of Maxilact Smart by DSM in 2018, which is a β-galactosidase that can reduce the hydrolysis time of lactose at 6°C by 33% when compared to Maxilact LGI 5000 from the same company (DSM, 2018). The period 2020–2021 is still in the 1.5-year period of secrecy granted by patent offices and applied patents in this period don't appear in the search; therefore, the reduction in the number of applied documents does not represent a lack of interest of the market. Besides, it is natural to expect a decline in the number of applied documents in 2020 due to the Sars-Cov2 pandemic.

Concerning classifications, the most recurrent IPC were C12N-009/38 (on 46% of the documents, referring to enzymes that act on β-galactose-glycosidic bonds, its compositions or production or

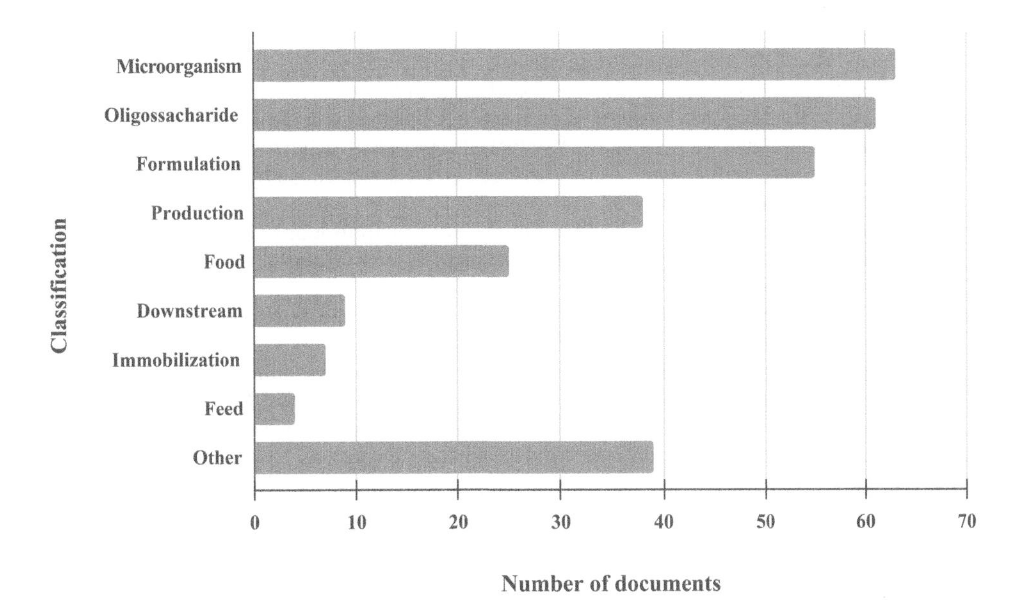

FIGURE 9.5 Number of applied patent documents about production and application of β-galactosidases by classification based on what technology the patent protects.

purification processes), C12P-019/14 (on 24% of the documents, referring to saccharides prepared by carbohydrases), and A23C-009/12 (on about 15% of the documents, relating to fermented milk preparations using microorganisms or enzymes in its production processes). These codes represent well the production and application of β-galactosidases, reiterating that new enzyme formulations development is also boosted by its potential applications. Besides, the IPC codes are also directly related to the classification elaborated by the authors (Figure 9.5). It is possible to observe that the majority of documents refer to new microorganism strains that can overproduce β-galactosidases or express this enzyme with a different configuration. These new strains are usually obtained with genetic engineering of microorganisms, mainly bacteria. Another great share of documents covers the production of oligosaccharides applying β-galactosidases, which is a rising market as previously discussed. Other patents discuss and protect new formulations and production, purification, and immobilization processes for β-galactosidases.

9.8 CONCLUSIONS

β-galactosidases have been produced and applied in different and well-established processes. Their hydrolytic potential and transgalactosylation activities make this important group of enzymes largely exploited not only by the dairy industry to produce compounds with low or no lactose content, but also to produce GOS through transgalactosylation activity. It is also possible to observe with this analysis that the development of processes involving production and application of β-galactosidases is surrounded by innovation and high technology. So, high quality new formulations, production and purification processes are expected with lower time and costs, microorganisms that can overproduce β-galactosidases and also new processes for the development of different lactose-free products and GOS.

REFERENCES

Abbasi, S., and Saeedabadian, A. 2015. Influences of lactose hydrolysis of milk and sugar reduction on some physical properties of ice cream. *Journal of Food Science and Technology* 52: 367–74.

Albuquerque, T. L., de Souza, M., Silva, N. C. G., Neto, C. A. C. G., Gonçalves, L. R. B, Fernandez-Lafuente, R., and Rocha, M. V. P. 2021. β-Galactosidase from *Kluyveromyces lactis*: characterization, production, immobilization and applications – A review. *International Journal of Biological Macromolecules* 191: 881–98.

Alikunju, A. P., Joy, S., Rahiman, M., Rosmine, E., Antony, A. C., Solomon, S., Manjusha, K., Saramma, A. V., Krishnan, K. P., and Hatha, A. A. M. 2018. A statistical approach to optimize cold active β-Galactosidase production by an arctic sediment pscychrotrophic bacteria, enterobacter ludwigii (MCC 3423) in cheese whey. *Catalysis Letters* 148(2): 712–24.

Al-Jazairi, M., Abou-Ghorra, S., Bakri, Y., and Mustafa, M. 2015. Optimization of β-galactosidase production by response surface methodology using locally isolated Kluyveromycesmarxianus. *International Food Research Journal* 22(4): 1361–67.

Am-Aiam, S., and Khanongnuch, C. 2015. Medium optimization for beta-galactosidase production by a thermotolerant yeast. *Chiang Mai Journal of Science* 42(4): 840–49.

Ansari, S. A., and Husain, Q. 2010. Lactose hydrolysis by β galactosidase immobilized on concanavalin A-cellulose in batch and continuous mode. *Journal of Molecular Catalysis B: Enzymatic* 63: 68–74.

Ansari, S. A., and Husain, Q. 2011. Immobilization of *Aspergillus oryzae* β galactosidase on concanavalin A-layered calcium alginate-cellulose beads and its application in lactose hydrolysis in continuous spiral bed reactors. *Polish Journal of Chemistry Technology* 13: 15–20.

Ansari, S. A., and Husain, Q. 2012. Lactose hydrolysis from milk/whey in batch and continuous processes by concanavalin A-Celite 545 immobilized *Aspergillus oryzae* β galactosidase. *Food and Bioproducts Processing* 90(2): 351–59.

Ansari, S. A., and Satar, R. 2012. Recombinant β-galactosidases – Past, present and future: A mini review. *Journal of Molecular Catalysis B: Enzymatic* 81: 1–6.

Argenta, A. B., Nogueira, A., and de P. Scheer, A. 2021. Hydrolysis of whey lactose: Kluyveromyces lactis β-galactosidase immobilisation and integrated process hydrolysis-ultrafiltration. *International Dairy Journal* 117: 105007.

Aulitto, M., Strazzulli, A., Sansone, F., Cozzolino, F., Monti, M., Moracci, M., Fiorentino, G., Limauro, D., Bartolucci, S., and Contursi, P. 2021. Prebiotic properties of *Bacillus coagulans* MA-13: Production of galactoside hydrolyzing enzymes and characterization of the transglycosylation properties of a GH42 β-galactosidase. *Microbial Cell Factories* 20(1): 1–17.

Becerra, M., Cerdán, E., and González Siso, M. I. 1997. Heterologous Kluyveromyces lactis β-galactosidase production and release by *Saccharomyces cerevisiae* osmotic-remedial thermosensitive autolytic mutants. *Biochimica et Biophysica Acta – General Subjects* 1335(3): 235–41.

Becerra, M., Díaz Prado, S., Cerdán, E., and González Siso, M. I. 2001a. Heterologous Kluyveromyces lactis β-galactosidase secretion by *Saccharomyces cerevisiae* super-secreting mutants. *Biotechnology Letters* 23(1): 33–40.

Beccerra, M., Díaz Prado, S., Rodríguez-Belmonte, E., Cerdán, M. E., and González Siso, M. I. 2002. Metabolic engineering for direct lactose utilization by Saccharomyces cerevisiae. *Biotechnology Letters* 24(17): 1391–96.

Becerra, M., Prado, S. D., Siso, M. I. G., and Cerdán, M. E. 2001b. New secretory strategies for *Kluyveromyces lactis* β-galactosidase. *ProteinEngineering* 14(5): 379–86.

Becerra, M., Rodríguez-Belmonte, E., Cerdán, M. E., and González Siso, M. I. (2004). Engineered autolytic yeast strains secreting *Kluyveromyces lactis* β-galactosidase for production of heterologous proteins in lactose media. *Journal of Biotechnology* 109(1–2): 131–37.

Benešová, E., Lipovová, P., Dvořáková, H., and Králová, B. 2009. β-D-Galactosidase from *Paenibacillusthiaminolyticus* catalyzing transfucosylation reactions. *Glycobiology* 20(4): 442–51.

Beniwal, A., Saini, P., Kokkiligadda, A., and Vij, S. 2018. Use of silicon dioxide nanoparticles for β-galactosidase immobilization and modulated ethanol production by co-immobilized *Kluyveromycesmarxianus* and *Saccharomyces cerevisiae* in deproteinized cheese whey. *LWT – Food Science and Technology* 87: 553–61.

Borrelli, G. M., and Trono, D. 2015. Recombinant lipases and phospholipases and their use as biocatalysts for industrial applications. *International Journal of Molecular Sciences* 16(9): 20774–840.

Boudjema, K., Fazouane-Naimi, F., Güven, K., Bekler, F. M., Acer, O., and Hellal, A. 2016. Production of intracellular β-galactosidase using a novel *Kluyveromycesmarxianus* DIV13-247 isolated from an Algerian dairy product. *Research Journal of Biotechnology* 11(6): 35–43.

Buchholz, K., and Klein, J. 1987. Characterization of immobilized biocatalysts. *Academic* 135: 3–30.

Bučko, M. Mislovičová, D., Nahálka, J., Vikartovská, A., Šefčovičová, J., Katrlík, J., JánTkáč, Gemeiner, P., Lacík, I., Štefuca, V., Polakovič, M., Rosenberg, M., Rebroš, M., Šmogrovičová, D., and Švitel, J. 2012. Immobilization in biotechnology and biorecognition: from macro-to nanoscale systems. *Chemical Papers* 66: 983–98.

Cardoso, B. B., Silvério, S. C., Abrunhosa, L., Teixeira, J. A., and Rodrigues, L. R. 2017. β-galactosidase from *Aspergillus lacticoffeatus*: A promising biocatalyst for the synthesis of novel prebiotics. *International Journal of Food Microbiology* 257: 67–74.

Carlson, J. L., Erickson, J. M., Lloyd, B. B., and Slavin, J. L. 2018. Health effects and sources of prebiotic dietary fiber. *Current Developments in Nutrition* 2: 1–8.

CAZy. 2019. Search Results for β-Galactosidase. http://www.cazy.org/search?page=recherche&recherche=3.2.1.23&tag=9 (accessed April 25, 2019).

Cho, K. M., Yoo, Y. J., and Kang, H. S. 1999. δ-Integration of endo/exo-glucanase and β-glucosidase genes into the yeast chromosomes for direct conversion of cellulose to ethanol. *Enzyme and Microbial Technology* 25(1–2): 23–30.

Cregg, J. M., Cereghino, J. L., Shi, J., and Higgins, D. R. 2000. Recombinant protein expression in *Pichia pastoris. Molecular Biotechnology* 16: 23–52.

Daabrowski, S. S., Sobiewska, G., Maciuńska, J., Synowiecki, J., and Kur, J. 2000. Cloning, expression, and purification of the His6-tagged thermostable β-galactosidase from Pyrococcuswoesei in Escherichia coli and some properties of the isolated enzyme. *Protein Expression and Purification* 19(1): 107–12.

Daavu, N. R., and Potireddy, P. S. D. 2020. *Beta galactosidase production from marine bacterial isolate* (Patent No. IN201941010345). Office of the Controller General of Patents, Designs and Trade Marks Department for Promotion of Industry and Internal Trade Ministry of Commerce and Industry. https://patentscope.wipo.int/search/pt/detail.jsf?docId=IN306692012&_cid=P12-KY03ND-00110-1 (accessed January 6, 2022).

Das, B., Roy, A. P., Bhattacharjee, S., Chakraborty, S., and Bhattacharjee, C., 2015. Lactose hydrolysis by β-galactosidase enzyme: Optimization using response surface methodology. *Ecotoxicology Environmental Safety* 121: 244–52.

DeCastro, M.-E., Escuder-Rodriguez, J.-J., Cerdan, M.-E., Becerra, M., Rodriguez-Belmonte, E., and Gonzalez-Siso, M.-I. 2018. Heat-loving β-Galactosidases from cultured and uncultured microorganisms. *Current Protein & Peptide Science* 19(12): 1224–34.

Dekker, P., Koenders, D., and Bruins, M. 2019. Lactose-free dairy products: Market developments, production, nutrition and health benefits. *Nutrients* 11(3): 551.

Domingues, L., Lima, N., and Teixeira, J. A. (2005). *Aspergillus niger* β-galactosidase production by yeast in a continuous high cell density reactor. *Process Biochemistry* 40(3–4): 1151–4.

Domingues, L., Oliveira, C., Castro, I., Lima, N., and Teixeira, J. A. 2004. Production of β-galactosidase from recombinant *Saccharomyces cerevisiae* grown on lactose. *Journal of Chemical Technology and Biotechnology* 79(8): 809–15.

Domingues, L., Onnela, M. L., Teixera, J. A., Lima, N., and Penttilä, M. 2000a. Construction of a flocculent brewer's yeast strain secreting *Aspergillus niger* β-galactosidase. *Applied Microbiology and Biotechnology* 54(1): 97–103.

Domingues, L., Teixeira, J., Penttilä, M., and Lima, N. 2002. Construction of a flocculent *Saccharomyces cerevisiae* strain secreting high levels of *Aspergillus niger* β-galactosidase. *Applied Microbiology and Biotechnology* 58(5): 645–50.

Domingues, L., Vicente, A. A., Lima, N., and Teixeira, J. A. 2000b. Applications of yeast flocculation in biotechnological processes. *Biotechnology and Bioprocess Engineering* 5(4): 288–305.

DSM. 2018. DSM launches Maxilact® Smart, a lactase enzyme setting the pace for production efficiency in lactose-free dairy | DSM Food Specialties. ([s.d.]). @dfs. Retrieved January 6, 2022, from https://www.dsm.com/food-specialties/en_US/insights/dairy/2018-03-15-dsm-launches-maxilact-smart-a-lactase-enzyme-setting-the-pace-for-production-efficiency-in-lactose-free-dairy.html.

El-Gindy, A., Ibrahim, Z., and Aziz, H. 2009. Improvement of extracellular β-galactosidase production by thermophilic fungi *Chaetomium thermophile* and *Thermomyces lanuginosus. Australian Journal of Basic and Applied Sciences* 3(3): 1925–32.

Genina, N., Räikkönen, H., Heinämäki, J., Veski, P., and Yliruusi, J. 2010. Nano-coating of β-galactosidase onto the surface of lactose by using an ultrasound-assisted technique. *AAPS Pharmaceutical Science and Technology* 11(2): 959–65.

González-Cataño, F., Tovar-Castro, L., Casta, E., Regalado-Gonzalez, C., Garc, B., and Amaya-Llano, S. 2017. Improvement of covalent immobilization procedure of β-Galactosidase from *Kluyveromyces lactis* for galactooligosaccharides production: Modeling andkinetic study. *Biotechnology Progress* 33: 1568–78.

Grosová, Z., Rosenberg, M., and Rebroš, M. 2008. Perspectives and applications of immobilised β-galactosidase in food industry – A review. *Czech Journal of Food Science* 26: 1–14.

Guisan, J. M. 2013. Immobilization of Enzymes and Cells, *Immobilization of Enzymes: A Literature Survey Beatriz.* Totowa, NJ: Humana.

Gustafsson, C., Govindarajan, S., and Minshull, J. 2004. Codon bias and heterologous protein expression. *Trends in Biotechnology* 22(7): 346–53.

Haider, T., and Husain, Q. 2009. Hydrolysis of milk/whey lactose by β-galactosidase: A comparative study of stirred batch process and packed bed reactor prepared with calcium alginate entrapped enzyme. *Chemical Engineering and Processing. Process Intensification* 48: 576–80.

Halbmayr, E., Mathiesen, G., Nguyen, T. H., Maischberger, T., Peterbauer, C. K., Eijsink, V. G. H., and Haltrich, D. 2008. High-level expression of recombinant β-galactosidases in Lactobacillus plantarum and Lactobacillus sakei using a sakacin p-based expression system. *Journal of Agricultural and Food Chemistry* 56(12): 4710–19.

Husain, Q. 2010. β Galactosidases and their potential applications: A review. *Critical Reviews in Biotechnology* 30(1): 41–62.

Husain, Q., Ansari, S. A., Alam, F., and Azam, A. 2011. Immobilization of *Aspergillus oryzae* β galactosidase on zinc oxide nanoparticles via simple adsorption mechanism. *International Journal of Biological Macromolecules* 49: 37–43.

Ingram, C. J., Mulcare, C. A., Itan, Y., Thomas, M. G., and Swallow, D. M. 2009. Lactose digestion and the evolutionary genetics of lactase persistence. *Human Genetics* 124: 579–91.

Jensen, T. Ø., Pogrebnyakov, I., Falkenberg, K. B., Redl, S., and Nielsen, A. T. 2017. Application of the thermostable β-galactosidase, BgaB, from *Geobacillus stearothermophilus* as a versatile reporter under anaerobic and aerobic conditions. *AMB Express* 7(1): 169.

Kamran, A., Bibi, Z., Aman, A., and Qader, S. A. U. 2016. Lactose hydrolysis approach: Isolation and production of β-galactosidase from newly isolated *Bacillus strain* B-2. *Biocatalysis and Agricultural Biotechnology* 5: 99–103.

Karasova, P., Spiwok, V., Mala, S., Kralova, B., and Russell, N. J. 2002. β-Galactosidase activity in psychrophic microorganisms and their potential use in food industry. *Czech Journal of Food Science* 20: 43–47.

Kardung, M., Cingiz, K., Costenoble, O., Delahaye, R., Heijman, W., Lovrić, M., van Leeuwen, M., M'Barek, R., van Meijl, H., Piotrowski, S., Ronzon, T., Sauer, J., Verhoog, D., Verkerk, P. J., Vrachioli, M., Wesseler, J. H. H., and Zhu, B. X. 2021. Development of the circular bioeconomy: Drivers and indicators. *Sustainability* 13(1): 413.

Katrolia, P., Min, Z., Qiaojuan, Y., Zhengqiang, J., Chunlei, S., and Lite, L. 2011. Characterisation of a thermostable family 42 β-galactosidase (BgalC) family from *Thermotoga maritima* showing efficient lactose hydrolysis. *Food Chemistry* 125: 614–21.

Kazemi, S., Khayati, G., and Faezi-Ghasemi, M. 2016. β-galactosidase production by *Aspergillus niger* ATCC 9142 using inexpensive substrates in solid-state fermentation: Optimization by orthogonal arrays design. *Iranian Biomedical Journal* 20(5): 287–94.

Khan, A., and Alzohairy, A. 2010. Recent advances and applications of immobilized enzyme technologies: a review. *Research Journal of Biological Sciences* 5: 565–75.

Klein, M. P., Fallavena, L. P., Schöffer, J. D. N., Ayub, M. A. Z., Rodrigues, R. C., Ninow, J. L., and Hertz, P. F. 2013. High stability of immobilized β-D-galactosidase for lactose hydrolysis and galactooligosaccharides synthesis. *Carbohydrate Polymers* 95: 465–70.

Kokkiligadda, A., Beniwal, A., Saini, P., and Vij, S. 2016. Utilization of cheese whey using synergistic immobilization of β-galactosidase and saccharomyces cerevisiae cells in dual matrices. *Applied Biochemistry and Biotechnology* 179: 1469–84.

Kosseva, M. R., Panesar, P. S., Kaur, G., and Kennedy, J. F. 2009. Use of immobilised biocatalysts in the processing of cheese whey. *International Journal of Biological Macromolecules* 45(5): 437–47.

Kumar, V., Ramakrishnan, S., Teeri, T. T., Knowles, J. K. C., and Hartley, B. S. 1992. Saccharomyces cerevisiae cells secreting an *Aspergillus niger* β-galactosidase grow on whey permeate. *Nature Biotechnology* 10: 667–74.

Kumar, V., Sharma, N., and Bhalla, T. C. 2014. In silico analysis of β -galactosidases primary and secondary structure in relation to temperature adaptation. *Journal of Amino Acids*: Article ID 475839.

Kumari, M., Padhi, S., and Sharma, S. et al. 2021. Biotechnological potential of psychrophilic microorganisms as the source of cold-active enzymes in food processing applications. *3 Biotech* 11: 479.

Markets and Markets. 2020. Lactose-free products market growth, analysis | industry insights & statistics | marketsandmarkets. ([s.d.]). https://www.marketsandmarkets.com/Market-Reports/lactose-free-products-market-4457397.html (accessed January 6, 2022).

Lang, M., Kamrat, T., and Nidetzky, B. 2006. Influence of ionic liquid cosolvent on transgalactosylation reactions catalyzed by thermostable β-glycosylhydrolaseCelB from *Pyrococcusfuriosus. Biotechnology and Bioengineering* 95: 1093–100.

Leahy, M., Vaughan, P., Fanning, L., Fanning, S., and Sheehan, D. 2001. Purification and some characteristics of a recombinant dimeric *Rhizobium meliloti* β-galactosidase expressed in *Escherichia coli*. *Enzyme and Microbial Technology* 28(7–8): 682–88.

Leonardi, M., Gerbault, P., Thomas, M. G., and Burger, J. 2012. The evolution of lactase persistence in Europe. A synthesis of archaeological and genetic evidence. *International Dairy Journal* 22: 88–97.

Li, L., Zhang, M., Jiang, Z., Tang, L., and Cong, Q. 2009. Characterisation of a thermostable family 42 β-galactosidase from *Thermotoga maritima*. *Food Chemistry* 112(4): 844–50.

Liu, D. M., Chen, J., and Shi, Y. P. 2018. Advances on methods and easy separated support materials for enzymes immobilization. *TrAC – Trends in Analytical Chemistry* 102: 332–42.

Lopes, T. S., Hakkaart, G. J. A. J., Koerts, B. L., Raué, H. A., and Planta, R. J. 1991. Mechanism of high-copy-number integration of pMIRY-type vectors into the ribosomal DNA of *Saccharomyces cerevisiae*. *Gene* 105(1): 83–90.

Market Data Forecast, 2021. Galacto-oligosaccharides (Gos) market growth and forecast (2022 - 2027). Market Data Forecast. https://www.marketdataforecast.com/market-reports/galacto-oligosaccharide-market. (accessed June 4, 2022).

Machado, J. R., Behling, M. B., Braga, A. R. C., and Kalil, S. J. 2015. β-Galactosidase production using glycerol and byproducts: Whey and residual glycerin. *Biocatalysis and Biotransformation* 33(4): 208–15.

Mahdian, S. M. A., Karimi, E., Tanipour, M. H., Parizadeh, S. M. R., Ghayour-Mobarhan, M., Bazaz, M. M., and Mashkani, B. 2016. Expression of a functional cold active β-galactosidase from *Planococcus* sp-L4 in *Pichia pastoris*. *Protein Expression and Purification* 125: 19–25.

Manera, A. P., Da Costa Ores, J., Ribeiro, V. A., André, C., Burkert, V., and Kalil, S. J. 2008. Optimization of the culture medium for the production of β-galactosidase from *Kluyveromyces marxianus* CCT 7082. *Food Technology and Biotechnology* 46(1): 66–72.

Manera, A. P., Zabot, G. L., Vladimir Oliveira, J., de Oliveira, D., Mazutti, M. A., Kalil, S. J., Treichel, H., and Filho, F. M. 2012. Enzymatic synthesis of galactooligosaccharides using pressurised fluids as reaction medium. *Food Chemistry* 133: 1408–13.

Mangiagalli, M., and Lotti, M. 2021. Cold-active β-galactosidases: Insight into cold adaptation mechanisms and biotechnological exploitation. *Marine Drugs* 19(1): 43.

Marim, A. V. C., Gabardo, S., and Ayub, M. A. Z. 2021. Porungo cheese whey: β-galactosidase production, characterization and lactose hydrolysis. *Brazilian Journal of Food Technology* 24: 1–11.

Martarello, R. D., Cunha, L., Cardoso, S. L., de Freitas, M. M., Silveira, D., Fonseca-Bazzo, Y. M., Homem-de-Mello, M., Filho, E. X. F., and Magalhães, P. O. 2019. Optimization and partial purification of beta-galactosidase production by *Aspergillus niger* isolated from Brazilian soils using soybean residue. *AMB Express* 9(1): 81.

Misselwitz, B., Pohl, D., Frühauf, H., Fried, M., Vavricka, S. R., and Fox, M. 2013. Lactose malabsorption and intolerance: pathogenesis, diagnosis and treatment. *United European Gastroenterology Journal* 1: 151–59.

Mlichová, Z., and M. Rosenberg. 2006. Current trends of β-galactosidase application in food technology. *Journal of Food and Nutrition Research* 45: 47–54.

Moloughney, S. 2014. Expectations for the enzymes market. *Nutraceuticals World*: 54–59.

Montalto, M., Nucera, G., Santoro, L., Curigliano, V., Vastola, M., Covino, M., Cuoco, L., Manna, R., Gasbarrini, A., and Gasbarrini, G. 2005. Effect of exogenous beta-galactosidase in patients with lactose malabsorption and intolerance: a crossover double-blind placebo-controlled study. *European Journal of Clinical Nutrition* 59: 489–93.

Mörschbächer, A. P., Volpato, G., and Souza, C. F. V. de. 2016. *Kluyveromyces lactis* β-galactosidase immobilization in calcium alginate spheres and gelatin for hydrolysis of cheese whey lactose. *Ciência Rural* 46: 921–26.

Movahedpour, A., Ahmadi, N., Ghalamfarsa, F., Ghesmati, Z., Khalifeh, M., Maleksabet, A., Shabaninejad, Z., Taheri-Anganeh, M., and Savardashtaki, A. 2021. β-Galactosidase: From its source and applications to its recombinant form. *Biotechnology and Applied Biochemistry February* 69(2): 1–17.

Murphy, J., and Walsh, G. 2019. Purification and characterization of a novel thermophilic β-galactosidase from *Picrophilustorridus* of potential industrial application. *Extremophiles* 23(6): 783–92.

Naiyu, C., Qingfang, Z., and Aiping, C. 2012. Method for producing low-temperature beta-galactosidase through microbial fermentation (Patent No. CN102653746). China National Intellectual Property Administration. https://patentscope.wipo.int/search/pt/detail.jsf?docId=CN85285808&_cid=P12-KY03PG-00792-1 (accessed June 4, 2022).

Nguyen, H. H., and Kim, M. 2017. An overview of techniques in enzyme immobilization. *Applied Science and Convergence Technology* 26: 157–63.

Nguyen, T. H., Splechtna, B., Yamabhai, M., Haltrich, D., and Peterbauer, C. 2007. Cloning and expression of the β-galactosidase genes from *Lactobacillus reuteri* in *Escherichia coli. Journal of Biotechnology* 129(4): 581–91.

Nobel, J. G. de, and Barnett, J. A. 1991. Passage of molecules through yeast cell walls: A brief essay-review. *Yeast 7*: 313–23.

Noh, B.-S., Seo, H.-Y., Park, W.-S., and Oh, S. 2016. Safety of kimchi. In *Regulating Safety of Traditional and Ethnic Foods.* V. Prakash, O. Martin-Belloso, L. Keener, S. B. Astley, S. Braun, H. McMahon, H. Lelieveld (eds), 369–380. Amsterdam, Elsevier.

Nout, R. 2015. Quality, safety, biofunctionality and fermentation control in soya. In *Advances in Fermented Foods and Beverages*, W. Holzapfel (ed.) 409–34. Amsterdam, Elsevier. https://doi.org/10.1016/B978-1-78242-015-6.00018-9

Olempska-Beer, Z. S., Merker, R. I., Ditto, M. D., and DiNovi, M. J. 2006. Food-processing enzymes from recombinant microorganisms-a review. *Regulatory Toxicology and Pharmacology* 45(2): 144–58.

Oliveira, C., Guimarães, P. M. R., and Domingues, L. 2011. Recombinant microbial systems for improved β-galactosidase production and biotechnological applications. *BiotechnologyAdvances* 29(6): 600–609.

Oliveira, C., Teixeira, J. A., Lima, N., Da Silva, N. A., and Domingues, L. 2007. Development of stable flocculent *Saccharomyces cerevisiae* strain for continuous Aspergillus niger β-galactosidase production. *Journal of Bioscience and Bioengineering* 103(4): 318–24.

Panesar, P. S., Panesar, R., Singh, R. S., Kennedy, J. F., and Kumar, H. 2006. Microbial production, immobilization and applications of β-D-galactosidase. *Journal of Chemical Technology and Biotechnology* 81(4): 530–43.

Park, A. R., and Oh, D. K. 2010. Effects of galactose and glucose on the hydrolysis reaction of a thermostable β-galactosidase from *Caldicellulosiruptor saccharolyticus. Applied Microbiology and Biotechnology* 85(5): 1427–35.

Pereira-Rodríguez, A., Fern´andez-Leiro, R., Gonz´alezSiso, M. I., Cerd´an, M. E., Becerra, M., and Sanz-Aparicio, J. 2010. Crystallization and preliminary X-ray crystallographic analysis of β-galactosidase from *Kluyveromyces lactis. Acta Crystallographica F* 66: 297–300.

Princely, S., Saleem Basha, N., Kirubakaran, J. J., and Dhanaraju, M. D. 2013. Biochemical characterization, partial purification, and production of an intracellular beta-galactosidase from *Streptococcus thermophilus* grown in whey. *Pelagia Research Library European Journal of Experimental Biology* 3(2): 242–51. www.pelagiaresearchlibrary.com.

Qingfang, Z., Naiyu, C., Shaohua, D., Xiaohui, Y., Shuang, Y., Aiping, C. 2012. Fermenting production method of low-temperature beta-galactosidase by marine microorganisms (Patent No. CN102653747). China National Intellectual Property Administration. https://patentscope.wipo.int/search/pt/detail.jsf?docId=CN85285849 (accessed June 4, 2022).

Ramakrishnan, S., and Hartley, B. S. 1993. Fermentation of lactose by yeast cells secreting recombinant fungal lactase. *Applied and Environmental Microbiology* 59(12): 4230–35.

Raol, G. G., Prajapati, V. S., and Raol, B. V. 2014. Formulation of low-cost, lactose-free production medium by response surface methodology for the production of β-galactosidase using halotolerant *Aspergillus tubengensis* GR-1. *Biocatalysis and Agricultural Biotechnology* 3(4): 181–87.

Raol, G. G., Raol, B. V., Prajapati, V. S., and Bhavsar, N. H. 2015. Utilization of agro-industrial waste for β-galactosidase production under solid state fermentation using halotolerant *Aspergillus tubingensis* GR1 isolate. *3 Biotech* 5(4): 411–21.

Regenhardt, S. A., Mammarella, E. J., Rubiolo, A. C., 2013. Hydrolysis of lactose from cheese whey using a reactor with β-galactosidase enzyme immobilised on a commercial UF membrane. *Chemical Process Engineering -InzynieriaChemicznaiProcesowa* 34: 375–85.

Rodríguez, Á. P., Leiro, R. F., Cristina, M. C., Cerdán, M. E., González Siso, M. I., and Becerra, M. 2006. Secretion and properties of a hybrid *Kluyveromyces lactis-Aspergillus niger* β-galactosidase. *Microbial Cell Factories* 5: 1–13.

Rossini, D., Porro, D., Brambilla, L., Venturini, M., Ranzi, B. M., Vanoni, M., and Alberghina, L. 1993. In *Saccharomyces cerevisiae,* protein secretion into the growth medium depends on environmental factors. *Yeast* 9(1): 77–84.

Rutkiewicz, M., Bujacz, A., Wanarska, M., Wierzbicka-Wos, A., and Cieslinski, H. 2019. Active site architecture and reaction mechanism determination of cold adapted 3-D-galactosidase from *Arthrobacter sp.* 32cB. *International Journal of Molecular Sciences* 20(17).

Sakai, A., Shimizu, Y., and Hishinuma, F. 1990. Integration of heterologous genes into the chromosome of Saccharomyces cerevisiae using a delta sequence of yeast retrotransposon Ty. *Applied Microbiology and Biotechnology* 33(3): 302–6.

Sangwan, V., Tomar, S. K., Ali, B., Singh, R. R. B., and Singh, A. K. (2015). Production of β-galactosidase from *Streptococcus thermophilus* for galactooligosaccharides synthesis. *Journal of Food Science and Technology* 52(7): 4206–15.

Schmidt, M., and Stougaard, P. (2010). Identification, cloning and expression of a cold-active β-galactosidase from a novel Arctic bacterium, *Alkalilactibacillusikkense. Environmental Technology* 31(10): 1107–14.

Sen, P., Nath, A., Bhattacharjee, C., Chowdhury, R., and Bhattacharya, P. 2014. Process engineering studies of free and micro-encapsulated β-galactosidase in batch and packed bed bioreactors for production of galactooligosaccharides. *Biochemical Engineering. Journal* 90: 59–72.

Sen, S., Ray, L., and Chattopadhyay, P. 2012. Production, purification, immobilization, and characterization of a thermostable β-galactosidase from *Aspergillus alliaceus. Applied Biochemistry and Biotechnology* 167(7): 1938–53.

Sharma, S. K., and Leblanc, R. M. 2017. Biosensors based on b-galactosidase enzyme: Recent advances and perspectives. *Analytical Biochemistry* 535: 1–11.

Shuanggui, X., Wenxuan, L., Lei, L. 2018. Method for producing high-yield high-temperature-resistant beta-galactosidase by carrying out magnetic field mutagenesis on bifidobacterium (Patent No. CN109055336). China National Intellectual Property Administration. https://patentscope.wipo.int/search/pt/detail.jsf?docId=CN235614840&_cid=P12-KY03MR-99925-1 (accessed June 4, 2022).

Singh, A. K., and Singh, K. 2012. Study on hydrolysis of lactose in whey by use of immobilized enzyme technology for production of instant energy drink. *Advance Journal of Food Science and Technology* 4: 84–90.

Soto, D., Escobar, S., Guzm´an, F., C´ardenas, C., Bernal, C., and Mesa, M. 2017. Structure-activity relationships on the study of β-galactosidase folding/unfolding due to interactions with immobilization additives: Triton X-100 and ethanol. *International Journal of Biological Macromolecules* 96: 87–92.

Suárez, S., Guerrero, C., Vera, C., and Illanes, A. 2018. Effect of particle size and enzyme load on the simultaneous reactions of lactose hydrolysis and transgalactosylation with glyoxyl-agarose immobilized β-galactosidase from *Aspergillus oryzae. Process Biochemistry* 73: 56–64.

Suopajärvi, T., Ricci, P., Karvonen, V., Ottolina, G., and Liimatainen, H. 2020. Acidic and alkaline deep eutectic solvents in delignification and nanofibrillation of corn stalk, wheat straw, and rapeseed stem residues. *Industrial Crops and Products* 145: 111956.

Szymańska, K., Bryjak, J., Mrowiec-Białoń, J., and Jarzebski, A. B. 2007. Application and properties of siliceous mesostructured cellular foams as enzymes carriers to obtain efficient biocatalysts. *Microporous and Mesoporous Material* 99: 167–75.

Thakur, M., Kumar Rai, A., and Singh, S. P. 2022. An acid-tolerant and cold-active β-galactosidase potentially suitable to process milk and whey samples. *Applied Microbiology and Biotechnology* 106: 3599.

Todea, A., Benea, I. C., Bîtcan, I., Péter, F., Klébert, S., Feczkó, T., Károly, Z., and Biró, E. 2021. One-pot biocatalytic conversion of lactose to gluconic acid and galacto-oligosaccharides using immobilized β-galactosidase and glucose oxidase. *Catalysis Today* 366: 202–11.

Trevan, M., 1980. Techniques of immobilization. In *Immobilized Enzymes. An introduction and applications in biotechnology*, M. D. Trevan (ed.), 138. Chichester, UK, John Wiley & Sons.

Tsanasidou, C., Kosma, I., Badeka, A., and Kontominas, M. 2021. Quality parameters of wheat bread with the addition of untreated cheese whey. *Molecules* 26: 7518.

Vasileva, N., Ivanov, Y., Damyanova, S., Kostova, I., and Godjevargova, T., 2016. Hydrolysis of whey lactose by immobilized β-galactosidase in a bioreactor with a spirally wound membrane. *International Journal of Biological Macromolecules* 82, 339–46. https://doi.org/10.1016/j.ijbiomac.2015.11.025

Vera, C., Guerrero, C., Aburto, C., Cordova, A., and Illanes, A. 2020. Conventional and non-conventional applications of β-galactosidases. *BBA – Proteins and Proteomics* 1868: 140271.

Vera, C., Guerrero, C., Conejeros, R., and Illanes, A. 2012. Synthesis of galacto-oligosaccharides by β-galactosidase from *Aspergillus oryzae* using partially dissolved and supersaturated solution of lactose. *Enzyme and Microbial Technology* 50: 188–94.

Vera, C., Guerrero, C., Wilson, L., and Illanes, A. 2017. Optimization of reaction conditions and the donor substrate in the synthesis of hexyl-β-D-galactoside. *Process Biochemistry* 58: 128–36.

Verma, M. L., Barrow, C. J., Kennedy, J. F., and Puri, M. 2012. Immobilization of β-D-galactosidase from Kluyveromyces lactis on functionalized silicon dioxide nanoparticles: Characterization and lactose hydrolysis. *International Journal of Biological Macromolecules* 50: 432–37.

Wang, G., Wang, H., Chen, Y., Pei, X., Sun, W., Liu, L., Wang, F., Umar Yaqoob, M., Tao, W., Xiao, Z., Jin, Y., Yang, S. T., Lin, D., and Wang, M. 2021. Optimization and comparison of the production of galactooligosaccharides using free or immobilized *Aspergillus oryzae* β-galactosidase, followed by purification using silica gel. *Food Chemistry* 362: 130195.

Wang, S.-D., Guo, G.-S., Li, L., Cao, L.-C., Tong, L., Ren, G.-H., and Liu, Y.-H. 2014. Identification and characterization of an unusual glycosyltransferase-like enzyme with β-galactosidase activity from a soil metagenomic library. *Enzyme and Microbial Technology* 57: 26–35.

Wei, Q., Shengping, Y., Qingdian, Y., and Rongxin, S. 2020. Integrated method for coproduction of beta-galactosidase enzymic preparation and ethanol product from lactose-enriched biomass (Patent No. CN111378636). China. https://patentscope.wipo.int/search/pt/detail.jsf?docId=CN299150652&_cid=P12-KY03OG-00411-1 (accessed June 4, 2022).

Wyrick, J. J., Aparicio, J. G., Chen, T., Barnett, J. D., Jennings, E. G., Young, R. A., Bell, S. P., and Aparicio, O. M. 2001. Genome-Wide distribution of ORC and MCM proteins in mapping of replication origins. *Science* 294(December): 2357–60.

Yamada, M., Chiba, S., Endo, Y., and Isobe, K. 2016. New alkalophilic β-galactosidase with high activity in alkaline pH region from *Teratosphaeriaacidotherma* AIU BGA-1. *Journal of Bioscience and Bioengineering* 123(1): 15–9.

Yuan, T., Yang, P., Wang, Y., Meng, K., Luo, H., Zhang, W., Wu, N., Fan, Y., and Yao, B. 2008. Heterologous expression of a gene encoding a thermostable β-galactosidase from *Alicyclobacillus acidocaldarius*. *Biotechnology Letters* 30(2): 343–8.

Yunus, F. un N., Nadeem, M., and Rashid, F. 2015. Single-cell protein production through microbial conversion of lignocellulosic residue (wheat bran) for animal feed. *Journal of the Institute of Brewing* 121(4): 553–7.

Zartha, J. W. S., Palop, F. M., Arango, B. A., Velez, F. M. S., and Avalos, A. G. P. 2016. S-Curve analysis and technology life cycle. Application in series of data of articles and patents *In Espacios* 31: 19.

Zolnere, K., and Ciprovica, I. 2017. The comparison of commercially available β-galactosidases for dairy industry: Review. *Food Science* 1: 215–22.

10 Microbial Enzymes for Reduction of Antinutritional Factors

Adenise Lorenci Woiciechowski[1], *Maria Giovana Binder Pagnoncelli*[1,2], *Thamarys Scapini*[1], *Fernanda Guilherme Prado*[1], *Fernanda Kelly Mezzalira*[1,2], *Carolina Mene Savian*[1], *and Carlos Ricardo Soccol*[1]

[1] Department do Bioprocess Engineering and Biotechnology, Federal University of Paraná, Curitiba, Paraná, Brazil

[2] Department of Chemistry and Biology, Universidade Tecnológica Federal do Paraná

CONTENTS

10.1 INTRODUCTION

Plant-based foods (seeds, leaves, cereals, and others) play an important role in human and animal nutrition, offering essential nutrients for supplying the daily biological functions, for being rich sources of fibre, vitamins, carbohydrates, and minerals (Awolu et al., 2017). Besides, plant-based foods are an environmentally sustainable alternative to protein intake, as production involves lower greenhouse gas emissions, presents high protein values, and can enhance system productivity by diversifying crop rotations, which can also naturally restore soil nitrogen (Margier et al., 2018; Fasolin et al., 2019).

During growth, plants synthesize many compounds, among them, antinutritional factors (ANFs) that are secondary metabolites responsible for preventing bacterial and fungal infections and insect attacks on plants (Zhang et al., 2015; Krogdahl et al., 2022). In lower concentrations, the ANFs do not exhibit deleterious effects on health; however, in high concentrations, those compounds are responsible for the reduction of nutrients' bioaccessibility, interfering in the human body's nutrient absorption, or even being potentially toxic (Robinson et al., 2019; Jeyakumar and Lawrence, 2022).

The reduction of ANF content in foods is carried out by different strategies, aiming to minimize the deleterious effects of high concentrations of these compounds. In some cases, easy and accessible processes are available, including cooking, germination, or soaking. In other cases, more complex

DOI: 10.1201/9781003311164-13

processes involving high temperature and pressure are necessary, such as hydrothermal treatments. Enzymatic technologies also present good results for ANF reduction and can be performed by fermentation and germination processes mediated by endogenous enzymes, or by industrial enzymes on food processing, boosting the enzyme market, or by enzymes in the digestive tract. The enzymatic methods are an important study field for ANF reduction in the food industry. The application of these methods for ANF reduction is advantageous for the consumer because they will increase the nutritional value of the food, increasing the nutrients quality and quantity, facilitating digestion.

Physical, thermal, and chemical methods will be discussed briefly, and the enzymatic will be detailed, describing the natural body processing with the digestive enzymes produced during digestion, and the enzymatic methods for food processing.

10.2 ANTINUTRITIONAL FACTORS (ANFs)

ANFs can be classified into proteinaceous, such as lectins, and protease inhibitors (e.g., trypsin and chymotrypsin inhibitors); or non-proteinaceous, such as phytic acid, tannins, saponins, α-galactosides, and alkaloids (Khattab and Arntfield, 2009; Bessada et al., 2019). Lectins are glycoproteins capable of interfering in the nutrient absorption, and these ANFs are found in cereals and some pulses that have sugar-binding activity, as shown on Table 10.1. Depending on the content, lectins can show agglutination activity in blood cells, known as hemagglutination (Salim-ur-Rehman et al., 2014; Bessada et al., 2019).

In the same group, protease inhibitors are present in many cultivated seeds as a defence-related protein (Roy and Dutta, 2009) and can inhibit the proteolytic enzymatic activity in the gastrointestinal tract. Therefore, when legumes and pulses are not properly prepared before being consumed, those ANFs can cause digestive functions disturbance (Thakur et al., 2019). As an example of the protease inhibitors, trypsin inhibitors are hydrolases that form an irreversible condition in the body, known as a complex of enzyme-trypsin inhibitor, which can reduce the trypsin content in the intestine and decrease protein digestibility (Robinson et al., 2019; Thakur et al., 2019).

Classified as non-proteinaceous, ingredients like phytic acids are secondary factors that are naturally concentrated in plant seeds, such as legumes, grains, and cereals, and can form complexes with dietary minerals (e.g., zinc, iron, calcium, and magnesium), reducing bioavailability and absorption of those minerals (García-Estepa et al., 1999; Parca et al., 2018; Rosa-Sibakov et al., 2018). Besides, it can link with other compounds, such as protein and digestive enzymes, like proteases and amylases, reducing the protein solubility and inhibiting hydrolysis (Parca et al., 2018).

In the same classification, tannins are phenolic compounds capable of precipitating proteins, amino acids, and alkaloid compounds, reducing their digestibility and amino acid availability (Robinson et al., 2019). Those ANFs also have astringent properties, resulting in complexes with salivary and glycoproteins that reduce palatability, which causes sensory limitation (Bessada et al., 2019). Tannins can also interfere in the dietary iron absorption (Rao and Deosthale, 1982), and are found in crops and legumes, such as sorghum (Thakur et al., 2019).

Saponins are secondary molecules produced by plants, containing a carbohydrate linked to an aglycone (Soetan, 2008), and capable of forming soap-like foams when shaken with aqueous solution. Besides, saponins have amphiphilic properties, which give them the capacity to hemolyse red blood cells (Shi et al., 2009). In addition, those ANFs can also reduce nutrients' bioavailability and decrease enzyme activity, affecting protein digestibility by inhibiting digestive enzymes, such as trypsin, as shown on Table 10.1. In high concentrations, they are responsible for bitter taste and astringency in plant-based food (Thakur et al., 2019).

α-Galactosides, such as stachyose, raffinose, and verbascose, are oligosaccharides found in legumes and are responsible for causing intestinal discomfort once cannot be digested by humans (Mulimani and Devendra, 1998). Other compounds are alkaloids, organic molecules synthesized from amino acids by plants, which can be composed of several carbon rings with side chains with nitrogen replacing one or more carbon atoms, which might lead to an anti-palatable effect

TABLE 10.1
Antinutritional Factors Classification and Nutritional Effects

Type	ANF	Examples of Food Source	Nutritional Effects	Reference
Proteinaceous	Lectins	Pulses andcereals	Sugar-biding activity, low nutrients absorption, hemagglutination	Gibson et al. (2006); Bessada et al. (2019); Samtiya et al. (2020)
	Protease inhibitors	Chickpeas, whitebeans, soybeans	Low protein digestibility, flatulence, upset digestive functions	Bessada et al. (2019); Thakur et al. (2019); Samtiya et al. (2020)
Non-proteinaceous	Phyticacid	Chickpeas, corn, lupins, sorghum, millet	Low mineral bioavailability and absorption, formation of insoluble complexes with metal ions	Gibson et al. (2006); Salim-ur-Rehman et al. (2014); Vashishth et al. (2017); Ojo (2018)
	Tannin	Berry fruits, cocoa beans, pomegranate, sorghum, barley	Low protein digestibility, reduction of essential amino acids absorption	Lampart-Szczapa et al. (2003); Serrano et al. (2009); Morzelle et al. (2019)
	Saponins	Pulses, kidney bean, chickpea, soybean, sunflower, lupin, groundnut	The compound binds to proteins. Hemolysis of red blood cells, low nutrients bioavailability, low digestive enzyme activity	Shi et al. (2009); Thakur et al. (2019)
	α-Galactosides	Pulses	Intestinal discomfort	Frias et al. (2000); Khattab and Arntfield (2009); Thirunathan and Manickavasagan (2019)
	Alkaloids	Lupins, solanine, leaves of coca plant	Unpalatability, neurologicaldisorder, gastrointestinal dysfunction	Jiménez-Martínez et al. (2007); Jezierny et al. (2010)
	Oxalates	Dark pepper, beets, cocoa, spinach	Reduce the calcium, sodium, and ammonium ions absorption. High intakes may form kidney stone (nephrolithiasis)	Gemede (2014); Mihrete (2019)

(Jadhav et al., 1981). These compounds also act on the nervous and digestive systems, increasing or disrupting electrochemical transmissions, with the potential to cause neurological disorder, gastrointestinal disfunction, and, in high concentrations, can lead to death (Jezierny et al., 2010; Thakur et al., 2019). In contrast, low concentrations of alkaloids have important pharmacological applications, such as analgesic, tumour cell killer, blood pressure reducer, respiration, and circulation stimulant (Muzquiz et al., 2012; Siah et al., 2012).

Oxalates are salt from the oxalic acid that can bind with calcium, magnesium, sodium, and potassium. Among these salts, calcium oxalate is insoluble, affecting the Ca absorption of the body and leading to the formation of kidney stones. Without Ca, a cofactor in enzymatic reactions and available in the human system, physiological and biochemical reactions can be compromised, such as nerve impulse transmission, clotting factors in blood, and, in high doses, it can also be fatal (Bhandari and Kawabata, 2006).

10.3 NON-ENZYMATIC TREATMENTS TO REDUCE ANTINUTRITIONAL FACTORS IN FOOD

Physical, thermal, and chemical methods can be used to reduce the negative effects of ANFs in foods. Among the most common methods used are soaking, cooking, autoclaving, pressure cooking, microwave cooking, extrusion, medium pH changes, in addition to enzymatic processes, which will be addressed in the next section. Table 10.2 provides an overview of results obtained using non-enzymatic methods to remove or decrease ANF content.

Among the processes, soaking or hydration in pure water or with chemical substances that change the pH solution are simple techniques used to reduce ANFs, and are often applied in the domestic consumption of legumes. The reduction of ANF content by soaking or hydration can occur by cooking under mild temperature conditions and are favoured when the compound has a low molecular weight and ionic character. Also, the ANF reduction can be associated with ANF solubility and leaching in water, and endogenous enzymes activity such as phytases (Ojo, 2018; Vashishth et al., 2021; Antoine et al., 2022).

Soaking strategies in water or chemical solutions for ANF removal are generally effective in the reduction of water-facilitated compounds – often observed for phytate reduction (see Table 10.2) – and combined with thermal processes can demonstrate satisfactory efficiencies. The combination of soaking or hydration followed by hydrothermal processes favours the hydrolysis of chemical structures into lower molecular weight compounds (Vijayakumari et al., 2007; Vashishth et al., 2021). This fact was observed in seeds of *Vigna unguiculata* (L.) Walp subsp. *Unguiculata* in which phytic acid content that was reduced by 27% and 31% when soaking in pure water or solutions with a slightly alkaline pH by the addition of sodium bicarbonate, and by 79% by the combination of soaking in water (12 h) and after cooking (30 min) (Kalpanadevi and Mohan, 2013).

Hydrothermal and/or physical processes can occur in different conditions of temperature, pressure, pH, contact time, etc. during the preparation and cooking of food, and processes can range from conventional cooking to the use of autoclaves, microwaves, thermal irradiation, among others (Vijayakumari et al., 2007; Saadi et al., 2022). Saadi et al. (2022) found that bean seed exposed to heat irradiation in autoclaves and heat ionization had significant reductions in the ANF content, which were attributed to changes in structural conformation. The tannin content reduction by heat ionization and heat irradiation was attributed to the exposure time of the seeds to the process and to the significant effect of the gallic acid content; in addition to the phytic acid removal observed for heat irradiation (Saadi et al., 2022).

Proteinaceous ANFs, such as trypsin inhibitors, have reduced activity or are completely inactivated during cooking processes (Ojo, 2018; Vashishth et al., 2021). For horse gram grains, two varieties were tested for different thermal processes for ANF reduction, and trypsin inhibitor activity was reduced in a range of 77–82% in thermal processes (autoclave, microwave, micro-ionization, extrusion), and compounds such as oxalate proved to be water-soluble and heat-sensitive, with more pronounced reductions in extrusion processes (Vashishth et al., 2021). After heat treatment in different processes (infrared, dry-heating, microwave), the trypsin inhibitors from the rice bran sample were reduced approximately 31% and were associated with protein denaturation, a result of the heat-labile characteristic of trypsin inhibitors (Irakli et al., 2020).

Generally, the reduction of ANFs in foods by processes involving thermal and hydrothermal conditions, or by chemical addition, are induced by chemical hydrolysis, degradation, conformational changes, and molecule isomerization (Vijayakumari et al., 2007; Kalpanadevi and Mohan, 2013; Ojo et al., 2018; Saadi et al., 2022). The phytate reduction content in hydrothermal processes may be associated with chemical degradation into inositol hexaphosphate hydrolysed into penta- and tetraphosphate, or by the formation of insoluble complexes with proteins and minerals (Vijayakumari et al., 2007; Kalpanadevi and Mohan, 2013). As well, the breaking of glycosidic bonds of the saponin structure can result in the content reduction of these compounds in hydrothermal processes (Ojo et al., 2018).

TABLE 10.2
Overview of Results Using Non-Enzymatic Processes to Reduce ANF Content in Foods

			Tannins		Phytates		Oxalates		Trypsin Inhibitor		Saponin		
Process	**Food**	**Conditions**	**Initial (mg g^{-1})**	**Removal (%)**	**Initial (mg g^{-1})**	**Removal (%)**	**Initial (mg g^{-1})**	**Removal (%)**	**Initial**	**Removal (%)**	**Initial (mg g^{-1})**	**Removal (%)**	**Reference**
Boiling	*Canavalia ensiformis*	Soaking + Atmospheric boiling	27.47	64.69	59.28	63.26	–	–	25.61 mg g^{-1}	100	5.29	78.83	Ojo et al. (2018)
	Cassia hirsutta	Soaking + Atmospheric boiling	62.3	78.46	56.26	62.14	–	–	10.88 mg g^{-1}	100	5.11	82.97	Ojo (2018)
Autoclave	Guar seed	Autoclaving for 15 min at 121°C	5.9	50	29.8	85	–	–	1.8 mg g^{-1}	66.7	27.5	~50	Nidhina and Muthukumar (2015)
	Vigna unguiculata (L.) Walp subsp. *unguiculata*	Autoclaving for 30 min at 103.04 KPa	3.8	82	3.98	70	–	–	26.48 IU mg^{-1}	92	–	–	Kalpanadevi and Mohan (2013)
Soaking	*Cassia hirsutta*	Hydration – 100% level (soaking in water up to 24 h at room temperature)	27.47	13.79	56.26	6.20	–	–	10.88 mg g^{-1}	14.06	5.11	4.30	Ojo (2018)
	Vigna unguiculata (L.) Walp subsp. *unguiculata*	Soaking for 12 horas in pure water	3.8	26	3.98	31	–	–	26.48 IU mg^{-1}	8	–	–	Kalpanadevi and Mohan (2013)

(Continued)

TABLE 10.2 *(Continued)*
Overview of Results Using Non-Enzymatic Processes to Reduce ANF Content in Foods

			Tannins		Phytates		Oxalates		Trypsin Inhibitor		Saponin		
Process	**Food**	**Conditions**	**Initial (mg g^{-1})**	**Removal (%)**	**Initial (mg g^{-1})**	**Removal (%)**	**Initial (mg g^{-1})**	**Removal (%)**	**Initial**	**Removal (%)**	**Initial (mg g^{-1})**	**Removal (%)**	**Reference**
Microwave	Horse gram grains (*Macrotyloma uniflorum*)	Soaking overnight + Microwave 800 W for 30 min	9.24	60.38	4.54	53.86	1.23	60.27	99.8 U g^{-1}	79.25	–	–	Vashishth et al. (2021)
	Rice bran	Microwave 650 W for 2 min (~160 °C) and up to 21% moisture)	3.39	0	27.08	26.0	6.52	34.81	14.93 UI g^{-1}	30.94	89.70	26.56	Irakli et al. (2020)
Extrusion	Lentil (*Lens culinarisMedik*)	180°C and moisture 22%	0.94	98.83	1.14	99.30	–	–	2.823 IU mg^{-1}	99.54	–	–	Rathod and Annapure (2016)
	Horse gram grains (*Macrotyloma uniflorum*)	160°C and moisture 21%	9.24	67.53	4.54	59.82	1.23	81.82	99.8 U g^{-1}	81.86	–	–	Vashishth et al. (2021)

The extrusion process is also considered an effective process for reducing ANFs and is used in food processing industries with different shapes, textures, and flavours, and it is also a promising technology for reducing tannins, phytates, oxalic acid, trypsin inhibitors, etc. (Rathod and Annapure, 2016; Vashishth et al., 2021). Extrusion is a process operating in an intense thermo-mechanical combined system, where the food is submitted at high temperatures, with short exposure times and intense shear force, which causes structural rupture of the food, changes the texture of the product, and can reduce the ANF compounds. As it is a process performed at high temperatures, the ANF removal by extrusion can cause conformational changes in the compounds and affect the nutritional composition of legumes and grains, because it is necessary to strictly control the operating conditions and the characteristics of the raw material to avoid losses of important nutritional compounds in the food (Nikmaram et al., 2017; Vashishth et al., 2021). Tannins may qualitatively change in the structure resulting in different chemical reactivity, change in the degree of polymerization, and even destruction of the structure under extreme operating conditions (Alonso et al., 1998). In addition, moisture is an important variable for the process. In lentils, increasing the moisture content from 14% to 22% positively impacted the phytates removal, tannins, and polyphenols (Rathod and Annapure, 2016).

Hydrothermal, thermal, and chemical processes are important to ANF reduction. However, reaction to extreme conditions (e.g., high temperature, pressure, and exposure time) can result in degradation of nutritional components important for food quality (Sheikh et al., 2021). Furthermore, the levels of ANFs do not need to be completely eliminated or destroyed, considering that the reduction to safe levels is enough to reduce the deleterious effect on health.

10.4 DIGESTIVE AND INDUSTRIAL ENZYMES TO REDUCE ANFs IN FOOD

Enzymes can be responsible for the ANF reduction in foods, and generally can be associated in two scenarios: (1) as a food processing method, as in germination or fermentation methods, or by the direct use of industrial enzymes; (2) via processes in the digestive tract, which can reduce the deleterious effect of ANFs by enzymatic pathways. In the first scenario, due to the wide variety of microorganisms present in nature, there are several enzymes that can be used by the food industry with different roles and abilities, improving production, processing, and other related components such as flavour, aroma, colour, texture, appearance, shelf life, and nutritional value (Hjort, 2007). Besides that, enzymatic processing becomes interesting as it is sustainable bioprocessing, which provides natural products, with a reduced amount of chemicals and attractive sensory properties (Adeyemo and Onilude, 2013; Glusac and Fishman, 2021). In the ANF removal, enzymes with different specificities can be applied directly by industrial enzymatic solutions, or through fermentation or germination processes. Endogenous enzymes have a great influence in germination and fermentation, which can be used to improve the nutritional quality of food before ingestion.

In the second scenario, the enzymatic processes for ANF reduction are related to digestive enzymes. The digestive process is mediated by enzymes which speed up chemical reactions by breaking down macronutrient molecules into smaller compounds to facilitate nutrient absorption in the digestive tract, playing a crucial role in its proper functioning and providing micronutrients (vitamins and minerals) (Maske et al., 2021; Assan et al., 2022). During digestion, enzymes can act synergistically to reduce the ANF content, and can be influenced by some factors such as the contact time between enzyme and substrate, pH, and the concentration of enzyme versus substrate (Yazici and Mazlum, 2019; Chakraborty et al., 2021).

Therefore, this section will address the biological processes mediated by enzymes to reduce ANF in the digestive tract and applied on industrial processes.

10.4.1 Digestive Enzymes

There are different types of digestive enzymes that act on the various components consumed daily proteins, carbohydrates, and fats) (Vazquez-Flores et al., 2018; Terra et al., 2019). Most digestive

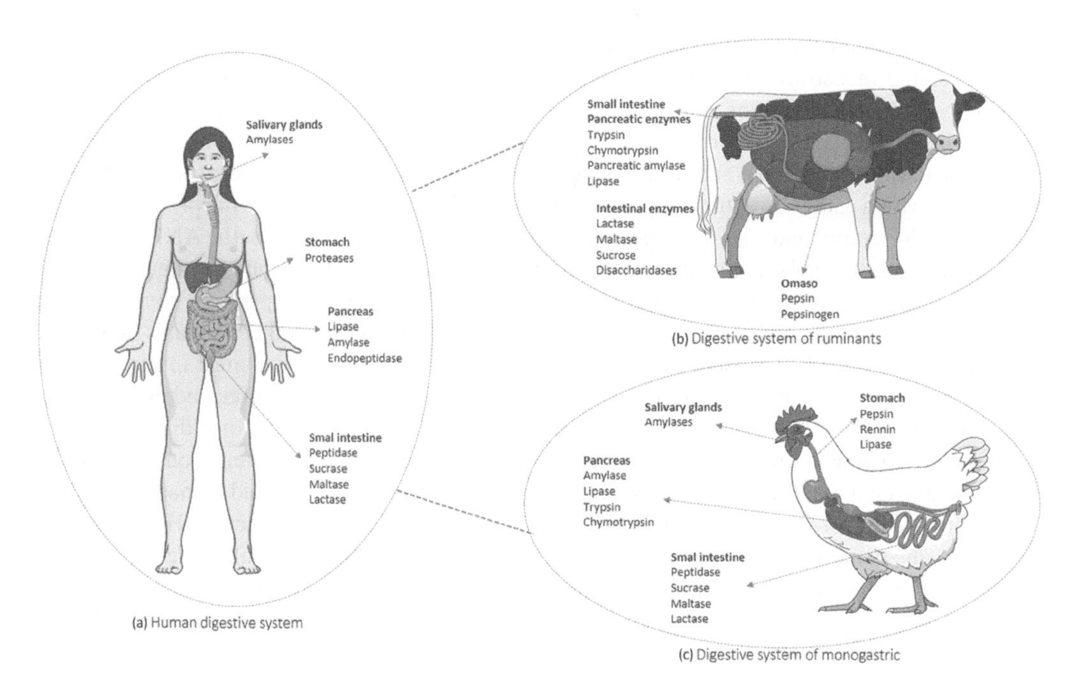

FIGURE 10.1 Digestive enzymes produced in the digestive system of humans, birds, and ruminants. (a) Human digestive system; (b) digestive system of ruminants; and (c) digestive system of monogastric (birds).

enzymes are produced by the organs of the digestive system, mainly the pancreas (Yazici and Mazlum, 2019; Larrosa and Otero, 2021) (Figure 10.1).

Starch, a molecule composed of amylose and amylopectin, is an important source of polysaccharide. Amylase enzymes hydrolyse starch to yield products such as dextrins, maltotriose, maltose, and eventually glucose (Vandenberghe et al., 2019). Disaccharides are digested to obtain simple sugars. The cleaved monosaccharides are then transported across the epithelial brush border for absorption and metabolism. Maltose is formed by two glucose molecules linked together and it is one of the main sources of glucose. Glucose is an essential nutrient for metabolism as it is the main source of energy. The maltase enzyme can hydrolyse maltose and release glucose (Nyambe-Silavwe and Williamson, 2018).

Sucrose is also a disaccharide, and it is naturally produced in plants. It is an enzyme which catalyses the hydrolysis of sucrose into fructose and glucose. This enzyme is present in the brush border membrane of small intestinal villi (Nichols et al., 2017). In the human gastrointestinal system (Figure 10.1a), the sucrase-isomaltase is synthasized and assembled in the rough endoplasmic reticulum as a homologous pro-enzyme dimer. It passes through the Golgi apparatus and is transported to the apical cell surface of villi. Then, the sucrase-isomaltase is cleaved by pancreatic proteases into its mature subunits, isomaltase and sucrase (Rose et al., 2018). Deficiency of this enzyme can cause a rare disease known as congenital sucrase-isomaltase deficiency. This deficiency causes malabsorption of carbohydrates by the intestine, thus causing abdominal distension, watery diarrhea, and/or discomfort. This is an inducible enzyme, and its synthesis is increased by stressful situations (Cohen, 2016; Husein et al., 2021).

Milk and dairy products are major dietary sources of protein, phosphorus, calcium, magnesium, and other nutrients. Lactose is the main carbohydrate present in this food group and is a disaccharide as well. Lactases are responsible for catalysing the lactose and release glucose and galactose as a product. These are produced in the intestinal mucosa, in the superficial zone of the microvilli of the small intestine (Figure 10.1) (Maske et al., 2021). Lactase is an enzyme which is vulnerable to any aggression to the intestinal mucosa, and therefore it may temporarily stop being produced until the mucosa recovers or may even stop irreversibly depending on the degree of injury (Ianiro et al., 2016; Maske et al., 2021). Deficiency of this enzyme has been the main cause of lactose intolerance, since its production in the intestinal villi and its function of breaking the bonds between glucose and galactose ensures that these sugars are digested normally (Ianiro et al., 2016). When unabsorbed, lactose is fermented by the colon microbiota, resulting in increased intestinal wall peristalsis, organic acids and gases production, and other problems caused by abnormal bowel flow (Qi and Tester, 2020). Lactose intolerance can be controlled by intaking manipulated compounds, depending on the degree of the intolerance. As it is responsible for the enzymatic glycosidic hydrolysis, lactose intolerance directly impacts the consumption of milk and dairy products (Ianiro et al., 2016; Qi and Tester, 2020).

The class of lipases corresponds to the enzymes responsible for the digestion of lipids, which are mostly derived from a diet rich in triacylglycerols or triglycerides. Gastric lipase is secreted by the tongue; however, the extremely acidic stomach pH does not allow the full action of this gastric lipase, reducing the speed of its enzymatic action, thus only the breaking of some ester bonds of short chain fatty acids occurs (Koca et al., 2018; Chen et al., 2021). Bile salts emulsify fat, so that the pancreatic lipase enzyme can act by breaking down triglycerides into diglycerides and free fatty acids; diglycerides undergo a new action of lipase giving rise to monoglycerides, fatty acids and glycerol; and about 70% of the diglycerides are absorbed by the intestinal mucosa, the remainder is converted into monoglycerides, glycerol, and fatty acids (Kong et al., 2019; Tian et al., 2019). Pancreatic juice has several digestive enzymes (mainly proteases and carbohydrates) (Figure 10.1). The pancreatic lipase is responsible for the hydrolysis of the ester bonds of lipids, releasing large amounts of cholesterol, fatty acids, glycerol, and some monoacylglycerol molecules (Yuan et al., 2021).

Peptidases hydrolyse the peptide bonds. The pepsin, a type of protease, works in the breakdown of proteins and trypsin, which acts in the digestion of the proteins that were not digested in the stomach (Figure 10.1) (Pujante et al., 2017; Vogt, 2021). The protein hydrolysis affects the food matrix properties, resulting in enhanced digestibility, decreased allergy, and greater modification of sensory quality (Tavano et al., 2018).

10.4.2 Biological Process and Enzymes Applied on ANF Removal in Food

The applicability of enzymes or microorganisms in processes is old. Although it goes back to ancient times, when used in baking, brewing, cheese making, etc., its use was as a fermenting microorganism or in crude preparations (Tang and Zhao, 2009). As technology advances the automation and control of industrial processes has enabled biological processes to be implemented to enhance the nutritional food quality, and promising results using enzymatic methods for ANF removal have been explored as a promising field of food science. These processes of applying enzymes can be conducted from different strategies.

Germination and fermentation are examples of biological processes applied to ANF reduction and they have the advantage of using mild conditions, thus preventing the degradation of the thermolabile substances present in food. During the processes of germination and fermentation, the endogenous enzymes degrade ANFs. Therefore, the nutrients in a germinated or fermented food are superior to ungerminated or unfermented (Nkhata et al., 2018).

Fermentation is carried out by enzymatic pathway and with the potential to reduce ANF, as well as the potential to improve the quality and nutritional profile of grains and seeds. This process can be applied by different strategies, the most applied being solid state fermentation (SSF) (occurs in a solid basis with low water activity) and submerged fermentation (SmF) (occurs in a medium with high water activity by submerging the substrate). In addition, strategies for growing the culture that carries out the fermentative process can also be applied, which can apply the endogenous microbiota of the food (also called spontaneous fermentation) or can be controlled by inoculating specific cultures (Saéz et al., 2018). Both processes are mediated by endogenous enzymes produced by the predominant cultures and are dependent on rigorous parameter controls for the effectiveness of the system.

Fermentation is a complex biological process, which, through the action of different mechanisms associated with the enzymes and metabolites involved, results in the alteration of the chemical structure of the food. Changes in nutritional and anti-nutritional aspects were evaluated for lupin grain using strains of *Rhyzopusoligosporus* as an inoculating agent, and significant reductions in nitrates (94.59%), tannins (82.10%), alkaloids (94%), urease activity (93.75%), phytic acid (70.06%), and trypsin activity (76.76%), and promoted an increase in antioxidant capacity, carotenoid, and phenolic contents (Villacrés et al., 2020). Soybean meal was also evaluated in two-step SSF using strains of *Bacillus subtilis* and *Aspergillus oryzae*, being in 24 hours for *B. subtilis*, subsequently the fungus *A. oryzae* was inoculated for 72 hours (Suprayogi et al., 2022). It is interesting to note that in addition to changes in the nutritional profile of amino acids, important results for the reduction of tannins, saponins, trypsin inhibitors, and phytic acid were achieved (Suprayogi et al., 2022). *Bacillus* and *Aspergillus*strains are known for enzymatic production during fermentation, mainly proteases and carbohydrases (Mukherjee et al., 2015; Cui et al., 2020; Li and Wang, 2021; Chen et al., 2022; Suprayogi et al., 2022). Strains of *Aspergillus* are especially known for displaying stability, selectivity, and properties that are industrially useful, including the ability to accumulate compounds and to efficiently produce metabolites and enzymes (e.g., amylases, cellulases, pectinases, peroxidases, lipases, xylanases, proteases, laccases) (Mukherjee et al., 2015; Soccol et al., 2017; Adsul et al., 2020; Colla et al., 2020).

Another biological process used for ANF reduction that involves an enzymatic process is germination. These are a traditional method used in grain and seed processing. ANF reduction during germination occurs naturally, without the need for the addition of chemicals. This process is dependent of rigorous control of operational parameters, such as humidity and temperature. Germination is a complex biological process that seeds and grains undergo as they physically recover after maturation. This process leads to the resumption of an intense metabolic activity that triggers cellular processes essential for plant growth. Most seeds have energy storage compounds such as oils, fat, starch, or protein. These compounds are converted into metabolically easily assimilated compounds by the plant during germination, and this process is mediated by endogenous enzymes such as lipases, amylases, and cellulases (Xu et al., 2019). Still considering the role of ANFs during grain ripening (secondary metabolites of plant protection), it is natural that, during germination, endogenous enzymes act in the ANFs, allowing the healthy development of the plant.

Reduction of ANF content in germination processes of a short duration (24–72h) is also associated with an increase in the mineral content and antioxidant activity, which are associated with the release of these molecules into the grain after the chemical bonds with ANF are broken, and by increasing the oxidant enzymes (Luo et al., 2014; Setia et al., 2019; Thakur et al., 2021). These structural changes induced by endogenous enzymes occur by hydrolysis of constituent bonds (e.g., nutrients and ANFs) and are responsible for initiating the grain germination process, triggering physiological and biomolecular changes in the food that will depend on the germination period and conditions (Aguilar et al., 2019; Bhinder et al., 2021). All enzymatic processes have particularities depending on the operating conditions, grains, and especially the germination time. And because it involves endogenous enzymes, it may vary depending on the plant cultivar, climate, humidity, and geographical conditions.

Different endogenous enzymes may be involved in the germination process. In lipid-rich grains, the enzymatic action of lipases is important for the hydrolysis of triacylglycerols into glycerol and fatty acids, resulting in energy release for plant growth. Likewise, the activity of α-amylase enzymes during the germination process may be responsible for the hydrolysis of starch grains to reducing sugars. On the other hand, protease enzymes can be associated with the hydrolysis of peptides into amino acids, as well as other proteins with and without catalytic activity that act synergistically in chain processes of oxidation and hydrolysis of compounds (Setia et al., 2019; Adetunji et al., 2021; Ozden et al., 2021).

Also, in ANF reduction by germination, α-galactosidases are highlighted because these enzymes catalyse the hydrolysis of the terminal α-galactosyl of oligosaccharides. The α-(1,6)-galactosidic links are found in oligosaccharides, such as stachyose, melibiose, and raffinose, and the intact oligosaccharides are not absorbed by the digestive tract (Hill, 2003; Gote et al., 2004). Raffinose family oligossacharides are an ANF present in high concentrations in legume seeds and grains and are accumulated during germination and maturation, being important carbohydrates for the plant's growth, and they are responsible for abdominal discomfort (Hill, 2003). This compound is naturally reduced during the germination stage of the grain by processes catalysed by endogenous enzymes with specificity for structure, such as glycosiltransferases (Sengupta et al., 2015).

The reduction of phytic acid during the germination processes is well reported and is generally associated with the action of phytase enzymes, which by hydrolysing chemical bonds increase the mineral content (Luo et al., 2014; Xu et al., 2019; Mesfin et al., 2021; Thakur et al., 2021). In the phytic acid removal in chickpeas conducted for 72 hours in an environmental test chamber, it was suggested that the action of endogenous phytase enzymes was responsible for the hydrolysis of the ANF compound (Mesfin et al., 2021). Phytase (myo-inositol-hexakis-phosphate-phosphohydrolase) is an enzyme known for its ability to release phosphate and other minerals from phytic acid and its salts (phytates) (Song et al., 2019). Complexes with minerals and proteins are formed, showing that the enzyme phytase increases the bioavailability of these important nutrients. Therefore, the functional role of phytase is to prevent the antinutritional effects of phytic acid and phytates (Shivanna and Venkateswaran, 2014).

In this scenario, it can be observed that microbial processes of ANF reduction are usually also associated with other biochemical processes, such as reduction of crude fibre content, and synthesis of carbohydrase enzymes, such as cellulases, xylanases, and glucanases (Jazi et al., 2019). These enzymes play a key role in glycosidic bonds hydrolysis and the release of reducing sugars. These processes, parallel to ANF reduction via enzymatic methods, are important for improving the quality and digestibility of grains and seeds.

Within the context of enzymatic processes associated with the ANF reduction and other biochemical aspects already discussed, the relevance of cellulase enzymes can be highlighted as important enzymes for industrial processes, for being produced on a large scale, and for being among the enzymes associated with the reduction of fibre content during fermentation processes. Cellulases are an important complex of enzymes for bioprocesses and are the third most used enzyme in the industrial sector, with numerous known applications in food processing (Bajaj and Mahajan, 2019). These are inducible enzymes synthesized by a range of microorganisms, such as fungi and bacteria, during their growth on cellulosic material (Lee and Koo, 2001). Cellulases are a group of enzymes that acting on distinct parts of cellulose and converts the pollysacharides into monomers of glucose, being composed by endoglucanase or endo-1,4-β-D-glucanase (EG; EC 3.2.1.4), exoglucanase or exo-1,4-β-cellobiohydrolase (CBH; EC .2.1.3.2.1.91), and β-glucosidase (BG; EC 3.2.1.21) (Toushik et al., 2017; Ejaz et al., 2021).

In the food industry, cellulases enzymes are used in combination with pectinases, amylases and hemicellulases. These enzymes are used in a variety of processes to increase extraction and clarification of fruit and vegetable juices, increase product yield, reduce juice viscosity, maintain a high filtration rate, inçrease aroma and volatile characteristics, and improve product properties (Juturu and Wu, 2014; Singh et al., 2021). In fruit juice extraction the presence of polysaccharides hinders the process. The addition of these enzymes decreases cell wall resistance and solubilizes polysaccharides, improving extraction and clarification, and increasing yield (Carvalho et al., 2008). Also, the fibres (which are of cellulosic nature) present in fruit juices, being insoluble and dense, cause problems to further extraction. The addition of enzymes acts by removing such fibres and facilitating juice filtration (Shariq and Sohail, 2019).

In winemaking the use of enzymes provides many advantages. Cellulosic biomaterials are converted into fermentable sugars by the action of the enzyme cellulase; the sugars are then converted into alcohol by the yeast. The use of β-glucosidases by hydrolysis of glycosylated precursors into glucose and aglycones improves the aroma of wines (Raveendran et al., 2018).

In the production of beverages, such as wine and beer, as well as in obtaining oligosaccharides that have functional activities, β-1,3-glucanases have different applications; these enzymes act on numerous natural substances that have β-1,3-glycosidic bonds (Blasco et al., 2006). β-Glucanases consist of multifunctional enzymes that degrade as β-glucans and hydrolyse polysaccharides such as cellulose. The β-1,3-D-glucanases or laminarinases consist of extracellular enzymes that perform the hydrolysis of β-1,3-D-glucan. These enzymes can be presented in two forms – endo and exo – where they synergistically perform the total hydrolysis of β-D-glucan. The exo form β-1,3-D-glucanase (EC.3.2.1.58) hydrolyses β-1,3 bonds in the non-reducing region of the β-glucan chain, while the endo form β-1,3-D-glucanase (EC.3.2.1.39) randomly hydrolyses β-1,3 bonds within the β-glucan chain. The β-1,3(4)-glucanases are capable of hydrolysing either the β- (1,3) or the β- (1,4) bonds of the β-D-glucan. Finally, β-1,3-1,4-D-glucanases exhibit a strict substrate particularity for breaking β-1,4 glycosidic bonds in 3-O-substituted glucopyranose units (Blasco et al., 2006; Chaari and Chaabouni, 2018). Therefore, the addition of β-glucanases to beverage production processes is important to promote the decrease of viscosity and increase the release of soluble sugars present in malt and other cereals.

Another growing demand for enzymes is associated with foods with a high lactose content, which can act as an anti-nutritional factor for intolerant organisms. Milk and dairy products are an important source of protein and minerals; however, around 70% of the adult population is not able to completely digest the lactose present in these foods, and this fact has discouraged the consumption of these products. The best enzyme known by the food industry to alleviate the situation and make it possible for lactose intolerant people to consume dairy products are β-galactosidases (lactases), that catalyse the hydrolytic process of β-1,4-D-galactosidic linkages found in lactose and releases D-glucose and D-galactose as product. Enzymes are produced by microorganisms, and when these enzymes are added directly to milk, a lactose-free product is obtained (Lomer et al., 2007; Mattar et al., 2012).

Enzymatic processes associated with food processing are a growing market and have presented important results for the food science field. Phytase enzymes, carbohydrases, proteases, and secondary metabolites have shown potential to improve the quality of vegetables, grains, and cereals, associated with the reduction of ANF content and improved nutrient bioavailability, whether endogenously via germination and fermentation processes, or by the addition of enzymes as a biological inoculant agent.

10.4.3 Industrial Producers of Enzymes to Reduce ANF in Food

The commercial production of enzymes has been gaining prominence due to its extensive applications in industrial sectors (Table 10.3). Enzymes have a wide range of industrial applications from food, feed, and beverages to pulp and paper, detergent, textile, and bioenergy industries. It has been a field of production whose market in 2015 was valued at 4.9 billion US dollars with growth prospects, by 2021, of 5% (Kumar, 2020). Industrial products that use enzymes in their formulation are more economical and efficient, since enzymes are highly specific, environmentally friendly, and act under mild process conditions. With the help of enzyme engineering, this sector has been evolving rapidly, providing better quality products, and optimizing enzyme production processes (Zhu et al., 2019; Kumar, 2020; Sharma et al., 2021). The advances in recombinant DNA technologies, together with enzyme engineering, have made the production process more accessible and facilitate the handling of microorganisms, so that the commercial demand can be met; enzymes which are generated by microorganisms can be genetically manipulated according to the needs of the industrial application (Sharma et al., 2021).

Industrial enzymes can be obtained from a wide range of sources such as plants, animals, and microorganisms. However, microbial production is preferable because microorganisms are more stable than plant and animal enzymes. Microbial enzymes play an important role in the food industry, whether by bacteria, yeasts, or fungi (Sharma et al., 2021; Tarafdar et al., 2021). Many recent works have cited the use of fungi in the production of enzymes due to their ability to secrete extracellular enzymes in great quantity and their fast growth under low-cost media; so, can be mentioned the fungi of the genus *Aspergillus* (Adsul et al., 2020; Kumar, 2020). Among the most produced enzymes via *Aspergillus* can be cited the digestive enzymes lipase, protease, keratinase, laccase, β-glucosidase, and many others (Kumar, 2020; Chen et al., 2021; Yahya et al., 2021). When referring to a large-scale production, it is important to remember the fungal cell factories such as *Aspergillus* sp., *Trichoderma* sp., *Candida* sp., *Penicillium* sp., and *Rhizopus* (Kumar, 2020). Proteases can also be biosynthesized by bacteria of the genera *Bacillus*, *Pseudomonas*, and *Clostridium*; amylases are also obtained with high commercial quality through *Bacillus* sp. (Sharma et al., 2021).

Most commercial enzyme production processes occur by SmF or SSF. The SmF system is easy to operate and monitor, making it the mode of choice to produce most industrial enzymes. However, some specific advantages of the SSF system, in particular the use of fungi, are preferred by some industries to produce various enzymes (Thomas et al., 2013; Patel et al., 2016). Many researches have used SSF because it is an environmentally correct methodology where the industrial solid residue can be directly applied as substrate in fermentation, such as wineries and breweries, olive mill residues, sunflower cake, sugarcane bagasse, fruit peels and pulp, cereals (Benabda et al., 2019). For SSF, the most suitable microorganisms for use are filamentous fungi due to their growth in myceli that colonizes the solid substrate (Benabda et al., 2019). A study using *Rhizopus oryzae* to produce two enzymes (amylase and protease) by SSF using wheat bran as substrate obtained positive results, and amylase production can be increased to bakery products contributing to the reduction of antinutritional factors (Benabda et al., 2019).

The class of lipases performs the hydrolysis of triacylglycerides into glycerol and free acids, as lipases are digestive enzymes and, for this reason, they are of great interest in the food industry as they can be used to reduce ANFs in foods. Extremophilic lipases have the characteristic of thermostability, which makes them even more desirable in the industrial environment. The new thermostable lipase from *Geobacillus stearothermophilus* T1.2RQ, is an α/β hydrolase that has excellent hydrolytic and transesterification activity and is promising for application in food industries. For large-scale production of this enzyme, and in an accessible way, *Bacillus subtilis* has shown great potential, especially for its extracellular secretion capacity that helps to simplify the purification process, making lipase an endotoxin-free enzyme and safe to be used in the food and drugs industries (Elemosho et al., 2021).

β-Galactosidase (EC 3.2.1.23), also called lactase, consists of an enzyme whose function is to catalyse the hydrolysis of terminal non-reducing β-d-galactose residues into β-D – galactosides. The main application of this enzyme is in the hydrolysis of lactose in dairy products processing (Oliveira et al., 2011). This enzyme is essential in the food industry, as it is used in the production of lactose-free products. Due to the growing number of people who cannot digest this carbohydrate, the development of several lactose-restricted products is increasing. Thus, the commercialization of β-galactosidases has presented great market potential (Albuquerque et al., 2021; Argenta et al., 2021).

Several sources can be used to obtain β-galactosidase, such as plants, animals, and microorganisms such as fungi, bacteria, and yeasts. Traditionally, the most used β-galactosidases in industries are obtained from *Aspergillus* spp. and *Kluyveromyces* spp. (Husain, 2010). Duringβ-galactosidase production from *Aspergillus* spp., the enzyme is secreted into the extracellular medium, facilitating enzyme recovery. Such fungal enzymes have an optimum pH in the acidic range (2.5–5.4) and a high optimum temperature, allowing their use at temperatures of up to 50°C. In contrast, with *Kluyveromyces* spp. β-galactosidase is produced and stored intracellularly; also the yeast enzyme has an almost neutral optimum pH (6.0–7.0), enabling a wider range of applications (Panesar et al., 2006; Domingues et al., 2010).

In addition to the cited microorganisms, bacteria like *Pyrococcuswoesei*, *Thermus* sp., and *Bacillus stearothermophilus* are mentioned to produce thermostable β-galactosidase. Furthermore,

TABLE 10.3
Producers and Suppliers of Commercial Enzymes to Reduce ANFs in Food

Enzyme	Use	Productname	Supplier
β-Galactosidase (lactase)	Hydrolyses lactose[a]	Maxilact®	DSM
		Fungal Lactase	American Laboratories Inc.
		Dairyzym Lactase	SternEnzym
		GODO-YNL2 Lactase	Dupont Nutrition & Health
		Lactase 14-DS	Amano Enzyme U.S.A. Co., Ltd.
		BioCore® Kids-I	National Enzyme Company
		BioCore® Dairy Ultra	
		BioCore® Dairy	
		BioCore® Kids	
		DairyIntolerance	
		BioCore® OptimumComplete-I	
		BioCore® Dairy-I	
		DENAZYME GY2	NAGASE & CO., LTD.
		Lactozym® Pure	Novozymes
α-Galactosidase	Hydrolyses indigestiblesugars[a]	α-Galactosidase DS	Amano Enzyme U.S.A. Co., Ltd.
		α-Galactosidase	Nutriteck
α-Galactosidaseand lactase	Hydrolyses lactose and indigestible sugars[a]	BioCore® Optimum Complete	National Enzyme Company
		BioCore® OptimumComplete-I	
Phytase	Hydrolyses the phosphate residues from phytic acid[b]	Phytase	American Laboratories Inc.
		Natuphos ®	Basf
		Ronozyme ® HiPhos	Novozymes
		Nutrizyme®PHY-P	Sunsonzymes
		PhytaseEnzyme	Starzyme
Endo-1,4- β-xylanase e endo-1,4- β-glucanase	Hydrolyse pentosans and beta-glucans[c]	Natugrain® TS	BASF
Hemicellulases, β-glucanase and phytase	Improve the bioavailability of nutrients from vegetable and other plant foods. Hydrolyses the phosphate residues from phytic acid[c]	ProCerelase®	National Enzyme Company
		CereCalase®	
Cellulases	Hydrolyse, or degrade the carbohydrate cellulose[c]	Cellulase	American Laboratories Inc.
		Cellulase	ProFood International, Inc.
		SQzyme CS P	Suntaq InternationalLtd.
Hemicellulases	Hydrolyse galactomannoglucans, mannans, and glucomannans[c]	PastazymHemicellulase	Stern Enzym
		Hemicellulase	American Laboratories Inc.
Cellulase and hemicellulases	Pectolytic properties with a high cellulosic and hemicellulosic activity[c]	Endozym® Rouge	AEB Group
Cellulase and β-glucosidase	Hydrolyses 1.4-beta-D-glycosidic linkages in cellulose, lichenin, and cereal beta-D-glucans[c]	Cellulase DS	Amano Enzyme U.S.A. Co., Ltd.
		SQzyme CS L	Suntaq InternationalLtd.

Notes:

[a] Ianiro et al. (2016); Vandenberghe et al. (2019).

[b] Song et al. (2019).

[c] Toushik et al. (2017); Bajaj and Mahajan (2019); and Ejaz et al. (2021).

Arthrobacter psychrolactophilus and *Pseudoalteromonas haloplanktis* have been reported for cold adaptive β-galactosidase (Chen et al., 2008; Van de Voorde et al., 2014). *Aspergillus oryzae, Aspergillus aculeatus, Penicillium expansum, Penicillium chrysogenum,* and *Paecilomyces aeruginеus* are the fungal sources mentioned to produce β-galactosidase (Katrolia et al., 2011; Frenzel et al., 2015). Comparing both production sources, β-galactosidase from bacterial sources is widely used for lactose hydrolysis because of their ease of fermentation, high enzymatic activity, and good stability (Anisha, 2017).

As mentioned, *Aspergillus* spp. and *Kluyveromyces* spp. are traditionally used in industry to obtain β-galactosidase. This is because the products obtained from these organisms have "generally recognized as safe" (GRAS) status for human consumption. Microorganisms like *Kluyveromyces lactis, Kluyveromyces fragilis, Kluyveromycesmarxianus, Candida kefyr, Saccharomyces cerevisiae*, and the fungi *Aspergillus niger* and *Aspergillus oryzae* are the main lactase producers (Ahmed and Satar, 2012; Saqib et al., 2017).

α-Galactosidase, also called melibiase (α-D-galactosidegalactohydrolase, EC 3.2.1.22), consists of an exoglycosidase whose function is to cleave the non-reducing terminal of galactose residues from α-D-galactooligosaccharides and polysaccharides. A-galactosisase is named melibiase because this enzyme cleaves the α-1,6 bond between galactose and glucose D-Gal-α (1→6)-D-Glc) into melibiose (Bhatia et al., 2020). Several sources and substrates are mentioned to produce α-galactosidase, such as animals, plants, bacteria, and fungi. However, microbial production is the most used. Microorganisms have great potential to produce α-galactosidases. Among these microbial sources of production, filamentous fungi are extensively used because they can grow in cheap sources of agricultural waste and secrete high levels of enzymes into the culture medium, a fact that contributes to reducing the cost of enzyme production (Aleksieva et al., 2010).

α-Galactosidase is produced by the SmF technique. Generally, extracellular α-galactosidases are produced by *Aspergillus* species (Manzanares et al., 1998; Liu et al., 2007; Ferreira et al., 2011). *Trichoderma* spp. and *Penicillium* spp. species also can be mentioned like producers of α-galactosidase (Golubev et al., 2004; Rezende et al., 2005). Intracellular production of α-galactosidase is mentioned within thermophilic fungus *Humicola* sp. (Kotwal et al., 1999). Phytases are commonly found in nature and can be obtained from a variety of sources, including plants, animals, and microorganisms; depending on the source they present different properties related to thermostability and optimal pH range due to the numerous different molecular forms produced (Jorquera et al., 2008; Chen et al., 2018). Phytases consist of a 350 kDa molecular enzyme, active within pH range of 4.5–6.0 and stability decreases dramatically when the pH value is less than 3.0 and greater than 7.5. The optimum temperature for phytases ranges from 25°C to 80°C. For example, a thermophilic fungus species, *Thermomyceslanuginosus*, has excellent phytase activity at 65°C. Mesophilic fungi such as *A. fumigatus* and *A. niger NRRL 3135* show optimal activity at 37°C and 55°C, respectively. The optimal temperature of the phytase produced by *Thermoascus auranticus* is 55°C (Shimizu, 1992; Pasamontes et al., 1997; Gupta et al., 2015).

10.5 CONCLUSIONS

Food production all over the world has always faced big challenges including, producing sufficient food for a population that is increasing day by day, of great quality in terms of nutritional value and taste, that is accessible, of a suitable cost, using a large amount of raw material, producing the smallest amount possible of residue, and with the best reuse possible. In this sense industries have put great efforts into developing processes that are increasingly efficient, more and more competitive, and using advanced technologies. Particularly regarding ANFs in food, removing in part or in total their concentration from the food is an important step to help industries reach their objective, since are substances that are present in the natural composition of many raw materials and makes it difficult the suitable use of the nutrients presents at these foods. Therefore, the reduction in the ANFs concentration contributes to the efficient use of the nutrients of consumed food. Another point discussed is the presence of some products in food that are not digested by humans and animals due to their stomach constitution or the lack of some digestive enzymes in their digestion. So, the food must be industrially processed before ingestion, and many industries can do this through adequate processing. Chemical, physical, thermal, and enzymatic processes are used for reduction of these compounds in foods, and enzymes, due to the great variety available, the specificity, and their biofriendly characteristics, play an important and fundamental role in this. Enzymes are generally extracted from the natural products, or produced by microorganisms classified as safe, in mild conditions, including low chemical products concentration. The market for enzymes involved in ANF removal from food is expanding and is a promising industrial sector all over the world. Besides, there is always demanded to find new and more efficient processes and products to remove ANFs from food for human and animal consumption.

REFERENCES

Adetunji, A. E., B. V. Sershen, and N. Pammenter. 2021. Effects of exogenous application of five antioxidants on vigour, viability, oxidative metabolism and germination enzymes in aged cabbage and lettuce seeds. *South African Journal of Botany* 137:85–97. https://doi.org/10.1016/j.sajb.2020.10.001.

Adeyemo, S. M., and A. A. Onilude. 2013. Enzymatic reduction of anti-nutritional factors in fermenting soybeans by *Lactobacillus plantarum* isolates from fermenting cereals. *Nigerian Food Journal* 31:84–90. https://doi.org/10.1016/s0189-7241(15)30080-1.

Adsul, M., S. K. Sandhu, R. R. Singhania, R. Gupta, S. K. Puri, and A. Mathur. 2020. Designing a cellulolytic enzyme cocktail for the efficient and economical conversion of lignocellulosic biomass to biofuels. *Enzyme and Microbial Technology* 133:109442. https://doi.org/10.1016/j.enzmictec.2019.109442.

Aguilar, J., A. C. Miano, J. Obregón, J. Soriano-Colchado, and G. Barraza-Jáuregui. 2019. Malting process as an alternative to obtain high nutritional quality quinoa flour. *Journal of Cereal Science* 90:102858. https://doi.org/10.1016/j.jcs.2019.102858.

Ahmed, S., and R. Satar. 2012. Recombinant β-galactosidases – past, present and future: a mini review. *Journal of Molecular Catalysis B: Enzymymatic* 81:1–6. https://doi.org/10.1016/j.molcatb.2012.04.012.

Albuquerque, T. L., M. Sousa, N. C. G. Silva, et al. 2021. β-galactosidase from *Kluyveromyces lactis*: characterization, production, immobilization and applications – A review. *Internacional Journal of Biological Macromoles* 191:881–98. https://doi.org/10.1016/j.ijbiomac.2021.09.133.

Aleksieva, P., B. Tchorbanov, and L. Nacheva. 2010. High-yield production of α-galactosidase excreted from *Penicillium chrysogenum* and *Aspergillus niger. Biotechnology & Biotechnological Equipment* 24:1620–23. https://doi.org/10.2478/V10133-010-0015-5.

Alonso, R., E. Orúe, and F. Marzo. 1998. Effects of extrusion and conventional processing methods on protein and antinutritional factor contents in pea seeds. *Food Chemistry* 63:505–12. https://doi.org/10.1016/S0308-8146(98)00037-5.

Anisha, G. S. 2017. β-Galactosidases. In *Current Developments in Biotechnology and Bioengineering*, ed. A. Pandey, S. Negi, and C. R. Soccol, 395–421. Elsevier. https://doi.org/10.1016/B978-0-444-63662-1.00017-8.

Antoine, T., S. Georgé, A. Leca, et al. 2022. Reduction of pulse "antinutritional" content by optimizing pulse canning process is insufficient to improve fat-soluble vitamin bioavailability. *Food Chemistry* 370:131021. https://doi.org/10.1016/j.foodchem.2021.131021.

Argenta, A. B., A. Nogueira, and A. P. Scheer. 2021. Hydrolysis of whey lactose: *Kluyveromyces lactis* β-galactosidase immobilisation and integrated process hydrolysis-ultrafiltration. *International Dairy Journal* 117:105007. https://doi.org/10.1016/j.idairyj.2021.105007.

Assan, D., F. K. A. Kuebutornye, V. Hlordzi, et al. 2022. Effects of probiotics on digestive enzymes of fish (finfish and shellfish); status and prospects: a mini review. *Comparative Biochemistry and Physiology Part B-Biochemistry and Molecular Biology* 257:110653. https://doi.org/10.1016/j.cbpb.2021.110653.

Awolu, O. O., O. S. Omoba, O. Olawoye, et al. 2017. Optimization of production and quality evaluation of maize-based snack supplemented with soybean and tiger-nut (*Cyperus esculenta*) flour. *Food Science and Nutrition* 5:3–13. https://doi.org/10.1002/fsn3.359.

Bajaj, P., and R. Mahajan. 2019. Cellulase and xylanase synergism in industrial biotechnology. *Applied Microbiology and Biotechnology* 103:8711–24. https://doi.org/10.1007/s00253-019-10146-0.

Benabda, O., S. M'hir, M. Kasmi, W. Mnif, and M. Hamdi. 2019. Optimization of protease and amylase production by *Rhizopus oryzae* cultivated on bread waste using solid-state fermentation. *Journal of Chemistry* 2019:1–9. https://doi.org/10.1155/2019/3738181.

Bessada, S. M. F., J. C. M. Barreira, and M. B. P. P. Oliveira. 2019. Pulses and food security: dietary protein, digestibility, bioactive and functional properties. *Trends in Food Science & Technology* 93:53–68. https://doi.org/10.1016/j.tifs.2019.08.022.

Bhandari, M. R., and J. Kawabata. 2006. Cooking effects on oxalate, phytate, trypsin and α-amylase inhibitors of wild yam tubers of Nepal. *Journal of Food Composition and Analysis* 19:524–30. https://doi.org/10.1016/j.jfca.2004.09.010.

Bhatia, S., A. Singh, N. Batra, and J. Singh. 2020. Microbial production and biotechnological applications of α-galactosidase. *International Journal of Biological Macromolecules* 150:1294–1313. https://doi.org/10.1016/j.ijbiomac.2019.10.140.

Bhinder, S., S. Kumari, B. Singh, A. Kaur, and N. Singh. 2021. Impact of germination on phenolic composition, antioxidant properties, antinutritional factors, mineral content and maillard reaction products of malted quinoa flour. *Food Chemistry* 346:128915. https://doi.org/10.1016/j.foodchem.2020.128915.

Blasco, L., P. Veiga-Crespo, M. Poza, and T. G. Villa. 2006. Hydrolases as markers of wine aging. *World Journal of Microbiology and Biotechnology* 22:1229–33. https://doi.org/10.1007/s11274-006-9165-x.

Carvalho, L. M. J., I. M. Castro, and C. A. B. da Silva. 2008. A study of retention of sugars in the process of clarification of pineapple juice (*Ananascomosus*, L. Merril) by micro- and ultra-filtration. *Journal of Food Engineering* 87:447–54. https://doi.org/10.1016/j.jfoodeng.2007.12.015.

Chaari, F., and S. E. Chaabouni. 2018. Fungal β-1,3-1,4-glucanases: production, proprieties and biotechnological applications. *Journal of the Science of Food Agriculture* 99:2657–64. https://doi.org/10.1002/jsfa.9491.

Chakraborty, A. J., S. Mitra, T. E. Tallei, et al. 2021. Bromelain a potential bioactive compound: a comprehensive overview from a pharmacological perspective. *Life* 11:317–43. https://doi.org/10.3390/life11040317.

Chen, K. I., C. Y. Chiang, C. Y. Ko, H. Y. Huang, and K. C. Cheng. 2018. Reduction of phytic acid in soymilk by immobilized phytase system. *Journal of Food Science* 83:2963–69. https://doi.org/10.1111/1750-3841.14394.

Chen, X., Y. Lu, A. Zhao, Y. Wu, Y. Zhang, and X. Yang. 2022. Quantitative analyses for several nutrients and volatile components during fermentation of soybean by *Bacillus subtilis natto. Food Chemistry* 374:131725. https://doi.org/10.1016/j.foodchem.2021.131725.

Chen, W., C. Zheng, Z. Jin, et al. 2021. Evaluation of *Yarrowia lipolytica* lipase 2 on growth performance, digestive enzyme activity and nutritional components of Russian sturgeon (*Acipensergueldenstaedtii*). *Iranian Journal of Fisheries Sciences* 20:396–409. https://doi.org/10.22092/ijfs.2021.123835.

Chen, W., H. Chen, Y. Xia, J. Zhao, F. Tian, and H. Zhang. 2008. Production, purification, and characterization of a potential thermostable galactosidase for milk lactose hydrolysis from *Bacillus stearothermophilus. Journal of Dairy Science* 91:1751–58. https://doi.org/10.3168/jds.2007-617.

Cohen, S. A. 2016. The clinical consequences of sucrase-isomaltase deficiency. *Molecular and Cellular Pediatrics* 3:3–6. https://doi.org/10.1186/s40348-015-0028-0.

Colla, L. M., N. E. Kreling, M. T. Nazari, et al. 2020. Use and applications of *Aspergillus niger* for the development of enzymes and products of biotechnological interest. In *Aspergillus niger: Pathogenicity, Cultivation and Uses*, ed. E. Baughan, 123–60. Nova Publishers.

Cui, J., P. Xia, L. Zhang, Y. Hu, Q. Xie, and H. Xiang. 2020. A novel fermented soybean, inoculated with selected *Bacillus*, *Lactobacillus* and *Hansenula* strains, showed strong antioxidant and anti-fatigue potential activity. *Food Chemistry* 333:127527. https://doi.org/10.1016/j.foodchem.2020.127527.

Domingues, L., P. M. R. Guimarães, and C. Oliveira. 2010. Metabolic engineering of *Saccharomyces cerevisiae* for lactose/whey fermentation. *Bioengineered Bugs* 1:164–71. https://doi.org/10.4161/bbug.1.3.10619.

Ejaz, U., M. Sohail, and A. Ghanemi. 2021. Cellulases: from bioactivity to a variety of industrial applications. *Biomimetics* 6:44. https://doi.org/10.3390/biomimetics6030044.

Elemosho, R., A. Suwanto, and M. Thenawidjaja. 2021. Extracellular expression in *Bacillus subtilis* of a thermostable *Geobacillus stearothermophilus* lipase. *Electronic Journal of Biotechnology* 53:71–9. https://doi.org/10.1016/j.ejbt.2021.07.003.

Fasolin, L. H., R. N. Pereira, A. C. Pinheiro, et al. 2019. Emergent food proteins – Towards sustainability, health and innovation. *Food Research International* 125:108586. https://doi.org/10.1016/j.foodres.2019.108586.

Ferreira, J. G., A. P. Reis, V. M. Guimarães, D. L. Falkoski, L. S. Fialho, and S. T. de Rezende. 2011. Purification and characterization of *Aspergillus terreus*α-galactosidases and their use for hydrolysis of soymilk oligosaccharides. *Applied Biochemistry and Biotechnology* 164:1111–25. https://doi.org/10.1007/s12010-011-9198-y.

Frenzel, M., K. Zerge, I. Clawin-Rädecker, and P. C. Lorenzen. 2015. Comparison of the galacto-oligosaccharide forming activity of different β-galactosidases. *Food Science and Technology* 60:1068–71. https://doi.org/10.1016/j.lwt.2014.10.064.

Frias, J., C. Vidal-Valverde, C. Sotomayor, C. Diaz-Pollan, and G. Urbano. 2000. Influence of processing on available carbohydrate content and antinutritional factors of chickpeas. *European Food Research and Technology* 210:340–45. https://doi.org/10.1007/s002170050560.

García-Estepa, R. M., E. Guerra-Hernández, and B. García-Villanova. 1999. Phytic acid content in milled cereal products and breads. *Food Research International* 32:217–21. https://doi.org/10.1016/S0963-9969(99)00092-7.

Gemede, H. F. 2014. Antinutritional factors in plant foods: Potential health benefits and adverse effects. *International Journal of Nutrition and Food Science* 3:284–89. https://doi.org/10.11648/j.ijnfs.20140304.18.

Gibson, R. S., L. Perlas, and C. Hotz. 2006. Improving the bioavailability of nutrients in plant foods at the household level. *Proceedings of the Nutrition Society* 65:160–68. https://doi.org/10.1079/PNS2006489.

Glusac, J., and A. Fishman. 2021. Enzymatic and chemical modification of zein for food application. *Trends in Food Science & Technology* 112:507–17. https://doi.org/10.1016/j.tifs.2021.04.024.

Golubev, A. M., R. A. P. Nagem, J. R. Brandão Neto, et al. 2004. Crystal structure of α-galactosidase from *Trichoderma reesei* and its complex with galactose: implications for catalytic mechanism. *Journal of Molecular Biology* 339:413–22. https://doi.org/10.1016/j.jmb.2004.03.062.

Gote, M., H. Umalkar, I. Khan, and J. Khire. 2004. Thermostable α-galactosidase from *Bacillus searothermophilus* (NCIM 5146) and its application in the removal of flatulence causing factors from soymilk. *Process Biochemistry* 39:1723–29. https://doi.org/10.1016/j.procbio.2003.07.008.

Gupta, R. K., S. S. Gangoliya, and N. K. Singh. 2015. Reduction of phytic acid and enhancement of bioavailable micronutrients in food grains. *Journal of Food Science and Technology* 52:676–84. https://doi.org/10.1007/s13197-013-0978-y.

Hill, G. D. 2003. Plant antinutritional factors: characteristics. *Encyclopedia of Food Sciences and Nutrition* (Second Edition), 4578–87. https://doi.org/10.1016/B0-12-227055-X/01318-3.

Hjort, C. 2007. Industrial enzyme production for food applications. In *Novel Enzyme Technology for Food Applications*, ed. R. Rastall, 43–59. Elsevier B.V. https://doi.org/10.1533/9781845693718.1.43.

Husain, Q. 2010. β-galactosidases and their potential applications: a review. *Critical Reviews in Biotechnology* 30:41–62. https://doi.org/10.3109/07388550903330497.

Husein, D. M., S. Rizk, and H. Y. Naim. 2021. Differential effects of sucrase-isomaltase mutants on its trafficking and function in irritable bowel syndrome: similarities to congenital sucrase-isomaltase deficiency. *Nutrients* 13:9–17. https://doi.org/10.3390/nu13010009.

Ianiro, G., S. Pecere, V. Giorgio, A. Gasbarrini, and G. Cammarota. 2016. Digestive enzyme supplementation in gastrointestinal diseases. *Current Drug Metabolism* 17:187–93. https://doi.org/10.2174/138920021702160114150137.

Irakli, M., A. Lazaridou, and C. G. Biliaderis. 2020. Comparative evaluation of the nutritional, antinutritional, functional, and bioactivity attributes of rice bran stabilized by different heat treatments. *Foods* 10:57. https://doi.org/10.3390/foods10010057.

Jadhav, S. J., E. P. Sharma, and D. K. Salunkhe. 1981. Naturally occurring toxic alkaloids in foods. *Critical Review in Toxicology* 9:21–104. https://doi.org/10.3109/10408448109059562.

Jazi, V., H. Mohebodini, A. Ashayerizadeh, A. Shabani, and R. Barekatain. 2019. Fermented soybean meal ameliorates *Salmonella typhimurium* infection in young broiler chickens. *Poultry Science* 98:5648–60. https://doi.org/10.3382/ps/pez338.

Jeyakumar, E., and Lawrence R. 2022. Microbial fermentation for reduction of antinutritional factors. In *Current Developments in Biotechnology and Bioengineering*, ed. A. K. Rai, S. P. Singh, A. Pandey, C. Larroche, C. R. Soccol, 239–60. Elsevier B.V. https://doi.org/10.1016/B978-0-12-823506-5.00012-6.

Jezierny, D., R. Mosenthin, and E. Bauer. 2010. The use of grain legumes as a protein source in pig nutrition: a review. *Animal Feed Science and Technology* 157:111–28. https://doi.org/10.1016/j.anifeedsci.2010.03.001.

Jiménez-Martínez, C., H. Hernández-Sánchez, and G. Dávila-Ortiz. 2007. Diminution of quinolizidine alkaloids, oligosaccharides and phenolic compounds from two species oflupinus and soybean seeds by the effect of *Rhizopus oligosporus*. *Journal of the Science of Food and Agriculture* 87:1315–22. https://doi.org/10.1002/jsfa.2851.

Jorquera, M., O. Martínez, F. Maruyama, P. Marschner, and M. L. Mora. 2008. Current and future biotechnological applications of bacterial phytases and phytase-producing bacteria. *Microbes Environ* 23:182–191. https://doi.org/10.1264/jsme2.23.182.

Juturu, V., and J. C. Wu. 2014. Microbial cellulases: engineering, production and applications. *Renewable and Sustainable Energy Reviews* 33:188–203. https://doi.org/10.1016/j.rser.2014.01.077.

Kalpanadevi, V., and V. R. Mohan. 2013. Effect of processing on antinutrients and in vitro protein digestibility of the underutilized legume, *Vigna unguiculata* (L.) Walp subsp. *unguiculata. Food Science and Technology* 51:455–61. https://doi.org/10.1016/j.lwt.2012.09.030.

Katrolia, P., Q. Yan, H. Jia, Y. Li, Z. Jiang, and C. Song. 2011. Molecular cloning and high-level expression of a β-galactosidase gene from *Paecilomycesaerugineus* in *Pichia pastoris. Journal of Molecular Catalysis B.: Enzymatic* 69:112–19. https://doi.org/10.1016/j.molcatb.2011.01.004.

Khattab, R. Y., and S. D. Arntfield. 2009. Nutritional quality of legume seeds as affected by some physical treatments 2. Antinutritional factors. *Food Science and Technology* 42:1113–18. https://doi.org/10.1016/j.lwt.2009.02.004.

Koca, S. B., E. Acar, and M. Naz. 2018. The effects of seasonal, sex and size on the digestive enzyme activities in freshwater crayfish (*Astacusleptodactylus* Eschscholtz, 1823). *Aquacture Research* 49:1598–1605. https://doi.org/10.1111/are.13615.

Kong, H., F. Wu, X. Jiang, et al. 2019. Nano-TiO_2 impairs digestive enzyme activities of marine mussels under ocean acidification. *Chemosphere* 237:124561. https://doi.org/10.1016/j.chemosphere.2019.124561.

Kotwal, S. M., M. M. Gote, M. I. Khan, and J. M. Khire. 1999. Production, purification and characterization of a constitutive intracellular β-galactosidase from the thermophilic fungus *Humicola* sp. *Journal of Industrial Microbiology and Biotechnology* 23:661–67. https://doi.org/10.1038/sj.jim.2900680.

Krogdahl, Å., T. M. Kortner, and R. W. Hardy. 2022. Antinutrients and adventitious toxins. In *Fish Nutrition*, ed. R. W. Hardy, and S. J. Kaushik, 775–821. Elsevier B.V. https://doi.org/10.1016/B978-0-12-819587-1.00001-X.

Kumar, A. 2020. *Aspergillus nidulans*: A potential resource of the production of the native and heterologous enzymes for industrial applications. *International Journal of Microbiology* 2020. https://doi.org/10.1155/2020/8894215.

Lampart-Szczapa, E., A. Siger, K. Trojanowska, M. Nogala-Kalucha, M. Malecka, and B. Pacholek. 2003. Chemical composition and antibacterial activities of lupin seeds extracts. *Nahrung/Food* 47:286–90. https://doi.org/10.1002/food.200390068.

Larrosa, A. P. Q., and D. M. Otero. 2021. Flour made from fruit by-products: characteristics, processing conditions, and applications. *Journal of Food Processing and Preservation* 45:e15398. https://doi.org/10.1111/jfpp.15398.

Lee, S., and Y. M. Koo. 2001. Pilot-scale production of cellulase using *Trichoderma reesei* RUT C-30 fed-batch mode. *Journal of Microbiology and Biotechnology* 11:229–33.

Li, W., and T. Wang. 2021. Effect of solid-state fermentation with *Bacillus subtilis* lwo on the proteolysis and the antioxidative properties of chickpeas. *International Journal of Food Microbiology* 338:108988. https://doi.org/10.1016/j.ijfoodmicro.2020.108988.

Liu, C., H. Ruan, H. Shen, et al. 2007. Optimization of the fermentation medium for α-galactosidase production from *Aspergillus foetidus* ZU-G1 using response surface methodology. *Journal of Food Science* 72:120–25. https://doi.org/10.1111/j.1750-3841.2007.00328.x.

Lomer, M. C. E., G. C. Parkes, and J. D. Sanderson. 2007. Review article: lactose intolerance in clinical practice – myths and realities. *Alimentary Pharmacology & Therapeutics* 27:93–103. https://doi.org/10.1111/j.1365-2036.2007.03557.x.

Luo, Y., W. H. Xie, X. X. Jin, Q. Wang, and Y. J. He. 2014. Effects of germination on iron, zinc, calcium, manganese, and copper availability from cereals and legumes. *CyTA – Journal of Food* 12:22–6. https://doi.org/10.1080/19476337.2013.782071.

Manzanares, P., L. H. Graaff, and J. Visser. 1998. Characterization of galactosidases from *Aspergillus niger*: Purification of a novel α-galactosidase activity. *Enzyme and Microbial Technology* 22:383–90. https://doi.org/10.1016/S0141-0229(97)00207-X.

Margier, M., S. Georgé, N. Hafnaoui, et al. 2018. Nutritional composition and bioactive content of legumes: characterization of pulses frequently consumed in France and effect of the cooking method. *Nutrients* 10:1668. https://doi.org/10.3390/nu10111668.

Maske, B. L., G. V. de M. Pereira, A. da S. Vale, et al. 2021. A review on enzyme-producing *Lactobacilli* associated with the human digestive process: from metabolism to application. *Enzyme and Microbial Technology* 149:109836. https://doi.org/10.1016/j.enzmictec.2021.109836.

Mattar, R., D. F. C. Mazo, and F. J. Carrilho. 2012. Lactose intolerance: Diagnosis, genetic, and clinical factors. *Clinical and Experimental Gastroenterolology* 2012:113–21. https://doi.org/10.2147/CEG.S32368.

Mesfin, N., A. Belay, and E. Amare. 2021. Effect of germination, roasting, and variety on physicochemical, techno-functional, and antioxidant properties of chickpea (*Cicer arietinum* L.) protein isolate powder. *Heliyon* 7:e08081. https://doi.org/10.1016/j.heliyon.2021.e08081.

Mihrete, Y. 2019. Review on antinutritional factors and their effect on mineral absorption. *Acta Scientific Nutritional Health* 3:84–9.

Morzelle, M. C., J. M. Salgado, A. P. Massarioli, et al. 2019. Potential benefits of phenolics from pomegranate pulp and peel in Alzheimer's disease: antioxidant activity and inhibition of acetylcholinesterase. *Journal of Food Bioactives* 5. https://doi.org/10.31665/JFB.2019.5181.

Mukherjee, R., R. Chakraborty, and A. Dutta. 2015. Role of fermentation in improving nutritional quality of soybean meal – a review. *Asian-Australasian Journal of Animal Sciences* 29:1523–29. https://doi.org/10.5713/ajas.15.0627.

Mulimani, V. H., and S. Devendra. 1998. Effect of soaking, cooking and crude α-galactosidase treatment on the oligosaccharide content of red gram flour. *Food Chemistry* 61:475–79. https://doi.org/10.1016/S0308-8146(97)00142-8.

Muzquiz, M., A. Varela, C. Burbano, C. Cuadrado, E. Guillamón, and M. M. Pedrosa. 2012. Bioactive compounds in legumes: pronutritive and antinutritive actions. Implications for nutrition and health. *Phytochemistry Reviews* 11:227–44. https://doi.org/10.1007/s11101-012-9233-9.

Nichols, B. L., S. E. Avery, R. Quezada-Calvillo, et al. 2017. Improved starch digestion of sucrase-deficient shrews treated with oral glucoamylase enzyme supplements. *Journal of Pediatric Gastroenterology and Nutrition* 65:e35–e42. https://doi.org/10.1097/MPG.0000000000001561.

Nidhina, N., and S. P. Muthukumar. 2015. Antinutritional factors and functionality of protein-rich fractions of industrial guar meal as affected by heat processing. *Food Chemistry* 173:920–26. https://doi.org/10.1016/j.foodchem.2014.10.071.

Nikmaram, N., S. Y. Leong, M. Koubaa, et al. 2017. Effect of extrusion on the anti-nutritional factors of food products: an overview. *Food Control* 79:62–73. https://doi.org/10.1016/j.foodcont.2017.03.027.

Nkhata, S. G., E. Ayua, E. H. Kamau, and J. B. Shingiro. 2018. Fermentation and germination improve nutritional value of cereals and legumes through activation of endogenous enzymes. *Food Science & Nutrition* 6:2446–58. https://doi.org/10.1002/fsn3.846.

Nyambe-Silavwe, H., and G. Williamson. 2018. Chlorogenic and phenolic acids are only very weak inhibitors of human salivary α-amylase and rat intestinal maltase activities. *Food Research International* 113:452–55. https://doi.org/10.1016/j.foodres.2018.07.038.

Ojo, M. A. 2018. Changes in some antinutritional components and in vitro multienzymes protein digestibility during hydrothermal processing of *Cassia hirsutta*. *Preventive Nutrition and Food Science* 23:152–59. https://doi.org/10.3746/pnf.2018.23.2.152.

Ojo, M. A., B. I. O. Ade-Omowaye, and P. O. Ngoddy. 2018. Processing effects of soaking and hydrothermal methods on the components and in vitro protein digestibility of *Canavalia ensiformis*. *International Food Research Journal* 25:720–29.

Oliveira, C., P. M. R. Guimarães, and L. Domingues. 2011. Recombinant microbial systems for improved β-galactosidase production and biotechnological applications. *Biotechnolology Advances* 29:600–09. https://doi.org/10.1016/j.biotechadv.2011.03.008.

Ozden, E., M. E. Light, and I. Demir. 2021. Alternating temperatures increase germination and emergence in relation to endogenous hormones and enzyme activities in aubergine seeds. *South African Journal of Botany* 139:130–39. https://doi.org/10.1016/j.sajb.2021.02.015.

Panesar, P. S., R. Panesar, R. S. Singh, J. F. Kennedy, and H. Kumar. 2006. Microbial production, immobilization and applications of β-D-galactosidase. *Journal of Chemical Technology & Biotechnology* 543:530–43. https://doi.org/10.1002/jctb.1453.

Parca, F., Y. O. Koca, and A. Unay. 2018. Nutritional and antinutritional factors of some pulses seed and their effects on human health. *International Journal of Secondary Metabolite* 5:331–42. https://doi.org/10.21448/ijsm.488651.

Pasamontes, L., M. Haiker, M. Wyss, M. Tessier, and A. P. van Loon. 1997. Gene cloning, purification, and characterization of a heat-stable phytase from the fungus *Aspergillus fumigatus*. *Applied and Environmental Microbiology* 63:1696–1700. https://doi.org/10.1128/aem.63.5.1696-1700.1997.

Patel, A. K., R. R. Singhania, and A. Pandey. 2016. Novel enzymatic processes applied to the food industry. *Current Opinion in Food Science* 7:64–72. https://doi.org/10.1016/j.cofs.2015.12.002.

Pujante, I. M., M. Díaz-López, J. M. Mancera, and F. J. Moyano. 2017. Characterization of digestive enzymes protease and α-amylase activities in the thick-lipped grey mullet (*Chelonlabrosus*, Risso 1827). *Aquaculture Research* 48:367–76. https://doi.org/10.1111/are.13038.

Qi, X., and R. F. Tester. 2020. Lactose, maltose, and sucrose in health and disease. *Molecular Nutrition & Food Research* 64:1901082. https://doi.org/10.1002/mnfr.201901082.

Rao, P. U., and Y. G. Deosthale. 1982. Tannin content of pulses: varietal differences and effects of germination and cooking. *Journal of the Science of Food and Agriculture* 33:1013–16. https://doi.org/10.1002/jsfa.2740331012.

Rathod, R. P., and U. S. Annapure. 2016. Effect of extrusion process on antinutritional factors and protein and starch digestibility of lentil splits. *Food Science and Technology* 66:114–23. https://doi.org/10.1016/j.lwt.2015.10.028.

Raveendran, S., B. Parameswaran, S. B. Ummalyma, et al. 2018. Applications of microbial enzymes in food industry. *Food Technology & Biotechnology* 56:16–30. https://doi.org/10.17113/ftb.56.01.18.5491.

Rezende, S. T., V. M. Guimarães, M. de C. Rodrigues, and C. R. Felix. 2005. Purification and characterization of an α-galactosidase from *Aspergillus fumigatus. Brazilian Archives of Biology and Technology* 48:195–202. https://doi.org/10.1590/s1516-89132005000200005.

Robinson, G. H. J., J. Balk, and C. Domoney. 2019. Improving pulse crops as a source of protein, starch and micronutrients. *Nutrition Bulletin* 44:202–15. https://doi.org/10.1111/nbu.12399.

Rosa-Sibakov, N., M. Re, A. Karsma, A. Laitila, and E. Nordlund. 2018. Phytic acid reduction by bioprocessing as a tool to improve the in vitro digestibility of faba bean protein. *Journal of Agricultural and Food Chemistry* 66:10394–399. https://doi.org/10.1021/acs.jafc.8b02948.

Rose, D. R., M. M. Chaudet, and K. Jones. 2018. Structural studies of the intestinal α-glucosidases, maltase-glucoamylase and sucrase-isomaltase. *Journal of Pediatric Gastroenterology and Nutrition* 66:S11–13. https://doi.org/10.1097/MPG.0000000000001953.

Roy, S., and S. K. Dutta. 2009. Genomic and cDNA cloning, expression, purification, and characterization of chymotrypsin-trypsin inhibitor from winged bean seeds. *Bioscience, Biotechnology & Biochemistry* 73:2671–2676. https://doi.org/10.1271/bbb.90519.

Saadi, S., N. Saari, H. M. Ghazali, and M. S. Abdulkarim. 2022. Mitigation of antinutritional factors and protease inhibitors of defatted winged bean-seed proteins using thermal and hydrothermal treatments: Denaturation/unfolding coupled hydrolysis mechanism. *Current Research in Food Science* 5:207–21. https://doi.org/10.1016/j.crfs.2022.01.011.

Saéz, G. D., L. Saavedra, E. M. Hebert, and G. Zárate. 2018. Identification and biotechnological characterization of lactic acid bacteria isolated from chickpea sourdough in northwestern Argentina. *Food Science and Technology* 93:249–56. https://doi.org/10.1016/j.lwt.2018.03.040.

Salim-ur-Rehman, J. A. A., F. M. Anjum, and M. A. Randhawa. 2014. Antinutrients and toxicity in plant-based foods. In *Practical Food Safety*, ed. R. Bhat and V. M. Gómez-López, 311–39. John Wiley & Sons, Ltda. https://doi.org/10.1002/9781118474563.ch16.

Samtiya, M., R. E. Aluko, and T. Dhewa. 2020. Plant food anti-nutritional factors and their reduction strategies: an overview. *Food Production, Processing and Nutrition* 2:6. https://doi.org/10.1186/s43014-020-0020-5.

Saqib, S., A. Akram, S. A. Halim, and R. Tassaduq. 2017. Sources of β-galactosidase and its applications in food industry. *3 Biotech* 7:79–86. https://doi.org/10.1007/s13205-017-0645-5.

Sengupta, S., S. Mukherjee, P. Basak, and L. Makumder. 2015. Significance of galactinol and raffinose family oligosaccharide synthesis in plants. *Frontiers in Plant Science* 6:656. https://doi.org/10.3389/fpls.2015.00656.

Serrano, J., R. Puupponen-Pimiä, A. Dauer, A. M. Aura, and F. S. Calixto. 2009. Tannins: current knowledge of food sources, intake, bioavailability and biological effects. *Molecular Nutrition & Food Research* 53:S310–329. https://doi.org/10.1002/mnfr.200900039.

Setia, R., Z. Dai, M. T. Nickerson, E. Sopiwnyk, L. Malcolmson, and Y. Ai. 2019. Impacts of short-term germination on the chemical compositions, technological characteristics and nutritional quality of yellow pea and faba bean flours. *Food Research International* 122:263–72. https://doi.org/10.1016/j.foodres.2019.04.021.

Shariq, M., and M. Sohail. 2019. Citrus limetta peels: a promising substrate for the production of multienzyme preparation from a yeast consortium. *Bioresource and Bioprocessing* 6:43–58. https://doi.org/10.1186/s40643-019-0278-0.

Sharma, A., G. Gupta, T. Ahmad, S. Mansoor, and B. Kaur. 2021. Enzyme engineering: current trends and future perspectives. *Food Reviews International* 37:121–54. https://doi.org/10.1080/87559129.2019.1695835.

Sheikh, M. A., C. S. Saini, and H. K. Sharma. 2021. Analyzing the effects of hydrothermal treatment on antinutritional factor content of plum kernel grits by using response surface methodology. *Applied Food Research* 1:100010. https://doi.org/10.1016/j.afres.2021.100010.

Shi, J., S. J. Xue, Y. Ma, D. Li, Y. Kakuda, and Y. Lan. 2009. Kinetic study of saponins B stability in navy beans under different processing conditions. *Journal of Food Engineering* 93:59–65. https://doi.org/10.1016/j.jfoodeng.2008.12.035.

Shimizu, M. 1992. Purification and characterization of phytase from *Bacillus subtilis* (*natto*) N–77. *Bioscience, Biotechnology & Biochemistry* 56:1266–69. https://doi.org/10.1271/bbb.56.1266.

Shivanna, G. B., and G. Venkateswaran. 2014. Phytase production by *Aspergillus niger* CFR 335 and *Aspergillus ficuum* SGA 01 through submerged and solid-state fermentation. *The Science World Journal* 2014:1–6. https://doi.org/10.1155/2014/392615.

Siah, S. D., I. Konczak, I., S. Agboola, J. A. Wood, and C. L. Blanchard. 2012. *In vitro* investigations of the potential health benefits of Australian-grown faba beans (*Vicia faba* L.): chemopreventative capacity and inhibitory effects on the angiotensin-converting enzyme, α-glucosidase and lipase. *British Journal of Nutrition* 108:S123–S134. https://doi.org/10.1017/S0007114512000803.

Singh, A., S. Bajar, A. Devi, and D. Pant. 2021. An overview on the recent developments in fungal cellulase production and their industrial applications. *Bioresource Technology Reports* 14:100652. https://doi.org/10.1016/j.biteb.2021.100652.

Soccol, C. R., E. S. F. da Costa, L. A. J. Letti, S. G. Karp, A. L. Woiciechowski, and L. P. de S. Vandenberghe. 2017. Recent developments and innovations in solid state fermentation. *Biotechnology Research and Innovation* 1:52–71. https://doi.org/10.1016/j.biori.2017.01.002.

Soetan, K.O. 2008. Pharmacological and other beneficial effects of antinutritional factors in plants – a review. *African Journal of Biotechnology* 7:4713–21.

Song, H., A. F. E. Sheikha, and D. M. Hu. 2019. The positive impacts of microbial phytase on its nutritional applications. *Trends in Food Science & Technology* 86:553–62. https://doi.org/10.1016/j.tifs.2018.12.001.

Suprayogi, W. P. S, A. Ratriyanto, N. Akhirini, R. F. Hadi, W. Setyono, and A. Irawan. 2022. Changes in nutritional and antinutritional aspects of soybean meals by mechanical and solid-state fermentation treatments with *Bacillus subtilis* and *Aspergillus oryzae*. *Bioresource Technology Reports* 17:100925. https://doi.org/10.1016/j.biteb.2021.100925.

Tang, W., and H. Zhao. 2009. Industrial biotechnology: tools and applications. *Biotechnology Journal* 4:1725–39. https://doi.org/10.1002/biot.200900127.

Tarafdar, A., R. Sirohi, V. K. Gaur, et al. 2021. Engineering interventions in enzyme production: lab to industrial scale. *Bioresource Technology* 326:124771. https://doi.org/10.1016/j.biortech.2021.124771.

Tavano, O. L., A. Berenguer-Murcia, F. Secundo, and R. Fernandez-Lafuente. 2018. Biotechnological applications of proteases in food technology. *Comprehensive Reviews in Food Science and Food Safety* 17:412–36. https://doi.org/10.1111/1541-4337.12326.

Terra, W. R., R. O. Dias, and C. Ferreira. 2019. Recruited lysosomal enzymes as major digestive enzymes in insects. *Biochemical Society Transactions* 47:615–23. https://doi.org/10.1042/BST20180344.

Thakur, P., K. Kumar, N. Ahmed, et al. 2021. Effect of soaking and germination treatments on nutritional, anti-nutritional, and bioactive properties of amaranth (*Amaranthus hypochondriacus* L.), quinoa (*Chenopodium quinoa* L.), and buckwheat (*Fagopyrum esculentum* L.). *Current Research in Food Science* 4:917–25. https://doi.org/10.1016/j.crfs.2021.11.019.

Thakur, A., V. Sharma, and A. Thakur. 2019. An overview of anti-nutritional factors in food. *International Journal of Chemical Studies* 7:12472–12479.

Thirunathan, P., and A. Manickavasagan. 2019. Processing methods for reducing α-galactosides in pulses. *Critical Reviews in Food Science Nutrition* 59:3334–48. https://doi.org/10.1080/10408398.2018.1490886.

Thomas, L., C. Larroche, and A. Pandey. 2013. Current developments in solid-state fermentation. *Biochemical Engineering Journal* 81:146–61. https://doi.org/10.1016/j.bej.2013.10.013.

Tian, H., Y. Meng, C. Li, et al. 2019. A study of the digestive enzyme activities in scaleless carp (*Gymnocypris przewalskii*) on the Qinghai-Tibetan Plateau. *Aquaculture Reports* 13:100174. https://doi.org/10.1016/j.aqrep.2018.10.002.

Toushik, S. H., K. T. Lee, J. S. Lee, and K. S. Kim. 2017. Functional applications of lignocellulolytic enzymes in the fruit and vegetable processing industries. *Journal of Food Science* 82:585–93. https://doi.org/10.1111/1750-3841.13636.

Van de Voorde, I., K. Goiris, E. Syryn, C. Van den Bussche, and G. Aerts. 2014. Evaluation of the cold-active *Pseudoalteromonas haloplanktis* β-galactosidase enzyme for lactose hydrolysis in whey permeate as primary step of D-tagatose production. *Process Biochemistry* 49:2134–40. https://doi.org/10.1016/j.procbio.2014.09.010.

Vandenberghe, L. P. S., S. G. Karp, M. G. B. Pagnoncelli, C. Rodrigues, A. B. P. Medeiros, and C. R. Soccol. 2019. Digestive enzymes: industrial applications in food products. In *Green Bio-processes: Enzymes in Industrial Food Processing*, ed. B. Parameswaran, S. Varjani, and S. Raveendran, 267–291. Springer. https://doi.org/10.1007/978-981-13-3263-0_14.

Vashishth, A., S. Ram, and V. Beniwal. 2017. Cereal phytases and their importance in improvement of micronutrients bioavailability. *3 Biotech* 7:42. https://doi.org/10.1007/s13205-017-0698-5.

Vashishth, R., A. D. Semwal, M. Naika, G. K. Sharma, and R. Kumar. 2021. Influence of cooking methods on antinutritional factors, oligosaccharides and protein quality of underutilized legume *Macrotyloma uniflorum*. *Food Research International* 143:110299. https://doi.org/10.1016/j.foodres.2021.110299.

Vazquez-Flores, A. A., A. I. Martinez-Gonzalez, E. Alvarez-Parrilla, et al. 2018. Proanthocyanidins with a low degree of polymerization are good inhibitors of digestive enzymes because of their ability to form specific interactions: a hypothesis. *Journal of Food Science* 83:2895–2902. https://doi.org/10.1111/1750-3841.14386.

Vijayakumari, K., M. Pugalenthi, and V. Vadivel. 2007. Effect of soaking and hydrothermal processing methods on the levels of antinutrients and in vitro protein digestibility of *Bauhinia purpurea l.* seeds. *Food Chemistry* 103:968–75. https://doi.org/10.1016/j.foodchem.2006.07.071.

Villacrés, E., M. B. Quelal, E. Fernández, G. Garcìa, G. Cueva, and C. M. Rosell. 2020. Impact of debittering and fermentation processes on the antinutritional and antioxidant compounds in *Lupinus mutabilis* sweet. *Food Science and Technology* 131:109745. https://doi.org/10.1016/j.lwt.2020.109745.

Vogt, G. 2021. Synthesis of digestive enzymes, food processing, and nutrient absorption in decapod crustaceans: a comparison to the mammalian model of digestion. *Zoology* 147:125945. https://doi.org/10.1016/j.zool.2021.125945.

Xu, M., Z. Jin, S. Simsek, C. Hall, J. Rao, and B. Chen. 2019. Effect of germination on the chemical composition, thermal, pasting, and moisture sorption properties of flours from chickpea, lentil, and yellow pea. *Food Chemistry* 295:579–87. https://doi.org/10.1016/j.foodchem.2019.05.167.

Yahya, S., F. Muhammad, M. Sohail, and S. A. Khan. 2021. Amylase production and growth pattern of two indigenously isolated Aspergilli under submerged fermentation: influence of physico-chemical parameters." *Pakistan Journal of Botany* 53:1147–55. https://doi.org/10.30848/PJB2021-3(31).

Yazici, M., and Y. Mazlum. 2019. Prebiotic applications in cultured crayfish and shrimps. *Kahramanmaraş Sütçü İmam Üniversitesi Tarım ve Doğa Dergisi* 22:153–63. https://doi.org/10.18016/ksutarimdoga.vi.471559.

Yuan, C., X. Ding, L. Jiang, W. Ye, J. Xu, and L. Qian. 2021. Effects of dietary phytosterols supplementation on serum parameters, nutrient digestibility and digestive enzyme of white feather broilers. *Italian Journal of Animal Science* 20:2102–09. https://doi.org/10.1080/1828051X.2021.2000895.

Zhang, G., Z. Xu, Y. Gao, X. Huang, Y. Zou, and T. Yang. 2015. Effects of germination on the nutritional properties, phenolic profiles, and antioxidant activities of buckwheat. *Journal of Food Science* 80:H1111–19. https://doi.org/10.1111/1750-3841.12830.

Zhu, D., Q. Wu, and L. Hua. 2019. Industrial enzymes. In *Comprehensive Biotechnology*, ed. M. Moo-Young, 1–13. Elsevier B.V. https://doi.org/10.1016/B978-0-444-64046-8.00148-8.

11 Microbial Enzymes for the Recovery of Nutraceuticals from Agri-Food Waste

Sri Charan Bindu Bavisetty[1], *Nur Maiyah*[1], *Shahrim Ab Karim*[2], *Dave Jaydeep Pinakin*[3], *Wahyu Haryati Maser*[3], *Kantiya Petsong*[4], *Theeraphol Senphan*[5], *Ali Muhammed Moula Ali*[3]

[1] Department of Fermentation Technology, School of Food-Industry, King Mongkut's Institute of Technology Ladkrabang, Bangkok, Thailand

[2] Department of Foodservice Management, Faculty of Food Science and Technology, University Putra Malaysia, Serdang, Selangor Darul Ehsan, Malaysia

[3] Department of Food Science and Technology, School of Food-Industry, King Mongkut's Institute of Technology Ladkrabang, Bangkok, Thailand

[4] Department of Food Technology, Faculty of Technology, Khon Kaen University, KhonKaen, Thailand

[5] Program in Food Science and Technology, Faculty of Engineering and Agro-Industry, Maejo University, Chiang Mai, Thailand

CONTENTS

DOI: 10.1201/9781003311164-14

11.1 INTRODUCTION

Food production is expanding with the world's growing population, increasing food waste. A considerable amount of environmental stress is generated by agriculture and food waste during various stages of the food supply chain (Mirabella et al., 2014). Approximately 221 million tons of food wastes are generated every year in the production and retail chain (FAO, 2021). Traditionally, only some of these by-products are used as livestock feed, and most of the remaining waste is dumped into composts. Dumping these wastes poses environmental risks due to the production of large quantities of greenhouse gases such as methane and carbon dioxide, leading to sustainability issues (El Sheikha and Ray, 2022). Therefore, researchers are compelled to search for alternative sustainable ways to deal with waste management. In this regard, the utilization of microbial enzymes for the recovery of nutraceuticals from agri-food waste is gaining attention.

The two major agro-industrial by-products are agriculture and the food processing industry. Agricultural residues include peels, seed pods, leaves, stems, roots, etc., from crop production. Aquaculture or fish processing waste contains skin, bones, heads, and offal (Ucak et al., 2021), while industrial residues mainly include peels and cake-like residues from the food processing (Alibardi and Cossu, 2016; Gençdağ et al., 2021). Food waste is chemically composed of by-products such as lipids, proteins, carbohydrates, and other components and is considered a rich source of nutraceuticals (Hadj Saadoun et al., 2021). Nutraceuticals are food or food products that confer health benefits and play a significant role in preventing and treating diseases. Nutraceutical product development is prompted by increased consumer demand for health-related food products (Aiello et al., 2020). Agri-food wastes are gaining much attention from the nutraceutical sector. They are considered a good source of bioactive compounds that can be quickly recovered for the dietary supplement formulation (Tenore et al., 2020). Bioactive compounds play a unique role in this process and can be used in various forms, such as antioxidants, positively impacting human health. Extraction of bioactive compounds using enzymes, mainly from agricultural and food wastes, is promising with multiple benefits, including conserving time and energy during the extraction process and boosting the reproducibility of the extraction process on a commercial level (Vilas-Boas et al., 2021). Traditionally, extraction of bioactive compounds was mainly based on solvent extraction, subcritical water extraction, supercritical fluid extraction, enzymes, ultrasound, microwaves, fermentation technology, and enzymatic hydrolysis. Among these, the extraction of nutraceuticals using enzymes produced by microorganisms is gaining interest as it aids in extracting nutraceuticals. Still, the transformation of the extracted products into bioactive compounds with enhanced activity is achieved.

Microbial maceration or fermentation has been commonly used to extract bioactive compounds from agricultural and food industry wastes (Hadj Saadoun et al., 2021). This low-cost process generates less waste and consumes considerably less energy. A variety of bacterial species (e.g., lactic acid bacteria (LAB), propionibacteria, chromobacteria, and acetobacteria) and fungal strains (e.g., *Aspergillus, Saccharomyces,* and *Rhodosporidium*) can grow on various agricultural and food industry wastes as substrates. These microorganisms can produce endogenous enzymes such as hydrolases, dehydrogenases, isomerases, and decarboxylases, which catalyse the conversion of biopolymer-containing food by-products into various nutraceuticals, including sugars, alcohols, peptides, fatty acids, or phenolic acid and their derivatives. In this way, waste from the agri-food industry can be a source of revenue. Therefore, extracting nutraceuticals through microbial enzymes is an environmentally friendly process that offers the potential to recover bioactive compounds. Figure 11.1 provides an overview of agri-food industrial waste and its nutraceutical potential.

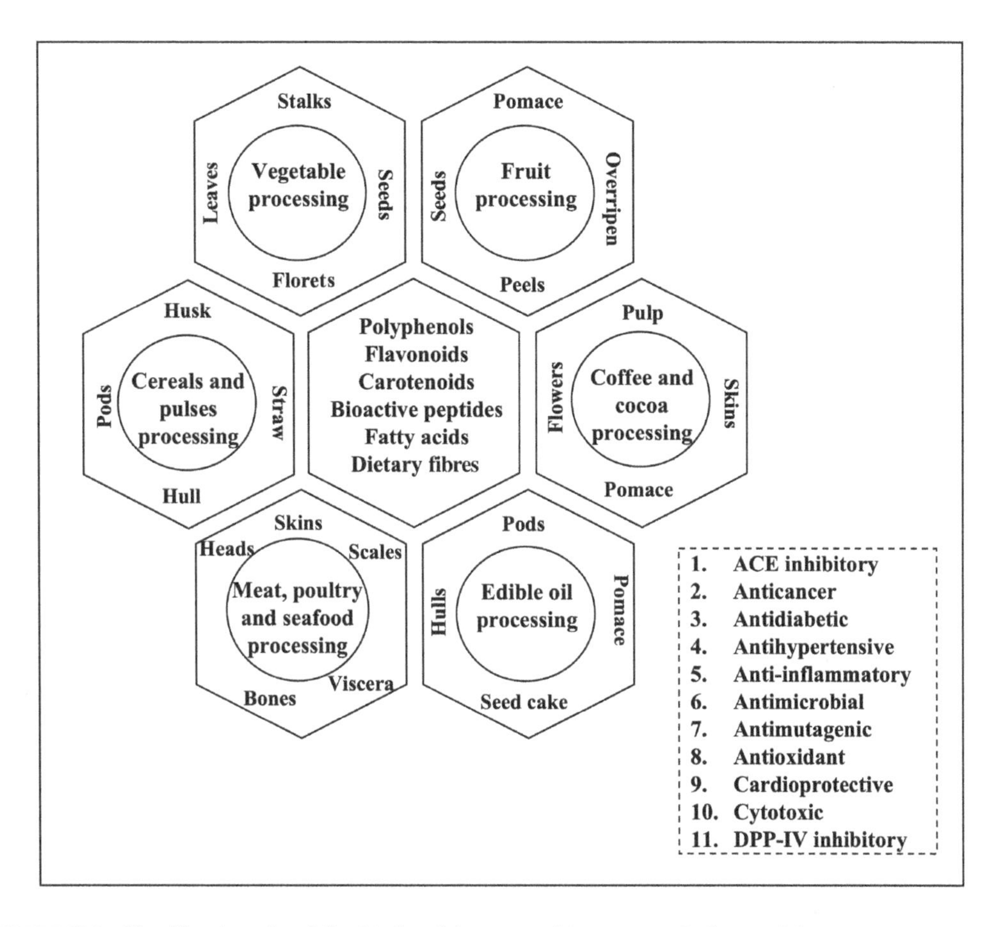

FIGURE 11.1 Classification of agri-food industrial waste and its nutraceutical potential.

This chapter provides an overview of industrial wastes from the agri-food industry, their nutraceutical value, and bioremediation as a strategy for converting by-products into bioactive compounds using microbial or fungal enzymes.

11.2 ENZYMES

Enzymes act as biocatalysts that accelerate chemical reactions in living organisms (Fasim et al., 2021). Enzymes in the form of extracts from plants and bacteria have long been used to produce wine, cheese, and other foods. Biotechnology has transformed traditional enzymes used in food, drinks, personal and home care, agriculture, bioenergy, pharmaceuticals, and various other industries (Fasim et al., 2021; Phukon et al., 2022). However, the application of purified enzymes has only been used for a few decades. Enzymes differ in their concentration, activity, availability, stability, and the presence of inhibitory factors. Animals, plants, and microbes have been replaced by recombinant DNA-based genetically developed creatures as enzyme suppliers (Ravindran and Jaiswal, 2016).

In the enzyme production process, microorganisms are an essential source. Among the available enzymes, proteases are most widely used in the industrial world and account for almost 60% of the total market for industrial enzymes (Haddar et al., 2010). Alkaline protease enzymes are increasingly used in animal feeds, foods, and detergent industries (Rebah, 2012). Based on the Enzyme Commission (EC) categorization, proteases are classified according to the source of their isolation, i.e., vegetables, animals, and microorganisms. In addition, they can also be distinguished according to their catalytic activity, i.e., active site, molecular size, exo- or endopeptidase, charge, and substrate specificity. Proteases are hydrolases that belong to group 3 and subgroup 4, which are responsible for the hydrolysis of the bond of peptides (dos Santos Aguilar and Sato, 2018).

Since the earliest times of civilization, people have used enzymes to produce food and beverages without knowing their nature, function, and activities. In the 19th century, with the development of biochemistry, enzymology began to develop. Wilhelm Kuhne coined the name enzyme for the first time in 1877. Due to their substrate specificity and consistent working settings, they are frequently preferred over chemical catalysis. Alpha-amylase, cellulase, xylanase, pectinase, protease, chitinase, and other enzymes are routinely utilized, depending on the unique makeup of the examined substances (Kumari et al., 2021; Torres-León et al., 2021).

Enzymes are biological catalysts in a specific biochemical reaction in metabolic processes secreted by almost all living cells (Ravindran and Jaiswal, 2016). Enzymes are categorized into the following classes based on their EC number: EC 1 (Oxidoreductases), EC 2 (Transferases), EC 3 (Hydrolases), EC 4 (Lyases), EC 5 (Isomerases), and EC 6 (Ligases), and EC 7 (Translocases) (McDonald and Tipton, 2021). These enzymes are beneficial as they are essential components in various products and production processes, as enzymes in the form of bacteria or plant extracts have been unknowingly used for centuries to produce several products, namely wine, cheese, bread, beer, and vinegar, as well as in the manufacture of goods such as leather and linen (Ravindran and Jaiswal, 2016). Lipases belong to the serine hydrolase family and are also known as carboxylic acid esterases (EC 3.1.1.3). There are two classifications of lipases based on the sources of the enzymes and their specificity. Based on the origins of lipases produced by animals, plants, and microbes (bacteria, fungi, and yeasts) and on their specificity, lipases are classified as substrate selective, enantioselective, and regioselective (nonspecific, 1,3 specific, and fatty acid specific) (Szymczak et al., 2021).

Proteases, also known as peptidases, are a complex group of proteolytic enzymes. They belong to class 3 (hydrolase) and class 4 (hydrolysis of peptide bonds) according to the EC classification and hydrolyse protein substrates into peptides and amino acids (Matkawala et al., 2021). Alkaline proteases (ATPases) account for more than 50% of global enzyme production. Marine microbes are a source of diverse alkaline proteases that play an important role in ecology and have applications in the industrial world (Barzkar, 2020). Proteolytic enzymes, also commonly referred to as proteases, can be classified as exopeptidases or endopeptidases (proteases) according to their site of action (Kieliszek et al., 2021).

11.2.1 Enzyme Classification

International Union of Biochemists classified the enzymes based on the reaction they are employed to catalyse; they are divided into six functional groups (Table 11.1). Enzyme technology is widely

TABLE 11.1
Enzymes Classification and Their Activity

Classes	Activity
Hydrolases	Hydrolases are hydrolytic enzymes that catalyse the hydrolysis reaction by cleaving the link and hydrolysing it with the addition of water.
Isomerases	Isomerase enzymes catalyse structural alterations in a molecule, changing its form.
Ligases	Ligases enzymes are known to catalyse the reaction of joining two large molecules by forming a new chemical bond.
Lyases	Catalyses the breaking of various chemical bonds by means other than hydrolysis and oxidation, often forming a new double bond or a new ring structure..
Oxidoreductases	Oxidation processes are catalysed by the enzyme Oxidoreductase, which involves electrons moving from one chemical form to another.
Transferases	Transferase enzymes aid in the movement of functional groups between acceptors and donors.

used in the food industry because of its advantages in safety, efficiency, and specificity (Pang et al., 2021).

11.3 NUTRACEUTICALS FROM AGRI-FOOD INDUSTRIAL WASTES

Food processing can lead to large amounts of by-products that can be used for further treatment or human consumption. There are many bioactive compounds in agri-food industrial wastewater, including those from fruits, vegetables, meats, and marine processing industries. These include flavonoids, polyphenols, dietary fibers and bioactive peptides. These by-products, if valorized well, can be used for nutraceuticals. Table 11.2 describes an overview of agro-industrial by-products, their bioactive compounds, and nutraceutical potential. Figure 11.2 provides a brief overview on the microbial enzyme production, nutraceutical extraction from waste and advantages.

11.3.1 Plant-Based Agri-Food Industrial Waste

Plant-based wastes from agri-food industries are perishable and require appropriate post-harvest processes to maintain their quality, resulting in high waste rates. Plant wastes include offcuts, hulls, crop residues, bran, seeds, husks, stems, starch, sugar, and leftovers after the extraction of juice and oil. Dietary fibres, bioactive polysaccharides, vitamins, carotenoids, and phenolic compounds are examples of potentially bioactive compounds found in plant waste from the agri-food industry (Table 11.2).

11.3.1.1 Cereals and Pulses Processing Industry

Bran, bark, and pods are by-products of cereals processing, which are rich in bioactive compounds. Rice and wheat are the most consumed cereals in the world. Wheat bran is the waste generated during the wheat milling process. Wheat bran is rich in phenolic acids like p-coumaric acid and ferulic acid, hence possesses strong antioxidant activity (Li et al., 2022). Ferulic acid is more readily absorbed by the body than other phenolic acids and stays in the bloodstream longer; therefore, it is considered a superior antioxidant (Yin et al., 2019). Rice bran is a by-product of removing outer layers from the grain during the dehulling process. The most abundant phenolic compound in rice bran is oryzanol, known for its cardioprotective effects (Sapna and Jayadeep, 2021).

Other interesting plant-based sources of bioactive substances is brewer spent grain which is generally obtained from barley and millets as by-products of beer processing. The spent grains are arich source of flavonoids like catechin and vanillin (Bonifácio-Lopes et al., 2022). Corn processing industries generate bran as a by-product rich in ferulic acid (Hussain et al., 2021). Ferulic acid and oligosaccharides combine to form feruloyl oligosaccharides, thereby possessing synergistic antioxidant capability (Vieira et al., 2020). Bran and husks are mainly composed of fibre (mainly arabinoxylans), known for their prebiotic effect (Verni et al., 2019).

Mung bean (*Vigna radiata*) is one of the most important legume crops. It can be found in the USA as well as many Asian countries. Mung bean peel and plum have high antioxidant capacity and resistance to α-amylase and α-glucosidase due totheabundance in phenolics such as gallic acid, ferulic acid, and flavonoids like isovitexin and vitexin (Zheng et al., 2020). Soybean (*Glycine max* L) is one the most important pulse crops globally, which is grown extensively to extract protein and oil. The main by-product of soy milk and tofu is an okara which is rich in isoflavones. Isoflavones provide nutraceutical benefits in terms of antioxidants as well as antimicrobial effects (Quintana et al., 2020).

Enzymatic recovery of Xylooligosaccharides (XOS) from lignocellulosic biomass waste from the cereal processing industry has gained huge interest in recent times. Jagtap et al. (2017) found a maximum yield (1 gm%) of XOS mixture after enzymatic hydrolysis of xylan rich wheat husk using the crude enzyme from *Aspergillus fumigatus*. The obtained XOS also showed great prebiotic potential and antioxidant activity. Enzyme type and concentration were crucial factors in improving

TABLE 11.2
Overview of Agri-Industrial By-Products, Main Bioactive Compounds, and Their Nutraceutical Potential

Agri-Food Industries	Particulars	Bioactive Compounds	Nutraceutical Potential	References
Plant Based Agri-Food Industrial Waste				
Cereal and pulse processing industry	Wheat bran	p-Coumaric acid, ferulic acid	Antioxidant, Antidiabetic	Li et al. (2022)
	Rice bran	p-Coumaric acid, ferulic acid, γ-oryzanol	Antioxidant	Sapna and Jayadeep (2022)
	Brewer spent grain	Catechin, vanilin	Antioxidant, antimicrobial	Bonifácio-Lopes et al. (2022)
	Maize bran	Ferulic acid	Antioxidant	Hussain et al. (2021)
	Mung bean skin	Polyphenols	Antioxidant	Zheng et al. (2020)
	Soybean okara	Isoflavones, flavonoids	Antioxidant	Quintana et al. (2020)
Edible oil industry	Walnut press cake	Phenolics acids, tannins, flavanols	Antioxidant	del Pilar Garcia-Mendoza et al. (2021)
	Pistachio shell	Tannins, flavonoids	Antioxidant	Cardullo et al. (2021)
	Cashew nut shell	Anacardic acid, cardanol	Anticancer	Venkatachalam et al. (2020)
	Pecan nut cake	Flavonoids	Antioxidant	Maciel et al. (2020)
	Flaxseed cake	Vanillic acid, caffeic acid, chlorogenic acid	Antioxidant	Kaur et al. (2021)
	Groundnut seed cake	Vanillic acid, chlorogenic acid, catechin, rutin	Antioxidant, antimicrobial	Duhan et al. (2021)
	Olive pomace	Hydroxytyrosol, tyrosol, oleuropein,	Antioxidant, anti-adipogenesis	Durante et al. (2019)
	Olive seed	Lutein, rutin, verbascoside, phytosterols	Antioxidant	Tamasi et al. (2019)
Fruits and vegetable processing industry	Cauliflower stalks and leaves	Lignin	Antioxidant	Majumdar et al. (2021)
	Beetroot pomace	Betacyanins, betaxanthins	Antioxidant	Fernando et al. (2021)
	Sweet potato leaves	Phenolic acids, flavonoids	Hypoglycaemic	Luo et al. (2021)
	Carrot pomace	β-Carotene	Antioxidant	Salehi and Taghian Dinani (2020)
	Tomato peel	Lycopene, lutein and β-carotene	Antioxidant	Szabo et al. (2021)
	Tomato seeds	Kaempferol, quercitin	Anti-inflammatory	Kumar et al. (2021)
	Grape skin	Flavan-3-ols, ethyl gallate	Antioxidant, anticancer	Milinčić et al. (2021)
	Grape seed	Catechin, epicatechin, proanthocyanidin, gallate, flavonols	Anti-cancer	Chen et al. (2020)
	Apple pomace	Gallic acid	Antioxidant	Radenkovs et al. (2018)

(Continued)

TABLE 11.2 ***(Continued)***
Overview of Agri-Industrial By-Products, Main Bioactive Compounds, and Their Nutraceutical Potential

Agri-Food Industries	Particulars	Bioactive Compounds	Nutraceutical Potential	References
Coffee and cocoa processing industries	Coffee silver skin	Phenolic acids, flavonoids, and secoiridoids	Antioxidant, antibacterial	Nzekoue et al. (2020)
	Cocoa husk	Flavan-3-ol	Antioxidant	Cádiz-Gurrea et al. (2020)
Animal Based Agri-Food Industrial Waste				
Red meat processing industries	Bovine hide	Gelatin	Anti-ageing	Ahmad et al. (2018)
	Pig skin	Gelatin	Anti-ageing	He et al. (2020)
	Camel skin	Gelatin	Anti-ageing	Al-Hassan (2020)
	Bovine blood	Bovine haemoglobin, Bovine plasma	Antioxidant, antimicrobial	Shirsath and Henchion (2021)
Poultry processing industries	Chicken feet	Collagen	Anti-inflammatory	Kodous (2020)
		Gelatin	Anti-ageing	Choe and Kim (2018)
	Chicken feathers	Keratin	–	Fagbemi et al. (2020)
Seafood processing industry	Pangasius fish skin	Collagen	Antioxidant	Azizah et al. (2020)
	Golden carp skin	Collagen	Antioxidant	Ali et al. (2018b)
	Yellow tuna skin	Gelatin	Antioxidant	Nurilmala et al. (2020)
	Carp fish scales	Collagen	Antioxidant	Chinh et al. (2019)
	Tilapia scales	Collagen and Gelatin	Antioxidant	Zhang et al. (2019)
	Aegla cholchol exoskeleton	Chitin and chitosan	Antioxidant, antifungal	Bernabe et al. (2020)
	Blue crab exoskeleton	Chitosan	Antioxidant	Hamdi et al. (2018)
	Yellow tail fish viscera	EPA and DHA	Anti-inflammatory, antihypertensive, anti-diabetic	Franklin et al. (2020)
	Squid viscera	PUFA	Anti-inflammatory, anti-arthritis	Guo et al. (2018)
	Penaeus monodon shells	Astaxanthin	Antioxidant, anticancer	Abidin (2021)
	Atlantic shrimp shells	Astaxanthin	Antioxidant, anti-inflammatory	Dave et al. (2020)
	Green tiger shrimp shells	Astaxanthin	Antioxidant	Sharayei et al. (2021)
	Coconut crab shells	Fucoxanthin	Antioxidant, anticancer	Rasulu et al. (2022)
	Fish liver	Squalene	Antioxidant	Ali et al. (2019)

EPA – eicosapentaenoic acid, DHA – docosahexaenoic acid, PUFA – polyunsaturated fatty acids

the release of bioactive compounds entrapped in a complex cell wall matrix. Sapna and Jayadeep (2021) investigated that *Bacillus subtilis* and *Trichoderma longibrachiatum* endo-xylanases treatment effectively enhanced the nutraceutical content and bioactivities of red rice bran. The fungal enzyme potently increased the quantity of total flavonoid and individual soluble phenolic components. The bacterial enzyme was more efficient in enhancing the number of total phenolics and individual bound phenolic components. Okara is a by-product of soybean processing produced in large quantities but underused by the food industry. Enzymatic hydrolysis by a β-glucosidase enzyme

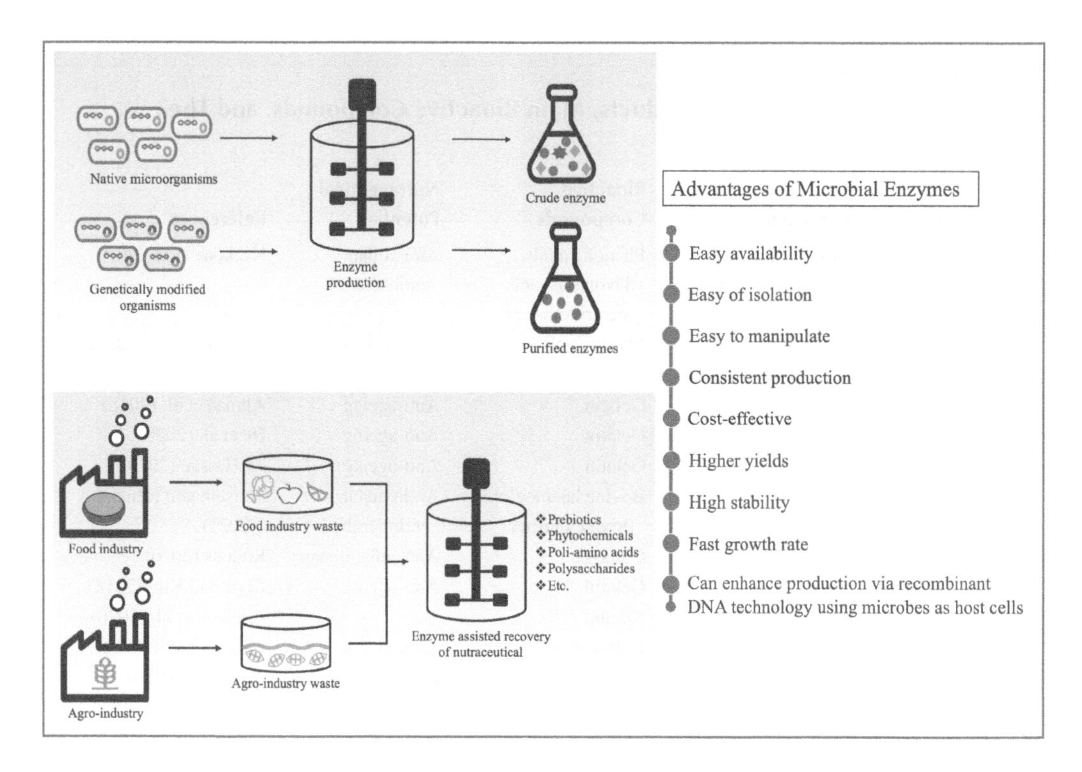

FIGURE 11.2 An overview of nutraceutical recovery using microbial enzymes and their advantages.

from *Saccharomyces cerevisiae* improved the bioconversion of isoflavone β-glucosides to the aglycones with a possible increase in their bioavailability (Santos et al., 2018).

11.3.1.2 Edible Oil Processing Industry

The edible oil processing industry includes several oil-bearing crops such as nuts, seeds, and olives. Nut by-products include skin, in volucre, shell, and defatted press cake. Walnut press cake is rich in phenolic acids, flavonoids, and tannins; hence, it possesses a strong antioxidant activity (del Pilar Garcia-Mendoza et al., 2021). The shells of cashew and pistachio nuts are a rich source of phenolic and flavonoids. Ancardic acid and cardanol in the cashew nutshell act as anticancer agents (Venkatachalam et al., 2020). Several researchers have reported the antioxidant and antimicrobial activities of groundnut seed cake (*Arachishypogaea* L.) and flaxseed cake (*Lecythispisonis* Cambess), which were hypothesized due tothe presence of phenolic acid profiles (vanillic acid, chlorogenic acid, catechin, rutin) (Duhan et al., 2021; Stodolak et al., 2017). The main by-products of olive oil extraction are olive pomace and seeds. The phenolic profile of olive oil pomace is mainly constituted oleanolic acid derivatives, hydroxytyrosol, comselogoside, hydroxytyrosol-4-glucoside, and minor amounts of salidroside, tyrosol, verbascoside, oleoside, riboside and other compounds, such as rutin and luteolin (Durante et al., 2019; Tamasi et al., 2019).

Researchers have investigated the effect of microbial enzymes on various properties of oil seeds press cake, like phenolic, antioxidant, and functional properties. The microbial enzymes are known to increase the phenolic and functional properties of the oilseed cake (Duhan et al., 2021; Sadh et al., 2018). The influence of hydrolysis by *Rhizopus oligosporus* enzymes on the antioxidant potential of flaxseed oil cake was studied by Stodolak et al. (2017). They found that enzymatic hydrolysis by carbohydrase of *Rhizopus oligosporus* resulted in a significant increase in phenolic content (13–85%), radical scavenging activity (about twofold higher against DPPH radical), and reducing power (by 20–30%). Sadh et al. (2018) also observed that *Aspergillus awamori* released the bound phenolic present in the groundnut seed cake by endogenous enzymes, which increases the total phenolic contents of the antioxidant properties. Tufariello et al. (2019) have observed the increased level of the highly valuable hydroxytyrosol during enzymatic hydrolysis of olive cake. This was most likely due to microbial glucosidase and esterase activity.

11.3.1.3 Fruits and Vegetable Processing Industry

Vegetables produce a wide range of by-products such as seeds, pomace, and peel. Vegetables in the Brassica family, such as cauliflower (*Brassica Oleracea* L. botrytis) and broccoli (*Brassica Oleracea* var. Parthenon), produce considerable amounts of by-products such as stalks (Lafarga et al., 2018). Cauliflower stalk is a promising source of lignin which acts as an antioxidant (Majumdar et al., 2021). The broccoli leaves contain higher levels of tocopherols and carotenoids than their stalks and florets (Liu et al., 2018).

According to the FAO (2019), the highest levels of food loss were contributed by roots, tubers, and oil crops, followed by fruits and vegetables. Beet, sweet potato, and carrots have high levels of underused biomass. The leaves of sweet potatoes are a rich source of phenolic and flavonoids (Luo et al., 2021). Betaxanthins and betacyanin (betalains), the water-soluble pigments containing nitrogen, are potential bioactive compounds in the beetroot pomace (Aiello et al., 2020; Fernando et al., 2021). Vegetables like tomatoes contain high levels of lycopene in tomato peels (Szabo et al., 2021). The tomato pomace is the main by-productof this vegetable. Tomato pomace is a mixture of tomato skin, seeds, and pulp. (Kumar et al., 2021; Mehta et al., 2018) reported antioxidant functions of bioactive compounds such as 3,4,5-tri-caffeoylquinic acid, quercetin-triglucoside, rutin, naringenin, and its chalcones from tomato pomace. Besides its antioxidant capacity, tomato pomace has also been reported to possess antimicrobial action against *S. aureus* and other Gram-positive bacteria (Mehta et al., 2018).

Peel and pomace are the major by-products from fruit processing industries obtained through juice and pulp manufacturing. Citrus by-products are rich in fibre, essential oils, vitamins, and flavonoids. Flavanones such as hesperidin and narirutin are found in citrus fruits (Louati et al., 2019). Grape (*Vitis Vinifera* Lin.) is famous for wine and juice production. Grape by-products include pomace, seed, and skin. Pomace, a mixture of grape skin, seed and pulp, is the most studied grape by-product and is rich in phenolic compounds, including flavonols, catechin, epicatechin, and procyanidins and phenolic acids (Ferri et al., 2017). The antioxidant activity of grape skin has been extensively studied for both red and white varieties (Milinčić et al., 2021). On the other hand, pome fruits such as apples and pears are commonly consumed as juices, puree, or unprocessed. The most common by-product of apple processing is pomace which has been reported to contain high levels of lipophilic and hydrophilic bioactive compounds (Radenkovs et al., 2018). Additionally, pomace oil from crab apple (*Malus Mill*) has been reported to contain significant amounts of nonacosane, nonacosan-10–ol, a-tocopherol and b-sitosterol (Radenkovs et al., 2018).

There is increasing attention on the exploration of vegetable processing waste as economically viable substrates to produce prebiotic oligosaccharides, especially xylooligosaccharides (XOS), as an excellent source for the maintenance and promotion of gut microbiota. Enzymatic hydrolysis of cauliflower stalks was carried out with produced crude endo-xylanase obtained from *Aspergillus niger* by Majumdar et al. (2021) to increase the production of XOS with potential prebiotic efficacy. Phenolic acids and flavonoids are critical antioxidant compounds. A considerable amount is bound to the cell wall of fruit and vegetable by-products (cellulose and pectin) like peels, seeds, and pomace leaves. Tomato pomace is a valuable by-product of the tomato processing industry and possesses considerable carotenoids and pigments with potential antioxidant properties. Researchers have observed that microbial hydrolytic enzymes can be an effective means for recovering powerful antioxidants from tomato by-products. Azabou et al. (2016) have observed higher lycopene recovery when extracted using *Fusarium solani* crude enzymes extract. Demirgül and Ozturk (2021) also reported the higher recovery of carotenoids from the tomato pomace during enzymatic hydrolysis using *A. niger.* Hydrolytic enzymes, i.e., cellulase, pectinase, and carbohydrase obtained from the

A. niger effectively released the polyphenols (gallic acid, catechin, epicatechin, and hydroxybenzoic acid) from the black grape pomace, apple pomace, and fig by-products (Buenrostro-Figueroa et al., 2017; Zambrano et al., 2018).

11.3.1.4 Coffee and Cocoa Processing Industry

The outer layer of green coffee beans, known as coffee silver skin, removed during roasting, is considered the main by-product. Coffee silver skin has been extensively studied for high antioxidant capacity due to high levels of phenolics such as caffeic acids, 5-O-caffeoylquinic, and caffeine (Nzekoue et al., 2020; Rochín-Medina et al., 2018). The cocoa husk is another by-product obtained during cocoa processing, rich in antioxidant activity (Cádiz-Gurrea et al., 2020). Spent coffee grounds are waste material generated during coffee beverage preparation. This by-product disposal causes a negative environmental impact and the loss of a rich source of nutrients and bioactive compounds. Rochín-Medina et al. (2018) have used the bacterial endo-cellulase and pectinase enzymes obtained from *Bacillus clausii* to achieve higher recovery of phenolic compounds from the spent coffee grounds. The accumulated cocoa shell can be used for bioconversion to obtain valuable compounds during cocoa processing. Lessa et al. (2018) have observed that the enzymatic hydrolysis of the cocoa shell by *Penicillium roqueforti* significantly increases fatty acids like oleic, linoleic, and gamma-linolenic and also increases the antioxidant activity.

11.3.2 Animal-Based Agri-Food Processing Waste

Meat, one of the most consumed foods globally, is an essential source of essential amino acids and minerals such as zinc and iron (Wu et al., 2014). Animal industry by-products include carcasses, blood, feathers, hooves, manure, hides, offal, heads, bones, fat, viscera, and seafood waste (Irshad and Sharma, 2015). Bioactive compounds like collagen, gelatin, chitosan, lipids, carotenoids, squalene, and bioactive peptides are extracted from the animal-based processing wastes, as mentioned in Table 11.2.

11.3.2.1 Red Meat Processing Industry

Butchers, meat processors, and wholesalers generate waste from red meat processing. The waste from the meat industry contains a high concentration of nitrogen, phosphorus, and fat, based on the type of waste. Bovine skin and hide are considered a rich source of bioactive substances like collagen and gelatin (Ahmad et al., 2018). Pig skin has commercial importance in cosmetic industries due to the gelatin production (He et al., 2020). Al-Hassan (2020) studied the anti-ageing properties of gelatin obtained from camel skin. Animal blood, a type of meat by-product, is an essential edible by-product as it contains high levels of iron and proteins (Borrajo et al., 2019). In Asia, blood is used to make blood curd, pudding, and cake. In Europe, animal blood is used for blood sausage, bread, and biscuits (Hsieh and Ofori, 2011). The animal blood collected in various slaughterhouses is a new source of different bioactive compounds like bovine haemoglobin and bovine plasma (Shirsath and Henchion, 2021).

11.3.2.2 Poultry Processing Industry

The poultry processing industry produces a large number of by-products, mainly in the form of heads, bones, legs, skin, feathers, and viscera, but also in the form of whole carcasses that are dead on arrival (Jayathilakan et al., 2012). The animal waste generated by the meat industry contains significant amounts of insoluble and poorly degradable structural proteins such as elastin, keratin, and collagen, which are significant components of the bone and hard tissue (Chakka et al., 2017). Kodous (2020) and Choe and Kim (2018) have studied the antioxidant and anti-inflammatory properties of gelatin and collagen obtained from the chicken feet. Chicken feathers are also a rich source of bioactive compounds like keratin and elastin (Alahyaribeik et al., 2020; Fagbemi et al., 2020).

11.3.2.3 Seafood Processing Industry

The increasing consumption of aquatic foods is due to their health benefits and the positive image that seafood products have acquired among consumers. The by-products of the fish industry, such as the bones, heads, skins and guts, account for 70% of the live weight of fish (FAO, 2019). These valuable ingredients (proteins and phospholipids, polyunsaturated fatty acids, polysaccharides, and bioactive compounds) are not fully utilized. Fish scales and skin are the sustainable sources for the gelatin and collagen. Various studies have found the antioxidant potential of collagen and gelatin obtained from the scales and skins of different fish species (Ali et al., 2018a; Azizah et al., 2020; Chinh et al., 2019; Nurilmala et al., 2020; Zhang et al., 2019). Fish viscera are a rich source of polyunsaturated fatty acids (PUFA) like EPA and DHA (Franklin et al., 2020; Guo et al., 2018, Rai et al., 2012, 2013). Ali et al. (2019) extracted squalene, a functional bioactive compound, from the fish liver tissues. Crustacean exoskeletons are a marine by-product that can be used as a biological matrix. They contain high bioactive substances such as chitin and chitosan (Bernabe et al., 2020; Hamdi et al., 2018). Crustacean shells are rich in carotenoids, mainly astaxanthin and their ester derivatives, which possess strong antioxidant and anticancer properties (Abidin, 2021; Montoya et al., 2021; Rasulu et al., 2022).

The development of the seafood processing industry has significantly accelerated seafood waste generation in the world. The disposal of seafood waste poses a severe environmental problem. The seafood waste like skin, scales, shells, and exoskeletons contains a substantial amount of high-quality protein, which is an excellent source of bioactive substance (Sasidharan and Venugopal, 2020). Endo-protease enzymes present in different microorganisms have shown the most promising hydrolytic activity for converting these wastes into valuable bioactive substances. Fang et al. (2017) have studied the hydrolytic effect of microbial protease obtained from *Aspergillus oryzae* on turbot fish skin and observed that the developed hydrolysates had a less fishy odour and high free amino acids along with increased antioxidant potential. Mirzapour Kouhdasht et al. (2018) produced gelatin from the skin of common carp fish by using the bacterial alkaline protease enzyme obtained from the *Bacillus licheniformis*. Shrimp processing wastes causing various environmental problems are considered the cheapest raw materials for recovering astaxanthin. LAB produce proteases that hydrolyse proteins of the shrimp shells and release astaxanthin free from the protein-chitin complex, thus facilitating its extraction (Dave et al., 2020). El-Bialy and Abd El-Khalek (2020) have recovered a higher amount of astaxanthin from shrimp shells by using *Lactobacillus acidophilus* and high antioxidant and antimicrobial potential.

11.4 RECOVERY OF BIOACTIVE COMPOUNDS FROM AGRI-FOOD INDUSTRIAL WASTE BY USING MICROBIAL ENZYMES

For bioremediation of agri-food industrial waste, the selection of microorganisms is crucial for their application in functional compound production. Bioremediation is used mainly to produce higher amounts of phenolic acid or other bioactive from by-products of the agro-industry. Hydrolysis of cell wall components by microbial enzymes stimulates the release of bound phenolics and other metabolites like organic acids, vitamins and peptides. Table 11.3 summarizes the recent application of microbial enzymes to extract bioactive compounds from agri-food industrial waste.

11.4.1 Protein Hydrolysate From Agri-Food Waste Using Microbial Enzyme

Protein hydrolysis is a chemical reaction involving the decomposition of proteins in the presence of water. Protein hydrolysate results from the hydrolysis of proteins using strong acids, bases, or enzymes (Chalamaiah et al., 2019). Many health benefits have been attributed to protein hydrolysate, including anti-obesity, hypocholesterolemic, anti-inflammatory, antihypertensive, anticancer, antioxidant, and immunomodulatory effects (Ashaolu, 2020; Chourasia et al., 2022; Padhi et al., 2021).

TABLE 11.3
Summary of Some Recent Bioremediation Applications by Utilizing Agri-Food Industrial Waste

Agri-Food Industry	Type of Waste	Nutraceutical Compounds	Microorganism Used for Bioremediation	Microbial Enzymes Used for Remediation	Incubation Conditions	Reference
Cereal and pulse processing industry	Wheat husk	Xylooligosaccharide	*Aspergillus fumigatus*	Xylanase	37°C for 12 h	Jagtap et al. (2017)
	Rice bran	Oryzanol	*Bacillus subtillis*	Endoxylanase	90°C for 90 min	Sapna and Jayadeep (2021)
	Oat husk	Polyphenols	*Monascus anka*	Cellulase, β-glucosidase	30°C for 14 days	Bei et al. (2017)
	Soybean okara	Polyphenols Isoflavones	*Saccharomyces cerevisiae*	β-Glucosidase	28°C for 24 h	Santos et al. (2018)
	Wheat straw Wheat bran	Xylooligosaccharide	*Aspergillus oryzae*	Xylanase	NA	Bhardwaj et al. (2019)
	Wheat bran	Ferulic acid	*Aspergillus niger*	Esterase	28°C for 7 days	Yin et al. (2019)
	Barley husk	Dietary fibres	*Pediococcus acidilactici*	NA	32°C for 72 h	Bartkiene et al. (2020)
	Mung bean husk	Torularhodin, β-carotene, torulene	*Rhodotorula mucilaginosa*	NA	25.8°C for 82 h	Sharma and Ghoshal (2020)
Edible oil processing industry	Peanut seed cake	Polyphenols	*Aspergillus awamori*	NA	30°C for 120 h	Sadh et al. (2018)
	Patè Olive Cake	Hydroxytyrosol	*Saccharomyces cerevisiae*	Microbial glucosidase, esterase	28°C for 60 days	Tufariello et al. (2019)
	Flaxseed cake	Polyphenols	*Rhizopus oligosporus*	Carbohydrase	30°C for 48 h	Stodolak et al. (2017)
Fruits and vegetable processing industries	Cauliflower stalk	Xylooligosaccharide	*Aspergillus niger*	Xylanase	NA	Majumdar et al. (2021)
	Red chicory leaves	Gallic acid Protocatechuic acids	*Saccharomyces cerevisiae*	Cellulase	30°C for 48 h	Kagkli et al. (2016)
	Tomato pomace	Carotenoids	*Aspergillus niger*	Cellulase	30°C for 72 h	Demirgül and Ozturk (2021)
	Tomato pomace	Lycopene	*Fusarium solani pisi*	Cellulase and pectinase	NA	Azabou et al. (2016)

(Continued)

TABLE 11.3 *(Continued)*
Summary of Some Recent Bioremediation Applications by Utilizing Agri-Food Industrial Waste

Agri-Food Industry	Type of Waste	Nutraceutical Compounds	Microorganism Used for Bioremediation	Microbial Enzymes Used for Remediation	Incubation Conditions	Reference
	Black grape pomace	Gallic acid Catechin Epicatechin	*Rhizomucor miehei*	Carbohydrase	37°C for 18 days	Zambrano et al. (2018)
	Apple pomace	Hydroxybenzoic acid	*Aspergillus niger*	Cellulase	37°C for 18 days	
	Plum pomace	Phenolic compounds	*Aspergillus niger* *Rhizopus oligosporus*	β-Glucosidases Carboxylesterases Feruloyl esterases	30°C for 14 days	Dulf et al. (2016)
	Apple by-products	Dietary fibres	*Weissella cibaria and Saccharomyces cerevisiae*	NA	30°C up to 72 h	Cantatore et al. (2019)
	Pomegranate waste	Ellagic acid	*Aspergillus niger and Saccharomyces cerevisiae*	NA	NA	Moccia et al. (2019)
	Fig by-products	Polyphenols	*Aspergillus niger*	Cellulase and pectinase	40°C for 36 h	Buenrostro-Figueroa et al. (2017)
Coffee and cocoa processing industry	Spent coffee ground	Polyphenols	*Bacillus clausii*	Cellulase and pectinase	37°C for 39 h	Rochín-Medina et al. (2018)
	Cocoa shell	Oleic acid Linoleic acid Gamma-linolenic acid	*Penicillium roqueforti*	NA	25°C for 7 days	Lessa et al. (2018)
Seafood processing industries	Turbot skin	Collagen	*Aspergillus oryzae*	Endoprotease	30°C for 48 h	Fang et al. (2017)
	Rohu head	EPA and DHA	NA	Endoprotease	55°C for 2 h	Bruno et al. (2019)
	Prawn shell	Chitin	*Lactobacillus plantarum subsp. Plantarum, Bacillus subtilis subsp.*	Protease	30°C for 5 days	Panzella et al. (2020)
	Common carp skin	Gelatin	*Bacillus licheniformis*	Alkaline protease	37°C for 48 h	Mirzapour et al. (2018)
	Nile tilapia head	Bioactive peptides	*Lactobacillus fermentum* *Lactobacillus plantarum*	Protease	37°C for 3 days	Gao et al. (2020)
	Shrimp waste	Astaxanthin	*Lactobacillus acidophilus*	NA	NA	El-Bialy and Abd El-Khalek (2020)

Bioactive peptides derived from dietary proteins can be incorporated into functional foods/nutraceuticals. Bioactive peptides are available in various formats, including beverages, meals, capsules, pills, liquids, and powders. Peptides have been isolated from foods such as eggs, milk, peanuts, whey, rice, chickpeas, corn, fish, algae, and other marine animals (Chalamaiah et al., 2019; Hathwar et al., 2011; Lasrado and Rai, 2018). Protein hydrolysate has been reported to possess numerous health benefits and can be produced from agricultural and food wastes using microbial enzymes. Some of the health benefits of protein hydrolysate are illustrated in Table 11.4.

Parameters that determine the degree of enzymatic hydrolysis include the time of hydrolysis (t), pH, temperature (T), and the ratio of enzyme and substrate (E/S). The most favourable conditions for protein hydrolysis are incubation temperature of 35–37°C, a 1 to 50 enzyme-to-substrate ratio, and a 24-hour incubation period. Spray drying is a standard method for obtaining a concentrated hydrolysed fraction (Sasidharan and Venugopal, 2020). The summary of several hydrolysis processes of agri-food waste by microbial enzymes is shown in Table 11.5.

11.4.2 Antioxidant

Bioactive peptides are the principal source of antioxidants in animal by-products, resulting from chemical or enzyme-assisted protein hydrolysis. Antioxidants are chemicals that can slow or stop other molecules from oxidizing. Antioxidants bond and stick together when they react with free radicals. In addition, other free radicals are produced, which are weak and harmless. Oxidation is a reaction of chemicals that results in the production of free radicals, which can set off a chain reaction that destroys cells. According to the study by Berraquero-García et al. (2022), molecular weight, amino acid sequence, secondary structure, and hydrophobic residues all influence the peptide fraction's potential ability to scavenge radicals.

Many protein hydrolysates obtained from food product residues possess antioxidant abilities due to aromatic residues such as tryptophan, tyrosine, and phenylalanine, which can give protons to electron-deficient radicals (Mechmeche et al., 2017). Besides aromatic residues, hydrophobic amino acid residues are also known to possess strong antioxidant potential due to their ability to trap free radicles and hydrogen transfer reactions. Protein hydrolysis in peach seeds with thermolysin produced hydrophobic peptides that could carry out hydrogen transfer reactions by trapping lipid peroxyl radicals because of their high solubility in hydrophobic amino acids residues (L, P, I) (Vásquez-Villanueva et al., 2016). Furthermore, metal ions can be bound by imidazole found in peptide chains with an H amino acid residue (García et al., 2015a). Kumar et al. (2012) studied the antioxidant potential through DPPH and reducing power of protein hydrolysate obtained from chicken feathers fermented using Bacillus sp. MPTK6. Berraquero-García et al., (2022) studied the metal chelating ability of the protein hydrolysate obtained from porcine blood by hydrolysis using protamex®. The IC_{50} for metal chelating ability using trypsin was 0.27 mg/mL. They concluded that the enzymatic valorization of blood protein increases the bioavailability of heme iron and produces antioxidant peptides. López-Pedrouso et al. (2020) studied the antioxidant potential of pork liver hydrolysates employing hydrolytic enzymes such as alcalase, bromelain, papain, and flavourzyme. Alcalase enzyme produced small peptides, whereas flavourzyme produced higher amounts of free amino acids. The peptides sequence with the highest antioxidant potential were found to be as follow: APAAIGPYSQAVLVDR from uncharacterized proteins; GLNQALVDLHALGSAR, ALFQDVQKPSQDEWGK, and LSGPQAGLGEYLFER from ferritin; and LGEHNIDVLEGNEQFINAAK from trypsinogen.

According to a study conducted by Bezus et al. (2021), chicken feather protein hydrolysate developed feather protein hydrolysate (FPH) with a strong antioxidant activity using enzymes obtained from *Bacillus* sp. Raw feathers are degraded into protein hydrolysate by this strain. The scavenging of ABTS radicals (2,20-azino-bis-(3-ethylbenzothiazoline)-6-sulphonic acid) by hydrolysate attained a value of around 95%, with an IC50 of 0.57 0.01 g^{-1}. FPH's lowering power to Fe+3 was also found (2.19 units Abs 700 in the presence of 2.03 g^{-1} proteins). With an IC_{50} of 0.53 0.01 g^{-1},

TABLE 11.4
Protein Hydrolysates from Agri-Food Waste and Their Nutraceutical Potential

Source	Sample	Active Peptides	Amino Acid Sequences	Nutraceutical Potential	References
Poultry and livestock	Porcine blood	Blood meal hydrolysates (BMH)	–	Antioxidant	Berraquero-García et al. (2022)
	Porcine liver	Porcine liver hydrolysates	APAAIGPYSQAVLVDR	Antioxidant	López-Pedrouso et al. (2020)
	Porcine liver	Ferritin	LSGPQAGLGEYLFER, ALFQDVQKPSQDEWGK, GLNQALVDLHALGSAR	Antioxidant	López-Pedrouso et al. (2020)
	Porcine liver	Trypsinogen	LGEHNIDVLEGNEQFINAAK	Antioxidant	López-Pedrouso et al. (2020)
	Bovine milk whey	Whey protein hydrolysates	–	Antimutagenic, antioxidant, antimicrobial	Halavach et al. (2020)
	Sheep milk whey	Whey protein hydrolysates	LAFNPTQLEGQCHV	Antioxidant, ACE inhibitor	Correa et al. (2019)
	Whey	Whey protein hydrolysates	LDAQSAPLR, LKGYGGVSLPEW, LKALPMH	Angiotensin-converting enzyme (ACE) inhibitory	Worsztynowicz et al. (2020)
	Whey	Whey protein hydrolysates	LKPTPEGDLEIL,LKALPMH, VLVLDTDYK, LKPTPEGDLE, LKGYGGVSLPE, ILDKVGINY	DPP-IV inhibitory	Worsztynowicz et al. (2020)
	Whey	Whey protein hydrolysates	AASDISLLDAQSAPLR, IIAEKTKIPAVF, IDALNENK, VLVLDTDYK	Antimicrobial	Worsztynowicz et al. (2020)
	Whey	Whey protein hydrolysates	IDALNENK	Proliferation	Worsztynowicz et al. (2020)
	Whey	Whey protein hydrolysates	LIVTQTMK	Cytotoxic	Worsztynowicz et al. (2020)
	Chicken feather	Feather protein hydrolysate (FPH)	RHILMFTYVANCEGPLPSY	Antioxidant	Bezus et al. (2021)
	Chicken visceral	Azocasein	–	Antioxidant, antihypertensive	dos Santos Aguilar et al. (2020)
	Chicken feather	Feather protein hydrolysate (FPH)	–	Antioxidant	Alahyaribeik et al. (2020)

(Continued)

TABLE 11.4 *(Continued)*
Protein Hydrolysates from Agri-Food Waste and Their Nutraceutical Potential

Source	Sample	Active Peptides	Amino Acid Sequences	Nutraceutical Potential	References
Aquatic and marine	Rainbow trout, whiting, anchovy (skeleton, fin, head, skin, and viscera)	Fish protein hydrolysates	–	Antioxidant	Korkmaz and Tokur (2022)
	Red tilapia scales	Fish scales protein hydrolysates	DENSHGTCRAYCVMFILKPP (Amino acid)	Antioxidant	Sierra-Lopera and Zapata-Montoya (2021)
	Nile tilapia (head, viscera)	Fish protein hydrolysates	DSEGHRTAPCYVMKILFW	Antioxidant	Gao et al. (2021)
	Shrimp shell	Shrimp shell protein hydrolysates (hyd)	–	Antioxidant, ACE inhibitory	Mechri et al. (2020)
	Rainbow trout skin	Fish skin protein hydrolysates	–	Anticancer, antioxidant	Yaghoubzadeh et al. (2020)
	Frogskin	Brevinin-2R	KLKNFAKGVAQSLLNKASCKLSGQ, CKLKNFAKGVAQSLLNKASKLSGQC	Anticancer	Hassanvand Jamadi et al. (2019)
	Shrimp (head, shell, tail)	Shrimp protein hydrolysates (hyd)	DESGHRTAPYVMILF	Detoxification	Dey and Dora (2014)
	Tuna dark muscle	Dark muscle protein hydrolysates	PTAEGGVYMVT	Anticancer	Hsu et al. (2011)
	Catla fish visceral	Fish visceral protein hydrolysates	HILKMFYTWGVDESGAPC		Bhaskar et al. (2008)
Plant	Cassava bagasse	Cassava bagasse hydrolysate	–	Antioxidant	Clerici et al. (2021)
	Cottonseed	Cottonseed protein hydrolysates (CPH)	DNSEQGAVMILYFKRHTCP	Antioxidant, antimicrobial	Song et al. (2020)
	Wheat bran	Wheat bran protein hydrolysates	NL, QL, FL, HAL, AAVL, AKTVF, TPLTR	Antihypertensive, antioxidant	Zou et al. (2020)
	Rice dregs	Rice dreg hydrolysates	GDMNP, LLLRW	Antioxidant	Chen et al. (2021)
	Olive seed	Olive seed protein hydrolysates	–	Antioxidant, antihypertensive	Esteve et al. (2015)
	Rice bran	Rice bran protein hydrolysates	–	Anti-inflammatory	Boonloh et al. (2015)
	Cherry seed	Cheery seed protein hydrolysates	–	Antioxidant, antihypertensive	García et al. (2015b)

TABLE 11.5
The Summary of Some Hydrolysis Process of Agri-Food Waste by Microbial Enzyme

Source	Sample	Waste	Hydrolysis Process					Reference
			pH	Hydrolysis Condition	Microorganism	Microbial Enzyme	Degree of Hydrolysis	
Poultry and livestock	Chicken	Feather	–	37°C for 1 h	*Pedobacter* sp.	Proteinase	–	Bezus et al. (2021)
	Chicken	Visceral	7.0	50°C for 2 h	*Aspergillus oryzae*	Flavourzyme	–	dos Santos Aguilar et al. (2020)
	Chicken	Visceral	7.0	50°C for 2 h	*Bacillus licheniformis*	Alcalase	–	dos Santos Aguilar et al. (2020)
	Chicken	Visceral	7.0	50°C for 2 h	*Bacillus amyloliquefaciens*	Neutrase	–	dos Santos Aguilar et al. (2020)
	Chicken	Feather	7.0	37°C	*Bacillus pumilis*	Protease	–	Alahyaribeik et al. (2020)
	Chicken	Feather	7.0–8.0	55°C	*Geobacillus stearothermophilus*	Protease	–	Alahyaribeik et al. (2020)
	Chicken	Feather	7.0	37°C	*Rhodococcus erythropolis*	Protease	–	Alahyaribeik et al. (2020)
	Sheep milk	Whey	–	45°C for 4 h	*Bacillus* sp.	Proteinase	–	Correa et al. (2019)
	Bovine milk	Whey	8.0	50°C for 3–4 h	*bacillus licheniformis*	Alcalase	–	Halavach et al. (2020)
	Bovine milk	Whey	7.0	50°C for 3–4 h	*Bacillus amyloliquefaciens*	Neutrase	–	Halavach et al. (2020)
	Cheese	Whey		37°C for 0–48 h	*Enterococcus faecalis*	Protease	–	Worsztynowicz et al. (2020)
	Porcine	Blood	8.5	60°C	*Bacillus subtilis*	Alcalase	24%	Berraquero-García et al. (2022)
	Porcine	Blood	7.8	55°C	*Bacillus subtilis*	Protamex	14%	Berraquero-García et al. (2022)
Aquatic and marine	Red tilapia	Scales	–	3 h	*Bacillus licheniformis*	Alcalase	16.7%	Sierra-Lopera and Zapata-Montoya (2021)
	Red tilapia	Scales	–	3 h	*Aspergillus oryzae*	Flavourzyme	5.9%	Sierra-Lopera and Zapata-Montoya (2021)
	Fish	Fish discards	–	–	*Pediococcus acidilactici*	Protease	–	Vázquez et al. (2020)
	Shrimp	Shrimp shell	–	–	*Anoxybacillus kamchatkensis*	Alkaline protease	32%	Mechri et al. (2020)
	Shrimp	Shrimp shell	–	–	*Aeribacillus pallidus*	Alkaline protease	27%	Mechri et al. (2020)
	Shrimp	Head, shell and tail	8.25	59.37°C for 84.42 min	*Bacillus lichenformis*	Alcalase	33.13%	Dey and Dora (2014)
	Rainbow trout	Skeleton, fin, head, skin, and viscera	8.0	60°C for 1 h	*Bacillus lichenformis*	Alkaline protease	64.35%	Korkmaz and Tokur (2022)

(Continued)

TABLE 11.5 *(Continued)*
The Summary of Some Hydrolysis Process of Agri-Food Waste by Microbial Enzyme

			Hydrolysis Process					
Source	**Sample**	**Waste**	**pH**	**Hydrolysis Condition**	**Microorganism**	**Microbial Enzyme**	**Degree of Hydrolysis**	**Reference**
	Rainbow trout	Skeleton, fin, head, skin, and viscera	7.0	60°C for 1 h	*Bacillus lichenformis*	Protamex	74.30%	Korkmaz and Tokur (2022)
	Whiting	Head and viscera	8.0	60°C for 1 h	*Bacillus lichenformis*	Alkaline protease	57.47%	Korkmaz and Tokur (2022)
	Whiting	Head and viscera	7.0	60°C for 1 h	*Bacillus lichenformis*	Protamex	54.43%	Korkmaz and Tokur (2022)
	Anchovy	Head and viscera	8.0	60°C for 1 h	*Bacillus lichenformis*	Alkaline protease	58.84%	Korkmaz and Tokur (2022)
	Anchovy	Head and viscera	7.0	60°C for 1 h	*Bacillus lichenformis*	Protamex	54.83%	Korkmaz and Tokur (2022)
	Frog	Skin	–	–	–	Alcalase, protamex	–	Hassanvand Jamadi et al. (2019)
	Catla fish	Visceral	8.5	50°C for 135 min	*Bacillus lichenformis*	Alcalase	50%	Bhaskar et al. (2008)
	Tuna	Dark muscle	7.5	37°C for 6 h	*Aspergillus melleus*	Protease XXIII	–	Hsu Li–Chan et al. (2011)
	Nile tilapia	Head, shell and tail	–	–	*Aspergillus oryzae*	Flavourzyme	–	Gao et al. (2021)
	Rainbow trout	Skin	8.5	58°C for 90 min	*Bacillus licheniformis*	Alcalase	28.38%	Yaghoubzadeh et al. (2020)
	Rainbow trout	Skin	7.0	50°C for 90 min	*Aspergillus oryzae*	Flavourzyme	43.83%	Yaghoubzadeh et al. (2020)
Plant	Rice	Dregs	8.0	60°C for 3 h	–	Alcalase	–	Chen et al. (2021)
	Cotton	Seed	9.0	55°C for 5 h	–	Alcalase	–	Song et al. (2020)
	Cassava	Bagasse	–	–	*Bacillus spp.*	Protease	–	Clerici et al. (2021)
	Wheat	Bran	8.0	50°C for 3 h	*Bacillus lobigii*	Alcalase	–	Zou et al. (2020)
	Olive	Seed		55°C for 2 h	–	Alcalase, Thermolycin	–	Esteve et al. (2015)
	Rice bran	Bran	8.0	55°C for 4 h	–	Protease	-	Boonloh et al. (2015)

the hydrolysate exhibited moderate nitric oxide scavenging action. Free amino acid concentration is high in FPH (almost 5 g^{-1}). The results suggest that this hydrolysate can be used to develop functional food.

11.4.3 ACE-Inhibitor/Antihypertensive

Protein hydrolysates with antihypertensive and antioxidant activities have gained attention due to their significant role in cardioprotective effects (Rai et al., 2017). Peptides with angiotensin-converting enzyme (ACE) inhibitory potential can range between 2 to 30 amino acids in size (Xu et al., 2016; Erdmann et al., 2008). The possibility of ACE inhibitory action is linked to amino acid sequences on the peptides, especially the sequence on the C-terminal end. Peptides with aromatic amino acids at the C-terminus (such as phenylalanine, tryptophan, and tyrosine), and branched-chain amino acids at the N-terminus (such as isoleucine, valine, and leucine) have been proven to exhibit significant ACE binding activity. Furthermore, the hydrophobic amino acids are also known to play an essential role in ACE active site binding. ACE inhibitory action is linked to amino acid sequences between them. The structure of the most potent and selective inhibitory peptides is comparable, and the arrangement in the C-terminal tripeptide sequence has a significant impact on ACE action. Due to interactions with three subsites in the active site of ACE, tripeptides with Trp, Tyr, Phe, Pro, and hydrophobic amino acids at the C-terminus have ACE inhibitory activity.

Correa et al. (2019) studied the properties of bioactive peptides from sheep whey using a novel proteinase from *Bacillus* sp. P7. The resulting peptides were reported to have significant antioxidant activity of 87% after 30 days of storage. Purified peptides of molecular weight 10kDa were encapsulated in phosphatidylcholine liposomes and were found to have an activity of 79% compared to 54% in unencapsulated forms.

11.4.4 Anticancer/Antiproliferative

Food-derived protein hydrolysates or peptides with immunomodulatory and anticancer activities reported from various food protein sources such as milk, egg, fish, oyster, and mussels. In vitro hydrolysis of food proteins using commercial proteolytic enzymes is the most common process for producing immunomodulatory and anticancer food protein hydrolysates. Food-derived protein hydrolysates or peptides' immunomodulatory and anticancer activities are related to the amino acid composition, sequence, and length.

Studies have demonstrated that C-terminal tripeptides containing Tyrosine and Tryptophan could bind free radicals, and polypeptides containing proline can synergize with polyphenols, reduce stress, and function as anti-tumour agents. The anti-tumour peptide by rapeseed RSP-4-3-3 inhibits the proliferation of HepG2 liver cancer cells in humans in vitro by participating in the cell-mediated mitochondrial apoptosis (Wang et al., 2016). You et al. (2011) studied the antiproliferative activity of protein hydrolysates produced from Loach protein which was digested using papain. They reported that the protein hydrolysate fractions could significantly suppress the proliferation of colon cancer cells (Caco-2) compared to liver cancer cells (HepG2 and MCF-7) and breast cancer cells. The antiproliferative activity of these three cancer cell lines was related to the oxygen radical absorbance capacity (ORAC) values of the protein hydrolysate fractions ($R2 > 0.86$, $p > 0.01$), indicating that the higher the antioxidant activity, the better the antiproliferative activity.

Additionally, several studies have reported the anticancer properties of seafood protein hydrolysates (Chi et al., 2015). Hsu et al. (2011) studied the antiproliferative action of peptides from tuna dark muscle using two commercial enzymes: protease and papain. The antiproliferative potential of the generated protein hydrolysates was tested against the MCF-7 human breast cancer cell line. The peptide fraction with a molecular weight ranging from 390 to 1400 Da possessed the highest antiproliferative efficacy. Amino acid sequences of the antiproliferative peptides identified from protease hydrolysate and papain were PTAEGGGVYMVT (1124 Da) and LPHVLTPEAGAT (1206 Da).

These peptides displayed dose-dependent inhibition of MCF-7 cells with IC_{50} values of 8.8 and 8.1 IM, respectively. In the other study, hydrolysate obtained from rainbow trout protein using flavourzyme and alcalase was a good source of bioactive peptides with antioxidant and anticancer properties (Yaghoubzadeh et al., 2020). The DPPH free radical activity and ferric-reducing antioxidant power of the hydrolysate fraction obtained from flavourzyme enzymes were more potent than alcalase fractions. Furthermore, hydrolysate fraction with less than 3 kDa demonstrated the strongest cellular inhibition in colon HCT-116 cancer cells.

11.4.5 Immunostimulant

Immunomodulatory protein hydrolysate consumption is hypothesized to delay or prevent western immune-related diseases. Wu et al. (2016) reported the immunomodulatory potential of defatted wheat germ globulin hydrolysate using alcalase, neutrase, papain, pepsin, or trypsin. Hydrolysates obtained from the alcalase enzyme were reported to possess the most muscular immunomodulatory activity regarding lymphocyte proliferation, phagocytosis of neutral red, and secretion of pro-inflammatory cytokines. The molecular weight of hydrolysate obtained from the alcalase enzyme ranged between 300 and 1450 Da.

11.4.6 Detoxification

Detoxification is essential for the body to maintain health and is a preventive measure by removing toxic substances from the body. The liver is primarily responsible for this function. Dey and Dora (2014) conducted research on shrimp head and shell waste with enzymatic hydrolysis of several types of enzymes, but found the best results were using alcalase for 90 minutes, using surface response methods, and a central composite design. The technology was applied to optimize hydrolysis settings for shrimp waste hydrolysis. In this study, Arginine is a semi-essential or conditionally essential amino acid involved in protein synthesis and other physiological processes such as detoxification and energy conversion.

11.4.7 Anti-Inflammatory

Inflammation is the body's naturally occurring mechanism to protect itself from infection by foreign microorganisms, in which white blood cells play a significant role. Anti-inflammatories are substances or compounds which can reduce inflammation. Several research studies have reported the anti-inflammatory properties of protein hydrolysates. Ahn et al. (2015) reported the anti-inflammatory properties of tripeptides Proline, Alanine, and Tyrosine with a molecular weight of 349.15 Da from salmon pectoral fin protein by-products. The hydrolysates were obtained using the pepsin enzyme. The tripeptide was found to exert anti-inflammatory properties by inhibiting NO/iNOS and PGE2/COX-2 pathways and inhibiting pro-inflammatory cytokine production, including TNF-α, IL-6, and IL-1β.

Da Rocha et al. (2018) studied the anti-inflammatory properties of protein hydrolysate obtained using the enzyme Protamex. The sequence of the peptides was found to be ARDCEGHILKMFPSTYV with a molecular weight of 1245 Da and 10% degree of hydrolysis. Boonloh et al. (2015) investigated the beneficial effects of rice bran protein hydrolysates in high carbohydrate, high-fat diet in rats suffering from metabolic syndrome. Metabolic syndrome is linked to chronic inflammation and insulin resistance. This study investigated the influence of rice bran protein hydrolysate on the expression of pro-inflammatory and anti-inflammatory genes in intra-abdominal fat cells. The mRNA Il-6, Tnf-, Mcp-1, and Nos2 levels in adipose tissue from the high carbohydrate, high-fat diet control group were higher. Pro-inflammatory gene expression was reduced considerably by rice bran protein hydrolysate. Furthermore, the anti-inflammatory gene Il-10 increased its expression in rats fed with a diet containing rice bran protein hydrolysates compared to a high carbohydrate, high fat diet, implying that rice bran protein hydrolysates can increase insulin resistance by reducing inflammatory cytokine-induced insulin resistance.

11.5 SAFETY OF BIOACTIVE COMPOUNDS ISOLATED FROM AGRI-FOOD WASTES USING MICROBIAL ENZYMES

The use of food waste-derived compounds presents several challenges, one of which compromises product safety, such as biological instability, potential pathogenic contamination, the potential for rapid auto-oxidation, high water activity, and high levels of active enzymes (Rosero-Delgado et al., 2021). The most commonly found contaminants are mycotoxins, microbial contaminants, and biogenic amines (Shukla et al., 2019). All these biological hazards can cause life-threatening diseases. Therefore, several factors must be considered when checking whether the agri-industrial waste is suitable for extracting bioactive components.

One of the critical contaminants which needs to be controlled are the mycotoxins (Skladanka et al., 2017). These are secondary metabolites of fungi that are known to be widely present in many food products worldwide. The smallest infestation of plant waste with microfungal can lead to contamination with mycotoxins, which remain in the end products and endanger their use for the production of nutraceuticals and the safety of their consumption (Jallow et al., 2021). Agri-food processing waste materials contain large numbers of microbes that lead to the breakdown of proteins, producing strong odours. Enzymes that accelerate or intensify the reactions associated with spoilage are still active in many waste processing methods (Russ and Schnappinger, 2007). The amount of water available determines whether a microorganism will grow or survive. Moulds are well adapted to low moisture conditions, while others produce spores or enter a survival state until moisture is high enough for bacterial activity (Johansson et al., 2012). Contamination of agricultural products or by-products is often attributed to foodborne pathogens, i.e., *Salmonella* and *Escherichia* spp.

Biogenic amines are essential nitrogen-containing compounds mainly formed by the decarboxylation of amino acids or by transamination of aldehydes and ketones (Zhang et al., 2021). They are found in various protein-rich food waste of plant or animal origin, e.g., meat and fish processing waste (Galanakis, 2020). They can cause adverse toxicological effects in humans, such as skin rash, oedema, headache, vomiting, hypotension, palpitations, and diarrhoea (Russo et al., 2010).

11.6 CONCLUSIONS

Agri-food industrial waste and by-products serve as vast reserves of nutraceuticals that offer their potential in the food and pharmaceutical industry. Utilization of this waste by microbial enzymes seems to be an economical and environmentally friendly alternative for producing nutraceuticals as an alternative to using toxic solvents. Although implementing microbial enzymes for the extraction of nutraceuticals in food by-products can offer benefits, these biomasses can also present their challenges, making the process of commercial implementation more complicated based on several factors that need to be considered. Some of the major obstacles are changes in structural and functional properties in processing, scaling up without compromising the bioactive properties of the target substances, and achieving high standards for safety and organoleptic properties. Moreover, further research is still needed on its direct application in the food manufacturing process, along with testing the content in the final food product and medicine ready for consumption. In addition, clinical trials related to the biological potential of extracted nutraceuticals in humans still need to be carried out to clarify the importance of these microbial enzyme-extracted nutraceuticals for humans.

REFERENCES

Abidin, Z. 2021. Total carotenoids, antioxidant and anticancer effect of *Penaeus monodon* shells extract. *Biointerface Research in Applied Chemistry 11*: 11293–11302.

Ahmad, T., Ismail, A., Ahmad, S.A., Khalil, K.A., Leo, T.K., Awad, E.A., Imlan, J.C., & Sazili, A.Q. 2018. Effects of ultrasound assisted extraction in conjugation with aid of actinidin on the molecular and physicochemical properties of bovine hide gelatin. *Molecules 23*: 730.

Ahn, C.-B., Cho, Y.-S., & Je, J.-Y. 2015. Purification and anti-inflammatory action of tripeptide from salmon pectoral fin by-product protein hydrolysate. *Food Chemistry 168*: 151–156.

Aiello, F., Restuccia, D., Spizzirri, U.G., Carullo, G., Leporini, M., & Loizzo, M.R. 2020. Improving kefir bioactive properties by functional enrichment with plant and agro-food waste extracts. *Fermentation 6*: 83.

Alahyaribeik, S., Sharifi, S.D., Tabandeh, F., Honarbakhsh, S., & Ghazanfari, S. 2020. Bioconversion of chicken feather wastes by keratinolytic bacteria. *Process Safety and Environmental Protection 135*: 171–178.

Al-Hassan, A. 2020. Gelatin from camel skins: Extraction and characterizations. *Food Hydrocolloids 101*: 105457.

Ali, A.M.M., Bavisetty, S.C.B., Prodpran, T., & Benjakul, S. 2019. Squalene from fish livers extracted by ultrasound-assisted direct in situ saponification: Purification and molecular characteristics. *Journal of the American Oil Chemists' Society 96*: 1059–1071.

Ali, A.M.M., Benjakul, S., Prodpran, T., & Kishimura, H. 2018a. Extraction and characterisation of collagen from the skin of golden carp (*Probarbus Jullieni*), a processing by-product. *Waste and Biomass Valorization 9*: 783–791.

Ali, A.M.M., Kishimura, H., & Benjakul, S. 2018b. Extraction efficiency and characteristics of acid and pepsin soluble collagens from the skin of golden carp (*Probarbus Jullieni*) as affected by ultrasonication. *Process Biochemistry 66*: 237–244.

Alibardi, L., & Cossu, R. 2016. Effects of carbohydrate, protein and lipid content of organic waste on hydrogen production and fermentation products. *Waste Management 47*: 69–77.

Ashaolu, T.J. 2020. Health applications of soy protein hydrolysates. *International Journal of Peptide Research and Therapeutics 26*: 2333–2343.

Azabou, S., Abid, Y., Sebii, H., Felfoul, I., Gargouri, A., & Attia, H. 2016. Potential of the solid-state fermentation of tomato by products by *Fusarium solani* pisi for enzymatic extraction of lycopene. *LWT-Food Science and Technology 68*: 280–287.

Azizah, N., Ochiai, Y., & Nurilmala, M. 2020. Collagen peptides from Pangasius fish skin as antioxidants. In IOP Conference Series: Earth and Environmental Science (Vol. 404, pp. 012055): IOP Publishing.

Bartkiene, E., Mozuriene, E., Lele, V., Zokaityte, E., Gruzauskas, R., Jakobsone, I., Juodeikiene, G., Ruibys, R., & Bartkevics, V. 2020. Changes of bioactive compounds in barley industry by-products during submerged and solid state fermentation with antimicrobial *Pediococcus acidilactici* strain LUHS29. *Food Science & Nutrition 8*: 340–350.

Barzkar, N. 2020. Marine microbial alkaline protease: An efficient and essential tool for various industrial applications. *International Journal of Biological Macromolecules 161*: 1216–1229.

Bei, Q., Liu, Y., Wang, L., Chen, G., & Wu, Z. 2017. Improving free, conjugated, and bound phenolic fractions in fermented oats (*Avena sativa* L.) with *Monascus anka* and their antioxidant activity. *Journal of Functional Foods 32*: 185–194.

Bernabe, P., Becheran, L., Cabrera-Barjas, G., Nesic, A., Alburquenque, C., Tapia, C.V., Taboada, E., Alderete, J., & De Los Ríos, P. 2020. Chilean crab (*Aegla cholchol*) as a new source of chitin and chitosan with antifungal properties against Candida spp. *International Journal of Biological Macromolecules 149*: 962–975.

Berraquero-García, C., Almécija, M.C., Guadix, E.M., & Pérez-Gálvez, R. 2022. Valorisation of blood protein from livestock to produce haem iron-fortified hydrolysates with antioxidant activity. *International Journal of Food Science & Technology 57*: 2479–2486.

Bezus, B., Ruscasso, F., Garmendia, G., Vero, S., Cavello, I., & Cavalitto, S. 2021. Revalorization of chicken feather waste into a high antioxidant activity feather protein hydrolysate using a novel psychrotolerant bacterium. *Biocatalysis and Agricultural Biotechnology 32*: 101925.

Bhardwaj, N., Kumar, B., Agarwal, K., Chaturvedi, V., & Verma, P. 2019. Purification and characterization of a thermo-acid/alkali stable xylanases from *Aspergillus oryzae* LC1 and its application in Xylo-oligosaccharides production from lignocellulosic agricultural wastes. *International Journal of Biological Macromolecules 122*: 1191–1202.

Bhaskar, N., Benila, T., Radha, C., & Lalitha, R.G. 2008. Optimization of enzymatic hydrolysis of visceral waste proteins of Catla (Catla catla) for preparing protein hydrolysate using a commercial protease. *Bioresource Technology 99*: 335–343.

Bonifácio-Lopes, T., Vilas-Boas, A., Machado, M., Costa, E.M., Silva, S., Pereira, R.N., Campos, D., Teixeira, J.A., & Pintado, M. 2022. Exploring the bioactive potential of brewers spent grain ohmic extracts. *Innovative Food Science & Emerging Technologies 76*: 102943.

Boonloh, K., Kukongviriyapan, V., Kongyingyoes, B., Kukongviriyapan, U., Thawornchinsombut, S., & Pannangpetch, P. 2015. Rice bran protein hydrolysates improve insulin resistance and decrease pro-inflammatory cytokine gene expression in rats fed a high carbohydrate-high fat diet. *Nutrients 7*: 6313–6329.

Borrajo, P., Pateiro, M., Barba, F.J., Mora, L., Franco, D., Toldrá, F., & Lorenzo, J.M. 2019. Antioxidant and antimicrobial activity of peptides extracted from meat by-products: A review. *Food Analytical Methods 12*: 2401–2415.

Bruno, S.F., Kudre, T.G., & Bhaskar, N. 2019. Impact of pretreatment-assisted enzymatic extraction on recovery, physicochemical and rheological properties of oil from Labeo rohita head. *Journal of Food Process Engineering 42*: e12990.

Buenrostro-Figueroa, J., Velázquez, M., Flores-Ortega, O., Ascacio-Valdés, J., Huerta-Ochoa, S., Aguilar, C., & Prado-Barragán, L. 2017. Solid state fermentation of fig (Ficus carica L.) by-products using fungi to obtain phenolic compounds with antioxidant activity and qualitative evaluation of phenolics obtained. *Process Biochemistry 62*: 16–23.

Cádiz-Gurrea, M.d.l.L., Fernández-Ochoa, Á., Leyva-Jiménez, F.J., Guerrero-Muñoz, N., Villegas-Aguilar, M.d.C., Pimentel-Moral, S., Ramos-Escudero, F., & Segura-Carretero, A. 2020. LC-MS and spectrophotometric approaches for evaluation of bioactive compounds from Peru cocoa by-products for commercial applications. *Molecules 25*: 3177.

Cantatore, V., Filannino, P., Gambacorta, G., De Pasquale, I., Pan, S., Gobbetti, M., & Di Cagno, R. 2019. Lactic acid fermentation to re-cycle apple by-products for wheat bread fortification. *Frontiers in Microbiology* 10: 2574.

Cardullo, N., Leanza, M., Muccilli, V., & Tringali, C. 2021. Valorization of agri-food waste from pistachio hard shells: Extraction of polyphenols as natural antioxidants. *Resources 10*: 45.

Chakka, A.K., Muhammed, A., Sakhare, P., & Bhaskar, N. 2017. Poultry processing waste as an alternative source for mammalian gelatin: Extraction and characterization of gelatin from chicken feet using food grade acids. *Waste and Biomass Valorization 8*: 2583–2593.

Chalamaiah, M., Ulug, S.K., Hong, H., & Wu, J. 2019. Regulatory requirements of bioactive peptides (protein hydrolysates) from food proteins. *Journal of Functional Foods 58*: 123–129.

Chen, M.-L., Ning, P., Jiao, Y., Xu, Z., & Cheng, Y.-H. 2021. Extraction of antioxidant peptides from rice dreg protein hydrolysate via an angling method. *Food Chemistry 337*: 128069.

Chen, Y., Wen, J., Deng, Z., Pan, X., Xie, X., & Peng, C. 2020. Effective utilization of food wastes: Bioactivity of grape seed extraction and its application in food industry. *Journal of Functional Foods 73*: 104113.

Chi, C.-F., Hu, F.-Y., Wang, B., Li, T., & Ding, G.-F. 2015. Antioxidant and anticancer peptides from the protein hydrolysate of blood clam (Tegillarca granosa) muscle. *Journal of Functional Foods 15*: 301–313.

Chinh, N.T., Manh, V.Q., Trung, V.Q., Lam, T.D., Huynh, M.D., Tung, N.Q., Trinh, N.D., & Hoang, T. 2019. Characterization of collagen derived from tropical freshwater carp fish scale wastes and its amino acid sequence. *Natural Product Communications 14*: 1934578X19866288.

Choe, J., & Kim, H. 2018. Effects of chicken feet gelatin extracted at different temperatures and wheat fiber with different particle sizes on the physicochemical properties of gels. *Poultry science 97*: 1082–1088.

Chourasia, R., Phukon, L. Chiring., Minhajul Abedin, M., Sahoo, D., & Rai, A.K. (2022). Production and characterization of bioactive peptides in novel functional soybean chhurpi produced using Lactobacillus delbrueckii WS4. *Food Chemistry 387*: 132889.

Clerici, N.J., Lermen, A.M., & Daroit, D.J. 2021. Agro-industrial by-products as substrates for the production of bacterial protease and antioxidant hydrolysates. *Biocatalysis and Agricultural Biotechnology 37*: 102174.

Correa, A.P.F., Bertolini, D., Lopes, N.A., Veras, F.F., Gregory, G., & Brandelli, A. 2019. Characterization of nanoliposomes containing bioactive peptides obtained from sheep whey hydrolysates. *Food Science and Technology 101*: 107–112.

Da Rocha, M., Alemán, A., Baccan, G.C., López-Caballero, M.E., Gómez-Guillén, C., Montero, P., & Prentice, C. 2018. Anti-inflammatory, antioxidant, and antimicrobial effects of underutilized fish protein hydrolysate. *Journal of Aquatic Food Product Technology 27*: 592–608.

Dave, D., Liu, Y., Pohling, J., Trenholm, S., & Murphy, W. 2020. Astaxanthin recovery from Atlantic shrimp (Pandalus borealis) processing materials. *Bioresource Technology Reports 11*: 100535.

del Pilar Garcia-Mendoza, M., Espinosa-Pardo, F.A., Savoire, R., Etchegoyen, C., Harscoat-Schiavo, C., & Subra-Paternault, P. 2021. Recovery and antioxidant activity of phenolic compounds extracted from walnut press-cake using various methods and conditions. *Industrial Crops and Products 167*: 113546.

Demirgül, K., & Ozturk, E. 2021. Changes in nutrients, energy, antioxidant and carotenoid levels of dried tomato (Lycopersicon esculentum) pomage treated with *Aspergillus niger* solid-state fermentation. *Turkish Journal of Agriculture-Food Science and Technology 9*: 701–708.

Dey, S.S., & Dora, K.C. 2014. Optimization of the production of shrimp waste protein hydrolysate using microbial proteases adopting response surface methodology. *Journal of Food Science and Technology 51*: 16–24.

dos Santos Aguilar, J.G., de Souza, A.K.S., & de Castro, R.J.S. 2020. Enzymatic hydrolysis of chicken viscera to obtain added-value protein hydrolysates with antioxidant and antihypertensive properties. *International Journal of Peptide Research and Therapeutics 26*: 717–725.

dos Santos Aguilar, J.G., & Sato, H.H. 2018. Microbial proteases: Production and application in obtaining protein hydrolysates. *Food Research International 103*: 253–262.

Duhan, J.S., Chawla, P., Kumar, S., Bains, A., & Sadh, P.K. 2021. Proximate composition, polyphenols, and antioxidant activity of solid state fermented peanut press cake. *Preparative Biochemistry & Biotechnology 51*: 340–349.

Dulf, F.V., Vodnar, D.C., & Socaciu, C. 2016. Effects of solid-state fermentation with two filamentous fungi on the total phenolic contents, flavonoids, antioxidant activities and lipid fractions of plum fruit (Prunus domestica L.) by-products. *Food Chemistry 209*: 27–36.

Durante, M., Bleve, G., Selvaggini, R., Veneziani, G., Servili, M., & Mita, G. 2019. Bioactive compounds and stability of a typical Italian bakery products "taralli" enriched with fermented olive paste. *Molecules 24*: 3258.

El-Bialy, H.A.A., & Abd El-Khalek, H.H. 2020. A comparative study on astaxanthin recovery from shrimp wastes using lactic fermentation and green solvents: An applied model on minced Tilapia. *Journal of Radiation Research and Applied Sciences 13*: 594–605.

El Sheikha, A.F., & Ray, R.C. 2022. Bioprocessing of horticultural wastes by solid-state fermentation into value-added/innovative bioproducts: A review. *Food Reviews International*: 1–56.

Esteve, C., Marina, M., & García, M. 2015. Novel strategy for the revalorization of olive (Olea europaea) residues based on the extraction of bioactive peptides. *Food Chemistry 167*: 272–280.

Fagbemi, O.D., Sithole, B., & Tesfaye, T. 2020. Optimization of keratin protein extraction from waste chicken feathers using hybrid pre-treatment techniques. *Sustainable Chemistry and Pharmacy 17*: 100267.

Fang, B., Sun, J., Dong, P., Xue, C., & Mao, X. 2017. Conversion of turbot skin wastes into valuable functional substances with an eco-friendly fermentation technology. *Journal of Cleaner Production 156*: 367–377.

FAO. 2019. The State of Food and Agriculture 2019. Moving forward on food loss and waste reduction. Rome. https://www.fao.org/3/ca6030en/ca6030en.pdf

FAO. 2021. Food Waste Index Report 2021. United Nations Environment Programme 2021. Nairobi. https://wedocs.unep.org/bitstream/handle/20.500.11822/35280/FoodWaste.pdf

Fasim, A., More, V.S., & More, S.S. 2021. Large-scale production of enzymes for biotechnology uses. *Current Opinion in Biotechnology 69*: 68–76.

Fernando, G.S.N., Wood, K., Papaioannou, E.H., Marshall, L.J., Sergeeva, N.N., & Boesch, C. 2021. Application of an ultrasound-assisted extraction method to recover betalains and polyphenols from red beetroot waste. *ACS Sustainable Chemistry & Engineering 9*: 8736–8747.

Ferri, M., Rondini, G., Calabretta, M.M., Michelini, E., Vallini, V., Fava, F., Roda, A., Minnucci, G., & Tassoni, A. 2017. White grape pomace extracts, obtained by a sequential enzymatic plus ethanol-based extraction, exert antioxidant, anti-tyrosinase and anti-inflammatory activities. *New Biotechnology 39*: 51–58.

Franklin, E.C., Haq, M., Roy, V.C., Park, J.S., & Chun, B.S. 2020. Supercritical CO2 extraction and quality comparison of lipids from Yellowtail fish (Seriola quinqueradiata) waste in different conditions. *Journal of Food Processing and Preservation 44*: e14892.

Galanakis, C.M. 2020. *Food Waste Recovery: Processing Technologies, Industrial Techniques, and Applications*. Academic Press.

Gao, P., Li, L., Xia, W., Xu, Y., & Liu, S. 2020. Valorization of Nile tilapia (Oreochromis niloticus) fish head for a novel fish sauce by fermentation with selected lactic acid bacteria. *LWT 129*: 109539.

Gao, R., Yu, Q., Shen, Y., Chu, Q., Chen, G., Fen, S., Yang, M., Yuan, L., McClements, D.J., & Sun, Q. 2021. Production, bioactive properties, and potential applications of fish protein hydrolysates: Developments and challenges. *Trends in Food Science & Technology 110*: 687–699.

Garcia, E.G., Marina, M.L., & Garcia, M.C. 2015a. Plum (Prunus domestica L.) by-product as a new and cheap source of bioactive peptides: Extraction method and peptides characterization. *Journal of Functional Foods 11*: 428–437.

García, M.C., Endermann, J., Gonzalez-Garcia, E., & Marina, M.L. 2015b. HPLC-Q-TOF-MS identification of antioxidant and antihypertensive peptides recovered from cherry (Prunus cerasus L.) subproducts. *Journal of Agricultural and Food Chemistry 63*: 1514–1520.

Gençdağ, E., Görgüç, A., & Yılmaz, F.M. 2021. Recent advances in the recovery techniques of plant-based proteins from agro-industrial by-products. *Food Reviews International 37*: 447–468.

Guo, Y., Huang, W.-C., Wu, Y., Qi, X., & Mao, X. 2018. Application of a low-voltage direct-current electric field for lipid extraction from squid viscera. *Journal of Cleaner Production 205*: 610–618.

Haddar, A., Fakhfakh-Zouari, N., Hmidet, N., Frikha, F., Nasri, M., & Kamoun, A.S. 2010. Low-cost fermentation medium for alkaline protease production by *Bacillus mojavensis* A21 using hulled grain of wheat and sardinella peptone. *Journal of Bioscience and Bioengineering 110*: 288–294.

Hadj Saadoun, J., Bertani, G., Levante, A., Vezzosi, F., Ricci, A., Bernini, V., & Lazzi, C. 2021. Fermentation of agri-food waste: A promising route for the production of aroma compounds. *Foods 10*: 707.

Halavach, T.M., Dudchik, N.V., Tarun, E.I., Zhygankov, V.G., Kurchenko, V.P., Romanovich, R.V., Khartitonov, V.D., & Asafov, V.A. 2020. Biologically active properties of hydrolysed and fermented milk proteins. *Journal of Microbiology, Biotechnology and Food Sciences 9*: 714–720.

Hamdi, M., Hajji, S., Affes, S., Taktak, W., Maâlej, H., Nasri, M., & Nasri, R. 2018. Development of a controlled bioconversion process for the recovery of chitosan from blue crab (Portunus segnis) exoskeleton. *Food Hydrocolloids 77*: 534–548.

Hassanvand Jamadi, R., Yaghoubi, H., & Sadeghizadeh, M. 2019. Brevinin-2R and derivatives as potential anticancer peptides: Synthesis, purification, characterization and biological activities. *International Journal of Peptide Research and Therapeutics 25*: 151–160.

Hathwar, S.C., Bijinu, B., Rai, A.K., & Narayan, B. 2011. Simultaneous recovery of lipids and proteins by enzymatic hydrolysis of fish industry waste using different commercial proteases. *Applied Biochemistry and Biotechnology 164*: 115–124.

He, L., Lan, W., Zhao, Y., Chen, S., Liu, S., Cen, L., Cao, S., Dong, L., Jin, R., & Liu, Y. 2020. Characterization of biocompatible pig skin collagen and application of collagen-based films for enzyme immobilization. *RSC Advances 10*: 7170–7180.

Hsieh, Y.-H.P., & Ofori, J.A. 2011. Blood-derived products for human consumption. *Revelation and Science 1*. https://journals.iium.edu.my/revival/index.php/revival/article/view/15

Hsu, K.-C., Li-Chan, E.C., & Jao, C.-L. 2011. Antiproliferative activity of peptides prepared from enzymatic hydrolysates of tuna dark muscle on human breast cancer cell line MCF-7. *Food Chemistry 126*: 617–622.

Hussain, M., Qamar, A., Saeed, F., Rasheed, R., Niaz, B., Afzaal, M., Mushtaq, Z., & Anjum, F. 2021. Biochemical properties of maize bran with special reference to different phenolic acids. *International Journal of Food Properties 24*: 1468–1478.

Irshad, A., & Sharma, B. 2015. Abattoir by-product utilization for sustainable meat industry: A review. *Journal of Animal Production Advances 5*: 681–696.

Jagtap, S., Deshmukh, R.A., Menon, S., & Das, S. 2017. Xylooligosaccharides production by crude microbial enzymes from agricultural waste without prior treatment and their potential application as nutraceuticals. *Bioresource Technology 245*: 283–288.

Jallow, A., Xie, H., Tang, X., Qi, Z., & Li, P. 2021. Worldwide aflatoxin contamination of agricultural products and foods: From occurrence to control. *Comprehensive Reviews in Food Science and Food Safety 20*: 2332–2381.

Jayathilakan, K., Sultana, K., Radhakrishna, K., & Bawa, A. 2012. Utilization of by-products and waste materials from meat, poultry and fish processing industries: A review. *Journal of Food Science and Technology 49*: 278–293.

Johansson, P., Ekstrand-Tobin, A., Svensson, T., & Bok, G. 2012. Laboratory study to determine the critical moisture level for mould growth on building materials. *International Biodeterioration & Biodegradation 73*: 23–32.

Kagkli, D.M., Corich, V., Bovo, B., Lante, A., & Giacomini, A. 2016. Antiradical and antimicrobial properties of fermented red chicory (Cichorium intybus L.) by-products. *Annals of Microbiology 66*: 1377–1386.

Kaur, M., Singh, B., Kaur, A., & Singh, N. 2021. Proximate, mineral, amino acid composition, phenolic profile, antioxidant and functional properties of oilseed cakes. *International Journal of Food Science & Technology 56*: 6732–6741.

Kieliszek, M., Pobiega, K., Piwowarek, K., & Kot, A.M. 2021. Characteristics of the proteolytic enzymes produced by lactic acid bacteria. *Molecules 26*: 1858.

Kodous, M. 2020. Physicochemical properties of hydrolyzed collagen produced from chicken feet. *Middle East Journal of Agriculture Research 9*: 81–89.

Korkmaz, K., & Tokur, B. 2022. Optimization of hydrolysis conditions for the production of protein hydrolysates from fish wastes using response surface methodology. *Food Bioscience 45*: 101312.

Kumar, D.M., Priya, P., Balasundari, S.N., Devi, G., Rebecca, A.I.N., & Kalaichelvan, P. 2012. Production and optimization of feather protein hydrolysate from *Bacillus* sp. MPTK6 and its antioxidant potential. *Middle-East Journal of Scientific Research 11*: 900–907.

Kumar, M., Tomar, M., Bhuyan, D.J., Punia, S., Grasso, S., Sa, A.G.A., Carciofi, B.A.M., Arrutia, F., Changan, S., & Singh, S. 2021. Tomato (Solanum lycopersicum L.) seed: A review on bioactives and biomedical activities. *Biomedicine & Pharmacotherapy 142*: 112018.

Kumari, M., Padhi, S., Sharma, S. et al. 2021. Biotechnological potential of psychrophilic microorganisms as the source of cold-active enzymes in food processing applications. *3 Biotech 11*: 479.

Lafarga, T., Bobo, G., Viñas, I., Zudaire, L., Simó, J., & Aguiló-Aguayo, I. 2018. Steaming and sous-vide: Effects on antioxidant activity, vitamin C, and total phenolic content of Brassica vegetables. *International Journal of Gastronomy and Food Science 13*: 134–139.

Lasrado, L.D., & Rai, A.K. 2018. Recovery of Nutraceuticals from Agri-Food Industry Waste by Lactic Acid Fermentation. In: Varjani, S., Parameswaran, B., Kumar, S., Khare, S. (eds) *Biosynthetic Technology and Environmental Challenges. Energy, Environment, and Sustainability*. Springer, Singapore. https://doi.org/10.1007/978-981-10-7434-9_11

Lessa, O.A., Reis, N.d.S., Leite, S.G.F., Gutarra, M.L.E., Souza, A.O., Gualberto, S.A., de Oliveira, J.R., Aguiar-Oliveira, E., & Franco, M. 2018. Effect of the solid state fermentation of cocoa shell on the secondary metabolites, antioxidant activity, and fatty acids. *Food Science and Biotechnology 27*: 107–113.

Li, Y., Li, M., Wang, L., & Li, Z. 2022. Effect of particle size on the release behavior and functional properties of wheat bran phenolic compounds during in vitro gastrointestinal digestion. *Food Chemistry 367*: 130751.

Liu, M., Zhang, L., Ser, S.L., Cumming, J.R., & Ku, K.-M. 2018. Comparative phytonutrient analysis of broccoli by-products: The potentials for broccoli by-product utilization. *Molecules 23*: 900.

López-Pedrouso, M., Borrajo, P., Pateiro, M., Lorenzo, J.M., & Franco, D. 2020. Antioxidant activity and peptidomic analysis of porcine liver hydrolysates using alcalase, bromelain, flavourzyme and papain enzymes. *Food Research International 137*: 109389.

Louati, I., Bahloul, N., Besombes, C., Allaf, K., & Kechaou, N. 2019. Instant controlled pressure-drop as texturing pretreatment for intensifying both final drying stage and extraction of phenolic compounds to valorize orange industry by-products (Citrus sinensis L.). *Food and Bioproducts Processing 114*: 85–94.

Luo, D., Mu, T., & Sun, H. 2021. Profiling of phenolic acids and flavonoids in sweet potato (Ipomoea batatas L.) leaves and evaluation of their anti-oxidant and hypoglycemic activities. *Food Bioscience 39*: 100801.

Maciel, L.G., Ribeiro, F.L., Teixeira, G.L., Molognoni, L., Dos Santos, J.N., Nunes, I.L., & Block, J.M. 2020. The potential of the pecan nut cake as an ingredient for the food industry. *Food Research International 127*: 108718.

Majumdar, S., Bhattacharyya, D., & Bhowal, J. 2021. Evaluation of nutraceutical application of xylooligosaccharide enzymatically produced from cauliflower stalk for its value addition through a sustainable approach. *Food & Function 12*: 5501–5523.

Matkawala, F., Nighojkar, S., Kumar, A., & Nighojkar, A. 2021. Microbial alkaline serine proteases: Production, properties and applications. *World Journal of Microbiology and Biotechnology 37*: 1–12.

McDonald, A.G., & Tipton, K.F. 2021. Enzyme nomenclature and classification: The state of the art. *The FEBS Journal*. https://doi.org/10.1111/febs.16274

Mechmeche, M., Kachouri, F., Chouabi, M., Ksontini, H., Setti, K., & Hamdi, M. 2017. Optimization of extraction parameters of protein isolate from tomato seed using response surface methodology. *Food Analytical Methods 10*: 809–819.

Mechri, S., Sellem, I., Bouacem, K., Jabeur, F., Chamkha, M., Hacene, H., Bouanane-Darenfed, A., & Jaouadi, B. 2020. Antioxidant and enzyme inhibitory activities of Metapenaeus monoceros by-product hydrolysates elaborated by purified alkaline proteases. *Waste and Biomass Valorization 11*: 6741–6755.

Mehta, D., Prasad, P., Sangwan, R.S., & Yadav, S.K. 2018. Tomato processing by-product valorization in bread and muffin: Improvement in physicochemical properties and shelf life stability. *Journal of Food Science and Technology 55*: 2560–2568.

Milinčić, D.D., Stanisavljević, N.S., Kostić, A.Ž., Bajić, S.S., Kojić, M.O., Gašić, U.M., Barać, M.B., Stanojević, S.P., Tešić, Ž.L., & Pešić, M.B. 2021. Phenolic compounds and biopotential of grape pomace extracts from Prokupac red grape variety. *LWT 138*: 110739.

Mirabella, N., Castellani, V., & Sala, S. 2014. Current options for the valorization of food manufacturing waste: A review. *Journal of Cleaner Production 65*: 28–41.

Mirzapour Kouhdasht, A., Moosavi-Nasab, M., & Aminlari, M. 2018. Gelatin production using fish wastes by extracted alkaline protease from *Bacillus licheniformis*. *Journal of Food Science and Technology 55*: 5175–5180.

Moccia, F., Flores-Gallegos, A.C., Chávez-González, M.L., Sepúlveda, L., Marzorati, S., Verotta, L., Panzella, L., Ascacio-Valdes, J.A., Aguilar, C.N., & Napolitano, A. 2019. Ellagic acid recovery by solid state fermentation of pomegranate wastes by *Aspergillus niger* and Saccharomyces cerevisiae: A comparison. *Molecules 24*: 3689.

Montoya, J.M., Mata, S.V., Acosta, J.L., Edgar, B., Cabrera, H., Valdez, L.G.L., Reyes, C., Jair, H., & Cureño, B. 2021. Obtaining of astaxanthin from crab exosqueletons and shrimp head shells. *Biointerface Research in Applied Chemistry 11*: 13516–13523.

Nurilmala, M., Hizbullah, H.H., Karnia, E., Kusumaningtyas, E., & Ochiai, Y. 2020. Characterization and antioxidant activity of collagen, gelatin, and the derived peptides from yellowfin tuna (Thunnus albacares) skin. *Marine Drugs 18*: 98.

Nzekoue, F.K., Angeloni, S., Navarini, L., Angeloni, C., Freschi, M., Hrelia, S., Vitali, L.A., Sagratini, G., Vittori, S., & Caprioli, G. 2020. Coffee silverskin extracts: Quantification of 30 bioactive compounds by a new HPLC-MS/MS method and evaluation of their antioxidant and antibacterial activities. *Food Research International 133*: 109128.

Padhi, S., Sanjukta, S., Chourasia, R., Labala, R.K., Singh, S.P., & Rai, A.K. 2021. A multifunctional peptide from *Bacillus* fermented soybean for effective inhibition of SARS-CoV-2 S1 receptor binding domain and modulation of toll like receptor 4: A molecular docking study. *Frontiers in Molecular Bioscience 31*: 636647.

Pang, C., Yin, X., Zhang, G., Liu, S., Zhou, J., Li, J., & Du, G. 2021. Current progress and prospects of enzyme technologies in future foods. *Systems Microbiology and Biomanufacturing 1*: 24–32.

Panzella, L., Moccia, F., Nasti, R., Marzorati, S., Verotta, L., & Napolitano, A. 2020. Bioactive phenolic compounds from agri-food wastes: An update on green and sustainable extraction methodologies. *Frontiers in Nutrition 7*: 60.

Phukon, L.C., Chourasia, R., Padhi, S. et al. 2022. Cold-adaptive traits identified by comparative genomic analysis of a lipase-producing *Pseudomonas* sp. HS6 isolated from snow-covered soil of Sikkim Himalaya and molecular simulation of lipase for wide substrate specificity. *Current Genetics*. https://doi.org/10.1007/s00294-022-01241-3

Quintana, G., Spínola, V., Martins, G.N., Gerbino, E., Gómez-Zavaglia, A., & Castilho, P.C. 2020. Release of health-related compounds during in vitro gastro-intestinal digestion of okara and okara fermented with Lactobacillus plantarum. *Journal of Food Science and Technology 57*: 1061–1070.

Radenkovs, V., Kviesis, J., Juhnevica-Radenkova, K., Valdovska, A., Püssa, T., Klavins, M., & Drudze, I. 2018. Valorization of wild apple (Malus spp.) by-products as a source of essential fatty acids, tocopherols and phytosterols with antimicrobial activity. *Plants 7*: 90.

Rai, A.K., Bhaskar, N., & Baskaran, V. 2013. Bioefficacy of EPA-DHA from lipids recovered from fish processing wastes through biotechnological approaches. *Food Chemistry 136(1)*: 80–86.

Rai, A.K., Sanjukta, S., & Jeyaram, K. 2017. Production of angiotensin I converting enzyme inhibitory (ACE-I) peptides during milk fermentation and its role in treatment of hypertension. *Critical Reviews in Food Science and Nutrition 57*: 2789–2800.

Rai, A.K., Swapna, H.C., Bhaskar, N., & Baskaran, V. 2012. Potential of seafood industry by-products as sources of recoverable lipids: Fatty acid composition of meat and non meat component of selected Indian marine fishes. *Journal of Food Biochemistry 36*: 441–448.

Rasulu, H., Praseptiangga, D., Joni, I.M., & Ramelan, A.H. 2022. Characterization of Physicochemical Properties of Powder Coconut Crab Shells (Birgus latro L.) from North Maluku. In *6th International Conference of Food, Agriculture, and Natural Resource (IC-FANRES 2021)* (pp. 357–360): Atlantis Press.

Ravindran, R., & Jaiswal, A.K. 2016. Microbial enzyme production using lignocellulosic food industry wastes as feedstock: A review. *Bioengineering 3*: 30.

Rebah, F.B.a.N.M. 2012. Fish processing wastes for microbial enzyme production: A review. *Biotech.*

Rochín-Medina, J.J., Ramírez, K., Rangel-Peraza, J.G., & Bustos-Terrones, Y.A. 2018. Increase of content and bioactivity of total phenolic compounds from spent coffee grounds through solid state fermentation by *Bacillus clausii*. *Journal of Food Science and Technology 55*: 915–923.

Rosero-Delgado, E.A., Zambrano-Arcentales, M.A., Gómez-Salcedo, Y., Baquerizo-Crespo, R.J., & Dustet-Mendoza, J.C. 2021. Biotechnology applied to treatments of agro-industrial wastes. In Maddela, N.R., García Cruzatty, L.C., Chakraborty, S. (eds) *Advances in the Domain of Environmental Biotechnology* (pp. 277–311): Springer, Singapore.

Russ, W., & Schnappinger, M. 2007. Waste related to the food industry: A challenge in material loops. In Oreopoulou, Vasso, Russ, Winfried (eds) *Utilization of by-products and treatment of waste in the food industry* (pp. 1–13): Springer, Boston, USA.

Russo, P., Spano, G., Arena, M., Capozzi, V., Grieco, F., & Beneduece, L. 2010. Are consumers aware of the risks related to biogenic amines in food. *Current Research, Technology and Education Topics in Applied Microbiology and Microbial Biotechnology*: 1087–1095.

Sadh, P.K., Chawla, P., & Duhan, J.S. 2018. Fermentation approach on phenolic, antioxidants and functional properties of peanut press cake. *Food Bioscience 22*: 113–120.

Sadh, P.K., Duhan, S., & Duhan, J.S. 2018. Agro-industrial wastes and their utilization using solid state fermentation: A review. *Bioresources and Bioprocessing 5*: 1–15.

Salehi, L., & Taghian Dinani, S. 2020. Application of electrohydrodynamic-ultrasonic procedure for extraction of β-carotene from carrot pomace. *Journal of Food Measurement and Characterization 14*: 3031–3039.

Santos, V.A.Q., Nascimento, C.G., Schmidt, C.A., Mantovani, D., Dekker, R.F., & da Cunha, M.A.A. 2018. Solid-state fermentation of soybean okara: Isoflavones biotransformation, antioxidant activity and enhancement of nutritional quality. *LWT 92*: 509–515.

Sapna, I., & Jayadeep, A. 2021. Role of endoxylanase and its concentrations in enhancing the nutraceutical components and bioactivities of red rice bran. *LWT 147*: 111675.

Sapna, I., & Jayadeep, A. 2022. Cellulolytic and xylanolytic enzyme combinations in the hydrolysis of red rice bran: A disparity in the release of nutraceuticals and its correlation with bioactivities. *LWT 154*: 112856.

Sasidharan, A., & Venugopal, V. 2020. Proteins and co-products from seafood processing discards: Their recovery, functional properties and applications. *Waste and Biomass Valorization 11*: 5647–5663.

Sharayei, P., Azarpazhooh, E., Zomorodi, S., Einafshar, S., & Ramaswamy, H.S. 2021. Optimization of ultrasonic-assisted extraction of astaxanthin from green tiger (Penaeus semisulcatus) shrimp shell. *Ultrasonics Sonochemistry 76*: 105666.

Sharma, R., & Ghoshal, G. 2020. Optimization of carotenoids production by Rhodotorula mucilaginosa (MTCC-1403) using agro-industrial waste in bioreactor: A statistical approach. *Biotechnology Reports 25*: e00407.

Shirsath, A.P., & Henchion, M.M. 2021. Bovine and ovine meat co-products valorisation opportunities: A systematic literature review. *Trends in Food Science & Technology 118*: 57–70.

Shukla, S., Lee, J.S., Bajpai, V.K., Khan, I., Huh, Y.S., Han, Y.-K., & Kim, M. 2019. Toxicological evaluation of lotus, ginkgo, and garlic tailored fermented Korean soybean paste (Doenjang) for biogenic amines, aflatoxins, and microbial hazards. *Food and Chemical Toxicology 133*: 110729.

Sierra-Lopera, L.M., & Zapata-Montoya, J.E. 2021. Optimization of enzymatic hydrolysis of red tilapia scales (Oreochromis sp.) to obtain bioactive peptides. *Biotechnology Reports 30*: e00611.

Skladanka, J., Adam, V., Zitka, O., Mlejnkova, V., Kalhotka, L., Horky, P., Konecna, K., Hodulikova, L., Knotova, D., & Balabanova, M. 2017. Comparison of biogenic amines and mycotoxins in alfalfa and red clover fodder depending on additives. *International Journal of Environmental Research and Public Health 14*: 418.

Song, W., Kong, X., Hua, Y., Li, X., Zhang, C., & Chen, Y. 2020. Antioxidant and antibacterial activity and in vitro digestion stability of cottonseed protein hydrolysates. *LWT 118*: 108724.

Stodolak, B., Starzyńska-Janiszewska, A., Wywrocka-Gurgul, A., & Wikiera, A. 2017. Solid-state fermented flaxseed oil cake of improved antioxidant capacity as potential food additive. *Journal of Food Processing and Preservation 41*: e12855.

Szabo, K., Dulf, F.V., Teleky, B.-E., Eleni, P., Boukouvalas, C., Krokida, M., Kapsalis, N., Rusu, A.V., Socol, C.T., & Vodnar, D.C. 2021. Evaluation of the bioactive compounds found in tomato seed oil and tomato peels influenced by industrial heat treatments. *Foods 10*: 110.

Szymczak, T., Cybulska, J., Podleśny, M., & Frąc, M. 2021. Various perspectives on microbial lipase production using agri-food waste and renewable products. *Agriculture 11*: 540.

Tamasi, G., Baratto, M.C., Bonechi, C., Byelyakova, A., Pardini, A., Donati, A., Leone, G., Consumi, M., Lamponi, S., & Magnani, A. 2019. Chemical characterization and antioxidant properties of products and by-products from Olea europaea L. *Food Science & Nutrition 7*: 2907–2920.

Tenore, G.C., Caruso, D., D'Avino, M., Buonomo, G., Caruso, G., Ciampaglia, R., Schiano, E., Maisto, M., Annunziata, G., & Novellino, E. 2020. A pilot screening of agro-food waste products as sources of nutraceutical formulations to improve simulated postprandial glycaemia and insulinaemia in healthy subjects. *Nutrients 12*: 1292.

Torres-León, C., Chávez-González, M.L., Hernández-Almanza, A., Martínez-Medina, G.A., Ramírez-Guzmán, N., Londoño-Hernández, L., & Aguilar, C.N. 2021. Recent advances on the microbiological and enzymatic processing for conversion of food wastes to valuable bioproducts. *Current Opinion in Food Science 38*: 40–45.

Tufariello, M., Durante, M., Veneziani, G., Taticchi, A., Servili, M., Bleve, G., & Mita, G. 2019. Patè olive cake: Possible exploitation of a by-product for food applications. *Frontiers in nutrition 6*: 3.

Ucak, I., Afreen, M., Montesano, D., Carrillo, C., Tomasevic, I., Simal-Gandara, J., & Barba, F.J. 2021. Functional and bioactive properties of peptides derived from marine side streams. *Marine Drugs 19*: 71.

Vásquez-Villanueva, R., Marina, M.L., & García, M.C. 2016. Identification by hydrophilic interaction and reversed-phase liquid chromatography–tandem mass spectrometry of peptides with antioxidant capacity in food residues. *Journal of Chromatography A 1428*: 185–192.

Vázquez, J.A., Fraguas, J., Mirón, J., Valcárcel, J., Pérez-Martín, R.I., & Antelo, L.T. 2020. Valorisation of fish discards assisted by enzymatic hydrolysis and microbial bioconversion: Lab and pilot plant studies and preliminary sustainability evaluation. *Journal of Cleaner Production 246*: 119027.

Venkatachalam, C.D., Sengottian, M., Lakshmanan, S., Thilagarajan, R.R.A., & Pugazanthi, S. 2020. Extraction of polyphenols from cashew nut (Anacardium occidentale) shells using rotocel extractor. In *AIP Conference Proceedings* (Vol. 2240, pp. 030001): AIP Publishing LLC.

Verni, M., Rizzello, C.G., & Coda, R. 2019. Fermentation biotechnology applied to cereal industry by-products: Nutritional and functional insights. *Frontiers in Nutrition 6*: 42.

Vieira, T.F., Corrêa, R.C., Peralta, R.A., Peralta-Muniz-Moreira, R.F., Bracht, A., & Peralta, R.M. 2020. An overview of structural aspects and health beneficial effects of antioxidant oligosaccharides. *Current Pharmaceutical Design 26*: 1759–1777.

Vilas-Boas, A.A., Pintado, M., & Oliveira, A.L. 2021. Natural bioactive compounds from food waste: Toxicity and safety concerns. *Foods 10*: 1564.

Wang, L., Zhang, J., Yuan, Q., Xie, H., Shi, J., & Ju, X. 2016. Separation and purification of an anti-tumor peptide from rapeseed (Brassica campestris L.) and the effect on cell apoptosis. *Food & Function 7*: 2239–2248.

Worsztynowicz, P., Białas, W., & Grajek, W. 2020. Integrated approach for obtaining bioactive peptides from whey proteins hydrolysed using a new proteolytic lactic acid bacteria. *Food Chemistry 312*: 126035.

Wu, G., Fanzo, J., Miller, D.D., Pingali, P., Post, M., Steiner, J.L., & Thalacker-Mercer, A.E. 2014. Production and supply of high-quality food protein for human consumption: Sustainability, challenges, and innovations. *Annals of the New York Academy of Sciences 1321*: 1–19.

Wu, W., Zhang, M., Sun, C., Brennan, M., Li, H., Wang, G., Lai, F., & Wu, H. 2016. Enzymatic preparation of immunomodulatory hydrolysates from defatted wheat germ (Triticum vulgare) globulin. *International Journal of Food Science & Technology 51*: 2556–2566.

Xu, Y., Bao, T., Han, W., Chen, W., Zheng, X., & Wang, J. 2016. Purification and identification of an angiotensin I-converting enzyme inhibitory peptide from cauliflower by-products protein hydrolysate. *Process Biochemistry 51*: 1299–1305.

Yaghoubzadeh, Z., Peyravii Ghadikolaii, F., Kaboosi, H., Safari, R., & Fattahi, E. 2020. Antioxidant activity and anticancer effect of bioactive peptides from rainbow trout (Oncorhynchus mykiss) skin hydrolysate. *International Journal of Peptide Research and Therapeutics 26*: 625–632.

Yin, Z.N., Wu, W.J., Sun, C.Z., Liu, H.F., Chen, W.B., Zhan, Q.P., Lei, Z.G., Xuan, X., Juan, J., & Kun, Y. 2019. Antioxidant and anti-inflammatory capacity of ferulic acid released from wheat bran by solid-state fermentation of *Aspergillus niger. Biomedical and Environmental Sciences 32*: 11–21.

You, L., Zhao, M., Liu, R.H., & Regenstein, J.M. 2011. Antioxidant and antiproliferative activities of loach (Misgurnus anguillicaudatus) peptides prepared by papain digestion. *Journal of Agricultural and Food Chemistry 59*: 7948–7953.

Zambrano, C., Kotogán, A., Bencsik, O., Papp, T., Vágvölgyi, C., Mondal, K.C., Krisch, J., & Takó, M. 2018. Mobilization of phenolic antioxidants from grape, apple and pitahaya residues via solid state fungal fermentation and carbohydrase treatment. *LWT 89*: 457–465.

Zhang, Y., Tu, D., Shen, Q., & Dai, Z. 2019. Fish scale valorization by hydrothermal pretreatment followed by enzymatic hydrolysis for gelatin hydrolysate production. *Molecules 24*: 2998.

Zhang, Z., Peng, Y., & Zheng, J. 2021. Amino acid and biogenic amine adductions derived from reactive metabolites. *Current Drug Metabolism 22*: 1076–1086.

Zheng, Y., Liu, S., Xie, J., Chen, Y., Dong, R., Zhang, X., Liu, S., Xie, J., Hu, X., & Yu, Q. 2020. Antioxidant, α-amylase and α-glucosidase inhibitory activities of bound polyphenols extracted from mung bean skin dietary fiber. *LWT 132*: 109943.

Zou, Z., Wang, M., Wang, Z., Aluko, R.E., & He, R. 2020. Antihypertensive and antioxidant activities of enzymatic wheat bran protein hydrolysates. *Journal of Food Biochemistry 44*: e13090.

Section IV

Technologies for Improved Enzyme Systems for Nutraceutical Production

12 Molecular Engineering of Microbial Food Enzymes

Ammini Naduvanthar Anoopkumar[1],
Embalil Mathachan Aneesh[1], *Aravind Madhavan*[2],
Parameswaran Binod[3,10], *Mukesh Kumar Awasthi*[4],
Mohammed Kuddus[5], *Ashok Pandey*[6,7,8],
Laya Liz Kuriakose[9], *and Raveendran Sindhu*[9]

[1] Centre for Research in Emerging Tropical Diseases (CRET-D), Department of Zoology, University of Calicut, Malappuram, Kerala, India

[2] Mycobacterium Research Group, Rajiv Gandhi Center for Biotechnology, Jagathy, Thiruvananthapuram, Kerala, India

[3] Microbial Processes and Technology Division, CSIR-National Institute for Interdisciplinary Science and Technology (CSIR-NIIST), Trivandrum, Kerala, India

[4] College of Natural Resources and Environment, Northwest A & F University, Yangling, Shaanxi, China

[5] Department of Biochemistry, University of Hail, Hail, Kingdom of Saudi Arabia

[6] Centre for Innovation and Translational Research, CSIR – Indian Institute for Toxicology Research (CSIR-IITR), Lucknow, India

[7] Centre for Energy and Environmental Sustainability, Lucknow, Uttar Pradesh, India

[8] Sustainability Cluster, School of Engineering, University of Petroleum and Energy Studies, Dehradun, Uttarakhand, India

[9] Department of Food Technology, TKM Institute of Technology, Kollam, Kerala, India

[10] Academy of Scientific and Innovative Research (AcSIR), Ghaziabad, Uttar Pradesh, India

CONTENTS

DOI: 10.1201/9781003311164-16

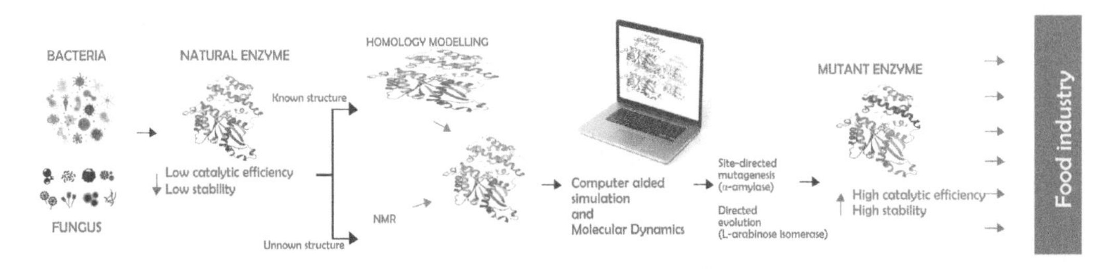

FIGURE 12.1 Integrating advanced molecular engineering approaches for enhanced enzyme stability and efficiency.

12.1 INTRODUCTION

The processing of food through the use of natural agents, especially with the aid of microbial consortia, is noted to be a well-known advanced method. The evidence concerning the aforesaid methods reminds us to go back to 6,000 BC when they were focused on bread baking, brewing of beer, then winemaking. However, the first determined microbial oxidation was witnessed through the production of vinegar in the 6,000 BC, indicating the historical significance of various processes associated with microbe-based approaches (Mitsutake et al., 2001).

In the modern epochs, specifically in the late 19th century, a mixture of pepsin and chymosin has been primarily used in various food processing approaches including in the production of bacterial amylases and in cheese making. Moreover, the 1930s witnessed the prominent use of pectinases in juice clarification. Besides the significance from the food perspective, the use of immobilized enzymes like invertase has illustrated its significance, since it also had advantageous effects in World War II for inverted sugar syrup production for the army (Liese et al., 2006). Generally, most classes of enzymes have significant application in food technology; however, many studies have suggested that hydrolases are likely to be the predominant class of enzymes.

As a general concept, many of the investigations have aimed to improve the performance of food technology approaches, with particular attention paid to augmenting the pH range, thermal stability, and catalytic activity. For instance, the following approaches have been shown to improve the performance of specific enzymes in the aforementioned ways: site-directed mutagenesis (α-amylase: thermostability and pH-activity profile), directed evolution (L-arabinose isomerase: pH-activity profile), and immobilization (Lactase: thermostability). From this, it is evident that recent trends have incorporated molecular engineering in food technology with special inference on microbial food enzymes (Figure 12.1) (Tomschy et al., 2002). This chapter focuses on how microbe-based approaches can be implemented to enhance various properties of enzymes for food processing technology.

12.2 RECENT TRENDS IN THE BIOPROSPECTING OF MICROBIAL FOOD ENZYMES

The production of enzymes (Anoopkumar et al., 2020; Rebello et al., 2018, 2019) at the industrial level is very popular due to its wide application in various fields including food technology. For this reason, rigorous researches are being directed worldwide to produce the enzymes in an easy and efficient way with special inference making them cost effective and sustainable in nature. As already introduced, microorganisms, especially the bacterial populations, are a favoured foundation for the production of enzymes since they are easily available with a fast growth rate (Illanes et al., 2012). Advancements in molecular biology, including the recombinant DNA technology (Figure 12.2), allow the research community to make specific effective genetic changes that can be favoured as a foundation for elevated enzyme production (Illanes et al., 2012).

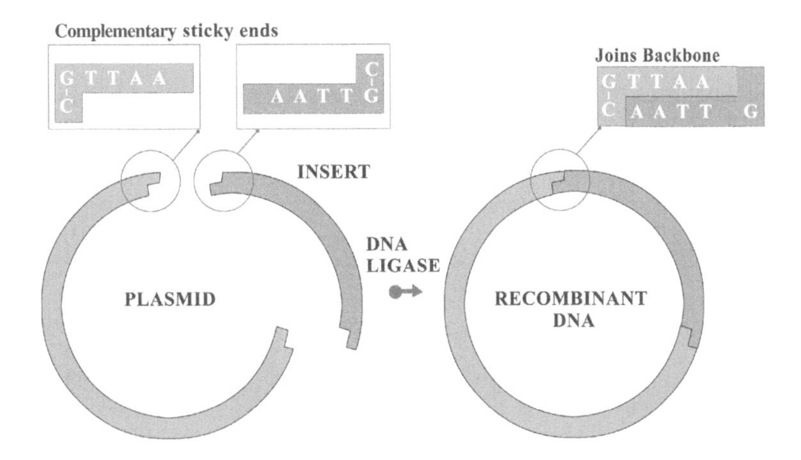

FIGURE 12.2 The recombinant DNA technology integrates the specific unit of genetic trait to the desired host.

The production of enzymes at an industrial level is primarily conducted in two significant ways: (1) screening to discover a microbial strain's potential and (2) the fermentation process. In addition to the process discussedseveral factors need to be considered for the efficacious output of the employed approach. For instance, the composition of growth media, especially the growth media comprising nitrogen sources (Fish and Lilly, 1984), carbon sources, and micronutrients, usually determines how the expected output level can be accomplished.

12.3 IMPLEMENTING MOLECULAR ENGINEERING FOR MICROBIAL ENZYMES

Nature usually offers a great diversity of microbial enzyme resources. The research community has explored and integrated the various traits of enzymes of microbial origin with food technological perspectives. While a vast diversity of microbial consortia inhabits the biosphere, only a minute percentage has been cultivated for research purposes (Adrio and Demain, 2014) through laboratory techniques. Integrating a particular genomic region with another host's genome offers a scientific way for exploring such microbes (Table 12.1), thereby producing a sufficient quantity of enzymes belonging to the Oxidoreductase, Lyase, Transferase, Hydrolase, and Isomerase classes of enzyme.

12.3.1 Glucose Oxidase

Glucose oxidase is found as a flavoprotein that is primarily known to catalyse the oxidation of β-D-glucose to produce hydrogen peroxide and D-glucono-delta-lactone by exploiting the molecular oxygen. From its original isolation, *Aspergillus niger* was found to be the most common source. The production of Glucose oxidase through molecular engineering especially through recombinant technology in *Hansenula polymorpha* and *Saccharomyces cerevisiae* results in hyperglycosylated, which opened the way for the possible use of the approach for enzyme production. However, effective results concerning the production of Glucose oxidase through recombinant expression have been observed from *Pichia pastoris*. When using *P. pastoris* as an effective host, it produced extensive quantities of overglycosylated enzymes that have different characteristics than the native enzyme.

As already mentioned, the main complication found during the production of Glucose oxidase at the industrial level is the low productivity rate. From a genetic engineering and eukaryotic expression perspective, the *P. pastoris* is famous for its ability to sustain as a prominent host. The ability to express the proteins at a high level without developing any chance of contamination would be the answer to the problem. Certain previous studies have validated the development of a system using glyceraldehyde-3-phosphate dehydrogenase gene promoter (pGAP) which facilitates the high expression of *Aspergillus niger* glucose oxidase (AnGOD) (Qiu et al., 2016).

TABLE 12.1
Various Enzymes, Their Microbial Origin, and Role in the Food Technology

Enzyme	Microbial Species as a Source	Role in Food Technology	References
α-Amylase	*Bacillus amyloliquefaciens* *Bacillus licheniformis* *Bacillus stearothermophilus* *Saccharomyces cerevisiae*	Brewing, starch liquefaction, baking, Bread quality	(Abd-Elhalem et al., 2015; Zhou et al., 2021)
Protease	*Aspergillus usamii* *Bacillus* sp. *Bacillus licheniformis* YP1A *Escherichia coli* *Bacillus firmus* 7728 *Bacillus megaterium* *Micrococcus* sp. *Bacillus amovivorus* *Streptomyces* sp. *CN902* *Streptomyces fungicidicus* MML1614 *Aspergillus clavatus* *Aspergillus parasiticus* *Pleurotus citrinopileatus*	Coagulation of milk Brewing	(Abdel-Fattah et al., 1974; Raveendran et al., 2018; Zhang et al., 2019)
Lipase	*Penicillium camemberti* *Aspergillus niger* *A. oryzae* *Candida antarctica* *Candida cylindracea* Ay30 *Helvina lanuginose* *Candida rugosa* *Pseudomonas* sp. *Geotrichumcandidum.*	Cheese production	(Raveendran et al., 2018)
Glucose oxidase	*Aspergillus niger* *Penicillium glaucum.* *Aspergillus niger* *Penicillium adametzii*	Food flavour	(Hanft and Koehler, 2006; Mikhailova et al., 2000)
Peroxidase	*Phanerochaete chrysosporium* *Streptomyces viridosporus*	Food colour	(Janusz et al., 2013)
Pectinase	*Aspergillus fumigatus* *Aspergillus foetidus* *Aspergillus niger* *Aspergillus awamori* *Penicillium restrictum* *Trichoderma viride* *Mucor piriformis* *Yarrowia lipolytica* *Saccharomyces fragilis* *Saccharomyces thermantitonum* *Torulopsiskefyr* *Candida pseudotropicalis*	Clarification of fruit juice	(Pasha et al., 2013; Raveendran et al., 2018; Sandri et al., 2011)

(Continued)

TABLE 12.1 *(Continued)*
Various Enzymes, Their Microbial Origin, and Role in the Food Technology

Enzyme	Microbial Species as a Source	Role in Food Technology	References
Laccase	*Azospirillum lipoferum* *Bacillus* sp. *Geobacillus* sp. *Streptomyces* sp. *Rhodococcus* sp. *Staphylococcus* sp. *Azospirillum* sp. *Lysinibacillus* sp. *Aquisalibacillus* sp.	Wine making	(Osma et al., 2010; Raveendran et al., 2018)
Phospholipase	*Fusarium oxysporum*	Cheese making	(Raveendran et al., 2018)
Glucoamylases	*Aspergillus niger* *Aspergillus awamori* *Saccharomyces cerevisiae* *Rhizopus oryzae*	Beer production	(Lilbaek et al., 2006; Pavezzi et al., 2008; Raveendran et al., 2018)
Lactase	*Kluyveromyces lactis* *Kluyveromyces fragilis*	Dairy foods including cheese, ice cream, and milk	(Cortés et al., 2005; Gupta et al., 2015)

Their findings prominently revealed that glucose oxidase activity achieved its maximum (107.18 U/mL) in the culture medium at pH 6 and 30°C, justifying the significance of molecular engineering in microbial food enzyme activity regulation. Due to the positive output, another investigation (Mu et al., 2019) focused on the development of recombinant enzymes through the expression of responsible genes, specifically the GOD gene from the *Penicillium chrysogenum* (75 mg/L) and *A. niger* (90 mg/L) in the *P. pastoris* expression system. The computer-aided mutation towards AnGOD revealed that T10K was found to be the most significant mutation since it encompasses a prominent increase in the apparent unfolding temperature (+ 6°C). Such mutations are known to stabilize the structure of the enzyme either through salt bridge interactions or through the addition of a new hydrogen bond. A drop in the conformational entropy, optimization of interactions (charge-charge), and enhancement of hydrophobic packing are also proposed to be stabilized by this mutation. Moreover, the R335T, S422N, R145N, E367Y, and H366Q mutations are counted among the most beneficial mutations associated with the improved hydrogen bond and salt bridge interactions, which justifies the significance of molecular engineering in microbial food enzymes.

12.3.2 Protease

Proteases are primarily known to catalyse the process in which the peptide bonds in proteins are hydrolysed. From 2014 to 2019 the demand for proteases in the world market was found to increase, with a compound annual growth rate of 5.3%. The demand is expected to continue to grow as the demand finds a plethora of industrial applications in food technology. Endopeptidases and exopeptidases are known to be the two major different classes of proteases classified according to the site of action (Rao et al., 1998). The production of protease with special inference with low cost and hyper-production along with superior quality usually requires technical assistance from the biotechnology fields, especially from genetic engineering. For instance, recent studies introduced the possible production of proteases through genetically modified *E. coli* and recombinant DNA technology. As mentioned in the previous section, the mutation may also possess a significant contribution to the protease isolation process from the microbial population.

The use of sequential in-vitro mutagenesis-based evolutionary engineering in certain situations allows the research community to isolate the enzyme easily and effectively. Mutation artificially developed in the subtilisin usually called "m-63" unveiled a prominent range of efficiency towards the catalytic activities. One of the major facts associated with the aforementioned approach is that it was twice as productive as the wild type at 10°C under N-succinyl-l-Ala-l-Ala-l-Pro-l-Phe-p-nitroanilide (Onaizi, 2018; Zhou et al., 2019). The cold-active proteases are principally found in microbial consortia from cold environments including polar regions, arctic regions, cold desert soil, sub-glacial water, permafrost, and deep sea. The advanced developments in molecular engineering, especially in cloning and expression studies, allow the scientific community to enhance the activity of the specific proteins, even in low temperatures for industrial needs. For instance, instead of needing high temperatures for conventional protease during the peeling process of leather, the same process can be achieved at the temperature of tap water with the aid of cold-active proteases (Joshi and Satyanarayana, 2013), which is a much more energy-efficient system. This same property of high activity at low temperatures is transferable to the food industry as it permits the alteration of heat labile foodstuffs, for example, cold-active enzymes can be used to quicken the ripening of slow-ripening cheeses (Sharma et al., 2022). In addition, they can find utility in softening food and developing the taste of frozen meat products.

The mutant enzymes such as Gin 19Ala, Glnl9His, and Glnl9Glu from the papain family revealed that the Glnl9His and Gln19Glu were involved in the acid-catalysed hydrolysis. The new way of pro-sequencing protein engineering unveiled the route towards investigating the protein-folding mechanisms thereby helping the researchers to make unique proteases with a plethora of valuable catalytic properties. The capacity of *Bacillussubtilis* to produce extracellular proteins has been considered by many researchers over the world for the extensive production of protease enzymes through recombinant technology. The mettaloproteases (npr) or subtilisin (apr) were witnessed to be the most important enzymes produced using the aforesaid strategy with *B. subtilis*. A study by Henner et al. (1985) substituted the npr and apr using the amylase promoter from *B. subtilis* and *Bacillus amyloliquefaciens* respectively, to enhance the expression. Such approaches can be used in cereal processing where it augments the volume of extract thereby elevating the number of certain elements including the α-amino nitrogen. Likewise, certain acidic aspartic proteases are able to clot the milk proteins for the preparation of curds, justifying the use of the aforesaid enzyme as a milk-clotting enzyme. Examining the caspase-7 (MEROPS: C14.004) for its peptide-bound structure validated the aforesaid concept by revealing the four binding subsites which change the substrate specificity once the site-directed mutagenesis is accomplished.

12.3.3 α-Amylase

The α-amylase enzymes are a significant group of amylolytic enzymes that are known to play a significant part in the processing of polysaccharides and oligosaccharides. Moreover, the ability of the enzymes belonging to this class to hydrolyse glycogen, starch, and various other saccharides to maltose and glucose make a significant contribution of the same to food technology (Rajagopalan and Krishnan, 2008; Souza et al., 2010). Different species of microbial consortia, especially the microbes belonging to the genus *Bacillus*, have commercial benefits in α-amylase production. In addition to this, microbial species such as *Bacillus stearothermophilus*, *B. amyloliquefaciens*, and *Bacillus licheniformis* find advanced applications in food technology (Konsoula and Liakopoulou-Kyriakides, 2007). For being necessary in food technology, the enzymes should have the ability to tolerate high temperatures. Hence many of the studies have employed computer-aided prediction approaches to reveal the thermostability of enzymes engineered through recombination technology and point mutation. The point mutation artificially created in certain amino acid sequence residues can alter the free energy of denaturation, which validates its prominent role in maintaining the thermostability of the studied enzymes. A recent attempt has also been made to produce α-amylase from several species of fungal genera, including *Penicillium* and *Aspergillus*. Certain mutations like Gln264Ser/Asn265Tyr can augment the thermal stability of enzymes retrieved from *B. licheniformis*. As mentioned in the previous section regarding the salt bridge bond, Asn188Thr mutation might be able to augment the thermal stability of the enzyme from *B. licheniformis*, through the aforesaid link Thr263.

Directed evolution can improve and alter the substrate specificity, activity, solubility, stability, and selectivity. However, electrostatic interactions, hydrophobic interactions, disulfide bonds, hydrogen bonds, reduced entropy of unfolding, salt bridges, and augmented incidence of proline residues also play a principal role in contributing to thermostability (Salminen et al., 1996). In addition to the above-discussed approaches, previous studies employed the deletion mutation method for increasing thermostability. For example, the three stabilizing mutations such as deletion of G177 and R176 followed by substitution of K269A in *B. amyloliquifaciens* caused prominent additive thermostabilization in bacterial α-amylases (Suzuki et al., 1989). Among the two discussed mutations, R176-G177 deletion also exhibited parallel effects on α-amylases as found in the R176. An investigation by Lévêque et al. (2000) cloned an α-amylase encoding gene of 1374 bp from *Thermococcus hydrothermalis* and the recombinant α-amylase was found to be active at pH 5.0 and at 75–85°C. A recent study by Zhou et al. (2019) validated the effectiveness of CRISPR/Cas9n-AID system in *Bacillus amyloliquefaciens* LB1ba02 and which allows multiple site editing instantaneously and simultaneously. They also validated the possibility of rapid gene knockout and integration. In order to verify the effectiveness of the developed strategy, they genetically altered the wild strain LB1ba02 and developed a mutant strain (LB1ba02△4) capable of secreting amylase. The mutant strain showed 1.25-fold more than the traditional strain validating the role of CRISPR/Cas9n-AID system in amylase production. have cloned eight genes with special inference on the development of novel α-amylases such as AmyF and AmyE.

To gain stability, certain studies have made amino acid substitutions using the tRNA suppressor method at H133. It is clear that there has been a noticeable upsurge and advancement in the research pertaining to microbial enzyme uses in food technology. Advanced and newer technologies like molecular must be continued for the exploration of α-amylases from novel microbial sources.

12.3.4 Lactase

The enzymes belonging to the Lactase (β-galactosidase) group usually have high stability, considerable activity, and ease of fermentation. Because of these characteristics, the enzymes are principally used in food processing industries. Primarily they convert the disaccharide lactose into glucose and galactose and then eventually enters glycolysis. Bacteria as well as fungi have received much more interest in the recent scenario as they are noted to be the primary source of β-galactosidase.

A study by Hsu et al. (2007) validated the extensive activity of the enzyme obtained from *Bifidobacterium longum* CCRC15708 and *B. longum* strain CCRC 15708. Besides the role of bacteria in β-galactosidase production, the yeast *Kluyveromyces lactis* was also found to produce the enzyme at the commercial level with special inference on the dairy environment. Isolation of the enzyme from such yeast could also be found as an added advantage since the process can offer lactose-free milk products (Pivarnik et al., 1995).

Lactococcus is reported by many researchers as being a model organism, making it a prominent candidate for metabolic engineering. Moreover, the easy availability of information concerning the complete genome sequence, the industrial significance, and the availability of effective genetic tools (de Vos, 2011) facilitates the use of *Lactococcus* for efficient cell factories. We can see the aforesaid perspective in another investigation done by Domingues et al. (2010), which revealed that the chromosomal integration inactivated the lactate dehydrogenase gene (ldhD) thereby pure L-lactic acid has been formed in *Lactobacillus helveticus*. Some of the microbial consortia are known to consume the lactose naturally using their lactose metabolization genes being potential candidates for cloning in *S. cerevisiae* cells. Moreover, the metabolic engineering strategies required for the β-galactosidase and lactose permease gene cloning has been discussed by Domingues et al. (2010).

12.3.5 Lipase

The lipases are always referred to as triacylglycerol acylhydrolase which do not necessitate any cofactor. The conversion of triglycerides into diglycerides and monoglycerides along with the formation of glycerol and fatty acids naturally happens with the help of lipases (Almeida et al., 2019). The lipases of microbial origin are industrially noted to be substantial since they offer superior stability, low manufacturing cost, and more availability than other resources including plant and animal resources. Among the various biocatalytic processes, the *Candida antartica* lipase B (CALB) is reported to be the most commonly found and used enzyme and has a large number of patents.

Advancements in genetic engineering technologies allowed the research community to investigate the lipase gene expression, structure, and function of proteins more in-depth. For instance, the application of various newer technologies including the molecular cloning technologies eased the approaches for heterologous expression of lipases derived from the *B. subtilis, P. pastoris, S. cerevisiae,* and *E. coli* (Borrelli and Trono, 2015). Despite increasing the variants of lipases that have been isolated from the microorganisms, the native enzymes are not suitable in certain circumstances. In such circumstances, the characteristics of the enzymes have thus to be improved. Here the authors suggest using genetic engineering to accomplish such targets in a tailor-made enzyme perspective. The directed evolution and rational design have been considered the most prominent approach to designing the proteins that are able to work under unusual circumstances. Currently, the following kits are used to incorporate an amino acid in a specific location for various genetic engineering perspectives; GenArt® by Life Technologies, QuickChange® by Agilent Technologies, and Q5® by NEB. The stability, recyclability, and activity of lipases can be improved through the immobilization process. The immobilization process can also control the various reaction parameters including the flow rate. Moreover, the insertion of a specific gene encoding the lipase enzyme made *E. coli* BL21 produce maximum activity in the wine making process (Chandra et al., 2020).

12.3.6 Pectinase

The pectinase usually referred to as pectinolytic enzymes are known to have the capacity to hydrolyse many multifaceted pectic elements. According to the classification based on the substrate specificity and mechanism of action, the different classes of pectinases are depolymerases, protopectinases, and esterases (Sakai, 1992). As already mentioned, genetic engineering can offer significant, efficient adaptations since the genetic changes can also be controlled completely using regulators. The process involves taking a specific responsible gene from a natural donor and introducing it to the host microorganism, which will have the ability to produce the pectinases at an industrial level.

Pectinases have various applications in food technology since they can reduce the viscosity and fermentation time, improve the clarity of juice, fand play a prominent role in the maceration of vegetables (Saharan and Sharma, 2018). The prominent role of the enzymes in the wine making process (facilitating filtration, augmenting the flavour and colour, and maximizing juice yield), coffee and tea fermentation, and oil extraction (olives, lax seed, and dates,) justifies the significance of focusing on the molecular engineering applications in food technology. Because of these many applications, certain investigations have focused on mutant strains of *Penicillium occitanis* developed through genetic engineering for the overproduction of many pectinases.

According to Bravo-Ruiz et al. (2017), the genome comparison of wild type with the mutant strain revealed that the mutation observed in a vastly conserved serine residue of protein comprising a Zn2Cys6 binuclear cluster DNA-binding domain. Likewise, the comparison of *P. occitanis* in terms of mutation accomplished through genetic engineering revealed the fact that the *P. occitanis* CT1 has the capacity to produce large quantities of pectinase. In addition to this, seven consistent single nucleotide polymorphisms (SNPs) were also found to be associated with the respective protein, indicating their role in possible pectinase production at the industrial level.

Given the detected deregulation of the enzymes found in the mutated strain, the authors give prime importance to non-synonymous SNPs influencing the transcription factor, which are responsible for the cysteine mutation of the amino acid sequence at a specific location, specifically at position 198. Likewise, targeted gene deletion and site-directed mutagenesis allowed Bravo-Ruiz et al. (2017) to replace the FOXG_08883 allele (wild type) with the mutant allele (Cys160Ser) in *Fusarium oxysporum*. In order to achieve the targeted deletion, they completely replaced the open reading frame with a hygromycin resistance cassette. Alazi et al. (2019) revealed that the W361R mutation formed in the regulator of D-galacturonic acid-responsive genes, specifically in GaaR and GaaX, results in pectinase production in *A. niger*. Based on these findings, the various regulatory mechanisms concerning the overproduction of different enzymes under inducing conditions have been revealed, which could be noted as a prominent contribution to food technology.

12.3.7 Laccase

Higher plants and fungi are recognized to be the most common source of laccase production. However, recent studies confirmed those microbial populations including the *Marinomonas mediterranea*, *Streptomyces cyaneus,* and *Streptomyces lavendulae* are novel sources for laccase production. The following organisms have been previously known to produce laccase at a considerable level: *Theiophora terrestris, Phlebia radiate, Phanerochaete chrysosporium* and *Lenzitesbetulina, Pleurotus ostreatus, Trametes versicolour, Trichoderma atroviride,* and *Trichoderma harzianum* (Shekher et al., 2011).

The activity of enzymes belonging to this class is lost after a particular number of cycles. For example, the enzymes were found to be inactive during the presence of free radicals produced by 1e- subtraction of mediators, phenols, or aromatic amines. However, studies concerning the mechanism behind this inactivation process (Kim et al., 2015) instigated by the formed free radicals are not well studied in previous investigations. The enzymes belonging to the class of laccase are not exempt from this, owing to their instability in the presence of various organic solvents including acetonitrile, ethanol, acetone, and methanol (Arregui et al., 2019). In order to overcome such situations, studies concerning the modification of pH at an optimum level for laccase are previously reported (Arregui et al., 2019).

Besides the approach that includes the specific steps of DNA recombination or random mutagenesis, studies that integrate the advantage from both combinatorial libraries and protein structures were found to be employed successfully. These strategies usually involve the mutation of a particular single amino acid codon to another one that will produce the 20 amino acids united to find the specific function. Moreover, it is also employed to augment the features of laccase at "hot-spot" residues which were previously identified through conventional strategies focused on random mutagenesis (Kunamneni et al., 2008). To obtain large quantity of laccases for industrial needs, novel fermentation approaches along with genetic modification can be employed. A study by Osma et al. (2010) cloned and heterologously expressed the Laccase genes for the aforesaid benefits in the following fungi: *Trichoderma reesei, Aspergillus oryzae* and *Aspergillus niger*. Despite the exciting potential enzymes belonging to the laccases bring, research concerning the implementation of advanced strategies for production levels should be undertaken.

12.4 CONCLUSIONS

Microbial enzymes are noteworthy candidates for many food research interventions. The enhanced production of microbial enzymes followed by its commercialization is recognized as a significant effort in the current scenario. However, many studies typically remain focused on experiments under laboratory conditions. Therefore, strategies should be set for increasing the product yields and efficiency. Here the authors suggest the advanced genetic engineering approaches including the site-directed mutagenesis and directed evolution for augmenting the product yield and efficiency of enzymes from microbial species having food technology applications.

REFERENCES

Abd-Elhalem, T. Basma, M. El-Sawy, R. F. Gamal, and K. A. Abou-Taleb. 2015. Production of amylases from *Bacillus amyloliquefaciens* under submerged fermentation using some agro-industrial by-products. *Annals of Agricultural Sciences* 60: 193–202.

Abdel-Fattah, A. F., N. M. El-Hawwary, and A. S. Amr. 1974. Milk-clotting enzymes of some streptomyces species. *Acta Microbiologica Polonica. Series B: Microbiologia Applicata* 6: 27–32.

Adrio, L. Jose, and A. L. Demain. 2014. Microbial enzymes: Tools for biotechnological processes. *Biomolecules* 4: 117–39.

Alazi, E., J. Niu, S. B. Otto, M. Arentshorst, T. T. M. Pham, A. Tsang, and A. F. J. Ram. 2019. W361r mutation in gaar, the regulator of d-galacturonic acid-responsive genes, leads to constitutive production of pectinases in *Aspergillus niger. MicrobiologyOpen* 8: e00732.

Almeida, J. M., V. P. Martini, J. Iulek, R. C. Alnoch, V. R. Moure, M. Müller-Santos, E. M. Souza, D. A. Mitchell, and N. Krieger. 2019. Biochemical characterization and application of a new lipase and its cognate foldase obtained from a metagenomic library derived from fat-contaminated soil. *International Journal of Biological Macromolecules* 137: 442–54.

Anoopkumar, A. N., S. Rebello, E. M. Aneesh, R. Sindhu, P. Binod, A. Pandey, and E. Gnansounou. 2020. Use of different enzymes in biorefinery systems. In *Biorefinery Production Technologies for Chemicals and Energy*, ed. A. Kuila and M. Mukhopadhyay, 357–68, Scrivener Publishing: Wiley.

Arregui, L., M. Ayala, X. Gómez-Gil, G. Gutiérrez-Soto, C. E. Hernández-Luna, M. H. de Los Santos, L. Levin, *et al.* 2019. Laccases: Structure, function, and potential application in water bioremediation. *Microbial Cell Factories* 18: 1–33.

Borrelli, G. M., and D. Trono. 2015. Recombinant lipases and phospholipases and their use as biocatalysts for industrial applications. *International Journal of Molecular Sciences* 16: 20774–840.

Bravo-Ruiz, G., A. H. Sassi, M. Marcet-Houben, A. D. Pietro, A. Gargouri, T. Gabaldon, and M. I. G. Roncero. 2017. Regulatory mechanisms of a highly pectinolytic mutant of penicillium occitanis and functional analysis of a candidate gene in the plant pathogen fusarium oxysporum. *Frontiers in Microbiology* 8: 1627.

Chandra, P., R. Singh, and P. K. Arora. 2020. Microbial lipases and their industrial applications: a comprehensive review. *Microbial Cell Factories* 19: 1–42.

Cortés, G., M. A. Trujillo-Roldán, O. T. Ramírez, and E. Galindo. 2005. Production of β-galactosidase by kluyveromyces marxianus under oscillating dissolved oxygen tension. *Process Biochemistry* 40: 773–78.

de Vos, W. M. 2011. Systems solutions by lactic acid bacteria: From paradigms to practice. *Microbial Cell Factories* 10: 1–13.

Domingues, L., P. M. R. Guimarães, and C. Oliveira. 2010. Metabolic engineering of saccharomyces cerevisiae for lactose/whey fermentation. *Bioengineered Bugs* 1: 164–71.

Fish, N. M., and M. D. Lilly. 1984. The interactions between fermentation and protein recovery. *Bio/technology* 2: 623–27.

Gupta, V. K., R. L. Mach, and S. Sreenivasaprasad. 2015. *Fungal Biomolecules: Sources, Applications and Recent Developments*, John Wiley & Sons.

Hanft, F., and P. Koehler. 2006. Studies on the effect of glucose oxidase in bread making. *Journal of the Science of Food and Agriculture* 86: 1699–704.

Henner, D. J., M. Yang, L. Band, and J. A. Wells. 1985. Expression of cloned protease genes in *Bacillus subtilis*. Paper presented at the Proceedings of the 9th International Spore Conference.

Hsu, C. A., S. L. Lee, and C. Chou. 2007. Enzymatic production of galactooligosaccharides by ß-Galactosidase from lactobacillus pentosus purification characterization and formation of galacto-oligosaccharides from bifidobacterium longum Bcrc 15708. *Journal Agriculture Food Chemistry* 55: 2225–30.

Illanes, A., A. Cauerhff, L. Wilson, and G. R. Castro. 2012. Recent trends in biocatalysis engineering. *Bioresource Technology* 115: 48–57.

Janusz, G., K. H. Kucharzyk, A. Pawlik, M. Staszczak, and A. J Paszczynski. 2013. Fungal laccase, manganese peroxidase and lignin peroxidase: Gene expression and regulation. *Enzyme and Microbial Technology* 52 (1): 1–12.

Joshi, S., and T. Satyanarayana. 2013. Biotechnology of cold-active proteases. *Biology* 2 (2): 755–83.

Kim, S. J., J. C. Joo, B. K. Song, Y. J. Yoo, and Y. H. Kim. 2015. Engineering a horseradish peroxidase C stable to radical attacks by mutating multiple radical coupling sites. *Biotechnology and Bioengineering* 112 (4): 668–76.

Konsoula, Z., and M. Liakopoulou-Kyriakides. 2007. Co-production of alpha-amylase and beta-galactosidase by *Bacillus subtilis* in complex organic substrates. *Bioresource Technology* 98 (1): 150–57.

Kunamneni, A., S. Camarero, C. García-Burgos, F. J Plou, A. Ballesteros, and M. Alcalde. 2008. Engineering and applications of fungal laccases for organic synthesis. *Microbial Cell Factories* 7 (1): 1–17.
Lévêque, E., B. Haye, and A. Belarbi. 2000. Cloning and expression of an alpha-amylase encoding gene from the hyperthermophilic archaebacterium thermococcus hydrothermalis and biochemical characterisation of the recombinant enzyme. *FEMS Microbiology Letters* 186 (1): 67–71.
Liese, A., K. Seelbach, and C. Wandrey. 2006. *Industrial Biotransformations*, John Wiley & Sons.
Lilbaek, H. M., M. L. Broe, E. Høier, T. M. Fatum, R. Ipsen, and N.K. Sørensen. 2006. Improving the yield of mozzarella cheese by phospholipase treatment of milk. *Journal of Dairy Science* 89 (11): 4114–25.
Mikhailova, R. V., Z. F. Shishko, M. I. Yasenko, and A. G. Lobanok. 2000. Effect of culture conditions on extracellular glucose oxidase production by strain penicillium adametzii BIM-90. *Mikologiya Fitopatologiya* 34 (4): 48–53.
Mitsutake, A., Y. Sugita, and Y. Okamoto. 2001. Generalized-ensemble algorithms for molecular simulations of biopolymers. *Peptide Science: Original Research on Biomolecules* 60 (2): 96–123.
Mu, Q. Y. Cui, M. Hu, Y. Tao, and B. Wu. 2019. Thermostability improvement of the glucose oxidase from *Aspergillus niger* for efficient gluconic acid production via computational design. *International Journal of Biological Macromolecules* 136: 1060–68.
Onaizi, Sagheer A. 2018. Enzymatic removal of protein fouling from self-assembled cellulosic nanofilms: Experimental and modeling studies. *European Biophysics Journal* 47 (8): 951–60.
Osma, J. F, J. L. Toca-Herrera, and S. Rodríguez-Couto. 2010. Uses of laccases in the food industry. *Enzyme Research* 2010: e918761.
Pasha, K. M., P. Anuradha, and D. Subbarao. 2013. Applications of pectinases in industrial sector. *International Journal of Pure and Applied Sciences and Technology* 16 (1): 89.
Pavezzi, F. C., E. Gomes, and R. da Silva. 2008. Production and characterization of glucoamylase from fungus Aspergillus awamori expressed in yeast Saccharomyces cerevisiae using different carbon sources. *Brazilian Journal of Microbiology* 39 (1): 108–14.
Pivarnik, L. F, A. G. Senecal, and A. G. Rand. 1995. Hydrolytic and transgalactosylic activities of commercial beta-galactosidase (lactase) in food processing. *Advances in Food and Nutrition Research* 38: 1–102.
Qiu, Z., Y. Guo, X. Bao, J. Hao, G. Sun, B. Peng, and W. Bi. 2016. Expression of *Aspergillus niger* glucose oxidase in yeast Pichia pastoris SMD1168. *Biotechnology & Biotechnological Equipment* 30 (5): 998–1005.
Rajagopalan, G., and C. Krishnan. 2008. Alpha-amylase production from catabolite derepressed Bacillus subtilis KCC103 utilizing sugarcane bagasse hydrolysate. *Bioresource Technology* 99 (8): 3044–50.
Rao, M. B., A. M. Tanksale, M. S. Ghatge, and V. V. Deshpande. 1998. Molecular and biotechnological aspects of microbial proteases. *Microbiology and Molecular Biology Reviews* 62 (3): 597–635.
Raveendran, S., B. Parameswaran, S. B. Ummalyma, A. Abraham, A. K. Mathew, A. Madhavan, S. Rebello, and A. Pandey. 2018. Applications of microbial enzymes in food industry. *Food Technology and Biotechnology* 56 (1): 16.
Rebello, S., A. N. Anoopkumar, S. Puthur, R. Sindhu, P. Binod, A. Pandey, and E. M. Aneesh. 2018. Zinc oxide phytase nanocomposites as contributory tools to improved thermostability and shelflife. *Bioresource Technology Reports* 3: 1–6.
Rebello, S., D. Balakrishnan, A. N. Anoopkumar, R. Sindhu, P. Binod, A. Pandey, and E. M. Aneesh. 2019. Industrial enzymes as feed supplements—Advantages to nutrition and global environment. In *Green Bio-Processes*, Binod Parameswaran, Sunita Varjani, and Sindhu Raveendran, eds., 293–304, Springer.
Saharan, R., and K. P. Sharma. 2018. Industrial applications of thermophilic pectinase: A review. *International Journal Current Research* 10: 70762–70.
Sakai, T. 1992. Degradation of pectins. *Microbial Degradation of Natural Products*. 1992: 57–81.
Salminen, T., A. Teplyakov, J. Kankare, B. S. Cooperman, R. Lahti, and A. Goldman. 1996. An unusual route to thermostability disclosed by the comparison of Thermus thermophilus and Escherichia coli inorganic pyrophosphatases. *Protein Science* 5 (6): 1014–25.
Sandri, I. G., R. C. Fontana, D. M. Barfknecht, and M. Moura da Silveira. 2011. Clarification of fruit juices by fungal pectinases. *LWT-food Science and Technology* 44 (10): 2217–22.
Sharma, S., V. Sharma, S. Chatterjee, and S. Kumar. 2022. Psychrophilic enzymes adaptations and industrial relevance. In *Extremophiles*, 166–82. CRC Press.
Shekher, R., S. Sehgal, M. Kamthania, and A. Kumar. 2011. Laccase: Microbial sources, production, purification, and potential biotechnological applications. *Enzyme Research* (2011): e217861.
Souza, R. B., R. G. Ferrari, M. Magnani, *et al.* 2010. Ciprofloxacin susceptibility reduction of salmonella strains isolated from outbreaks. *Brazilian Journal of Microbiology* 41: 497–500.

Suzuki, Y., N. Ito, T. Yuuki, H. Yamagata, and S. Udaka. 1989. Amino acid residues stabilizing a bacillus alpha-amylase against irreversible thermoinactivation. *Journal of Biological Chemistry* 264 (32): 18933–38.

Tomschy, A., R. Brugger, M. Lehmann, *et al.* 2002. Engineering of phytase for improved activity at low Ph. *Applied and Environmental Microbiology* 68 (4): 1907–13.

Zhang, Y., Y. Xia, P. F. H Lai, X. Liu, Z. Xiong, J. Liu, and L. Ai. 2019. Fermentation conditions of serine/alkaline milk-clotting enzyme production by newly isolated *Bacillus licheniformis* Bl312. *Annals of Microbiology* 69 (12): 1289–300.

Zhou, C., H. Liu, F. Yuan, *et al.* 2019. Development and application of a CRISPR/Cas9 system for *Bacillus licheniformis* genome editing. *International Journal of Biological Macromolecules* 122: 329–37.

Zhou, N., T. Semumu, and A. Gamero. 2021. Non-conventional yeasts as alternatives in modern baking for improved performance and aroma enhancement. *Fermentation* 7(3): 102.

13 Metagenomics for the Identification of Microbial Enzymes for Nutraceutical Production

Pratyusha Patidar and Tulika P. Srivastava
School of Biosciences and Bioengineering, Indian Institute of Technology Mandi, Kamand, Mandi, Himachal Pradesh, India

CONTENTS

13.1 INTRODUCTION

In an era of changing lifestyle where the consumption of processed food that may harm our health is increasing, it is important to look for a dietary option that is not just enriched in nutrient-rich food but also has some pharmaceutical value. Consumption of unhealthy processed food can lead to many life-threatening diseases. With the increasing awareness of these aspects, a rapid upsurge has been observed in the consumption of products having both nutritional and physiological benefits. In 1989, Dr. Stephen DeFelice, MD, founder, and chairman of the Foundation for Innovation in Medicine (FIM), Cranford, NJ, coined the term "nutraceutical" (Kalra, 2003). Dr. DeFelice defined nutraceutical as "a food (or part of a food) that provides medical or health benefits, including the prevention and/or treatment of a disease" (Kalra, 2003). The term nutraceutical essentially combines the two terms – nutrition and pharmaceutical. Hence, a nutraceutical is any substance that provides physiological and health benefits along with nutrition.

Recently, an increasing demand for nutraceuticals has been observed owing to several advantages associated with them. These include properties such as non-toxicity, safety, their source (i.e., natural), and the obvious health and therapeutic benefits associated with them. Some of the examples of nutraceuticals include carotenoids, vitamins, fatty acids, flavonoids, enzymes, and probiotics

DOI: 10.1201/9781003311164-17

(Kumar et al., 2019). Mostly these compounds are derived from fruits and vegetables; however, recently microbes are emerging as a potential source of such nutraceuticals. These compounds could be very effective in the potential treatment of diseases including Alzheimer's, Parkinson's, diabetes, obesity, and inflammatory responses (Nasri et al., 2014).

Food processing industries and the nutraceuticals producing industry have achieved great success with the application of microbial enzymes. The application of microbial enzymes has increased with the development and advancements in the field of metagenomics. In this chapter, the role of microbial enzymes in nutraceuticals production and how metagenomics can facilitate the role of these enzymes in various nutraceutical applications will be discussed.

13.2 METAGENOMICS – WHAT, WHY, AND HOW

13.2.1 What is Metagenomics?

Metagenomics is a term derived from "meta", which means all or whole and "genomics" meaning the study of genes. The term metagenomics was first introduced by Jo Handelsman, Jon Clardy, Robert M. Goodman, Sean F. Brady, and others (Handelsman et al., 1998). Metagenomics is a promising molecular tool that is used to study the genetic aspect of microbes present in a given environment in a culture-independent manner. Unlike the traditional microbiological techniques requiring culturing of the microbes, metagenomics can provide an insight into the compositional, metabolic, and functional aspects of microbes without the need of culturing them (Coughlan et al., 2015). Figure 13.1 represents a general schematic for the metagenomic technique. Briefly, in order to explore the taxonomic and functional diversity of any environment, the DNA from an environment sample is directly extracted. The source of the sample varies based on the environment being explored; for example, to explore gut microbiome, faeces is the most preferred choice and to explore the microbial diversity of hot springs, DNA is directly isolated from water obtained from the hot springs. After the isolation of DNA, a metagenomic library is constructed which involves quality analysis and quantification of the libraries and attaching sequencing technology-specific adaptors followed by sequencing. In order to pool multiple samples in a single run sample, barcoding is implemented using specific index sequences per sample. After obtaining the sequence data, bioinformatics analysis is performed to estimate the complete microbial diversity and abundance. The subsequent sections provide more detailed information about the two different metagenomics approaches.

13.2.2 Why Metagenomics?

Microorganisms produce a wide variety of metabolites that have many industrial applications. Microbes are an excellent source of enzymes that can act as nutraceuticals or can be used in the production of other nutraceutical substances. However, at present, only as few as 1% of microbes are known due to their culturable nature. The significance of metagenomics is from the fact that the other 99% of the microorganisms remain unculturable (Locey & Lennon, 2016). Hence, this major chunk of microbes, that remains unknown, can act as a novel source of genetic diversity. This microbial diversity can be of invaluable significance to industries working in the domains of health, environment, and agriculture. For example, metagenomics can be used to identify novel enzymes having nutraceutical applications. Metagenomic approaches provide a culture-independent genomic analysis of environmental microorganisms, thus giving us access to gene resources for uncultivable microorganisms. Moreover, with the decreasing cost of sequencing, metagenomics is becoming a favourable tool in the field of microbiology.

13.2.3 How Do Metagenomic Approaches Work?

Several different metagenomic approaches are presently being used to address various problems and applications in industrial microbiology. With a wide variety of sequencing techniques and reduced

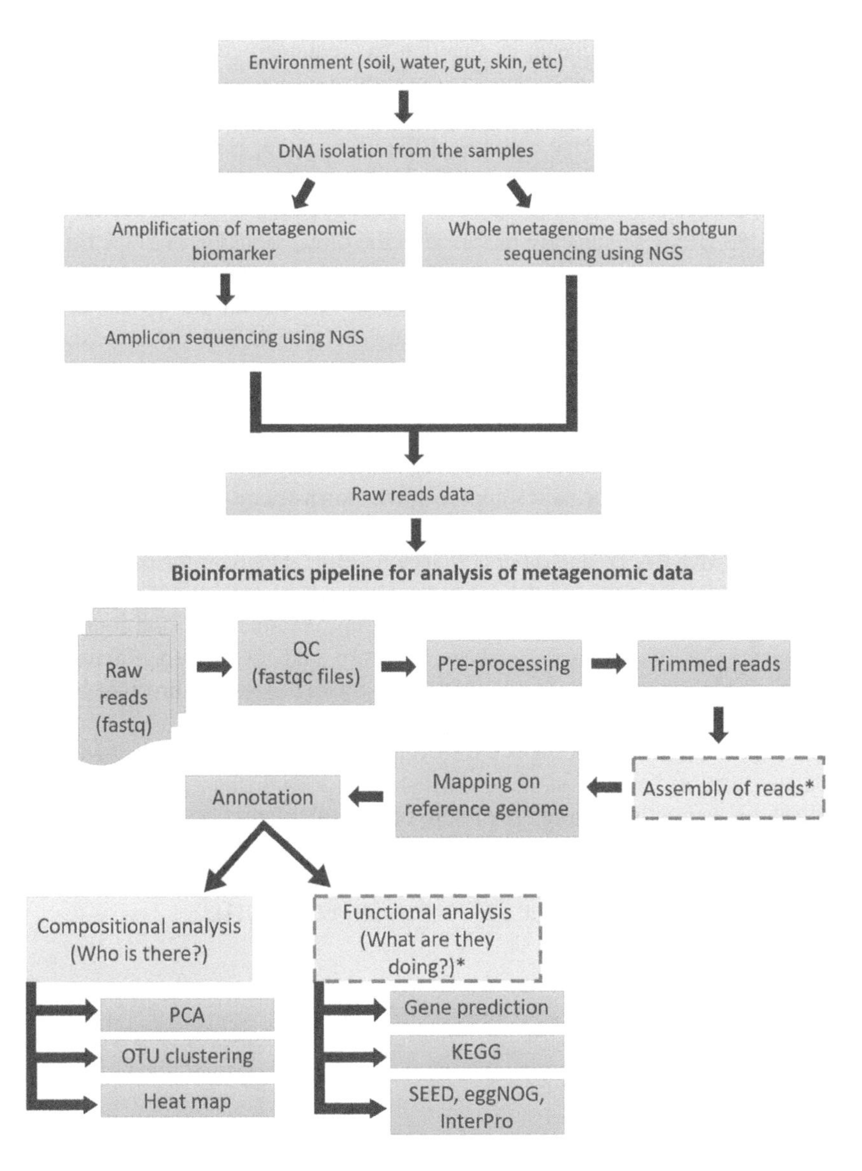

FIGURE 13.1 A general schematic for metagenomic sequencing techniques and downstream bioinformatics analyses. *Analysis steps applicable to shotgun whole genome metagenomic approach only.

sequencing cost, it has now become possible to sequence a large number of samples at the same time. Some of the popular sequencing platforms include Illumina, SOLiD, Ion Torrent, 454 pyrosequencing, PacBio, and Nanopore. There are two types of metagenomic sequencing approaches, namely amplicon-based, which includes 16S rRNA (for prokaryotes genomes) and 18S rRNA (for eukaryotic genomes) amplicons, and whole metagenomic shotgun sequencing-based methods. These two types of metagenomic techniques are discussed and reviewed by Ghosh and colleagues (Ghosh et al., 2018).

13.2.3.1 Marker-Based or Amplicon-Based Sequencing

Amplicon sequencing approach is based on the variable region (V1 to V9) found in the 16S rRNA gene sequence. 16S rRNA is an RNA component of the 30S subunit of prokaryotic RNA. 16S rRNA sequence is used for metagenomic sequencing due to several advantages (Patel, 2001), which include its presence in all the bacteria, its function has remained constant over the years, size of 16S rRNA is large enough to provide statistically significant sequence information, it is evolutionary conserved, and despite being evolutionary conserved, the 16S rRNA consists of nine variable regions that are considerably diverse amongst different bacteria. Hence, this helps in classifying bacteria in a population. Likewise for fungi, 18S rRNA is used as a genetic maker in metagenomic studies (Liu et al., 2015).

16S rRNA or 18S rRNA sequencing begins with the direct isolation of DNA from the metagenomic sample obtained from an environment. The basic steps involved in the 16S rRNA metagenomic sequencing (Pichler et al., 2018) are isolation of DNA from the given environment (soil, water, gut, skin, etc.), followed by quality and quantity check of the DNA (carried out at each step), amplification of the 16S or 18S rRNA gene (amplicon PCR), removal of primers and primer dimers (clean-up PCR), attachment of dual indices and adaptors (index PCR), and finally a second PCR clean-up to clean up the final library before quantification. Post quantification, normalization, and pooling of the library, the metagenomic DNA is sequenced using next-generation sequencing (NGS) technique. Finally, using bioinformatics approach the downstream compositional and functional analyses is performed. Various pipelines can be used for metagenomic data analysis, including QIIME, MOTHUR, MG-RAST, and others (Sharpton, 2014). The bioinformatics analysis involves pre-processing, mapping of reads on the reference genomes or 16S rRNA database, followed by Operational Taxonomic Unit (OTU) building for taxonomic classification, and compositional analysis of the mapped reads (phylogenetic analysis, diversity analysis microbial abundance analysis, and others). Figure 13.2 shows the upstream and downstream processing involved in the 16S rRNA based metagenomic sequencing technique.

16S rRNA marker-based sequencing is a very valuable technique, however, there are a few limitations associated with it. Some of its major limitations are its difficulty in strain-level identification, its applicability being limited only to prokaryotes (particularly bacteria), it fails to distinguish two different organisms with the same 16S rRNA (may be classified as the same species), it does not enable functional analysis, it has limited resolution, and has primer bias in PCR amplification (Gupta et al., 2019; Muhamad Rizal et al., 2020; Poretsky et al., 2014).

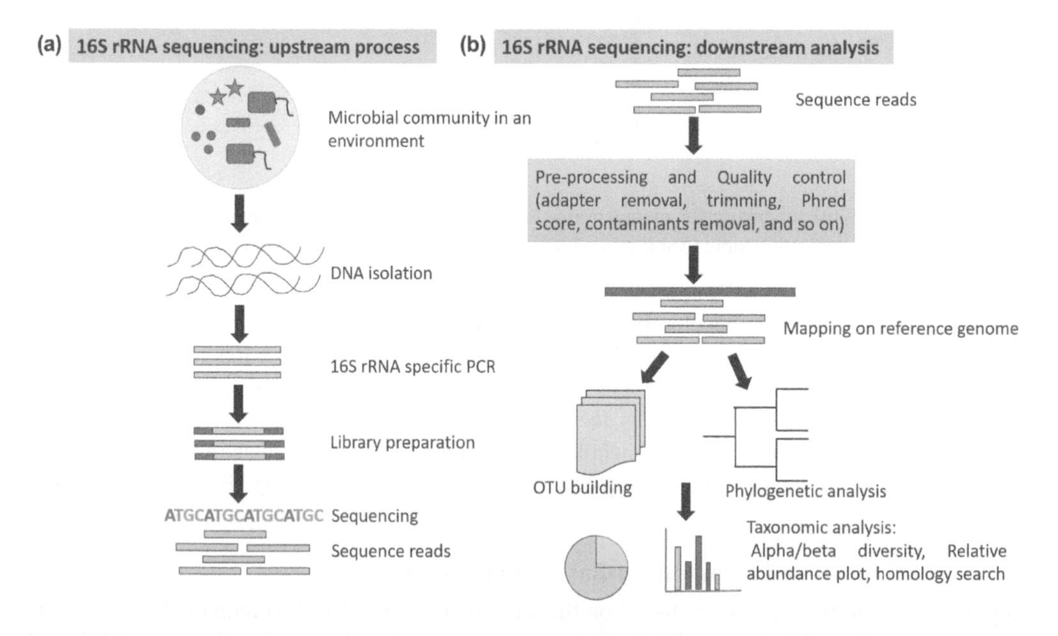

FIGURE 13.2 A general schematic for the (a) upstream process of 16S rRNA metagenomic sequencing and (b) downstream processing of the 16S rRNA metagenomics sequencing data.

13.2.3.2 Shotgun Metagenomic Sequencing

The shotgun metagenomic approach involves whole-genome sequencing, unlike the marker-based approach. Shotgun metagenomic sequencing is useful in taxonomic as well as functional analysis of the microbial population found in an environment. Though both 16S rRNA and shotgun metagenomic sequencing are metagenomic approaches, there are differences in their methodology. The common steps followed in the shotgun metagenomic sequencing include isolation of DNA from an environment (soil, water, gut, skin, etc.) using kit-based or manual methods, followed by quality and quantity checking of the isolated DNA using techniques like Gel electrophoresis, Qubit, Bioanalyser, and qPCR. The metagenomic DNA is then subjected to random fragmentation, adaptor ligation, and PCR amplification using random primers. Finally, the DNA fragments are sequenced using NGS (Brumfield et al., 2020).

The reads obtained from NGS sequencing are then subjected to assembly and mapping. If the reference genome is available then the reads are directly mapped on it, while in case of unavailability of a reference genome, *de novo* assembly is carried out. Assembly and mapping results in the annotation of the reads. These annotated reads are further analysed using different bioinformatics tools. Both compositional (taxonomic classification, phylogenetic analysis, principal component analysis (PCA), alpha- and beta-diversity analysis, microbial abundance, and other analyses) and functional analysis (gene prediction, protein family classification, pathway mapping, etc.) can be performed for the shotgun metagenomic data. Functional analysis is one of the major advantages of the shotgun approach over the amplicon-based approach. Figure 13.3 provide the overview of the shotgun metagenomic sequencing technique.

Shotgun metagenomic sequencing offers several advantages over the 16S rRNA approach. Unlike 16S rRNA, shotgun metagenomic sequencing enables functional analysis for the microbial communities, and it provides more comprehensive information as compared to the marker-based approach. Further, besides bacteria, it can be used for other organisms including viruses, fungi, and protozoa. However, the shotgun metagenomic approach is also associated with a few limitations. With the shotgun approach, there is a greater probability of contamination in the microbial reads creeping in, it follows a tedious assembly process as a large number of reads are generated, and also

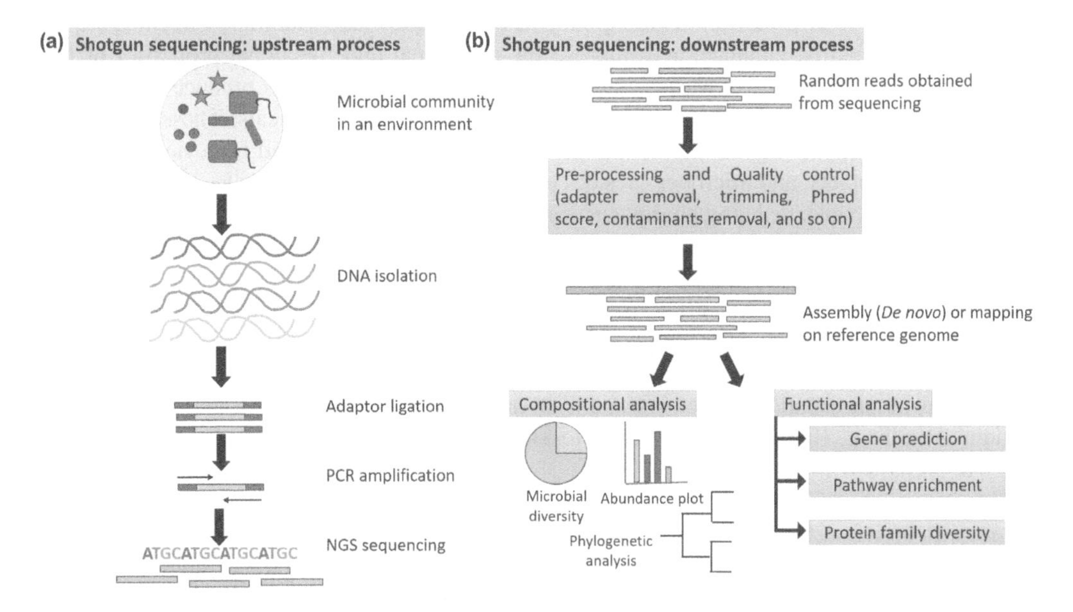

FIGURE 13.3 A general schematic for the (a) upstream process of shotgun whole-genome metagenomic sequencing and (b) downstream processing of the shotgun whole-genome metagenomics sequencing data.

TABLE 13.1
List of Commonly Used Computational Tools in Metagenomic Analyses

Sr. no	Analysis	Bioinformatics Tools/Databases
1.	Pre-processing and quality control (QC) trimming/removal of low-quality regions and contaminating sequences	FASTQC (https://www.bioinformatics.babraham.ac.uk/projects/fastqc/), Cutadapt (Martin, 2011), Trimmomatic (Bolger et al., 2014)
2.	Assembly and mapping	Bowtie (Langmead et al., 2009), MetaVelvet (Namiki et al., 2012), SOAPdenovo (Li et al., 2010)
3.	Taxonomic analysis	QIIME (Bolyen et al., 2019), Mothur (Schloss et al., 2009), BLAST (Altschul et al., 1990), RDP classifier (Wang et al., 2007), SILVA (Quast et al., 2013), Greengenes (DeSantis et al., 2006), MEGAN (Huson et al., 2007)
4.	Functional analysis	KEGG (Kanehisa, 2000), COG (Tatusov, 2000), SEED (Overbeek et al., 2005), eggNOG (Huerta-Cepas et al., 2019), MG-RAST (Meyer et al., 2008), UniRef (Suzek et al., 2015), MetaCyc (Caspi et al., 2016), PICRUSt (Langille et al., 2013), Pfam (Finn et al., 2009) (Mistry et al., 2021), TIGRFAMs (Haft et al., 2003) (Li et al., 2021), RDP (Cole et al., 2014), MEGAN (Huson et al., 2007), GeneMark (Borodovsky & McIninch, 1993), InterPro (Blum et al., 2021)

it is relatively more expensive. Table 13.1 provides the list of the commonly used bioinformatics tools for the analysis of metagenomic data.

13.3 APPLICATIONS OF METAGENOMIC APPROACHES TO EXPLORE MICROBIAL ENZYMES IN NUTRACEUTICALS PRODUCTION

Nutraceuticals are obtained from different sources, including plants and microbes. Microbes produce a large variety of products including enzymes, antibiotics, and vitamins. Enzymes are frequently used to modify, enhance, and improve the nutritional and functional properties of food items. Enzymatic hydrolysis of various substrates is used as a method to produce nutraceutical products. Protein-rich resources are an important category of nutraceuticals. These protein-rich resources can be produced, modified, and improved using enzymes obtained from various sources, such as animals, plants, and microbes. Besides enzymes, some bioactive peptides have been found to have health, physiological, and pharmaceutical benefits (Tapal & Tiku, 2019).

Enzymes are an important category of nutraceuticals. There are different natural sources available for enzyme production. The three primary natural sources include plants, animals, and microbes. Over half of the industrial enzymes are made by yeasts and moulds. Bacteria provide about 30%, animals provide 8%, and plants provide around 4% of the industrial enzymes (Thapa et al., 2019). Pepsin, lysozyme, α–Amylase, Pancrelipase, trypsin, and Chymotrypsin are some of the enzymes of nutraceutical value obtained from animals. Plants can act as a source of enzymes, including Pectinase, Hemicellulase, α-Galactosidase, Biodiastase, β-Amylase, and Glucoamylase (Singh & Sinha, 2012).

There are several advantages of microbes over plants and animals for harvesting the useful enzymes. In a review, Singh et al. have discussed several advantages of microbes over eukaryotic systems. As discussed in the review, the production processes are relatively simple as microbes are less complex as compared to eukaryotic systems in terms of their physiology and anatomy. In contrast to eukaryotes, microbes are easy to grow and maintain as their growth requirements are simple, their culturing is less time-consuming, they offer high reproducibility and consistency in the production process, isolation and purification of microbial enzymes is easier, production processes are cost-effective and less labour-intensive, and it is easier to perform genetic manipulations of microbes to enhance the production of a desirable product. Furthermore, some microbial enzymes are relatively more thermostable than their plant and animal counterparts, and it is noteworthy that thermostability is an important characteristic in the industrial application of enzymes. Processes involving microbes offer better controllability and ease in optimization. And, most importantly, the ethical concerns associated with plant and animal subjects are more complex (Singh et al., 2016).

Microbial enzymes can be of immense value as they have a wide range of applications across various fields. One of the important applications of these microbial enzymes is in nutraceutical production. Microbial enzymes themselves can be directly used as nutraceuticals or can be used to produce hydrolysates that have various physiological benefits. Digestive aid or digestive enzyme syrups contain several microbial enzymes that promote digestion in case of indigestion, stomach illness, and other chronic digestive disorders. For example, most of the digestive enzyme syrups contain a combination of diastase (also known as amylase) and pepsin (Thorat et al., 2022). Diastase hydrolyses polysaccharides (mainly starch) into simpler monosaccharides. On the other hand, pepsin is an endopeptidase that breaks down proteins into simpler peptides.

Recently, several metagenomic studies have reported the presence of amylase (i.e., diastase) in the metagenomes of rumen (Motahar et al., 2021), marine sediments (Nair et al., 2017), mangrove soil (Alves et al., 2020), and other extreme environments (Gupta et al., 2021). Though not much has been explored in case of microbial pepsin sources and its sources had been assumed to be restricted to eukaryotes only, homologues of pepsin have been reported in the prokaryotes as well (Rawlings & Bateman, 2009). Nonetheless, several fungi, including *Aspergillus*, *Penicillium*, *Rhizopus*, and *Neurospora*, have been reported to produce pepsin-like enzymes (Mamo & Assefa, 2018).

In a recent study by Ibacache-Quiroga et al., the use of 16S rRNA amplicon and shotgun metagenomic approaches revealed the presence of caseinase activity in the Kefir beverage, a probiotic food (Ibacache-Quiroga et al., 2022). In their review, Escuder-Rodríguez et al., have discussed the discovery of cellulase from thermophiles using metagenomics (Escuder-Rodríguez et al., 2018). Other thermophilic enzymes discovered through metagenomics and have nutraceutical application as reported in this study include cellobiosidase, β-glucosidases, and endoglucanases. Metagenomics can be used to identify and isolate the microbial enzymes of nutraceutical value. Table 13.2 contains the list of microbial enzymes with their metagenomics sources and potential nutraceutical application.

The microbial diversity in fermented foods can be explored using metagenomics to identify and extract microbial enzymes. Since these enzymes have important biological and pharmaceutical properties, they can be used in nutraceutical industries. Some of the enzymes, identified from fermented foods, using metagenomic approaches have been listed in the Table 13.3.

13.4 APPLICATION OF METAGENOMICS-ASSISTED APPROACHES IN IDENTIFYING MICROBIAL ENZYMES FOR NUTRACEUTICAL PRODUCTION

Certain microbial enzymes of nutraceutical interest are difficult and expensive to maintain. Traditional chemical synthesis of enzymes that can truly mimic the complexity of biological enzymes is difficult. Also, an environment-friendly alternative is necessary as industrial or chemical synthesis increases the burden on the environment. Hence, substituting traditional chemical processes with naturally sourced enzymatic pathways is a preferred approach (Coughlan et al., 2015). The invention of recombinant DNA technology has led to the development of genetically modified microbial strains having the potential to produce desirable proteins and enzymes. Sometimes the enzymes isolated from natural sources might not satisfy the requirements of the food processing conditions, as they require them to be functional under extreme environmental conditions such as

TABLE 13.2
List of Enzymes, Their Microbial and Metagenomic Sources, and Their Potential Nutraceutical Applications

Enzyme	Microbial Source	Metagenomic Source	Potential Nutraceutical Applications
Hemicellulase	*Clostridia*, *Ruminococcus flavefaciens*, *Trichoderma*, *Aspergillus*, and mushrooms (Shallom & Shoham, 2003)	Termite gut (Liu et al., 2019; Tokuda et al., 2018)	Extraction of dietary fibre from wastes of *Cynara cardunculus* (Domingo et al., 2015)
Alcalase	*Bacillus* spp. (Tacias-Pascacio et al., 2020)	Soil (Gabor et al., 2012)	Production of hydrolysates rich in peptides with anti-microbial, anti-oxidant, ACE[a] inhibition, anti-inflammatory, and immunomodulatory properties (Tacias-Pascacio et al., 2020)
Alkaline protease	*Bacillus licheniformis* (Bezawada et al., 2011)	Alkaline hot spring (Choure et al., 2021)	Production of hydrolysates containing ACE[a] inhibitory peptides (Shu et al., 2018)
Amyloglucosidase	*Aspergillus niger*, *Saccharomycopsis fibuligera*, Ascomycetes (Amaral-Fonseca et al., 2018) www.specialtyenzymes.com	Rumen (Shen et al., 2020)	Synthesis of α-glucosylated derivative of pterostilbene having anti-oxidant activity and nutraceutical application (González-Alfonso et al., 2018)
Catalase	Ascomycetes (Hansberg et al., 2012)	Water, soil, human skin, milk (Juwita et al., 2022; Trifonova et al., 2021; Xie et al., 2018)	Diminishes H_2O_2 level and modulate the release of a biological active agent that could be beneficial in irritable bowel syndrome (IBD) and related conditions (Alothman et al., 2021)
Cellulase	*Thermobifida fusca*, *Thermobispora bispora* *Streptomyces* sp., *Thermonospora* sp., *Erwinia chrysanthemi*, *Ruminococcus albus*, *Cellulomonas* sp., *Clostridium* sp., *Bacillus* sp., and *Acetivibrio cellulolyticus* (Ejaz et al., 2021)	Paper industry waste (Cui et al., 2019)	Cellulase improves extractability of nutraceutical components from rice bran (Sapna & Jayadeep, 2022)
Chymosin	*Escherichia coli* (Kawaguchi et al., 1987)	*Tenebrio molitor* larvae gut (Wang & Zhang, 2015)	Production of biologically active peptides with immunomodulating properties, from milk components (Lucarini, 2017; Séverin & Wenshui, 2005)
Glucose oxidase	*Aspergillus niger*, *Penicillium* spp. (Dubey et al., 2017)	Soil (Bao et al., 2021; Pan et al., 2021)	Extraction of Resveratrol, a natural phenol having therapeutic use (Chen et al., 2016)
Invertase	Yeast (Kulshrestha et al., 2013)	Pulque, water kefir beverage (Escobar-Zepeda et al., 2020; Verce et al., 2019)	Production of a fermented beverage (kombucha) (Wang et al., 2020)

(Continued)

TABLE 13.2 ***(Continued)***
List of Enzymes, Their Microbial and Metagenomic Sources, and Their Potential Nutraceutical Applications

Enzyme	Microbial Source	Metagenomic Source	Potential Nutraceutical Applications
Lactase or β-galactosidase	*Bifidobacterium infantis*, *Bifidobacterium longum*, *Bifidobacterium longum* *Lactobacillus* (Saqib et al., 2017)	Human gut, dairy wastewater (Eberhardtet al., 2020; Mulualem et al., 2021)	Exogeneous supply of lactase for treating lactose intolerance (Ojetti et al., 2009)
Proteinase	*Lactobacillus helveticus*, *Leuconostoc* spp., *Streptococcus thermophilus*, and *Lactobacillus delbrueckii* (Settanni & Moschetti, 2010; Yamamoto et al., 1994)	Soil (Calderon et al., 2019; Souza et al., 2018)	ACE[a] inhibition (Rai et al., 2017)
Thermolysin	*Bacillus thermoproteolyticus* (Khan et al., 2009)	Grassland soils (Séneca et al., 2021).	ACE[a] inhibition (Chourasia et al., 2020) and synthesis of aspartame, an artificial non-caloric sweetner (Birrane et al., 2014)
Xylanase	*Fusarium venenatum* (Sibbeseb & Sorensen, 2014)	Termite gut, hot spring (Joshi et al., 2020; Wu et al., 2021)	Endo-1,4-beta-xylanase (xylanolytic enzyme) in combination with cellulolytic enzyme improves the nutraceutical profile and anti-oxidant potential of red rice bran (Sapna & Jayadeep, 2022)
β-Glucosidase	*Lactobacillus kimchi* (Ko et al., 2014)	Soil (Matsuzawa & Yaoi, 2017)	Production of Resveratrol, having anti-inflammatory, anti-oxidant, antiallergic, cardioprotective, and anti-tumour properties (Ko et al., 2014)

Note:
[a] Angiotensin-I-converting enzyme.

pH, temperature, and pressure. Hence, they need to be genetically modified to enhance their specificity, purity, multifunctionality, catalytic activity, and yield. Such genetic changes promote their potential utilization in industrial applications (Chourasia et al., 2020). Recombinant DNA technology can be applied to microbes to produce nutraceuticals or enzymes that facilitate the production of other types of nutraceuticals.

The recombinant DNA techniques can be used along with metagenomics techniques to identify, isolate, produce, and modify enzymes of nutraceutical value. An amalgamation of these two techniques can help in the identification and production of microbial enzymes for unculturable microbes as well. This approach is known as functional metagenomics. Metagenomics can help address certain questions: "Who is there?", "What are they doing there?", and "Who is doing what?" The approaches for the first two questions have been discussed in previous sections of the chapter. The third question, "Who is doing what?", can be addressed though functional metagenomics which enables the study of the functional role of microbial genes. Figure 13.4 describes a brief workflow

TABLE 13.3
List of Microbial Enzymes Identified from Fermented Foods Using Metagenomics and Their Nutraceutical Applications

Enzyme	Metagenomic Source	Nutraceutical Applications	References
Fibrinolytic enzymes	Indonesian traditional fermented foods (*Tauco*, yellow *Oncom*, *Tempoyak*, and *Terasi*	Dissolves blood clots and protects against stroke, thrombosis, myocardial infarction, and other cardiovascular diseases	Purwaeni et al. (2018); Zakaria et al. (2015)
Caseinase, β-galactosidase, lactate dehydrogenases	Kefir beverage	Probiotic, increases protein metabolism, glucose absorption, and prevents lactose intolerance by lactose hydrolysis	Ibacache-Quiroga et al. (2022)
LAB enzymes (amylase, cellulase, protease)	Malaysian fermented foods (*tapai pulut*, *tempeh*, *tempoyak*, and *fu yu*	Probiotic, anti-inflammatory, anti-oxidative, anti-cancer, anti-microbial, anti-diabetic, anti-hypertensive, and anti-hypercholesterolemic	Azizi et al. (2021); Ho & Yin Sze (2018)
Proteolytic enzymes	Himalayan fermented soybean food (*Kinema*)	Improves digestibility of complex proteins	Kharnaior & Tamang (2021)
Fungal amylase	Himalayan alcoholic producing starters (*marcha*, *thiat*, *humao*, *hamei*, *chowan*, *phut*, *dawdim*, and *khekhrii*)	Bioactive peptides with immunomodulatory, anti-thrombic, and anti-hypertensive	Anupma & Tamang (2020); Tamang et al. (2016)
Proteolytic enzymes, amino acid lyases, threonine aldolase	Mongolian traditional fermented mare's milk (*Koumiss*)	Rich source of essential amino acids, vitamins, and has ACE[a] inhibitory activity	Chen et al. (2010); Yao et al. (2017)

Note:
[a] Angiotensin-I-converting enzyme.

to identify microbial enzymes, having nutraceutical potential, using recombinant DNA technology and metagenomics. Several studies involving the use of metagenomics in the nutraceutical industry have been reported. A general approach involves identification of the gene of interest, followed by the cloning of the putative gene in a suitable host. The cloned protein product is isolated and tested for its functional role. A few examples of these studies are discussed in the following paragraphs.

Functional metagenomics have been used in nutraceutical production, particularly microbial enzymes. In 2005, Walter et al. used the functional metagenomics method and developed a metagenomic library, consisting of 5,760 bacterial artificial chromosome (BAC) clones, prepared in Escherichia coli DH10B from DNA extracted from the large-bowel microbiota of BALB/c mice. They were able to identify three clones with ß-glucanase activity (Walter et al., 2005). Glucans cannot be digested by humans and so their hydrolysis relies on bacterial fermentation. As glucans are associated with health benefits in humans, glucan-hydrolysing enzymes isolated from bowel-dwelling microbiota may be of interest to nutraceutical-related industries (Abumweis et al., 2010).

A recent study by Patel and colleagues led to the discovery of a novel D-allulose 3-epimerase gene (*daeM*) using metagenomic sequencing of the sample from the hot-spring site at Tattapani geothermal field, Surguja district, Chhattisgarh, India. The putative *daeM* gene was then cloned in

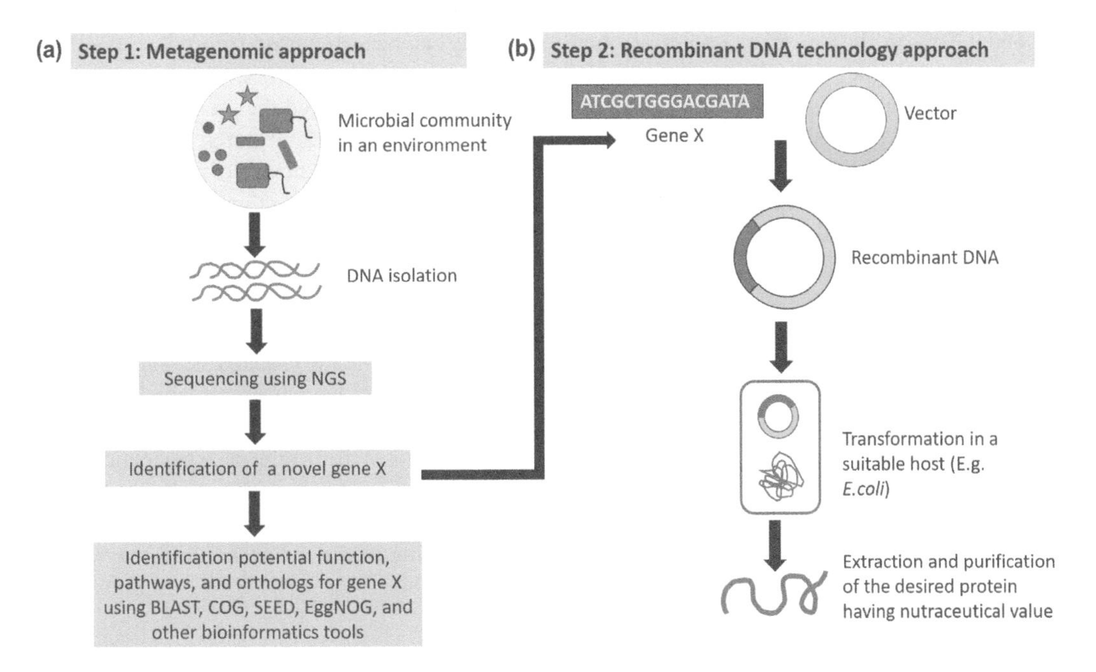

FIGURE 13.4 Identification of microbial enzymes for nutraceutical applications using metagenomics and recombinant DNA.

Bacillus subtilis to produce D-allulose. D-allulose is a sugar substitute with anti-diabetes and anti-obesity properties and therefore it is a useful nutraceutical (Patel et al., 2020). Another metagenomic study was performed on the sample collected from a warm aquatic site, located in Chhattisgarh, India. The study involved identification of a putative amylosucrase gene (AS_{met}) from the metagenome sequence data, followed by gene cloning of AS_{met} into *Escherichia coli*, extraction and purification of the protein, and biochemical characterization of the enzyme. Amylosucrase, the catalytic product of AS_{met}, is involved in production of turanose production from sucrose biomass. Further, turanose is considered to be a functional sweetener and an anti-inflammatory disaccharide (Agarwal et al., 2019).

A similar study led to identification of a novel putative trehalose synthase gene (*treM*) from a thermal spring metagenome followed by gene cloning of *treM*, heterologous expression in *Escherichia coli*, and biochemical characterization for trehalose and trehalulose. Being a thermostable enzyme, it has greater preference for industrial applications (Agarwal & Singh, 2021). The trehalose produced using trehalose synthase is a known potent ingredient that could diminish the progression of insulin resistance (Arai et al., 2010). Trehalose had also been reported to have anti-oxidant properties (Mizunoe et al., 2018). Furthermore, trehalose stabilizes the protein structures and reduces peptide aggregation, and therefore it is a potential therapeutic candidate for the treatment of neurodegenerative diseases (Khalifeh et al., 2019).

In 2020, Joshi et al. reported the presence of a novel xylanase gene in the extreme temperature hot spring metagenome. Xylanase is an enzyme that is known to produce xylooligosaccharide, a short-chain carbohydrate with prebiotic potential, as it promotes the growth of intestinal microbiota, prevents constipation, enhances mineral absorption, modulates immune responses, and has additional health benefits (Joshi et al., 2020). In the past, the xylanase gene has been reported to be present in the cattle rumen microbiota metagenome. It was observed that the novel xylanase had the potential to produce xylooligosaccharides and ferulic acid from wheat straw (Cheng et al., 2012). The thermal aquatic habitat (hot spring) metagenome led to discovery of a novel pullulanase, a potent enzyme for starch debranching. The identified enzyme caused a significant increase of about 80% in the amylose content of potato starch. It has been reported to produce resistant starch having higher crystallinity, heat-stability, and resistance to α-amylase digestion (Thakur et al., 2021a). Subsequently, another study led by the same group reported the production of resistant starch from pullulanase (Thakur et al., 2021b). This resistant fraction of starch is called slowly digestible starch and this makes it an important nutraceutical. It is known to prevent diseases associated with the consumption of glycaemic carbohydrates and promotes growth of probiotics (Magallanes-Cruz et al., 2017).

Fibrinolytic enzymes are a group of enzymes that can degrade the fibrin mesh formed during blood clotting. These enzymes are therefore useful in blood clot removal in individuals with cardiovascular problems. Besides blood clot removal, these enzymes also act as blood pressure regulators and dental caries preventers, and they are antimicrobial, anti-inflammatory, analgesic, and anti-bacterial agents (Kotb, 2017). Microbial fibrinolytic enzyme producers have been screened using the metagenomics approach (Purwaeni et al., 2018). Chitinases, which act on chitin, have been reported to produce chitin derivatives (chitosan and chitooligosaccharides) with high pharmaceutical and nutritional potential. In 2017, Berini et al. performed an experiment to clone two metagenome-sourced chitinases using metagenomic sequencing of soil sample (Berini et al., 2017). Cretoiu and co-workers discovered chitobiosidase through genetic screening of the metagenomic library from agricultural soil (Cretoiu et al., 2015).

13.5 FUTURE PERSPECTIVES

Presently, the nutraceutical industry is at a nascent stage and is still evolving, but with increasing awareness, the industry is emerging rapidly amongst the other industrial sectors. However, there are still some challenges that need to be addressed: (1) reduction of the potential allergenicity and other side effects of the microbial enzymes, (2) cost, (3) extended shelf-life, (4) stability, (5) regulation of nutraceutical products and others. Moreover, there are several limitations associated with metagenomic techniques: (1) lack of sufficient tools for metagenomic analyses, (2) limitations associated with NGS techniques such as read length, insufficient sequence coverage, and so on, and (3) bias in data processing and analysis, among others.

Several obstacles are often encountered while using metagenomics approach in nutraceutical research such as (1) difficulty in novel enzyme discovery due to unavailability of suitable hosts for fungal genes as the promoter and intron regions cannot be recognized by a bacterial host; (2) genes in environmental samples might not be expressed efficiently in the host owing to codon bias in translation; (3) need of high amino acid homology (of novel microbial enzymes) with already known enzymes, and hence enzymatic activities must be verified before assigning a function; (4) in soil studies, DNA samples are often contaminated by DNA of other organisms or soil component like humic acids which interfere with the enzymatic activity (methods used to remove the humic acids can further reduce the yield of metagenomic DNA); and (5) lack of enzyme-specific substrates for metagenomically identified novel enzymes make it difficult to do functional screening.

However, with further advancements in the fields of molecular biology, genetic engineering, and genomics, among others, these problems will be resolved. One of the prospects, towards which research studies are going on presently, is "orthologs" discovery. Presently, this approach is being used, but it needs to be extended for other newly discovered novel enzymes and unknown enzymes that are yet to be discovered. The known or novel genes identified from a microorganism through metagenomic techniques can be further subjected to computational tools to find their orthologs. Orthologs of the gene of interest (encoding the desirable enzyme) can be used in industrial applications. This will help in the identification of culturable microbes (discovered through gene ortholog study) that can replace the unculturable ones in industries. These orthologous genes can also be further cloned into other culturable microbial species to produce the desired enzyme having some potential nutraceutical applications.

Further, identification of suitable hosts for gene expression of novel genes of nutraceutical interest is another challenge which needs to be addressed in future. This will enable the utilization of their gene-products in nutraceutical industries. Nonetheless, this problem can be overcome to some extent through whole-genome metagenomics. Furthermore, a multi-OMICS approach involving Metatranscriptomics, Metaproteomics, and Metabolomics can help in addressing the gaps in the nutraceuticals industry. In future, exploring the lesser known ecological niches, including extreme environments, can lead to identification of useful nutraceutical microbial sources. Various fermented foods have been reported to have immense physiological and biological benefits. Additionally, more extensive metagenomic studies on indigenous fermented foods will help in exploring the nutritional and health benefits associated with fermented food products.

13.6 CONCLUSIONS

Modern lifestyle and food habits have made us more prone to various health problems. However, increasing consumer awareness and the health benefits associated with functional food has further led to several advancements in the production of nutraceuticals. Microbial enzymes have been proven to be of immense value in the production of nutraceuticals. Recent advancements in the field of metagenomics have brought a paradigm shift across various fields. Metagenomic techniques have become indispensable across different sectors including, health, agriculture, and environment. Hence, metagenomics is also considered to be a game-changer in several fields. It has brought a revolution in the field of microbiology and its application. It has enabled the scientific community to address many present-day problems. It provides a culture-independent method for the identification of microbial enzymes. These microbial enzymes can be directly used as nutraceuticals or to facilitate the production of other nutraceutical products such as carbohydrates, vitamins, hormones, proteins, and prebiotics.

REFERENCES

Abumweis, S. S., Jew, S., & Ames, N. P. 2010. β-Glucan from Barley and Its Lipid-Lowering Capacity: A Meta-Analysis of Randomized, Controlled Trials. *European Journal of Clinical Nutrition* 64:1472–1480. https://doi.org/10.1038/ejcn.2010.178

Agarwal, N., Narnoliya, L. K., & Singh, S. P. 2019. Characterization of a Novel Amylosucrase Gene from the Metagenome of a Thermal Aquatic Habitat, and Its Use in Turanose Production from Sucrose Biomass. *Enzyme and Microbial Technology* 131: 109372. https://doi.org/10.1016/j.enzmictec.2019.109372

Agarwal, N., & Singh, S. P. 2021. A Novel Trehalose Synthase for the Production of Trehalose and Trehalulose. *Microbiology Spectrum* 9: 1–14. https://doi.org/10.1128/Spectrum.01333-21

Alothman, M., Ispas-Szabo, P., & Mateescu, M. A. 2021. Design of Catalase Monolithic Tablets for Intestinal Targeted Delivery. *Pharmaceutics* 13(1): 69. https://doi.org/10.3390/pharmaceutics13010069

Altschul, S. F., Gish, W., Miller, W., Myers, E. W., & Lipman, D. J. 1990. Basic Local Alignment Search Tool. *Journal of Molecular Biology* 215(3): 403–410. https://doi.org/10.1016/S0022-2836(05)80360-2

Alves, K. J., da Silva, M. C. P., Cotta, S. R., Ottoni, J. R., van Elsas, J. D., de Oliveira, V. M., & Andreote, F. D. 2020. Mangrove Soil as a Source for Novel Xylanase and Amylase as Determined by Cultivation-Dependent and Cultivation-Independent Methods. *Brazilian Journal of Microbiology* 51(1): 217–228. https://doi.org/10.1007/s42770-019-00162-7

Amaral-Fonseca, M., Kopp, W., Giordano, R., Fernández-Lafuente, R., & Tardioli, P. 2018. Preparation of Magnetic Cross-Linked Amyloglucosidase Aggregates: Solving Some Activity Problems. *Catalysts* 8(11): 496. https://doi.org/10.3390/catal8110496

Anupma, A., & Tamang, J. P. 2020. Diversity of Filamentous Fungi Isolated from Some Amylase and Alcohol-Producing Starters of India. *Frontiers in Microbiology* 11: 905. https://doi.org/10.3389/fmicb.2020.00905

Arai, C., Arai, N., Mizote, A., *et al.* 2010. Trehalose Prevents Adipocyte Hypertrophy and Mitigates Insulin Resistance. *Nutrition Research* 30(12): 840–848. https://doi.org/10.1016/j.nutres.2010.10.009

Azizi, N. F., Kumar, M. R., Yeap, S. K., *et al.* 2021. Kefir and Its Biological Activities. *Foods* 10(6): 1210. 10.3390/foods10061210

Bao, T., Deng, S., Yu, K., Li, W., & Dong, A. 2021. Metagenomic Insights into Seasonal Variations in the Soil Microbial Community and Function in a *Larix gmelinii* Forest of Mohe, China. *Journal of Forestry Research* 32(1): 371–383. https://doi.org/10.1007/s11676-019-01090-w

Berini, F., Casciello, C., Marcone, G. L., & Marinelli, F. 2017. Metagenomics: Novel Enzymes from Non-Culturable Microbes. *FEMS Microbiology Letters* 364(21): fnx211.

Bezawada, J., Yan, S., John, R. P., Tyagi, R. D., & Surampalli, R. Y. 2011. Recovery of *Bacillus licheniformis* Alkaline Protease from Supernatant of Fermented Wastewater Sludge Using Ultrafiltration and Its Characterization. *Biotechnology Research International.*

Birrane, G., Bhyravbhatla, B., & Navia, M. A. 2014. Synthesis of Aspartame by Thermolysin: An X-Ray Structural Study. *ACS Medicinal Chemistry Letters* 5(6): 706–710. https://doi.org/10.1021/ml500101z

Blum, M., Chang, H.-Y., Chuguransky, S., *et al.* 2021. The InterPro Protein Families and Domains Database: 20 Years On. *Nucleic Acids Research* 49(D1): D344–D354. https://doi.org/10.1093/nar/gkaa977

Bolger, A. M., Lohse, M., & Usadel, B. 2014. Trimmomatic: A Flexible Trimmer for Illumina Sequence Data. *Bioinformatics* 30(15): 2114–2120. https://doi.org/10.1093/bioinformatics/btu170

Bolyen, E., Rideout, J. R., Dillon, M. R., *et al.* 2019. Reproducible, Interactive, Scalable and Extensible Microbiome Data Science Using QIIME 2. *Nature Biotechnology* 37: 852–857. https://doi.org/10.1038/s41587-019-0209-9

Borodovsky, M., & McIninch, J. 1993. GENMARK: Parallel Gene Recognition for Both DNA Strands. *Computers & Chemistry* 17(2): 123–133. https://doi.org/10.1016/0097-8485(93)85004-V

Brumfield, K. D., Huq, A., Colwell, R. R., Olds, J. L., & Leddy, M. B. 2020. Microbial Resolution of Whole Genome Shotgun and 16S Amplicon Metagenomic Sequencing Using Publicly Available NEON Data. *PLoS One* 15(2): e0228899. https://dx.plos.org/10.1371/journal.pone.0228899

Calderon, D., Peña, L., Suarez, A., Villamil, C., *et al.* 2019. Recovery and Functional Validation of Hidden Soil Enzymes in Metagenomic Libraries. *Microbiology Open* 8(4): e00572. https://doi.org/10.1002/mbo3.572

Caspi, R., Billington, R., Ferrer, L., Foerster, H., *et al.* 2016. The MetaCyc Database of Metabolic Pathways and Enzymes and the BioCyc Collection of Pathway/Genome Databases. *Nucleic Acids Research* 44(D1): D471–D480. https://doi.org/10.1093/nar/gkv1164

Chen, H., Deng, Q., Ji, X., Zhou, X., Kelly, G., & Zhang, J. 2016. Glucose Oxidase-Assisted Extraction of Resveratrol from Japanese Knotweed (*Fallopia japonica*). *New Journal of Chemistry* 40(9): 8131–8140. https://doi.org/10.1039/C6NJ01294A

Chen, Y., Wang, Z., Chen, X., Liu, Y., Zhang, H., & Sun, T. 2010. Identification of Angiotensin I-Converting Enzyme Inhibitory Peptides from Koumiss, a Traditional Fermented Mare's Milk. *Journal of Dairy Science* 93(3): 884–892. https://doi.org/10.3168/jds.2009-2672

Cheng, F., Sheng, J., Dong, R., Men, Y., Gan, L., & Shen, L. 2012. Novel Xylanase from a Holstein Cattle Rumen Metagenomic Library and Its Application in Xylooligosaccharide and Ferulic Acid Production from Wheat Straw. *Journal of Agricultural and Food Chemistry* 60(51): 12516–12524. https://doi.org/10.1021/jf302337w

Chourasia, R., Phukon, L. C., Singh, S. P., Rai, A. K., & Sahoo, D. 2020. Role of Enzymatic Bioprocesses for the Production of Functional Food and Nutraceuticals. *In* Biomass, Biofuels, Biochemicals (pp. 309–334). Elsevier. https://doi.org/10.1016/B978-0-12-819820-9.00015-6

Choure, K., Parsai, S., Kotoky, R., *et al.* 2021. Comparative Metagenomic Analysis of Two Alkaline Hot Springs of Madhya Pradesh, India and Deciphering the Extremophiles for Industrial Enzymes. *Frontiers in Genetics* 12. 10.3389/fgene.2021.643423.

Cole, J. R., Wang, Q., Fish, J. A., *et al.* 2014. Ribosomal Database Project: Data and Tools for High Throughput rRNA Analysis. *Nucleic Acids Research* 42(Database issue): D633–D642. https://doi.org/10.1093/nar/gkt1244

Coughlan, L. M., Cotter, P. D., Hill, C., & Alvarez-Ordóñez, A. 2015. Biotechnological Applications of Functional Metagenomics in the Food and Pharmaceutical Industries. *Frontiers in Microbiology* 6: 1–22. https://doi.org/10.3389/fmicb.2015.00672

Cretoiu, M. S., Berini, F., Kielak, A. M., Marinelli, F., & van Elsas, J. D. 2015. A Novel Salt-Tolerant Chitobiosidase Discovered by Genetic Screening of a Metagenomic Library Derived from Chitin-Amended Agricultural Soil. *Applied Microbiology and Biotechnology* 99(19): 8199–8215. https://doi.org/10.1007/s00253-015-6639-5

Cui, J., Mai, G., Wang, Z., Liu, Q., Zhou, Y., Ma, Y., & Liu, C. 2019. Metagenomic Insights into a Cellulose-Rich Niche Reveal Microbial Cooperation in Cellulose Degradation. *Frontiers in Microbiology* 10: 618. https://doi.org/10.3389/fmicb.2019.00618

DeSantis, T. Z., Hugenholtz, P., Larsen, N., *et al.* 2006. Greengenes, a Chimera-Checked 16S rRNA Gene Database and Workbench Compatible with ARB. *Applied and Environmental Microbiology* 72(7): 5069–5072. https://doi.org/10.1128/AEM.03006-05

Domingo, C. S., Soria, M., Rojas, A. M., Fissore, E. N., & Gerschenson, L. N. 2015. Protease and Hemicellulase Assisted Extraction of Dietary Fiber from Wastes of Cynara Cardunculus. *International Journal of Molecular Sciences* 16(3): 6057–6075. https://doi.org/10.3390/ijms16036057

Dubey, M. K., Zehra, A., Aamir, M., *et al.* 2017. Improvement Strategies, Cost Effective Production, and Potential Applications of Fungal Glucose Oxidase (GOD): Current Updates. *Frontiers in Microbiology* 8: 1–22. https://doi.org/10.3389/fmicb.2017.01032

Eberhardt, M. F., Irazoqui, J. M., & Amadio, A. F. 2020. β-Galactosidases from a Sequence-Based Metagenome: Cloning, Expression, Purification and Characterization. *Microorganisms* 9(1): 55. https://doi.org/10.3390/microorganisms9010055

Ejaz, U., Sohail, M., & Ghanemi, A. 2021. Cellulases: From Bioactivity to a Variety of Industrial Applications. *Biomimetics* 6(3): 44. https://doi.org/10.3390/biomimetics6030044

Escobar-Zepeda, A., Montor, J., Olvera, C., Sanchez-Flores, A., & Lopez-Munguia, A. 2020. An Extended Taxonomic Profile and Metabolic Potential Analysis of Pulque Microbial Community Using Metagenomics. *Journal of Food Science and Technology* 5: 83–97. 10.25177/JFST.5.2.RA.10637

Escuder-Rodríguez, J.-J., DeCastro, M.-E., Cerdán, M.-E., Rodríguez-Belmonte, E., Becerra, M., & González-Siso, M.-I. 2018. Cellulases from Thermophiles Found by Metagenomics. *Microorganisms* 6(3): 66. https://doi.org/10.3390/microorganisms6030066

Finn, R. D., Mistry, J., Tate, J., *et al.* 2009. The Pfam Protein Families Database. *Nucleic Acids Research* 38(Database issue): D211–D222. https://doi.org/10.1093/nar/gkp985

Gabor, E., Niehaus, F., Aehle, W., & Eck, J. 2012. Zooming in on Metagenomics: Molecular Microdiversity of Subtilisin Carlsberg in Soil. *Journal of Molecular Biology* 418(1–2): 16–20. https://doi.org/10.1016/j.jmb.2012.02.015

Ghosh, A., Mehta, A., & Khan, A. M. 2018. Metagenomic Analysis and Its Applications. *Encyclopedia of Bioinformatics and Computational Biology: ABC of Bioinformatics* 1–3: 184–193. https://doi.org/10.1016/B978-0-12-809633-8.20178-7

González-Alfonso, J. L., Rodrigo-Frutos, D., Belmonte-Reche, E., *et al.* 2018. Enzymatic Synthesis of a Novel Pterostilbene α-Glucoside by the Combination of Cyclodextrin Glucanotransferase and Amyloglucosidase. *Molecules* 23(6): 1271. https://doi.org/10.3390/molecules23061271

Gupta, N., Beliya, E., Paul, J. S., Tiwari, S., Kunjam, S., & Jadhav, S. K. 2021. Molecular Strategies to Enhance Stability and Catalysis of Extremophile-Derived α-Amylase Using Computational Biology. *Extremophiles* 25(3): 221–233. https://doi.org/10.1007/s00792-021-01223-2

Gupta, S., Mortensen, M. S., Schjørring, S., *et al.* 2019. Amplicon Sequencing Provides More Accurate Microbiome Information in Healthy Children Compared to Culturing. *Communications Biology* 2(1): 1–7. https://doi.org/10.1038/s42003-019-0540-1

Haft, D. H., Selengut, J. D., & White, O. 2003. The TIGRFAMs Database of Protein Families. *Nucleic Acids Research* 31(1): 371–373. https://doi.org/10.1093/nar/gkg128

Handelsman, J., Rondon, M. R., Brady, S. F., Clardy, J., & Goodman, R. M. 1998. Molecular Biological Access to the Chemistry of Unknown Soil Microbes: A New Frontier for Natural Products. *Chemistry and Biology* 5(10). https://doi.org/10.1016/S1074-5521(98)90108-9

Hansberg, W., Salas-Lizana, R., & Domínguez, L. 2012. Fungal Catalases: Function, Phylogenetic Origin and Structure. *Archives of Biochemistry and Biophysics* 525(2): 170–180. https://doi.org/10.1016/j.abb.2012.05.014

Ho, J. C. K., & Yin Sze, L. 2018. Isolation, Identification and Characterization of Enzyme-Producing Lactic Acid Bacteria from Traditional Fermented Foods. *Bioscience Horizons: The International Journal of Student Research* 11. https://doi.org/10.1093/biohorizons/hzy004

Huerta-Cepas, J., Szklarczyk, D., Heller, D., *et al.* 2019. eggNOG 5.0: A Hierarchical, Functionally and Phylogenetically Annotated Orthology Resource Based on 5090 Organisms and 2502 Viruses. *Nucleic Acids Research* 47(D1): D309–D314. https://doi.org/10.1093/nar/gky1085

Huson, D. H., Auch, A. F., Qi, J., & Schuster, S. C. 2007. MEGAN Analysis of Metagenomic Data. *Genome Research* 17(3): 377–386. https://doi.org/10.1101/gr.5969107

Ibacache-Quiroga, C., González-Pizarro, K., Charifeh, M., Canales, C., Díaz-Viciedo, R., Schmachtenberg, O., & Dinamarca, M. A. 2022. Metagenomic and Functional Characterization of Two Chilean Kefir Beverages Reveals a Dairy Beverage Containing Active Enzymes, Short-Chain Fatty Acids, Microbial β-Amyloids, and Bio-Film Inhibitors. *Foods* 11(7): 900. https://doi.org/10.3390/foods11070900

Joshi, N., Sharma, M., & Singh, S. P. 2020. Characterization of a Novel Xylanase from an Extreme Temperature Hot Spring Metagenome for Xylooligosaccharide Production. *Applied Microbiology and Biotechnology* 104(11): 4889–4901. https://doi.org/10.1007/s00253-020-10562-7

Juwita, S., Indrawati, A., Damajanti, R., Safika, S., & Mayasari, N. L. P. I. 2022. Genetic Relationship of Staphylococcus Aureus Isolated from Humans, Animals, Environment, and Dangke Products in Dairy Farms of South Sulawesi Province, Indonesia. *Veterinary World* 15(3): 558. 10.14202/vetworld.2022.558-564

Kalra, E. K. 2003. Nutraceutical – Definition and Introduction. *AAPS Journal* 5(3): 1–2.

Kanehisa, M. 2000. KEGG: Kyoto Encyclopedia of Genes and Genomes. *Nucleic Acids Research* 28(1): 27–30. https://doi.org/10.1093/nar/28.1.27

Kawaguchi, Y., Kosugi, S., Sasaki, K., Uozumi, T., & Beppu, T. 1987. Production of Chymosin in Escherichia coli Cells and Its Enzymatic Properties. *Agricultural and Biological Chemistry* 51(7): 1871–1877. https://doi.org/10.1080/00021369.1987.10868318

Khalifeh, M., Barreto, G. E., & Sahebkar, A. 2019. Trehalose as a Promising Therapeutic Candidate for the Treatment of Parkinson's Disease. *British Journal of Pharmacology* 176(9): 1173–1189. https://doi.org/10.1111/bph.14623

Khan, M. T. H., Fuskevåg, O.-M., & Sylte, I. 2009. Discovery of Potent Thermolysin Inhibitors Using Structure Based Virtual Screening and Binding Assays. *Journal of Medicinal Chemistry* 52(1): 48–61. https://doi.org/10.1021/jm8008019

Kharnaior, P., & Tamang, J. P. 2021. Bacterial and Fungal Communities and Their Predictive Functional Profiles in Kinema, a Naturally Fermented Soybean Food of India, Nepal and Bhutan. *Food Research International* 140: 110055. https://doi.org/10.1016/j.foodres.2020.110055

Ko, J.-A., Park, J. Y., Kwon, H. J., *et al.* 2014. Purification and Functional Characterization of the First Stilbene Glucoside-Specific β-Glucosidase Isolated from Lactobacillus Kimchi. *Enzyme and Microbial Technology* 67: 59–66. https://doi.org/10.1016/j.enzmictec.2014.09.001

Kotb, E. 2017. Microbial Fibrinolytic Enzyme Production and Applications. *Microbial Functional Foods and Nutraceuticals* 13: 175. https://doi.org/10.1002/9781119048961.ch8

Kulshrestha, S., Tyagi, P., Sindhi, V., & Yadavilli, K. S. 2013. Invertase and Its Applications – A Brief Review. *Journal of Pharmacy Research* 7(9): 792–797. https://doi.org/10.1016/j.jopr.2013.07.014

Kumar, S., Patel, R., & Seshadri, S. 2019. The Role of Nutraceuticals in Preventing and Managing Mental Disorders. *In* Functional Foods and Mental Health (1st ed.): 50–71.

Langille, M. G. I., Zaneveld, J., Caporaso, J. G., *et al.* 2013. Predictive Functional Profiling of Microbial Communities Using 16S rRNA Marker Gene Sequences. *Nature Biotechnology* 31(9): 814–821. https://doi.org/10.1038/nbt.2676

Langmead, B., Trapnell, C., Pop, M., & Salzberg, S. L. 2009. Ultrafast and Memory-Efficient Alignment of Short DNA Sequences to the Human Genome. *Genome Biology* 10(3): 1–10. https://doi.org/10.1186/gb-2009-10-3-r25

Li, W., O'Neill, K. R., Haft, D. H., *et al.* 2021. RefSeq: Expanding the Prokaryotic Genome Annotation Pipeline Reach With Protein Family Model Curation. *Nucleic Acids Research* 49(D1): D1020–D1028. https://doi.org/10.1093/nar/gkaa1105

Li, R., Zhu, H., Ruan, J., *et al.* 2010. De Novo Assembly of Human Genomes with Massively Parallel Short Read Sequencing. *Genome Research* 20(2): 265–272. https://doi.org/10.1101/gr.097261.109

Liu, N., Li, H., Chevrette, M. G., *et al.* 2019. Functional Metagenomics Reveals Abundant Polysaccharide-Degrading Gene Clusters and Cellobiose Utilization Pathways Within Gut Microbiota of a Wood-Feeding Higher Termite. *The ISME Journal* 13(1): 104–117. https://doi.org/10.1038/s41396-018-0255-1

Liu, J., Yu, Y., Cai, Z., Bartlam, M., & Wang, Y. 2015. Comparison of ITS and 18S rDNA for Estimating Fungal Diversity Using PCR–DGGE. *World Journal of Microbiology and Biotechnology* 31(9): 1387–1395. https://doi.org/10.1007/s11274-015-1890-6

Locey, K. J., & Lennon, J. T. 2016. Scaling Laws Predict Global Microbial Diversity. *Proceedings of the National Academy of Sciences of the United States of America* 113(21): 5970–5975. https://doi.org/10.1073/pnas.1521291113

Lucarini, M. 2017. Bioactive Peptides in Milk: From Encrypted Sequences to Nutraceutical Aspects. *Beverages* 3: 41. https://doi.org/10.3390/beverages3030041

Magallanes-Cruz, P. A., Flores-Silva, P. C., & Bello-Perez, L. A. 2017. Starch Structure Influences Its Digestibility: A Review. *Journal of Food Science* 82(9): 2016–2023. https://doi.org/10.1111/1750-3841.13809

Mamo, J., & Assefa, F. 2018. The Role of Microbial Aspartic Protease Enzyme in Food and Beverage Industries. *Journal of Food Quality* 2018. https://doi.org/10.1155/2018/7957269

Martin, M. 2011. Cutadapt Removes Adapter Sequences from High-Throughput Sequencing Reads. *EMBnet. Journal* 17(1): 10–12. https://doi.org/10.14806/ej.17.1.200

Matsuzawa, T., & Yaoi, K. 2017. Screening, Identification, and Characterization of a Novel Saccharide-Stimulated β-Glycosidase from a Soil Metagenomic Library. *Applied Microbiology and Biotechnology* 101(2): 633–646. https://doi.org/10.1007/s00253-016-7803-2

Meyer, F., Paarmann, D., D'Souza, M., *et al.* 2008. The Metagenomics RAST Server – A Public Resource for the Automatic Phylogenetic and Functional Analysis of Metagenomes. *BMC Bioinformatics* 9(1): 386. https://doi.org/10.1186/1471-2105-9-386

Mistry, J., Chuguransky, S., Williams, L., *et al.* 2021. Pfam: The Protein Families Database in 2021. *Nucleic Acids Research* 49(D1): D412–D419. https://doi.org/10.1093/nar/gkaa913

Mizunoe, Y., Kobayashi, M., Sudo, Y., *et al.* 2018. Trehalose Protects Against Oxidative Stress by Regulating the Keap1–Nrf2 and Autophagy Pathways. *Redox Biology* 15: 115–124.

Motahar, S. F. S., Ariaeenejad, S., Salami, M., Emam-Djomeh, Z., & Mamaghani, A. S. A. 2021. Improving the Quality of Gluten-Free Bread by a Novel Acidic Thermostable α-Amylase from Metagenomics Data. *Food Chemistry* 352: 129307. https://doi.org/10.1016/j.foodchem.2021.129307

Muhamad Rizal, N. S., Neoh, H. M., Ramli, R., *et al.* 2020. Advantages and Limitations of 16S rRNA Next-Generation Sequencing for Pathogen Identification in the Diagnostic Microbiology Laboratory: Perspectives from a Middle-Income Country. *Diagnostics* 10(10): 816. https://doi.org/10.3390/diagnostics10100816

Mulualem, D. M., Agbavwe, C., Ogilvie, L. A., Jones, B. V., Kilcoyne, M., O'Byrne, C., & Boyd, A. 2021. Metagenomic Identification, Purification and Characterisation of the Bifidobacterium Adolescentis BgaC β-Galactosidase. *Applied Microbiology and Biotechnology* 105(3): 1063–1078. https://doi.org/10.1007/s00253-020-11084-y

Nair, H. P., Vincent, H., Puthusseri, R. M., & Bhat, S. G. 2017. Molecular Cloning and Characterization of a Halotolerant α-Amylase from Marine Metagenomic Library Derived from Arabian Sea Sediments. *3 Biotech* 7(1): 1–9. https://doi.org/10.1007/s13205-017-0674-0

Namiki, T., Hachiya, T., Tanaka, H., & Sakakibara, Y. 2012. MetaVelvet: An Extension of Velvet Assembler to De Novo Metagenome Assembly from Short Sequence Reads. *Nucleic Acids Research* 40(20): e155. https://doi.org/10.1093/nar/gks678

Nasri, H., Baradaran, A., Shirzad, H., & Kopaei, M. R. 2014. New Concepts in Nutraceuticals as Alternative for Pharmaceuticals. *International Journal of Preventive Medicine* 5(12): 1487–1499.

Ojetti, V., Gigante, G., Ainora, M. E., *et al.* 2009. S1213 The Effect of Oral Supplementation with Lactobacillus Reuteri or Tilactase in Lactose-Intolerant Patients: A Placebo Controlled Study. *Gastroenterology* 136(5): A-214. https://doi.org/10.1016/S0016-5085(09)60962-8

Overbeek, R., Begley, T., Butler, R. M., *et al.* 2005. The Subsystems Approach to Genome Annotation and Its Use in the Project to Annotate 1000 Genomes. *Nucleic Acids Research* 33(17): 5691–5702. https://doi.org/10.1093/nar/gki866

Pan, Y., Zheng, X., & Xiang, Y. 2021. Structure-Function Elucidation of a Microbial Consortium in Degrading Rice Straw and Producing Acetic and Butyric Acids via Metagenome Combining 16S rDNA Sequencing. *Bioresource Technology* 340: 125709. https://doi.org/10.1016/j.biortech.2021.125709

Patel, J. B. 2001. 16S rRNA Gene Sequencing for Bacterial Pathogen Identification in the Clinical Laboratory. *Molecular Diagnosis: A Journal Devoted to the Understanding of Human Disease through the Clinical Application of Molecular Biology* 6(4): 313–321. https://doi.org/10.1054/modi.2001.29158

Patel, S. N., Kaushal, G., & Singh, S. P. 2020. A Novel D-Allulose 3-Epimerase Gene from the Metagenome of a Thermal Aquatic Habitat and D-Allulose Production by *Bacillus subtilis* Whole-Cell Catalysis. *Applied and Environmental Microbiology* 86(5): 02605–02619. https://doi.org/10.1128/AEM.02605-19

Pichler, M., Coskun, Ö. K., Ortega-Arbulú, A.-S., Conci, N., Wörheide, G., Vargas, S., & Orsi, W. D. 2018. A 16S rRNA Gene Sequencing and Analysis Protocol for the Illumina MiniSeq Platform. *MicrobiologyOpen* 7(6): 2. https://doi.org/10.1002/mbo3.611

Poretsky, R., Rodriguez-R, L. M., Luo, C., Tsementzi, D., & Konstantinidis, K. T. 2014. Strengths and Limitations of 16S rRNA Gene Amplicon Sequencing in Revealing Temporal Microbial Community Dynamics. *PLoS One* 9(4): e93827. https://doi.org/10.1371%2Fjournal.pone.0093827

Purwaeni, E., Darojatin, I., Riani, C., & Retnoningrum, D. S. 2018. Bacterial Fibrinolytic Enzyme Coding Sequences from Indonesian Traditional Fermented Foods Isolated Using Metagenomic Approach and Their Expression in *Escherichia coli*. *Food Biotechnology* 32(1): 47–59. https://doi.org/10.1080/08905436.2017.1413986

Quast, C., Pruesse, E., Yilmaz, P., *et al.* 2013. The SILVA Ribosomal RNA Gene Database Project: Improved Data Processing and Web-Based Tools. *Nucleic Acids Research* 41(Database issue): D590–D596. https://doi.org/10.1093/nar/gks1219

Rai, A. K., Sanjukta, S., & Jeyaram, K. 2017. Production of Angiotensin I Converting Enzyme Inhibitory (ACE-I) Peptides During Milk Fermentation and Their Role in Reducing Hypertension. *Critical Reviews in Food Science and Nutrition* 57(13): 2789–2800. https://doi.org/10.1080/10408398.2015.1068736

Rawlings, N. D., & Bateman, A. 2009. Pepsin Homologues in Bacteria. *BMC Genomics* 10(1): 1–10.

Sapna, I., & Jayadeep, A. 2022. Cellulolytic and Xylanolytic Enzyme Combinations in the Hydrolysis of Red Rice Bran: A Disparity in the Release of Nutraceuticals and Its Correlation With Bioactivities. *Lwt* 154: 112856. https://doi.org/10.1016/j.lwt.2021.112856

Saqib, S., Akram, A., Halim, S. A., & Tassaduq, R. 2017. Sources of β-Galactosidase and Its Applications in Food Industry. *3 Biotech* 7(1): 79. https://doi.org/10.1007/s13205-017-0645-5

Schloss, P. D., Westcott, S. L., Ryabin, T., *et al.* 2009. Introducing Mothur: Open-Source, Platform-Independent, Community-Supported Software for Describing and Comparing Microbial Communities. *Applied and Environmental Microbiology* 75(23): 7537–7541. https://doi.org/10.1128/AEM.01541-09

Séneca, J., Söllinger, A., Herbold, C. W., *et al.* 2021. Increased Microbial Expression of Organic Nitrogen Cycling Genes in Long-Term Warmed Grassland Soils. *ISME Communications* 1(1): 1–9. https://doi.org/10.1038/s43705-021-00073-5

Settanni, L., & Moschetti, G. 2010. Non-Starter Lactic Acid Bacteria Used to Improve Cheese Quality and Provide Health Benefits. *Food Microbiology* 27(6): 691–697. https://doi.org/10.1016/j.fm.2010.05.023

Séverin, S., & Wenshui, X. 2005. Milk Biologically Active Components as Nutraceuticals: Review. *Critical Reviews in Food Science and Nutrition* 45(7–8): 645–656. https://doi.org/10.1080/10408690490911756

Shallom, D., & Shoham, Y. 2003. Microbial Hemicellulases. *Current Opinion in Microbiology* 6(3): 219–228. https://doi.org/10.1016/s1369-5274(03)00056-0

Sharpton, T. J. 2014. An Introduction to the Analysis of Shotgun Metagenomic Data. *Frontiers in Plant Science* 5: 209. https://doi.org/10.3389/fpls.2014.00209

Shen, J., Zheng, L., Chen, X., Han, X., Cao, Y., & Yao, J. 2020. Metagenomic Analyses of Microbial and Carbohydrate-Active Enzymes in the Rumen of Dairy Goats Fed Different Rumen Degradable Starch. *Frontiers in Microbiology* 11: 1003. https://doi.org/10.3389/fmicb.2020.01003

Shu, G., Huang, J., Bao, C., Meng, J., Chen, H., & Cao, J. 2018. Effect of Different Proteases on the Degree of Hydrolysis and Angiotensin I-Converting Enzyme-Inhibitory Activity in Goat and Cow Milk. *Biomolecules* 8(4): 101. https://doi.org/10.3390/biom8040101

Sibbeseb, O., & Sorensen, F. 2014. Polypeptides with Xylanase Activity (Patent No. US 8,765438 B2). *WIPO Patent Application.*

Singh, J., & Sinha, S. 2012. Classification, Regulatory Acts and Applications of Nutraceuticals for Health. *International Journal of Pharmacy and Biological Sciences* 2(1): 177–187.

Singh, R., Kumar, M., Mittal, A., & Mehta, P. K. 2016. Microbial Enzymes: Industrial Progress in 21st Century. *3 Biotech* 6(2): 1–15. 10.1007/s13205-016-0485-8

Souza, R. C., Cantão, M. E., Nogueira, M. A., Vasconcelos, A. T. R., & Hungria, M. 2018. Outstanding Impact of Soil Tillage on the Abundance of Soil Hydrolases Revealed by a Metagenomic Approach. *Brazilian Journal of Microbiology* 49: 723–730. https://doi.org/10.1016/j.bjm.2018.03.001

Suzek, B. E., Wang, Y., Huang, H., McGarvey, P. B., Wu, C. H., & UniProt Consortium. 2015. UniRef Clusters: A Comprehensive and Scalable Alternative for Improving Sequence Similarity Searches. *Bioinformatics (Oxford, England)* 31(6): 926–932. https://doi.org/10.1093/bioinformatics/btu739

Tacias-Pascacio, V. G., Morellon-Sterling, R., Siar, E.-H., Tavano, O., Berenguer-Murcia, Á., & Fernandez-Lafuente, R. 2020. Use of Alcalase in the Production of Bioactive Peptides: A Review. *International Journal of Biological Macromolecules* 165: 2143–2196. https://doi.org/10.1016/j.ijbiomac.2020.10.060

Tamang, J. P., Shin, D. H., Jung, S. J., & Chae, S. W. 2016. Functional Properties of Microorganisms in Fermented Foods. *Frontiers in Microbiology* 7: 578.

Tapal, A., & Tiku, P. K. 2019. Nutritional and Nutraceutical Improvement by Enzymatic Modification of Food Proteins. *In* Enzymes in Food Biotechnology (pp. 471–481). Elsevier. https://doi.org/10.1016/B978-0-12-813280-7.00027-X

Tatusov, R. L. 2000. The COG Database: A Tool for Genome-Scale Analysis of Protein Functions and Evolution. *Nucleic Acids Research* 28(1): 33–36. https://doi.org/10.1093/nar/28.1.33

Thakur, M., Rai, A. K., Mishra, B. B., & Singh, S. P. 2021a. Novel Insight into Valorization of Potato Peel Biomass Into Type III Resistant Starch and Maltooligosaccharide Molecules. *Environmental Technology & Innovation* 24: 101827. https://doi.org/10.1016/j.eti.2021.101827

Thakur, M., Sharma, N., Rai, A. K., & Singh, S. P. 2021b. A Novel Cold-Active Type I Pullulanase from a Hot-Spring Metagenome for Effective Debranching and Production of Resistant Starch. *Bioresource Technology* 320: 124288. https://doi.org/10.1016/j.biortech.2020.124288

Thapa, S., Li, H., OHair, J., *et al.* 2019. Biochemical Characteristics of Microbial Enzymes and Their Significance from Industrial Perspectives. *Molecular Biotechnology* 61(8): 579–601. https://doi.org/10.1007/s12033-019-00187-1

Thorat, V., Kirdat, K., Tiwarekar, B., *et al.* 2022. Paenibacillus albicereus sp. nov. and Niallia alba sp. nov., Isolated from Digestive Syrup. *Archives of Microbiology* 204(2): 1–11. https://doi.org/10.1007/s00203-021-02749-x

Tokuda, G., Mikaelyan, A., Fukui, C., Matsuura, Y., Watanabe, H., Fujishima, M., & Brune, A. 2018. Fiber-Associated Spirochetes are Major Agents of Hemicellulose Degradation in the Hindgut of Wood-Feeding Higher Termites. *Proceedings of the National Academy of Sciences* 115(51): E11996–E12004. https://doi.org/10.1073/pnas.1810550115

Trifonova, T., Kosmacheva, A., Sprygin, A., Chesnokova, S., & Byadovskaya, O. 2021. Enzymatic Activity and Microbial Diversity of Sod-Podzolic Soil Microbiota Using 16S rRNA Amplicon Sequencing following Antibiotic Exposure. *Antibiotics* 10(8): 970. https://doi.org/10.3390/antibiotics10080970

Verce, M., De Vuyst, L., & Weckx, S. 2019. Shotgun Metagenomics of a Water Kefir Fermentation Ecosystem Reveals a Novel Oenococcus Species. *Frontiers in Microbiology* 10: 479. https://doi.org/10.3389/fmicb.2019.00479

Walter, J., Mangold, M., & Tannock, G. W. 2005. Construction, Analysis, and β-Glucanase Screening of a Bacterial Artificial Chromosome Library from the Large-Bowel Microbiota of Mice. *Applied and Environmental Microbiology* 71(5): 2347–2354. https://doi.org/10.1128/AEM.71.5.2347-2354.2005

Wang, Q., Garrity, G. M., Tiedje, J. M., & Cole, J. R. 2007. Naïve Bayesian Classifier for Rapid Assignment of rRNA Sequences into the New Bacterial Taxonomy. *Applied and Environmental Microbiology* 73(16): 5261–5267. https://doi.org/10.1128/AEM.00062-07

Wang, S., Zhang, L., Qi, L., *et al.* 2020. Effect of Synthetic Microbial Community on Nutraceutical and Sensory Qualities of Kombucha. *International Journal of Food Science & Technology* 55(10): 3327–3333. https://doi.org/10.1111/ijfs.14596

Wang, Y., & Zhang, Y. 2015. Investigation of Gut-Associated Bacteria in Tenebrio Molitor (Coleoptera: Tenebrionidae) Larvae Using Culture-Dependent and DGGE Methods. *Annals of the Entomological Society of America* 108(5): 941–949. https://doi.org/10.1093/aesa/sav079

Wu, H., Ioannou, E., Henrissat, B., Montanier, C. Y., Bozonnet, S., O'Donohue, M. J., & Dumon, C. 2021. Multimodularity of a GH10 Xylanase Found in the Termite Gut Metagenome. *Applied and Environmental Microbiology* 87(3): e01714–20. https://doi.org/10.1128/AEM.01714-20

Xie, W., Luo, H., Murugapiran, S. K., *et al.* 2018. Localized High Abundance of Marine Group II Archaea in the Subtropical Pearl River Estuary: Implications for Their Niche Adaptation. *Environmental Microbiology* 20(2): 734–754. https://doi.org/10.1111/1462-2920.14004

Yamamoto, N., Akino, A., & Takano, T. 1994. Antihypertensive Effect of the Peptides Derived from Casein by an Extracellular Proteinase from *Lactobacillus helveticus* CP790. *Journal of Dairy Science* 77(4): 917–922. https://doi.org/10.3168/jds.S0022-0302(94)77026-0

Yao, G., Yu, J., Hou, Q., *et al.* 2017. A Perspective Study of Koumiss Microbiome by Metagenomics Analysis Based on Single-Cell Amplification Technique. *Frontiers in Microbiology* 8: 165. https://doi.org/10.3389/fmicb.2017.00165

Zakaria, Z., Salleh, M. M., & Rashid, N. A. A. 2015. Screening and Identification of Fibrinolytic Bacteria from Malaysian Fermented Seafood Products. *Journal of Applied Pharmaceutical Science* 5(10): 022–031. 10.7324/JAPS.2015.501005.

14 CRISPR Technology for Probiotics and Nutraceuticals Production

Jalaja Vidya[1], *Yesodharan Vysakh*[2], *and Anand Krishnan*[2]

[1] Department of Botany and Biotechnology, Milad E-Sherif Memorial College (MSM), Kayamkulam, Alappuzha, Kerala, India

[2] Cardiology Department, Sree Chitra Tirunal Institute for Medical Sciences and Technology, Thiruvananthapuram, Kerala, India

CONTENTS

DOI: 10.1201/9781003311164-18

14.1 INTRODUCTION

The term probiotics was first introduced by Vergin (1954) in his study about the detrimental effects of antibiotics and other microbial substances on the gut microbial population where he observed that "probiotika" are beneficial to the gut microflora. The term was derived from the Greek language which means "for life". "Probiotic" was originally used to describe compounds produced by microbes that can stimulate the growth and development of other organisms. Later it was used to describe extracts from tissue that can stimulate microbial growth and animal feed supplements exerting a beneficial effect on animals by enhancing their intestinal flora balance. The definitions for probiotics have expanded over time concurrently with the developing interest in the use of live bacterial supplements and with the progress in understanding their mechanisms of action. The present-day definition of probiotics put forward by the US Food and Drug Administration (FDA) and the World Health Organization (WHO) jointly is "Live microorganisms which when administered in adequate amounts confer a health benefit to the host" (Hamilton-Miller, 1999).

Trillions of microorganisms live on the exposed surfaces of the human body and live in a symbiotic association with the host and play many essential roles in the production of metabolites, maturation of the host-immune response, brain–gut axis, and much more (Gilbert et al., 2018; Martens et al., 2018). The association between humans and probiotics for well-being has a long history. It was earlier observed that in breastfed infants, the gut microbiota was dominated by bifidobacteria which were absent in infants fed with formula and those infants who suffered from diarrhoea. This observation spread light to a concept that microbes present in the gut play a key role in maintaining health (Kechagia et al., 2013) As research in the field progressed with time, the probiotic successfully evolved by accumulating more relevant and substantial evidence that probiotic bacteria can contribute to human health. Probiotics and food containing probiotics gained more popularity due to their significant contributions to good health (Thantsha et al., 2012).

14.1.1 Ideal Characteristics of a Probiotic Strain

Functional foods can be described as foods or food ingredients that exhibit beneficial effects on the consumer's health beyond their nutritive value. Amicrobial strain should fulfil a number of criteria or specific properties to be regarded as a probiotic. There are different categories of criteria such as safety, technological aspects, and performance (Gibson and Fuller, 2000). Based on the safety, a probiotic strain should be of human origin and has to be isolated from the gastrointestinal tract of healthy individuals. They should have "generally recognized as safe" (GRAS) status; be non-pathogenic; not cause any negative impact such as infective gastrointestinal disorders or endocarditis; should not deconjugate bile salts; should not carry any antibiotic resistance genes that can be transferred to pathogens; must be immuno-tolerant and genetically stable; and should not possess any plasmid transfer mechanism. The strain or its cell components, products, or after its properties after death, must be non-toxic, non-pathogenic, non-allergic, or non-carcinogenic even in individuals with less immune capability. A probiotic strain should not promote inflammation and must have a desirable antibiogram profile (Collins et al., 1998; Havenaar and Huis In't Veld, 1992; Saarela et al., 2000; Ziemer and Gibson, 1998).

14.1.2 Mechanism of Action

There are various mechanisms of action for probiotics to enhance the growth and development of the host, although the exact mechanism by which they exert their effects is still under study and not fully untangled. Their activities range from production of bacteriocin and short chain fatty acid, lowering the pH of gut, and nutrient competition to stimulate the function of mucosal barrier and immunomodulation. Probiotics also helps in the induction of several aspects of the innate and acquired immune response by inducing IgA secretion and phagocytosis, modifying T-cell responses, attenuating Th2 responses, and enhancing Th1 responses (Guarner and Malagelada, 2003; Isolauri et al., 2001; McNaught and MacFie, 2001). One of such mechanisms by which probiotics inhibit the pathogens is by competition for sites adhesion, where probiotics fight with many pathogenic organisms for cellular attachments with the epithelium of GI tract to colonize effectively and thereby preventing pathogens from adhering to the mucosa acting as "colonization barriers" (Fuller, 1991). Another mechanism by which the probiotics attack pathogenic strains is by the modification of the microbial flora of the specific region through the synthesis of antimicrobial compounds (Rolfe, 2000). Different types of bifidobacteria and Lactobacilli produce bacteriocins and other antimicrobial compounds that inhibit the growth of pathogenic organisms.

14.1.3 Nutraceuticals

Nutraceuticals can be defined as biologically active substances that may be consumed as food or part of a food (Yapijakis, 2009). Nutraceuticals play a very important role in promoting good health as they help in the prevention of disease by strengthening the defence mechanism of the body and refine the immune system of body against diseases. Nutraceutical can be consumed in different forms such as capsule, tablet, or liquid form. They are also taken as dietary fibre, probiotics, prebiotics, polyunsaturated fatty acids, and antioxidants (Padhi et al. 2022; Rai et al. 2012). As natural foods, nutraceuticals can be categorized into different groups depending on various characteristics such as foods available in the market, mechanism of action, and their chemical nature (Sachdeva et al., 2020).

Based on the nutraceuticals' availability in the market they are categorized into traditional nutraceuticals and non-traditional nutraceuticals. Traditional nutraceuticals are natural substances with no changes to the foodstuff whereas non-traditional nutraceuticals are artificial foods prepared with the aid of biotechnology. Again, based on their constituents, they are grouped into different categories such as nutrients, herbals, phytochemicals, and probiotic microorganisms.

14.2 COMMON SOURCES OF MICROBIAL NUTRACEUTICALS AND PROBIOTICS

14.2.1 Bacteria as Probiotic

Under normal or "equilibrium" conditions, friendly bacteria in the gut exceed the unfriendly ones and they develop a physical barrier against unfriendly bacteria thereby becoming gut-beneficial bacteria. Intake of excess amounts of antibiotics can lead to bacterial imbalance in the gut that will lead to gas, cramping, or diarrhoea. Probiotics have been used for prevention or treatment of many conditions such as diarrhoea, ulcerative colitis, irritable bowel syndrome, and Crohn's disease. Even though there are many varieties of bacteria which can be used as probiotics, the two types that are commonly used as probiotics which can be found in stores are *Lactobacillus* and *Bifidobacterium*. For the preparation of pharmaceutical and probiotic foods, different genera of lactic acid bacteria (LAB) and bifidobacteria are used (Kneifel and Domig, 2014). Other than these two, bacterial species which are used as probiotics include *Saccharomyces, Streptococcus, Enterococcus, Bacillus, Pediococcus, Leuconostoc,* and *Escherichia coli*. Bacteria as probiotic are available as food mainly as fermented foods such as cheeses, kimchi, pickles, and raw unfiltered apple cider vinegar and also as dietary supplements like powders capsules, liquids, and other forms which contain a wide variety of strains and doses. Some of the probiotics also come with a mixed culture of live microorganisms (Kailasapathy and Chin, 2000).

14.2.2 Lactobacillus

Lactobacillus is rod-shaped gram positive facultative anaerobic or microaerophilic bacteria. They form a major part of the lactic acid bacteria (LAB) which metabolize hexose sugars to lactic acid.

Due to the production of lactic acid an acidic environment is created, which inhibits the growth and survival of various types of harmful bacteria (Makarova et al., 2006). Normally Lactobacilli are present in the gastrointestinal tract and vagina of human beings. Lactobacilli along with Bifidobacterium are one of the first bacteria that colonize in the gut of infants after delivery (Walter, 2008). Lactobacilli species express several key properties such as adherence to intestinal surfaces, tolerance of high acid and bile concentration; they can withstand gastric juice and low pH; inhibit pathogenic species by its antimicrobial activity; produce exopolysaccharides; resist antibiotics; and remove cholesterol since they are commonly selected as probiotics (Lee et al., 2014; Ruiz et al., 2013; Tulumoglu et al., 2013). *Lactobacillus rhamnosus* CRL1505, through controlling immune-coagulative responses, has been effective in clearing respiratory viruses and reducing viral-associated pulmonary damage (Zelaya et al., 2014). Lactobacillus is commonly added to fermented foods such as yoghurts. It is also commonly taken in dietary supplements. In adults, lactobacillus has most often been taken by mouth, alone or together with other probiotics, in doses of 50 million to 100 billion colony-forming units (CFUs) daily, for up to 6 months. In children, lactobacillus has most often been taken by mouth in doses of 100 million to 50 billion CFUs daily, for up to 3 months (Allen et al., 2013).

14.2.3 Bifidobacterium

The genus Bifidobacterium are non-motile, gram-positive anaerobic bacteria. They are found as endosymbiotic inhabitants in the gastrointestinal tract and vagina of humans and other mammals (Chen et al., 2007). The genus Bifidobacterium are widely used as probiotic bacteria due to their different resistance mechanisms to bile salts, since they are able to produce beneficial effects of probiotic bacteria in the presence of this biological fluid. Different strains of bifidobacteria that are widely used as probiotics include *B. adolescentis, B. infantis, B. animalissubsplactis, B. animalissubspanimalis, B. bifidum, B. longum,* and *B. breve.* Bifidobacterium species along with Lactobacillus and other probiotics are used for the treatment of travellers' diarrhoea, constipation, antibiotic-associated diarrhoea, suspension of disease activity of gut inflammation, and moderate ulcerative colitis (Aloisio et al., 2012; Chmielewska and Szajewska, 2010; Hempel et al., 2012). Bifidobacterium is also used to reduce diarrhoea induced by radiation and for the treatment of necrotizing enterocolitis in newborns, for food allergies, to reduce the risk of development of disease for eczema, and it also has cholesterol-lowering capacities (Demers et al., 2014; Di Gioiaet al., 2014; Isolauri et al., 2012).

14.2.4 Lactococcus

The genus Lactococcus are gram-positive LAB that are commonly used for the manufacturing of fermented products in the dairy industry. They convert sugars into lactic acid, developing an acidic condition, thereby preventing the growth of bacteria which spoil in milk products. Certain strains of *Lactococcus lactis* subsp. *lactis* grow attached to vaginal epithelial cells and are responsible for the production of nisin (*Lactococcus lactis subsp. lactis* CV56) and are also used for the treatment of antibiotic-associated diarrhoea along with other probiotics bacteria (Gao et al., 2011; Johnston et al., 2011; Yang et al., 2013).

14.2.5 Bacillus

Bacillus is spore-forming gram-positive aerobic or facultative aerobic bacteria. Different strains of Bacillus are *B. subtilis, B. subtilis, B. coagulans, B. cereus* etc. *Bacillus coagulans* possess probiotic properties along with other microorganisms and are very successful in the prevention and treatment of antibiotic-associated diarrhoea (Doron et al., 2008; Hempel et al., 2012). The spores of *Bacillus subtilis* are used as probiotics for consumption of animal and also used for the treatment of diarrhoea and *H. pylori* eradication in humans (Larsen et al., 2014; Tompkins et al., 2010; Zokaeifar et al., 2014).

14.2.6 Escherichia

The genus Escherichia is non-spore-forming, rod-shaped facultative anaerobic bacteria from the family Enterobacteriaceae. *Escherichia coli* are commonly found in the lower intestine and even a probiotic strain is known: *Escherichia coli* Nissle 1917 (EcN). *Escherichia coli* together with other probiotics can be used for the treatment of constipation and inflammatory bowel disease. *E. coli* could also reduce ulcerative colitis, gastrointestinal disorder, and Crohn's disease. Research is going on in the field of treatment of colon cancer using *E. coli* (Behnsen et al., 2013; Chmielewska and Szajewska, 2010).

14.3 CRISPR

Genetic manipulation is an impressive tool to study probiotic mechanisms of action and to potentially create improved strains (Lebeer et al., 2018; Petrova et al., 2018). Prokaryotes are continuously under the threat of large varieties of viruses and to escape from these menaces they have developed diverse defensive strategies. From the initial studies about phages which infect bacteria, scientists have identified and characterized bacteria that escape and are resistant to phage infection. These investigations about the bacterial resistance against viral infections led to the discovery of different defence mechanisms. Out of these, clustered regularly interspaced short palindromic repeat (CRISPR)-Cas system give adaptive immunity to fight foreign elements (Gilbert et al., 2018; Kechagia et al., 2013). CRISPR and CRISPR-associated (cas) genes give a special defence mechanism that provides robust adaptation to viruses that rapidly evolve. The CRISPR-cas and its function in anti-plasmid and antiviral defence was first analysed by in-silico studies. The first illustration of CRISPR loci was in 1987, following the sequencing of the gene iap of *Escherichia coli* (Ishino et al., 1987). Then back in 2002, small non-coding RNA was isolated and sequenced from the species of *Archaeoglobusfulgidus,* which revealed that CRISPR loci were transcribed to small RNAs. Also in the same year, cas genes were identified as a gene family associated with CRISPR loci (Jansen et al., 2002). The major advantage of CRISPR-Cas is that it is a programmable form of immunity which can be used to target any sequence as enormous diversity of specificities (Pingoud et al., 2014). There are three types (I–III) of CRISPR systems that have been identified among a wide range of bacterial and archaeal hosts. Every CRISPR system consists of a cluster of CRISPR-Cas genes, noncoding RNAs, and a distinctive array of repetitive elements (direct repeats). These repetitive elements are interspaced by short variable sequences which are obtained from exogenous DNA targets called protospacers, and together they constitute the CRISPR RNA (crRNA) array (Makarova et al., 2011). The two major units of the CRISPR-Cas systems are, first, a suite of genes that are encoding proteins involved in adaptation or spacer acquisition and effector functions, which include pre-crRNA processing and targeting of recognition and cleavage. The adaptation module is more or less uniform in almost all the different type of the CRISPR-Cas systems. The adaptation module consists of an endonuclease Casl and the structural subunit Cas2. Second, in contrast to the adaptation module, the effector modules are extremely variable between CRISPR-Cas types and subtypes (Amitai and Sorek, 2016). Cas9 is a nuclease guided by small RNAs which is highly specific, easier to design, highly efficient and well-suited for high-throughput and multiplexed gene editing. This can be used for a variety of cell types and organisms (Ran et al., 2013).

This close association of the microbes with humans makes the microbiome an exciting target for therapies with an aim to induce desired responses including metabolic, immunological, or even neurological changes, out of which probiotics is currently a promising field for food manufacturers, with exceptional potential for growth. The uses of probiotics involve intake of live probiotic cultures to improve intestinal microflora for the overall health of consumers. CRISPR-Cas systems are widespread in about 40% of bacteria and 90% of archaea even though the distribution and classification among these vary greatly within the phylogenetic clades (Crawley et al., 2018). The first proof that CRISPR-Cas9 can be implemented for making precise and scar-less genome editing in bacteria came from a study in which the Cas9 protein from *Streptococcus pyogenes* was integrated in the chromosome of *Streptococcus pneumonia* which opened a new era of gene editing in microbes using CRISPR-Cas systems (Jiang et al., 2013; Weiser et al., 2018). Nowadays the CRISPR-Cas systems are widely employed for genome editing and transcriptional regulation in a diverse variety of organisms. For the editing in probiotics the endogenous CRISPR-Cas systems can in some cases can be used, or otherwise engineered CRISPR-Cas systems can be introduced into target bacteria. These systems are now widely used to modify the genomes of microbiome-associated or probiotic bacteria, yeast, and bacteriophages. These systems can also be used to kill specific strains based on their sequence without touching the rest of the microbiome.

14.4 GENOME EDITING IN MICROBES BY CRISPR

Genome editing permits genetic material to be added, removed, or altered at particular locations in the genome. It enables scientists to scrutinize the genetic basis of physiological and metabolic processes in any living organism. As the word means, the process uses the enzymes to "cut" DNA creating a double-strand break (DSB). DSB repair occurs by non-homologous end joining (NHEJ) or homology-directed repair (HDR) (Carroll, 2014). NHEJ produces random mutations (gene knock-out), while HDR uses additional DNA to create a desired sequence within the genome (gene knock-in). Microbial genome engineering is achieved through a number of different techniques each with varying degrees of specificity, efficiency, and applicability. Recently, these tools and techniques have been expanded to a great extent for a diverse range of bacterial host organisms other than the typical *E. coli* cells. Among these genome editing methods, the breakthrough discovery of CRISPR technology emerged as a new generation genome-editing tool for efficient and robust strategy regardless of the kind of organism being modified.

The occurrence of genetic variations through spontaneous mutations was revealed earlier by the classical genetic studies which were followed by mutagenesis induced by physical or chemicals means. This random mutagenesis approach was further advanced through the discovery of jumping genes or transposons in which the induced transposon insertions cause arbitrary mutations in the genome of some organisms. Since these strategies are non-specific and random, the targeted editing of genome at specific locations was achieved through the process of homologous recombination (HR) with significant precision, but the strategy is limited by its least efficiency and laborious characterization and selection process. Also the application of this technique is limited to laboratory animals like mice.

Almost all of the genome editing approaches are based on the cellular mechanism of repairing DNA damage. Hence the targeted DSB induced by the sequence specific endonucleases in DNA backbone trigger repair and this forms the basement for all genome editing methods. This editing mechanism by repair is broadly classified into two types: homology dependent and homology independent methods. The repair systems such as NHEJ and microhomology-mediated end joining (MMEJ) facilitate joining of exogenous DNA in break points without any precision and foreign DNA may be integrated in either orientation (Figure 14.1). On the other hand, the precise modification of the genome can be achieved by another approach of homology-directed repair (HDR) or even by HR to exchange several kilobases of DNA. The most common nucleases which are used to make DSBs in DNA arose from investigation into natural biological processes and are zinc-finger nucleases (ZFNs) (Miller et al., 1985; Pavletich and Pabo, 1991), transcription activator-like effector nucleases (TALENs) (Boch et al., 2009; Moscou and Bogdanove, 2009) and CRISPR-Cas (Ishino et al., 1987) of which the latter dominates nowadays worldwide.

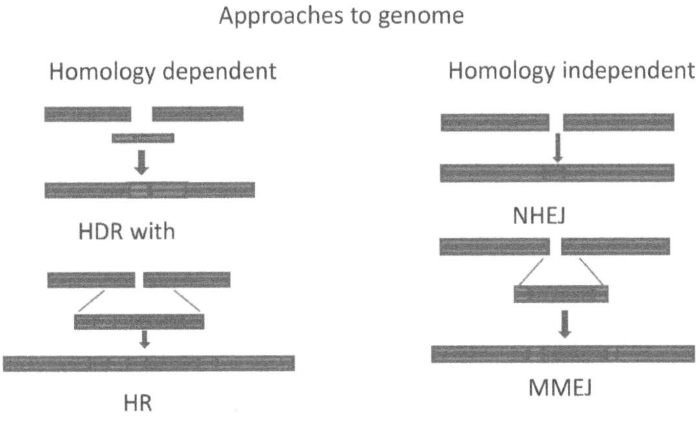

FIGURE 14.1 Double strand DNA breakage and the repair and recombination events. In a homology dependent event, homologous recombination or HDR takes place when a single-stranded oligodeoxynucleotides (ssODN) with close homology to the site are available. The absence of template DNA leads to homology independent repair either by NHEJ or MMEJ pathways.

14.5 COMMON METHODS FOR GENOME EDITING IN MICROBES

In prokaryotes, other than the CRISPR tools, the genome editing is done through several other methods; however, it faces several limitations such as least efficiency, inconsistency, and laborious operating methods. An organism's genetic makeup can be modified by introducing new genetic elements by genetic engineering tools. This enables insertion of genes into specific sites as well as the editing of specific sequence within the genome like directed mutagenesis. Some of these methods are listed below.

14.5.1 Genome Editing by Suicide Plasmids

This method is one of the earliest approaches based on a plasmid with selective replication ability and is having a homologous sequence in frame with a selectable marker and a transposon sequence (Selvaraj and Iyer, 1983). Since the plasmid cannot replicate in recipient cells, the strain with successful integration of the target DNA was selected by antibiotic selection and screening. The plasmid also contains a secI site with 18bp asymmetric sequence in between the mutant allele and selectable marker gene which is cleaved by secI homing endonuclease expressed by *E.coli* from another plasmid. This method is particularly useful in organisms in which advanced methods of editing are poorly working with large insertions or deletions and for protein purification studies after site directed mutagenesis (Konishi et al., 1991).

14.5.2 Clos Tron Method

Due to the lack of conventional methods knock-in/knock-out mutation in Clostridium, a recombination independent procedure is adopted for editing the genome of Clostridia typically by Clostron method. The method uses group II intron with transposon activity and a retro-transposition-activated marker to precisely insert the foreign DNA into defined sites within the genome (Kuehne et al., 2011). The procedure of gene insertion involves a plasmid containing a cassette of introns, having transposon activity and antibiotic resistance is transformed into a host, disrupting the desired gene, leaving behind a plasmid with a functional antibiotic marker (Heap et al., 2007).

14.5.3 Recombineering/Lambda Red System

This is a widely used method based on bacteriophage λ. Lambda Red consists of three proteins: alpha (α), beta (β), and lambda (γ). α is an exonuclease, which digests the 5'-ended strand of a dsDNA end. β binds to ssDNA and promotes strand annealing. Finally, γ binds to the bacterial RecBCD enzyme and inhibits its activities. These proteins induce a "hyper-recombination" state in *E. coli* and other bacteria, in which recombination occurs at high frequency regardless of the sequence length (Sharan et al., 2009). The antibiotic selection and secI site can be also included to enable counter selection eliminate the resistance marker by HR (Tas et al., 2015).

14.6 CRISPR-BASED GENOME EDITING METHODS

Since the discovery of CRISPR-Cas9 system the genetic engineers have developed a number of methods across different organisms based on different DNA repair mechanisms using a variety of heterologous recombinases and a number of plasmids, as outlined in the following sections.

14.6.1 Homologous Directed Repair

In this approach the major components of the editing such as linear DNA template, recombinase, and the cas9/gRNA are either incorporated in a one-plasmid system or in two- or three-plasmid systems and co-transformed into the host bacterium. Here the heterologous recombinases such as λ red or RecT are delivered via plasmid or phage which is then co-transformed with DNA template and Cas9 nuclease in plasmid with antibiotic resistance genes. In one plasmid system, the larger size of the plasmid makes it harder to transform and it restricts the cloning of the gRNA targets cloning strain. The recombinase T enhances the editing by binding to the ssDNA and protects it from degradation in *Corynebacterium glutamicum* (Wang et al., 2018), *Lactobacillus plantarum, Lactobacillus brevis,* and *Lactoccocus lactis* (Guo et al., 2019).

14.6.2 Non-Homologous End Joining

The NHEJ machinery naturally occurs in very few bacterial genera and in case of others it must be heterogeneously encoded in the CRISPR plasmid. The NHEJ repair machinery basically consists of two proteins, Ku and LigD, of which the former binds to the ends of cleaved DNA and the latter joins the ends together, often introducing insertions/deletions or non-specific mutations (Arroyo-Olarte et al., 2021).

14.6.3 Alternative End-Joining (A-EJ) Pathway

Also known as MMEJ, alternative end-joining involves microhomologies of 1–9 nt near the Cas9 cut site after resection of DNA ends by RecBCD being ligated by LigA, leaving behind deletions of variable sizes after repair. This approach, combined with CRISPR-Cas9, has been successfully reported in several species such as *E. coli, Streptomyces coelicolor,* and *Pectobacterium atroseptícum.* Both of these methods, NHEJ as well as A-EJ, are not at all specific as they cannot be used for specific editing but are effective for gene knock-outs (Chen et al., 2017).

14.7 GENOME EDITING IN FOOD GRADE MICROBES

The use of omics technologies like genomics, proteomics, and transcriptomics has entirely changed the understanding of microbiomes. CRISPR-Cas technologies have recently redefined the gene-editing toolbox and demonstrated our potential to alter the genome of food microbes. An integrated method based on CRISPR-based genome editing has the potential to enable the engineering of micro-organisms and the creation of microbiomes that will have an impact on the food supply chain (Pan and Barrangou, 2020). Genome editing technologies are needed to create strains for new applications and products, and will be useful in established ones like food and probiotics, as a research tool for acquiring mechanistic insights and identifying novel properties (Börner et al., 2018).

CRISPR-Cas systems are widely found in archaea and bacteria and are well-known acquired immunity mechanisms. CRISPR-Cas is the RNA-guided nucleases system that is now widely considered the most dependable tool for genome editing. The first evidence of their existence was discovered in the *Escherichia coli* genome in 1987, as an unusual repeating DNA sequence, later described as a cluster of regularly interspersed short palindromic repeats (CRISPR), was discovered during the investigation of genes involved in phosphate metabolism (Ishino et al., 2018). CRISPR-Cas is a bacterial adaptive immune system that protects them from intrusive elements like bacteriophages and plasmids. When foreign DNA is introduced into a bacterial cell, tiny DNA fragments are incorporated into the CRISPR locus as novel spacers between CRISPR repeats. The CRISPR locus is transcribed and processed to produce small CRISPR RNAs (crRNAs) that direct the Cas effector nuclease to complementary target sequences. Following further exposure to foreign DNA for which the cell possesses a complementary crRNA, a sequence-specific DNA-RNA duplex forms. The crRNA thus directs Cas nucleases to cleave the foreign double-stranded DNA, resulting in sequence-specific targeting. The invasive DNA is then destroyed by host nucleases, which interrupts and prevents targeted DNA replication (van Pijkeren and Barrangou, 2017).

The CRISPR-fuelled genome editing mania has largely focused on engineering complex eukaryotes and there are numerous opportunities to apply in bacteria in general and food microorganisms in particular (Pan and Barrangou, 2020). The CRISPR-Cas system was first employed as a counter selection tool to assist ssDNA mediated recombineering in *Lactobacillus reuteri* (Zuo and Marcotte, 2021). It helps detect and remove recombinant cells with low recombination efficiency and unedited cells (Oh and van Pijkeren, 2014). In *Tatumellacitrea*, a member of the Enterobacteriaceae family recognized for producing 2-keto-D-gluconicacid, a precursor to vitamin C, the CRISPR-Cas9 system showed 100% efficacy in targeted chromosomal deletions (Dey, 2020).

14.8 GENOME EDITING IN LACTIC ACID BACTERIA

LAB are GRAS status organisms, so are safe for human consumption. As they comprise an important group of microbes in food biotechnological applications, genome editing of LAB for probiotics and nutraceuticals is a major focus of interest in nowadays. The limiting factor in molecular editing of these organisms is because of its poor acceptance by the public due to its "genetically modified organism" (GMO) label. The strain improvement in this category of organisms by random mutagenesis is laborious and time consuming, and to overcome these drawbacks, the strains must be edited with specificity for particular probiotic or nutraceutical applications.

Although the primary application of LAB is fermented food production, several strains gaining attraction as probiotic food supplements and have huge market value. There exist a number of methods for editing the genetic makeup of LAB without introducing any marker genes and these methods should come up with an edited microbe which can be labelled as non-GMOs. Endogenous CRISPR-based tool is used in *P. acidilactici* species to strengthen the lactic acid production by nicotinamide adenine dinucleotide (NADH)-dependent ldh gene expression and it enhanced both lactic acid production and cell growth (Kasai et al., 2019). A probiotic strain *L. Plantarum* producing GlcNAc yielding 797.3 mg/litre was constructed using CRISPR editing tool without introducing any exogenous genes (Zhou et al., 2019). The team inactivated nagB to block the reverse reaction of fructose-6-phosphate (F6P) to glucosamine 6-phosphate (GlcN-6P) and the feedback repression is relieved by riboswitch replacement and point mutation in glmS1.

14.9 GENOME EDITING IN STREPTOMYCES

Streptomyces is an important genus in bacteria well known for its production of secondary metabolites having wide application as antibiotics, immunosuppressants, insecticides, anti-cancerous agents etc. The genome sequencing of streptomyces revealed a large number of biosynthetic gene clusters (BGCs) for natural products (NPs). Surprisingly most of these BGCs in streptomyces are not expressed or poorly expressed under laboratory conditions, and hence remain an unexplored pool of novel bioactive compounds. The activation of this silent BGC for the synthesis of these natural products can be achieved in a number of ways: by induction of BGC expression by knock-out of negative regulatory genes and replacement of native promoters with strong ones, or by over expression of positive regulatory genes. Moreover, activation of silent BGCs by CRISPR-mediated knock-in tactic to modify the promoters or activators for production of NPs in diverse species of streptomyces. There are reports that triggered the biosynthetic pathways for NPs by CRISPR knock-in strategy by inserting the constitutive upstream of BGC operons or activators in some streptomyces species (Cobb et al., 2015).

14.10 GENOME EDITING IN CYANOBACTERIA

Cyanobacteria, the group of photosynthetic bacteria maintains the energy balance of the ecosystem by assimilating the CO2 utilizing solar energy and are able to produce a succession of products such as biofuels, pharmaceuticals, and nutrients. But the genetic complexity of these bacteria compared to others makes the genetic engineering process tedious because of this ploidy (Li et al., 2016). Hence multiple rounds of segregation steps are required to get a final fully edited version of the host which is laborious and time consuming. This difficulty is addressed by CRISPR editing in a number of species in cyanobacteria by transient expression of Cas9 to minimize the toxicity attributed by high-level expression in the cyanobacterial host system (Griese et al., 2011; Xiao et al., 2018). Also like some other prokaryotes, cyanobacteria lack an NHEJ pathway and the CRISPR system introduces mutations at specific locations by HR. Cyanobacteria capable of producing free fatty acids by incorporation of the thioesterase gene from *E. coli* in a fastest growing strain *Synechococcus elongatus* UTEX 2973 was generated through CRISPR-Cas9 tool by Racharaks and team in 2021 (Racharacks et al., 2021). The CRISPR-based gene manipulation for improved production of succinate in cyanobacteria through simultaneous glgc knock-out and gltA/ppc knock-in strategy resulted in an 11-fold increase in succinate production compared to the wild type control (Li et al., 2016).

14.11 GENOME EDITING IN LACTOBACILLUS AND BIFIDOBACTERIUM

The group of probiotic bacteria, such as Lactobacilli and Bifidobacteria, serves as therapeutic delivery systems to deliver proteins and peptides with utmost safety and precision (Plavec and Berlec, 2020). Among the probiotic microbes, CRISPR-Cas9 system is widely distributed in Lactobacilli due to its genetic diversity contributed by horizontal gene transfer, extensive genome remodeling, and phage infections (Sun et al., 2015). Next generation probiotics based on Lactobacilli focus on its immunomodulatory properties and the cell surface components play a key role in host –microbe interactions. For example, teichoic acid (TA) found on the gram-positive cell wall interacts with mammalian toll-like receptors and induce cytokine response. Hence alteration of these cell wall components by genome editing augment the probiotic profiles of strains (Grangette et al., 2005). The presence of TA in lactobacillus is highly conserved except in some species like *L. reuteri*, *Lactobacillus rhamnosus*, *Lactobacillus fermentum*, and *Lactobacillus casei*. The engineered strain of *Lactobacillus acidophilus* NCFM with phosphoglycerol transferase knock-outs lacks TAs in the cell wall and showed improved anti-inflammatory prolife compared to wild type (Khazaie et al. 2012).

Like that the probiotic, Lactobacilli inhabiting the human vagina have been engineered to express the VHH antibody against HIV infection, and Lactobacilli and bifidobacterium capable of tolerating the harsh environment of the gastrointestinal tract have been successfully established recently. Mutations in genes or regulatory elements introduced by genome editing enables high-level expression of effector molecules compared to native promoter/enhancers (Zuo and Marcotte, 2021).

Similarly, genome editing in multiple target sites in *L. crispatus* NCK1350 by native endogenous CRISPR-Cas3 system has proved the potential of *L. crispatus* strain to develop into next generation probiotics and vaccines against common infections and sexually transmitted diseases. The gene encoding exopolysaccharide priming-glycosyl transferase (p-gtf) was mutated in a number of ways by deletions, insertions, and substitutions to flexibly alter the therapeutic and probiotic potential for mucosal vaccine delivery (Hidalgo-Cantabrana et al., 2019).

14.12 GENOME EDITING IN YEAST

Yeasts are considered probiotic only when they can survive the temperature and acidity of the stomach. *Saccharomyces cereviceae* is the most preferred probiotic yeast which is proved to be beneficial for gut health. Other strains like *Saccharomyces boulardii* also have been reported with similar beneficial phenotypes and has the ability to stay long in gut compared to the former one. Yeasts have long been widely used for bioprocesses in industries and research due to its ability to tolerate harsh living conditions and convenient genetic manipulation. Out of all other types, *Saccharomyces cerevisiae* becomes the most popular cell factory for the production of various biofuels, chemicals, and other natural products (de Jong et al., 2012; Kavscek et al., 2015). Even though *S. Cerevisiae* is mostly used there are many other non-conventional yeasts which have different evolutionary distance from *S. cerevisiae* and are also increasingly gaining attention for their ability to produce various recombinant proteins, fine chemicals, and oils (Nombela et al., 2006). Despite many excellent properties, it is still arduous to engineer these non-conventional yeasts because of the lack of genome editing tools when compared with *S. cerevisiae* which has numerous advanced genetic tools and biological devices (Jensen and Keasling, 2015). Metabolic engineering of the *Saccharomyces boulardii* genome for secretion of human lysozyme and introduction of xylose assimilating pathway was successfully engineered through CRISPR editing for food grade applications (Liu et al., 2016).

The multiplex genome editing based CRISPR-Cas 9 aimed at reconstructing complex metabolic pathways and focusing on integrating multiple genes of interest in a single transformation, thereby simplifying the reconstruction of complex pathways. As the metabolites from plant origin usually have complex multigene biosynthetic pathways, the multiplex CRISPR-Cas9 system in yeast is well suited to functional genomics research in plant specialized metabolism (Utomo et al., 2021). This trusty editing method enabled the construction of cell factories for improved production of chemicals by speeding up the genetic writing. Its efficiency was illustrated by the significantly high yield in production of resveratrol by CRISPR–Cas9 based multiplex genome editing. The biosynthetic pathway of resveratrol (three genes) was integrated into rDNA repeats cluster of *O. polymorpha* which resulted in a 21-fold higher resveratrol production (97 mg/L) compared to the single copy pathway (Wang et al., 2018). *Scheffersomycesstipitis* is one of the most distinguished microorganisms for biomass refinery due to its excellent native capacity for catabolizing xylose and it has immense potential for producing shikimate pathway derived molecules. Establishing the CRISPR–Cas9 system encountered challenges such as the lack of the stable and useful plasmid to express the CRISPR components. Recently, a 500 bp minimal fragment of centromere (CEN) was identified to significantly stabilize the autonomously replicating sequences (ARS)-containing vector and enable exogenous gene expression (Cao et al., 2017). The CRISPR–Cas9 system is currently much less efficient in non-conventional yeasts as compared to mammal and yeasts *S. cerevisiae*, which might be attributed to comparatively low HR repair efficiency.

Another class of nutraceutical, L-(+)-Ergothioneine (ERG), belonging to the class of antioxidants, was successfully produced in yeast *S. cerevisiae* by CRISPR-Cas9 market free method by Van der Hoek et al. (2019). ERG has a dual role in the human body both as a nutraceutical and as vitamin for health protection, acting against cellular oxidative damage. The components of the biosynthetic pathways of this particular molecule were engineered in baker's yeast along with the transporter proteins.

TABLE 14.1
Summary of Genome Edited Probiotics and Their Products

Microorganism	Relevant Feature	Reference
Lactobacillus reuteri	Probiotic strain and producer of biotherapeutics	Oh and van Pijkeren (2014)
Streptococcus thermophilus	Probiotic and industrial fermentation strains	Barrangou et al. (2007)
Streptomyces albus	Producer of heterologous secondary metabolites	Cobb et al. (2015)
Streptomyces coelicolor	Source of pharmacologically active secondary metabolites	Tong et al. (2015)
Streptomyces lividans	Source of pharmacologically active secondary metabolites	Cobb et al. (2015)
Streptomyces viridochromogenes	Source of pharmacologically active secondary metabolites	Cobb et al. (2015)
Tatumellacitrea	Producer of vitamin C precursor (2-keto-D-gluconic acid)	Jiang et al. (2015)
S. elongates	Succinate production	Li et al. (2016)
S. coelicolor	Production of secondary metabolites	Li et al. (2018)
E. coli Nissle	Probiotic strain against inflammatory bowel disease (IBD)	Praveschotinunt et al. (2019)
E. coli Nissle 1917	Polydopamine coated *E. coli* against intestinal inflammation	Praveschotinunt et al. (2019)
E. coli SYNB8802	Protection against kidney damage by reducing oxalic acid	Zheng et al. (2020)
E. coli LS31T	Production of riboflavin	Liu et al. (2021)
Leuconostoc citreum	Riboflavin biosynthesis	Son et al. (2020)
Saccharomyces	Self-regulating engineered yeast to treat IBD	Scott et al. (2021)

14.13 GENOME EDITING IN FUNGI

Filamentous fungi are well known for their ability to synthesize an infinite number of bioactive metabolites from enzymes and organic acids to therapeutic molecules. These biosynthetic gene clusters are being discovered by the emerging genome editing tools which support the exploration of hidden gene clusters (Satish et al., 2020). Site-specific gene insertions and deletions in the genome of endophytic filamentous fungi *Pestalotiopsisfici* was successfully experimented by Xu et al (2021). The probiotic and nutraceuticals produced by the modified strains and their relevance are summarized in Table 14.1.

14.14 CRISPR ENGINEERED PROBIOTIC START-UPS

The CRISPR genome editing in the field of probiotics and nutraceuticals introduced designer probiotic bacterial strains capable of eliminating pathogenic and harmful microbial community through special lytic phages. It can be used precisely to eliminate particular genes related to pathogenicity or can modulate the gene expression without altering its genome. The Nobel laureates Jennifer Doudna and Emmanuel Charpentier apply their invention to contribute to the health wellness of mankind by leveraging this tool for therapeutic applications. They formed a group of companies which focus on medical diagnostics, treatment, and probiotic products for therapeutic purposes. Some of these are Mammoth Biosciences, Caribou Biosciences, Editas Medicine, and CRISPR therapeutics, of which we have specific interest in CRISPR engineered probiotics start-ups which are still in the nascent stage and include SPIPR Biome, Van Pijkeren Laboratory, Eligo Bioscince, Locus Bioscience, du Pont, and Novome Biotechnologies (www.patent art.com). The aims and major focus of these start-ups are summarized in Table 14.2.

TABLE 14.2
Summary of CRISPR-Based Start-Ups in Probiotics and Nutraceuticals

Start-Up Companies	Aim and Focus	Applications
SNIPR Biome	Editing of endogenous microbiome to kill pathogens	Engineered probiotics as an add-on therapy along with standard therapy
Van Pijkeren Laboratory	Probiotics as delivery vehicles for bacteriophages to target pathogens at site	Address antibiotic resistance of pathogens by engineered probiotic strains
Eligo Bioscience	CRISPR system to genetically disable inflammation-inducing gene in healthy skin bacteria	Developing a topical cream to treat a root cause of acne
Locus Biosciences	CRISPR-Cas3-enhanced bacteriophage (crPhage™) specifically targeting *Escherichia coli (E. coli)* bacteria that causes urinary tract infections	Recombinant bacteriophage therapy for Urinary Tract Infections
Novome Biotechnologies	Engineering Bacteroides, which can effectively compete with other resident microbes to durably colonize the gut	Hyperoxaluria treatment with Genetically Engineered Microbial Medicines (GEMMs) platform
DuPont	Collection of phages for immunization of bacterial cultures	Production of CRISPRized dairy products

14.15 CONCLUSIONS

Just like gene therapy, the advent of CRISPR technology gives immense scope and technological knowledge to the scientific community to precisely tailor the genome of almost all living entities for the benefit of mankind. CRISPR technology for next generation probiotics aims to create designer probiotic microbes to act precisely and specifically to eliminate pathogenic invaders through bacteria or phage particles either by the endogenous Cas9 system or by engineered strains. As we have discussed, CRISPR technology is rapidly advancing in almost all the genera of beneficial microbes or natural commensals for nutraceutical and probiotic application. These crazy molecular scissors can be adapted for editing any organism in a food chain such as beneficial microbes, agricultural crops, and live stocks much faster than traditional gene editing technologies for the production probiotics, prebiotics, and functional foods for the promotion of human health. More research into microalgae, yeast, fungi, and even into agricultural crops is needed in future as it will have a huge impact on food as well as pharmaceutical industries to assure a better lifestyle for a better life.

REFERENCES

Allen, S.J., Wareham, K., Wang, D., et al. 2013. A high-dose preparation of lactobacilli and bifidobacteria in the prevention of antibiotic-associated and *Clostridium difficile* diarrhoea in older people admitted to hospital: A multicentre, randomised, double-blind, placebo-controlled, parallel arm trial (PLACIDE). *Health Technology Assessment* 17(57): 1–140.

Aloisio, I., Santini, C., Biavati, B., et al. 2012. Characterization of Bifidobacterium spp. strains for the treatment of enteric disorders in newborns. *Applied Microbiology and Biotechnology* 96: 1561–1576.

Amitai, G., and Sorek, R. 2016. CRISPR–Cas adaptation: Insights into the mechanism of action. *Nature Reviews Microbiology* 14: 67–76. https://doi.org/10.1038/nrmicro.2015.14

Arroyo-Olarte, R.D., Bravo Rodríguez, R., and Morales-Ríos, E. 2021. Genome editing in bacteria: CRISPR-Cas and beyond. *Microorganisms* 9(4): 844. doi:10.3390/microorganisms9040844

Barrangou, R., Fremaux, C., Deveau, H., et al. 2007. CRISPR provides acquired resistance against viruses in prokaryotes. *Science* 315: 1709–1712.

Behnsen, J., Deriu, E., Sassone-Corsi, M., and Raffatellu, M. 2013. Probiotics: Properties, examples, and specific applications. *Cold Spring Harbor Perspectives in Medicine* 3: a010074–a010074.

Boch, J., Scholze, H., Schornack, S., et al. 2009. Breaking the code of DNA binding specificity of TAL-type III effectors. *Science* 326(5959): 1509–1512. doi:10.1126/science.1178811

Börner, R.A., Kandasamy, V., Axelsen, A.M., Nielsen, A.T., and Bosma, E.F. 2018. Genome editing of lactic acid bacteria: opportunities for food, feed, pharma and biotech. *FEMS Microbiology Letters* 36(1): 291. doi:10.1093/femsle/fny291

Cao, M., Gao, M., Lopez-Garcia C.L., et al. 2017. Centromeric DNA facilitates nonconventional yeast genetic engineering. *ACS Synthetic Biology* 18: 1545–1553. doi: 10.1021/acssynbio.7b00046.

Carroll, D. 2014. Genome engineering with targetable nucleases. *Annual Review of Biochemistry* 83: 409–439. doi:10.1146/annurev-biochem-060713-035418

Chen, J., Cai, W., and Feng, Y. 2007. Development of intestinal bifidobacteria and lactobacilli in breast-fed neonates. *Clinical Nutrition* 26: 559–566.

Chen, W., Zhang, Y., Yeo, W.-S., Bae, T., and Ji, Q. 2017. Rapid and efficient genome editing in *Staphylococcus aureus* by using an engineered CRISPR/Cas9 system. *Journal of the American Chemical Society* 139: 3790–3795. doi: 10.1021/jacs.6b13317

Chmielewska, A., and Szajewska, H. 2010. Systematic review of randomised controlled trials: Probiotics for functional constipation. *World Journal of Gastroenterology* 16(1): 69–75. doi.org/10.3748/wjg.v16.i1.69

Cobb, R.E., Wang, Y., and Zhao, H. 2015. High-efficiency multiplex genome editing of Streptomyces species using an engineered CRISPR/Cas system. *ACS Synthetic Biology* 4(6): 723–728. doi:10.1021/sb500351f

Collins, J.K., Thornton, G., and Sullivan, G.O. 1998. Selection of probiotic strains for human applications. *International Dairy Journal* 8: 487–490.

Crawley, A.B., Henriksen, J.R., and Barrangou, R. 2018. CRISPRdisco: An automated pipeline for the discovery and analysis of CRISPR-Cas systems. *CRISPR Journal* 1: 171–181.

de Jong, B., Siewers, V., and Nielsen, J. 2012. Systems biology of yeast: Enabling technology for development of cell factories for production of advanced biofuels. *Current Opinion in Biotechnology* 23(4): 624–630. doi:10.1016/j.copbio.2011.11.021

Demers, M., Dagnault, A., and Desjardins, J. 2014. A randomized double-blind controlled trial: Impact of probiotics on diarrhea in patients treated with pelvic radiation. *Clinical Nutrition* 33: 761–767.

Dey, A. 2020. CRISPR/Cas genome editing to optimize pharmacologically active plant natural products. *Pharmacological Research* 164: 105359. doi:10.1016/j.phrs.2020.105359

Di Gioia, D., Aloisio, I., Mazzola, G., and Biavati, B. 2014. Bifidobacteria: Their impact on gut microbiota composition and their applications as probiotics in infants. *Applied Microbiology and Biotechnology* 98: 563–577.

Doron, S.I., Hibberd, P.L., and Gorbach, S.L. 2008. Probiotics for prevention of antibiotic-associated diarrhea. *Journal of Clinical Gastroenterology* 42: S58–S63.

Fuller, R. 1991. Probiotics in human medicine. *Gut* 32: 439–442.

Gao, Y., et al. 2011. Complete genome sequence of *Lactococcus lactis* subsp. lactis CV56, a probiotic strain isolated from the vaginas of healthy women. *Journal of Bacteriology* 193: 2886–2887.

Gibson, G.R., and Fuller, R. 2000. Aspects of in vitro and in vivo research approaches directed toward identifying probiotics and prebiotics for human use. *The Journal of Nutrition* 130: 391S–395S.

Gilbert, J.A., et al. 2018. Current understanding of the human microbiome. *Nature Medicine* 24: 392–400.

Grangette, C., Nutten, S., Palumbo, E., et al. 2005. Enhanced anti-inflammatory capacity of a *Lactobacillus plantarum* mutant synthesizing modified teichoic acids. *Proceedings of the National Academy of Sciences of the United States of America* 102(29): 10321–10326. doi:10.1073/pnas.0504084102

Griese, M., Lange, C., and Soppa, J. 2011. Ploidy in cyanobacteria. *FEMS Microbiology Letters* 323(2): 124–131. doi:10.1111/j.1574-6968.2011.02368.x

Guarner, F., and Malagelada, J.R. 2003. Gut flora in health and disease. *The Lancet* 361: 512–519.

Guo, T., Xin, Y., Zhang, Y., Gu, X., and Kong, J. 2019. A rapid and versatile tool for genomic engineering in *Lactococcus lactis*. *Microbial Cell Factories* 18(1): 22. doi:10.1186/s12934-019-1075-3

Hamilton-Miller, J.M.T. 1999. Probiotics: A critical review. *Journal of Antimicrobial Chemotherapy* 43(6): 849. https://doi.org/10.1093/jac/43.6.849

Havenaar, R., and Huis In't Veld, J.H.J. 1992. Probiotics: A general view. In The Lactic Acid Bacteria Volume 1 (ed. Wood, B. J. B.) 151–170 (Springer US). doi:10.1007/978-1-4615-3522-5_6.

Heap, J.T., Pennington, O.J., Cartman, S.T., Carter, G.P., and Minton, N.P. 2007. The ClosTron: A universal gene knock-out system for the genus Clostridium. *Journal of Microbiological Methods* 70(3): 452–464. doi: 10.1016/j.mimet.2007.05.021

Hempel, S., Newberry, S.J., Maher, A.R., et al. 2012. Probiotics for the prevention and treatment of antibiotic-associated diarrhea: A systematic review and meta-analysis. *Critical Reviews* 307(18): 1959–1969. doi:10.1001/jama.2012.3507

Hidalgo-Cantabrana, C., Goh, Y.J., Pan, M., Sanozky-Dawes, R., and Barrangou, R. 2019. Genome editing using the endogenous type I CRISPR-Cas system in *Lactobacillus crispatus*. *Proceedings of the National Academy of Sciences of the United States of America* 116(32): 15774–15783. doi:10.1073/pnas.1905421116

Ishino, Y., Krupovic, M., and Forterre, P. 2018. History of CRISPR-Cas from encounter with a mysterious repeated sequence to genome editing technology. *Journal of Bacteriology* 200(7). doi:10.1128/JB.00580-17

Ishino, Y., Shinagawa, H., Makino, K., Amemura, M., and Nakata, A. 1987. Nucleotide sequence of the iap gene, responsible for alkaline phosphatase isozyme conversion in *Escherichia coli*, and identification of the gene product. *Journal of Bacteriology* 169(12): 5429–5433. doi:10.1128/jb.169.12.5429-5433.1987

Isolauri, E., Rautava, S., and Salminen, S. 2012. Probiotics in the development and treatment of allergic disease. *Gastroenterology Clinics of North America* 41: 747–762.

Isolauri, E., Sütas, Y., Kankaanpää, P., Arvilommi, H., and Salminen, S. 2001. Probiotics: Effects on immunity. *The American Journal of Clinical Nutrition* 73: 444s–450s.

Jansen, R., van Embden, J.D.A., Gaastra, W., and Schouls, L.M. 2002. Identification of genes that are associated with DNA repeats in prokaryotes. *Molecular Microbiology* 43: 1565–1575.

Jensen, M.K., and Keasling, J.D. 2015. Recent applications of synthetic biology tools for yeast metabolic engineering. *FEMS Yeast Research* 15(1): 1–10. doi:10.1111/1567-1364.12185

Jiang, W., Bikard, D., Cox, D., Zhang, F., and Marraffini, L.A. 2013. RNA-guided editing of bacterial genomes using CRISPR-Cas systems. *Nature Biotechnology* 31: 233–239.

Jiang, Y., Chen, B., Duan, C., Sun, B., Yang, J., and Yang, S. 2015. Multigene editing in the *Escherichia coli* genome via the CRISPR-Cas9 system. *Applied and Environmental Microbiology* 81(7): 2506–2514.

Johnston, B.C., Goldenberg, J.Z., Vandvik, P.O., Sun, X., and Guyatt, G.H. 2011. Probiotics for the prevention of pediatric antibiotic-associated diarrhoea. In Cochrane Database of Systematic Reviews (ed. The Cochrane Collaboration) CD004827.pub3 (John Wiley & Sons, Ltd). doi:10.1002/14651858.CD004827.pub3.

Kailasapathy, K., and Chin, J. 2000. Survival and therapeutic potential of probiotic organisms with reference to *Lactobacillus acidophilus* and Bifidobacterium spp. *Immunology & Cell Biology* 78(1): 80–88.

Kasai, T., Suzuki, Y., Kouzuma, A., and Watanabe, K. 2019. Roles of D-lactate dehydrogenases in the anaerobic growth of *Shewanellaoneidensis* MR-1 on sugars. *Applied and Environmental Microbiology* 85(3): e02668–18. doi:10.1128/AEM.02668-18

Kavscek, M., Stražar, M., Curk, T., Natter, K., and Petrovič, U. 2015. Yeast as a cell factory: Current state and perspectives. *Microbial Cell Factories* 14: 94. doi: 10.1186/s12934-015-0281-x

Kechagia, M., et al. 2013. Health benefits of probiotics: A review. *ISRN Nutrition* 2013: 1–7.

Khazaie, K., Zadeh, M., Khan, M.W., et al. 2012. Abating colon cancer polyposis by *Lactobacillus acidophilus* deficient in lipoteichoic acid. *Proceedings of the National Academy of Sciences of the United States of America* 109(26): 10462–10467. doi:10.1073/pnas.1207230109

Kneifel, W., and Domig, K.J. 2014. Probiotic bacteria. In Encyclopedia of Food Microbiology 154–157 (Elsevier). doi:10.1016/B978-0-12-384730-0.00271-8.

Konishi, K., Van Doren, S.R., Kramer, D.M., Crofts, A.R., and Gennis, R.B. 1991. Preparation and characterization of the water-soluble heme-binding domain of cytochrome c1 from the *Rhodobacter sphaeroides* bc1 complex. *The Journal of Biological Chemistry* 266(22): 14270–14276.

Kuehne, S.A., Heap, J.T., Cooksley, C.M., Cartman, S.T., and Minton, N.P. 2011. ClosTron-mediated engineering of Clostridium. *Methods in Molecular Biology* 765: 389–407. doi:10.1007/978-1-61779-197-0_23

Larsen, N., et al. 2014. Characterization of Bacillus spp. strains for use as probiotic additives in pig feed. *Applied Microbiology and Biotechnology* 98: 1105–1118.

Lebeer, S., et al. 2018. Identification of probiotic effector molecules: present state and future perspectives. *Current Opinion in Biotechnology* 49: 217–223.

Lee, S.J., et al. 2014. The effects of co-administration of probiotics with herbal medicine on obesity, metabolic endotoxemia and dysbiosis: A randomized double-blind controlled clinical trial. *Clinical Nutrition* 33: 973–981.

Li, H., Shen, C.R., Huang, C.H., Sung, L.Y., Wu, M.Y., and Hu, Y.C. 2016. CRISPR Cas9 for the genome engineering of cyanobacteria and succinate production. *Metabolic Engineering* 38: 293–302. doi: 10.1016/j.ymben.2016.09.006

Li, L., Wei, K., Zheng, G., et al. 2018. CRISPR-Cpf1 assisted multiplex genome editing and transcriptional repression in Streptomyces. *Applied and Environmental Microbiology* 84(18); pii: e00827–18.

Liu, S., Hu, W., Wang, Z., and Chen, T. 2021. Rational engineering of *Escherichia coli* for high-level production of riboflavin. *Journal of Agricultural and Food Chemistry* 69 (41): 12241–12249.

Liu, J.J., Kong, I.I., Zhang, G.C., et al. 2016. Metabolic engineering of probiotic *Saccharomyces boulardii*. *Applied and Environmental Microbiology* 82(8): 2280–2287. doi: 10.1128/AEM.0005716.

Makarova, K., et al. 2006. Comparative genomics of the lactic acid bacteria. *Proceedings of the National Academy of Sciences of the United States of America* 103: 15611–15616.

Makarova, K.S., et al. 2011. Evolution and classification of the CRISPR–Cas systems. *Nature Reviews Microbiology* 9: 467–477.

Martens, E.C., Neumann, M., and Desai, M.S. 2018. Interactions of commensal and pathogenic microorganisms with the intestinal mucosal barrier. *Nature Reviews Microbiology* 16: 457–470.

McNaught, C.E., and MacFie, J. 2001. Probiotics in clinical practice: A critical review of the evidence. *Nutrition Research* 21: 343–353.

Miller, J., McLachlan, A.D., and Klug, A. 1985. Repetitive zinc-binding domains in the protein transcription factor IIIA from *Xenopus oocytes. The EMBO Journal* 4(6): 1609–1614.

Moscou, M.J., and Bogdanove, A.J. 2009. A simple cipher governs DNA recognition by TAL effectors. *Science* 326(5959): 1501. doi:10.1126/science.1178817

Nombela, C., Gil, C., and Chaffin, W.L. 2006. Non-conventional protein secretion in yeast. *Trends in Microbiology* 14(1): 15–21. doi:10.1016/j.tim.2005.11.009

Oh, J.H., and van Pijkeren, J.P. 2014. CRISPR-Cas9-assisted recombineering in *Lactobacillus reuteri. Nucleic Acids Research* 42(17): e131–e131.

Padhi, S., Chourasia, R., Kumari, M., Singh, S.P., and Rai, A.K. 2022. Production and characterization of bioactive peptides from rice beans using *Bacillus subtilis. Bioresource Technology* 351: 126932.

Pan, M., and Barrangou, R. 2020. Combining omics technologies with CRISPR-based genome editing to study food microbes. *Current Opinion in Biotechnology* 61: 198–208. doi:10.1016/j.copbio.2019.12.027

Pavletich, N.P., and Pabo, C.O. 1991. Zinc finger-DNA recognition: Crystal structure of a Zif268-DNA complex at 2.1 A. *Science* 252(5007): 809–817. doi:10.1126/science.2028256

Petrova, M.I., et al. 2018. Engineering *Lactobacillus rhamnosus* GG and GR-1 to express HIV-inhibiting griffithsin. *International Journal of Antimicrobial Agents* 52: 599–607.

Pingoud, A., Wilson, G.G., and Wende, W. 2014. Type II restriction endonucleases—A historical perspective and more. *Nucleic Acids Research* 42: 7489–7527.

Plavec, T.V., and Berlec, A. 2020. Safety aspects of genetically modified lactic acid bacteria. *Microorganisms* 8(2): 297. doi:10.3390/microorganisms8020297

Praveschotinunt, P., Duraj-Thatte, A.M., Gelfat, I., Bahl, F., Chou, D.B., and Joshi, N.S. 2019. Engineered *E. coli* Nissle 1917 for the delivery of matrix-tethered therapeutic domains to the gut. *Nature Communications* 10(1): 5580.

Racharaks, R., Arnold, W., and Peccia, J. 2021. Development of CRISPR-Cas9 knock-in tools for free fatty acid production using the fast-growing cyanobacterial strain *Synechococcus elongatus* UTEX 2973. *Journal of Microbiological Methods* 189: 106315. doi:10.1016/j.mimet.2021.106315

Rai, A.K., Swapna, H.C., Bhaskar, N., and Baskaran, V. 2012. Potential of seafood industry byproducts as sources of recoverable lipids: fatty acid composition of meat and non meat component of selected Indian marine fishes. *Journal of Food Biochemistry* 36: 441–448.

Ran, F.A., et al. 2013. Genome engineering using the CRISPR-Cas9 system. *Nature Protocols* 8: 2281–2308.

Rolfe, R.D. 2000. The role of probiotic cultures in the control of gastrointestinal health. *The Journal of Nutrition* 130: 396S–402S.

Ruiz, L., Margolles, A., and Sánchez, B. 2013. Bile resistance mechanisms in Lactobacillus and Bifidobacterium. *Frontiers in Microbiology* 4. https://doi.org/10.3389/fmicb.2013.00396

Saarela, M., Mogensen, G., Fondén, R., Mättö, J., and Mattila-Sandholm, T. 2000. Probiotic bacteria: Safety, functional and technological properties. *Journal of Biotechnology* 84: 197–215.

Sachdeva, V., Roy, A., and Bharadvaja, N. 2020. Current prospects of nutraceuticals: A review. *Current Pharmaceutical Biotechnology* 21(10): 884–896. doi:10.2174/1389201021666200130113441

Satish, L., Shamili, S., Muthubharathi, B.C., et al. 2020. CRISPR-Cas9 system for fungi genome engineering toward industrial applications. In Genome Engineering via CRISPR-Cas9 System 69–81 (Amsterdam: Elsevier Science).

Scott, B.M., Gutierrez-Vazquez, C., Sanmarco, L.M., et al. 2021. Self-tunable engineered yeast probiotics for the treatment of inflammatory bowel disease. *Nature Medicine* 27 (7): 1212–1222.

Selvaraj, G., and Iyer, V.N. 1983. Suicide plasmid vehicles for insertion mutagenesis in *Rhizobium meliloti* and related bacteria. *Journal of Bacteriology* 156(3): 1292–1300. doi:10.1128/jb.156.3.1292-1300.1983

Sharan, S.K., Thomason, L.C., Kuznetsov, S.G., and Court, D.L. 2009. Recombineering: A homologous recombination-based method of genetic engineering. *Nature Protocols* 4(2): 206–223. doi:10.1038/nprot.2008.227

Son, J., Jang, S.H., Cha, J.W., and Jeong, K.J. 2020. Development of CRISPR interference (CRISPRi) platform for metabolic engineering of *Leuconostoc citreum* and its application for engineering riboflavin biosynthesis. *International Journal of Molecular Sciences* 21: 16.

Sun, Z., Harris, H.M., McCann, A., et al. 2015. Expanding the biotechnology potential of lactobacilli through comparative genomics of 213 strains and associated genera. *Nature Communications* 6: 8322. doi:10.1038/ncomms9322

Tas, H., Nguyen, C.T., Patel, R., Kim, N.H., and Kuhlman, T.E. 2015. An integrated system for precise genome modification in *Escherichia coli*. *PLoS One* 10(9): e0136963. doi:10.1371/journal.pone.0136963

Thantsha, M.S., Mamvura, C.I., and Booyens, J. 2012. Probiotics – What they are, their benefits and challenges. In New Advances in the Basic and Clinical Gastroenterology (ed. Brzozowski, T.) (InTech). doi:10.5772/32889.

Tompkins, T., Xu, X., and Ahmarani, J. 2010. A comprehensive review of post-market clinical studies performed in adults with an Asian probiotic formulation. *Beneficial Microbes* 1: 93–106.

Tong, Y., Charusanti, P., Zhang, L., Weber, T., and Lee, S.Y. 2015. CRISPR-Cas9 based engineering of actinomycetal genomes. *ACS Synthetic Biology* 4(9): 1020–1029. doi:10.1021/acssynbio.5b00038

Tulumoglu, S., et al. 2013. Probiotic properties of lactobacilli species isolated from children's feces. *Anaerobe* 24: 36–42.

Utomo, J.C., Hodgins, C.L., and Ro, D.K. 2021. Multiplex genome editing in yeast by CRISPR/Cas9 - A potent and agile tool to reconstruct complex metabolic pathways. *Frontiers in Plant Science* 12: 719148. doi: 10.3389/fpls.2021.719148

Van der Hoek, S.A., Darbani, B., Zugaj, K.E., et al. 2019. Engineering the yeast *Saccharomyces cerevisiae* for the production of L-(+)-ergothioneine. *Frontiers in Bioengineering and Biotechnology* 7: 262. doi: 10.3389/fbioe.2019.00262.

Van Pijkeren, J.P., and Barrangou, R. 2017. Genome editing of food-grade lactobacilli to develop therapeutic probiotics. *Microbiology Spectrum* 5(5). doi:10.1128/microbiolspec.BAD-0013-2016

Vergin F. 1954. Anti- und Probiotika [Antibiotics and probiotics]. *Hippokrates* 25(4): 116–119.

Walter, J. 2008. Ecological role of lactobacilli in the gastrointestinal tract: Implications for fundamental and biomedical research. *Applied and Environmental Microbiology* 74: 4985–4996.

Wang, B., Hu, Q., Zhang, Y., et al. 2018. A RecET-assisted CRISPR-Cas9 genome editing in *Corynebacterium glutamicum*. *Microbial Cell Factories* 17(1): 63. doi:10.1186/s12934-018-0910-2

Weiser, J.N., Ferreira, D.M., and Paton, J.C. 2018. *Streptococcus pneumoniae*: transmission, colonization and invasion. *Nature Reviews Microbiology* 16: 355–367.

Xiao, Y., Wang, S., Rommelfanger, S., et al. 2018. Developing a Cas9-based tool to engineer native plasmids in *Synechocystis sp.* PCC 6803, *Biotechnology and Bioengineering* 115: 2305–2314

Xu, X., Huang, R., and Yin, W.B. 2021. An optimized and efficient CRISPR/Cas9 system for the endophytic fungus *Pestalotiopsisfici*. *Journal of Fungi* 7(10): 809. doi:10.3390/jof7100809

Yang, X., Wang, Y., and Huo, G. 2013. Complete genome sequence of *Lactococcus lactis* subsp. lactis KLDS4.0325. *Genome Announcement* 1: e00962–13.

Yapijakis, C. 2009. Hippocrates of Kos, the father of clinical medicine, and Asclepiades of Bithynia, the father of molecular medicine. *In Vivo* 23: 507.

Zelaya, H., et al. 2014. Immunobiotic lactobacilli reduce viral-associated pulmonary damage through the modulation of inflammation–coagulation interactions. *International Immunopharmacology* 19: 161–173.

Zheng, R., Li, Y., Wang, L., Fang, X., et al. 2020. CRISPR/Cas9-mediated metabolic pathway reprogramming in a novel humanized rat model ameliorates primary hyperoxaluria type 1. *Kidney International* 98(4): 947–957.

Zhou, D., Jiang, Z., Pang, Q., Zhu, Y., Wang, Q., and Qi, Q. 2019. CRISPR/Cas9-assisted seamless genome editing in *Lactobacillus plantarum* and its application in N-Acetylglucosamine production. *Applied and Environmental Microbiology* 85(21): e01367–19. doi: 10.1128/AEM.01367-19.

Ziemer, C.J., and Gibson, G.R. 1998. An overview of probiotics, prebiotics and synbiotics in the functional food concept: Perspectives and future strategies. *International Dairy Journal* 8: 473–479.

Zokaeifar, H., et al. 2014. Administration of Bacillus subtilis strains in the rearing water enhances the water quality, growth performance, immune response, and resistance against *Vibrio harveyi* infection in juvenile white shrimp, Litopenaeusvannamei. *Fish & Shellfish Immunology* 36: 68–74.

Zuo, F., and Marcotte, H. 2021. Advancing mechanistic understanding and bioengineering of probiotic lactobacilli and bifidobacteria by genome editing. *Current Opinion in Biotechnology* 70: 75–82. doi:10.1016/j.copbio.2020.12.015.

Index